# INTEROPERATING GEOGRAPHIC INFORMATION SYSTEMS

THE KLUWER INTERNATIONAL SERIES IN ENGINEERING AND COMPUTER SCIENCE

# INTEROPERATING GEOGRAPHIC INFORMATION SYSTEMS

*edited by*

**Michael Goodchild**
*University of California, Santa Barbara*

**Max Egenhofer**
*University of Maine*

**Robin Fegeas**
*U. S. Geological Survey*

**Cliff Kottman**
*Open GIS Consortium Inc.*

Springer Science+Business Media, LLC

ISBN 978-0-7923-8436-6 ISBN 978-1-4615-5189-8 (eBook)
DOI 10.1007/978-1-4615-5189-8

**Library of Congress Cataloging-in-Publication Data**

A C.I.P. Catalogue record for this book is available
from the Library of Congress.



*Printed on acid-free paper.*

# TABLE OF CONTENTS

## *Education, Institutions, and Markets*

# LIST OF CONTRIBUTORS

Gennady L. Andrienko
GMD - German National Research
Center for Information Technology
Schloss Birlinghoven
Sankt-Augustin
D-53754 Germany
*gennady.andrienko@gmd.de*

Natalia V. Andrienko
GMD - German National Research
Center for Information Technology
Schloss Birlinghoven
Sankt-Augustin
D-53754 Germany
*gennady.andrienko@gmd.de*

Edith Au
Pencom Web Works
40 Fulton Street
New York, NY 10038, USA
*edith@pencom.com*

Oleg Balovnev
Institute of Computer Science III
University of Bonn
Roemerstr. 164
53117 Bonn, Germany
*oleg@cs.uni-bonn.de*

Chumki Basu
Bellcore
MCC 1E-322B
445 South Street
Morristown, NJ 07960-6438, USA
*cbasu@bellcore.com*

Yvan Bédard
Centre for Research in Geomatics
Laval University
Ste-Foy, Quebec G1K 7P4, Canada
*yvan.bedard@scg.ulaval.ca*

Clifford Behrens
Bellcore
MCC 1J314R
445 South Street
Morristown, NJ 07960-6438, USA
*cliff@bellcore.com*

David A. Bennett
Department of Geography
Southern Illinois University,
Carbondale, IL 62901-4514, USA
*dbennett@siu.edu*

Yaser A. Bishr
International Institute for Aerospace
Survey and Earth Sciences (ITC)
Hengelosestraat 99
PO Box 6
7500 AA Enschede, The Netherlands
*yaser@itc.nl*

Patrice Boursier
L3i
Université de La Rochelle
Av. Norillac
17042 La Rochelle, France
*patrice.boursier@univ-lr.fr*

Martin Bruenig
Institute of Computer Science III
University of Bonn
Roemerstr. 164
53117 Bonn, Germany
*martin@cs.uni-bonn.de*

António Câmara
Departamento de Ciencias e
Engenharia do Ambiente
Campus da Faculdade de Ciências e
Tecnologia
Universidade Nova de Lisboa
2825 Monte de Caparica, Portugal
*asc@mail.fct.unl.pt*

Sang Kyun Cha
Knowledge and Data Engineering
Laboratory
School of Electrical Engineering
Seoul National University
San 56-1 Shinrim-dong Kwanak-ku
Seoul 151-742, Korea
*chask@kdb.snu.ac.kr*

Gilles Clément
Logiciels et Applications
Scientifiques (L.A.S.) Inc.
1600, Boul. St-Martin Est, Tour A,
bureau 650
Laval, Quebec H7G 4R8, Canada
*cl@las.com*

Armin B. Cremers
Institute of Computer Science III
University of Bonn
Roemerstr. 164
53117 Bonn, Germany
*abc@cs.uni-bonn.de*

Xavier Culos
SILOGIC Société d'ingénierie
informatique
78 chemin des Sept Derniers
31200 Toulouse, France
*xavier.culos@silogic.fr*

Clodoveu A. Davis Jr.
PRODABEL Empresa de Informática
e Informação do Municipo do Belo
Horizonte S.A.
Av Presidente Carlos Luz, 1275
31200-000 Belo Horizonte – MG
Brazil
*clodoveu@pbh.gov.br*

Thierry Deltheil
Département Recherche et
Développement
SILOGIC Société d'ingénierie
informatique
78 chemin des 7 deniers
31200 Toulouse, France
*tdeltheil@silogic.fr*

Liping Di
Raytheon STX Corporation
7701 Greenbelt Road, Suite 400
Greenbelt, MD 20770, USA
*lpd@rattler.gsfc.nasa.gov*

Khoa Doan
Best Software Corporation
11413 Isaac Newton Square
Reston, VA 20190, USA
*Khoa_Doan@bestsoftware.com*

Elaine R. Dobinson
Physical Oceanography Distributed
Active Archive Center (PODAAC)
Jet Propulsion Laboratory
Pasadena, CA 91109, USA
*elaine.dobinson@jpl.nasa.gov*

John D. Evans
Massachusetts Institute of
Technology
77 Massachusetts Ave., Room 9-514
Cambridge MA 02139, USA
*jdevans@mit.edu*

Ian Finch
Connect
3, Brownlow Street
University of Liverpool
Liverpool L69 3GL, UK
*I.Finch@csc.liv.ac.uk*

Frederico T. Fonseca
PRODABEL
Rua Contria 1098
30430-460 Belo Horizonte MG
Brazil
*fred@pbh.gov.br*

Andrew Frank
Technical University Vienna
Gusshaustr 27-29
A-1050 Vienna, Austria
*frank@geoinfo.tuwein.ac.at*

Volker Gaede
IC-Parc, Imperial College
London SW7 2AZ, UK
*v.gaede@icparc.ic.ac.uk*

Mark N. Gahegan
Department of Geographic
Information Science
Curtin University of Technology
PO Box U 1987
Perth 6845, Australia
*mark@cs.curtin.edu.au*

Pedro Pereira Gonçalves
Grupo de Análise de Sistemas
Ambientais
Faculdade de Ciências e Tecnologia
Universidade Nova de Lisboa
2825 Portugal
*pmg@uninova.pt*

Denis Gouin
Defence Research Establishment
Valcartier (DREV)
2459 Pie-XI Blvd. North
Val-Bélair
Quebec G3J 1X5, Canada
*denis.gouin@drev.dnd.ca*

Oliver Günther
Institut fur Wirtschaftsinformatik
Humboldt-Universität zu Berlin
Spandauer Str. 1
10178 Berlin, Germany
*guenther@wiwi.hu-berlin.de*

Francis Harvey
Department of Geography
Patterson Office Tower
University of Kentucky
Lexington, KY 40506-0027, USA
*fharv1@pop.uky.edu*

D Ian Heywood
The Open and Distance Learning
Centre
Robert Gordon University
School Hill
Aberdeen AB10 FR, UK
*ian_heywood@compuserve.com*

Doug Ilg
Raytheon STX Corporation
7701 Greenbelt Road, Suite 400
Greenbelt, MD 20770, USA
*dilg@rattler.gsfc.nasa.gov*

Sylvie Iris
SILOGIC
78 chemin des Sept Deniers
31200 Toulouse, France
*sylvie.iris@silogic.fr*

Steven R. Johnson
Brigham Young University
1348 E 900 South
Provo, Utah 84606-6307, USA
*srj2@email.byu.edu*

Marcus Jürgens
Institute of Statistics and Econometrics
Freie Universität Berlin
Garystr. 21
D-14195 Berlin, Germany
*juergens@inf.fu-berlin.de*

Karen K. Kemp
National Center for Geographic Information and Analysis
University of California
Santa Barbara, CA 93106-4060, USA
*kemp@ncgia.ucsb.edu*

Joo Kwan Kim
Knowledge and Data Engineering Laboratory
School of Electrical Engineering
Seoul National University
San 56-1 Shinrim-dong Kwanak-ku
Seoul 151-742, Korea
*jkkim@kdb.snu.ac.kr*

Ki Hong Kim
Knowledge and Data Engineering Laboratory
School of Electrical Engineering
Seoul National University
San 56-1 Shinrim-dong Kwanak-ku
Seoul 151-742, Korea
*next@kdb.snu.ac.kr*

Clifford A. Kottman
Vice President, Technology Development
Open GIS Consortium
6614 Rockland Dr
Clifton, VA 20124-2414, USA
*ckottman@opengis.org*

Werner Kuhn
Institute for Geoinformatics (IFGI)
University of Muenster
Robert–Koch-Str 26
48149 Muenster, Germany
*kuhn@ifgi.uni-muenster.de*

Donatas Kvedarauskas
LRI, Batiment 490
Université Paris-Sud
91405 Orsay Cedex, France
*dkvedar@lri.fr*

Yong Sik Kwon
Knowledge and Data Engineering Laboratory
School of Electrical Engineering
Seoul National University
San 56-1 Shinrim-dong Kwanak-ku
Seoul 151-742, Korea
*yskwon@kdb.snu.ac.kr*

Christian Larouche
Logiciels et Applications Scientifiques (L.A.S.) Inc.
1600, Boul. St-Martin Est, Tour A, bureau 650
Laval, Quebec H7G 4R8, Canada
*gc@las.com*

Xavier Lopez
University of California, Berkeley
102 South Hall #4600
Berkeley, CA 94720, USA
*xavier@sims.berkeley.edu*

Zakaria Maamar
Pattern Analysis and Machine Intelligence Lab
Department of Systems Design Engineering
University of Waterloo
Waterloo, Ontario N2L 3G1, Canada
*maamar@watfast.uwaterloo.ca*

Ken McDonald
NASA/GSFC Code 505
Greenbelt, MD 20771, USA
*ken.mcdonald@gsfc.nasa.gov*

Co Meijer
Cadastre
P.O.Box 9046
7300 GH Apeldoorn, Netherlands
*meijer@kadaster.nl*

Paul Morin
J2 Geomatics
615 Booth Street
Ottawa, Ontario K1A 0K2, Canada
*pmorin@ncs.dnd.ca*

Bernard Moulin
Computer Science Department, and
Centre for Research in Geomatics
Laval University
Ste-Foy, Quebec G1K 7P4, Canada
*moulin@ift.ulaval.ca*

Joaquim Muchaxo
Departamento de Ciencias e
Engenharia do Ambiente
Campus da Faculdade de Ciências e
Tecnologia
Universidade Nova de Lisboa
2825 Monte de Caparica, Portugal
*jmlm@mail.fct.unl.pt*

Rudolf Müller
Institut fur Wirtschaftsinformatik
Humboldt-Universitat zu Berlin
Spandauer Str. 1
10178 Berlin, Germany
*rmueller@wiwi.hu-berlin.de*

Nelson Neves
Departamento de Ciencias e
Engenharia do Ambiente
Campus da Faculdade de Ciências e
Tecnologia
Universidade Nova de Lisboa
2825 Monte de Caparica, Portugal
*jnn@mail.fct.unl.pt*

Silvia Nittel
University of California, Los Angeles
Computer Science Department
4801 Boelter Hall
Los Angeles, CA 90095-1596, USA
*silvia@cs.ucla.edu*

Brandon S. Plewe
Department of Geography
690 SWKT
Brigham Young University
Provo, Utah 84602-5526, USA
*brandon_plewe@byu.edu*

David Pullar
University of Queensland
Brisbane
Queensland 4072, Australia
*d.pullar@mailbox.uq.edu.au*

Hardy Pundt
Institute for Geoinformatics (IFGI)
University of Muenster
Robert-Koch-Str 26
48149 Muenster, Germany
*pundt@ifgi.uni-muenster.de*

M. Radwan
International Institute for Aerospace
Survey and Earth Sciences (ITC)
Hengelosestraat 99
PO Box 6
7500 AA Enschede, Netherlands
*radwan@itc.nl*

Robert G. Raskin
Physical Oceanography Distributed
Active Archive Center (PO.DAAC)
Jet Propulsion Laboratory
Pasadena, CA 91109, USA
*rob.raskin@jpl.nasa.gov*

Derek E. Reeve
Department of Geographical and Environmental Sciences
The University of Huddersfield
Queens Gate
Huddersfield HD1 3DH, UK
*d.e.reeve@hud.ac.uk*

Raja Sengupta
Department of Geography
Southern Illinois University
Carbondale, IL 62901-4514, USA
*sarojsen@siu.edu*

Amit P. Sheth
Large Scale Distributed Information Systems Lab
Computer Science Department
University of Georgia
Athens, GA 30602-7404, USA
*amit@cs.uga.edu*

Leon Shklar
Pencom Web Works
40 Fulton St
New York, NY 10038, USA
*leon@pencom.com*

João Pedro Silva
Departamento de Ciencias e Engenharia do Ambiente
Faculdade de Ciências e Tecnologia
Universidade Nova de Lisboa
2825 Monte de Caparica, Portugal
*jps@mail.fct.unl.pt*

Eleanor Small
Connect
3, Brownlow Street
University of Liverpool
Liverpool L69 3GL, UK
*eleanor.small@gs.com*

Chang Bin Song
Knowledge and Data Engineering Laboratory
School of Electrical Engineering
Seoul National University
San 56-1 Shinrim-dong Kwanak-ku
Seoul 151-742, Korea
*tsangbi@kdb.snu.ac.kr*

Kristin Stock
Queensland University of Technology
Brisbane
Queensland 4001, Australia
*k.stock@qut.edu.au*

R. Suresh
Raytheon STX Corporation
7701 Greenbelt Road, Suite 400
Greenbelt, MD 20770, USA
*suresh@rattler.gsfc.nasa.gov*

Frank Tuijnman
Professional GEO Systems
Damrak 44
1012 LK Amsterdam, Netherlands
*tom@pgs.nl*

Harry Uitermark
Cadastre
P.O.Box 9046
7300 GH Apeldoorn, Netherlands
*uitermark@kadaster.nl*

Carel van den Berg
Professional GEO Systems
Damrak 44
1012 LK Amsterdam, Netherlands
*frank@pgs.nl*

Peter van Oosterom
Cadastre
P.O.Box 9046
7300 GH Apeldoorn, Netherlands
*oosterom@kadaster.nl*

Andrej Vckovski
Netcetera AG
Zypressenstrasse 71
Postfach
CH-8040 Zurich, Switzerland
*vckovski@netcetera.ch*

Tom Vijlbrief
Professional GEO Systems
Damrak 44
1012 LK Amsterdam, Netherlands
*tom@pgs.nl*

Agnès Voisard
Institute of Computer Science
Freie Universität Berlin
Takustr. 9
D-14195 Berlin, Germany
*voisard@inf.fu-berlin.de*

Greg A. Wade
Department of Computer Science
Southern Illinois University
Carbondale, IL 62901-4514, USA
*wade@cs.siu.edu*

Chuanchu Wang
Kent Ridge Digital Labs
21 Heng Mui Keng Terrace
Singapore 119613
*ccwang@krdl.org.sg*

Jiankang Wu
Kent Ridge Digital Labs
21 Heng Mui Keng Terrace
Singapore 119613
*jiankang@krdl.org.sg*

Nancy Yeager
National Computational Science
Alliance
University of Illinois
Urbana-Champaign, IL 61820, USA
*nyeager@ncsa.uiuc.edu*

# PREFACE

Since the first commercial products became available around 1980, geographic information systems (GISs) have become a significant area of application of computing technology. They developed in response to practical needs, in the handling and analysis of digital geographic information, and without any strong theoretical framework. As a result, the GIS software industry that exists today is fragmented: each vendor has adopted a distinct terminology and approach; a technician trained on one GIS has a difficult time adapting to a different one; and a database built for one system is not necessarily usable in another. Just as the science of statistics provides a common mathematically-based language and framework for much statistical software, so a strong theoretical framework in geographic information science can provide principles and terminology for an easier-to-use, more coherent, and more *interoperable* GIS industry. The research community has made great strides in recent years in developing that fundamental theory, building on work in computer science, cognitive science, linguistics, information science, applied mathematics, and statistics. At the same time data sharing, exchange, and integration in GIS have become significant issues, where progress offers substantial benefits in reduced costs, increased flexibility, and speedier projects. This book captures the state of progress, places it in the context of fundamental theory, and expands on the institutional implications of greater interoperation among GISs.

Interoperation means many things to people. It means *openness* in the software industry, because open publication of internal data structures allows GIS users to build applications that integrate software components from different developers, and it allows new vendors to enter the market with competing products that are interchangeable with existing components, just as the concept of interchangeable parts helps competition in the automobile industry. In the past few years the Open GIS Consortium (OGC) has emerged as a major force in the trend to openness, as a consortium of GIS vendors, agencies, and academic institutions (*http://www.opengis.org*). Interoperation also means the ability to *exchange* data freely between systems, because each system would have knowledge of other systems' formats. Exchange standards such as the Spatial Data Transfer Standard (also known as Federal Information Processing Standard 173; Arctur et al. 1998) have had a significant impact on the ease with which data can be transferred between systems. They allow a user of one vendor's products to make use of data prepared using another vendor's products, because data can be transferred in a standard format. Interoperability also means *commonality in user interaction*, as system designers build interfaces that can be customized to a 'look and feel' familiar to the user.

*Simplification* is a common theme in discussions of interoperability—simplification in the complex collections of formats and standards in the industry, simplification in the interaction between user and system, simplification in the knowledge a user requires to be effective. In an interoperable world the user would have to *know less* in order to achieve the same outcome. Training on ARC/INFO would not be wasted if the user transferred to an Intergraph platform, and there would be no need to master the complex details of data formats in order to assemble a project database from different sources. From an educational perspective, progress in interoperability would be measured by what it was no longer necessary to teach.

The term *transparency* is used when a user no longer needs to be aware of the details of a computer implementation to use it effectively. A database management system offers transparency to its users, who need to know nothing about the actual implementation of a data model, or about the physical locations of data and software, but can work instead at a conceptual level. Transparency implies that certain things are no longer important to the user, and no longer intrude upon the user's conceptualization of the problem. It implies a uniform view of multiple, heterogeneous, distributed, and autonomous participating systems.

Another term with particular relevance to interoperation is *similarity*, a measure of the degree to which two data sets, software systems, disciplines, or agencies use the same vocabulary, follow the same conventions, and thus find it easy to interoperate. Currently, interoperation is possible only over the narrowest of domains. The effort to achieve interoperation is thus an effort to extend domains, or to raise the threshold of similarity below which interoperation is possible.

The current architecture of GISs requires its users to be specialists, who must learn a terminology that is largely system-specific, a user interface that is similarly dominated by details of implementation, and a world of data that is riddled with convention. In order to make use of today's GIS one must be a *spatially-aware professional* (SAP). The ability to decode acronyms is one of the tests of an SAP, who must know, for example, the meaning of all of the Dxx acronyms—DEM, DTM, DLG, DRG, DOQ, DCW (digital elevation model, digital terrain model, digital line graph, digital raster graphic, digital ortho-photo quadrangle, digital chart of the world respectively)—and their general characteristics. SAPs hold much of the *metadata* of the common data sets in their heads, and thus are able to locate necessary data and assess their fitness for use without use of the apparatus normally required to support information retrieval, such as directories, catalogs, and libraries. SAPs will have taken courses in GIS, or may have acquired their awareness through the use of software, attendance at conferences, or from the published literature.

SAPs know the conventions of the geographic information community, and its language. They know, for example, the conventions that allow the producers of DOQs to assert that the representative fraction of their product is 1:12,000 (Goodchild and Proctor 1997). This has no relationship to representative fraction as normally defined for paper maps, since there is no distance in a digital database that can be compared to distance on the ground. Rather, a DOQ has a scale of 1:12,000 because its positional accuracy, which is well-defined, matches that of a map at that scale, according to national map accuracy standards (*http://mapping.usgs.gov/www/ti/DOQ/doqpt1.html*).

Perhaps most importantly, SAPs know the conventions of GIS discretization, which maps real-world objects and fields to their digital equivalents. The object conceptualized by a user as a continuous line is discretized as a *polyline*, consisting of mathematically straight connections between discrete points. Similarly, by convention an area is discretized as a *polygon*, and may even be referred to as such by an SAP. A field is discretized in many different ways that are embedded within distinct suites of software. Thus the same concept, a continuous surface of elevation, that is widely understood across many disciplines and professional cultures, may be represented in GIS as a TIN (triangulated irregular network, or triangular mesh), the digitized contours of a DLG, or the regular grid of a DEM. While the conceptual schema is the same, the implementations are entirely different, and are never hidden from the GIS user, ensuring that GIS is essentially inaccessible as a tool to anyone other than an SAP. Vckovski (1998a) has argued that an interoperable world would have just one conceptualization of a continuous surface, and that many aspects of the actual implementation can be made transparent to the user.

Interoperability conceived in this way is clearly relevant at many levels (see, for example, the collection of papers edited by Vckovski 1998b), and in many different aspects of GIS. Many different types of detail can be made transparent to the user, and many aspects of implementations can be hidden. Efforts are needed on many fronts, and many conceptual and technical problems will have to be solved, before much progress can be made towards the goal of interoperability. That progress might be measured by ease of use, represented by the amount of training needed to accomplish a certain task, or by the number of user actions required. It might be measured less directly in terms of redundancy of instruction, as items in the GIS curriculum that now must be covered before students can make effective use of GIS become redundant, or at least relegated to classes that focus on the technical details of GIS, rather than its applications. Other suitable metrics might be based on the transferability of knowledge, measuring the effort required by someone trained on System A to achieve the same productivity on System B. Progress might also be measured by comparing across disciplines, or application domains.

Many recent developments in information technology and GIS are immediately and obviously relevant to interoperability, either by motivating interest in achieving its objectives, or by providing the technical means to do so. The Internet and its applications, particularly the World Wide Web (WWW), are driving much of the interest in interoperability, because they make transfer of data and software possible, but fail to resolve many of the more difficult issues that transfer raises. Developments in distributed systems, client–server architectures, digital libraries, and other related areas are also high on the technical agenda at this time, as the contents of this book make clear.

The idea of a book on Interoperating GISs arose from an initiative of the U.S. National Center for Geographic Information and Analysis (NCGIA), a consortium of the University of California, Santa Barbara; the University at Buffalo; and the University of Maine; and under the auspices of its Project Varenius, NCGIA's project to advance geographic information science. Varenius is funded by the U.S. National Science Foundation through Cooperative Agreement SBR 9600465. NCGIA sponsored an international conference on interoperating GISs, Interop '97, in Santa Barbara in early December 1997, and held an invitational workshop immediately following the conference to develop a research agenda in the field. The results of that workshop are available as a printed report and on the WWW (Goodchild et al. 1997). Following the conference, the authors of the papers judged best by the conference program committee were invited to submit expanded versions for this book.

The book is divided into four major sections. In the first, Max Egenhofer introduces the need for theory, and nine chapters that present various theoretical frameworks. The second contains six chapters on modeling in distributed environments. In the third section, Cliff Kottman introduces a series of chapters on prototypes and functional systems, and on the lessons learned from them. In the final section, Robin Fegeas introduces a set of chapters on the institutional context of interoperation, issues of electronic commerce and GIS, and education.

The editors hope that this book will lay a firm foundation for research on interoperating GISs, and help push the field towards a more coherent, simpler, and more integrated future. We thank all of the authors for their contributions, and especially the National Science Foundation and the Open GIS Consortium for their collaboration in making this effort possible. Many of the software, system, and corporate names that appear throughout the book are registered as trade marks or otherwise protected.

## References

Arctur D, Hair D, Timson G, Martin E P, Fegeas R 1998 Issues and prospects for the next generation of the spatial data transfer standard (SDTS). *International Journal of Geographical Information Science* 12(4): 403–426

Goodchild M F, Proctor J 1997 Scale in a digital geographic world. *Geographical and Environmental Modelling* 1(1): 5–23

Goodchild M F, Egenhofer M J, Fegeas R 1997 *Interoperating GISs: Report of the Specialist Meeting*. National Center for Geographic Information and Analysis, Santa Barbara. *www.ncgia.org*

Vckovski A 1998a *Interoperable and Distributed Processing in GIS*. Taylor and Francis, London

Vckovski A 1998b Guest editorial: Special Issue, Interoperability in GIS. *International Journal of Geographical Information Science* 12(4): 297–298

# ABOUT THE EDITORS

**Dr. Michael F. Goodchild** is Professor of Geography at the University of California, Santa Barbara; Chair of the Executive Committee, National Center for Geographic Information and Analysis (NCGIA); Associate Director of the Alexandria Digital Library Project; and Director of NCGIA's Varenius project. He received his BA degree from Cambridge University in Physics in 1965 and his PhD in Geography from McMaster University in 1969. After 19 years at the University of Western Ontario, he moved to Santa Barbara in 1988. He was Director of NCGIA from 1991 to 1997. In 1990 he was given the Canadian Association of Geographers Award for Scholarly Distinction, and in 1996 the Association of American Geographers award for Outstanding Scholarship; he has won the American Society of Photogrammetry and Remote Sensing Intergraph Award and twice won the Horwood Critique Prize of the Urban and Regional Information Systems Association. He was Editor of *Geographical Analysis* between 1987 and 1990, and serves on the editorial boards of ten other journals and book series. His major publications include *Geographical Information Systems: Principles and Applications* (1991); *Environmental Modeling with GIS* (1993); *Accuracy of Spatial Databases* (1989); *GIS and Environmental Modeling: Progress and Research Issues* (1996); and *Scale in Remote Sensing and GIS* (1997); in addition he is author of some 300 scientific papers. He is currently Chair of the National Research Council's Mapping Science Committee. His research interests center on geographic information science, spatial analysis, the future of the library, and uncertainty in geographic data.

**Dr. Max J. Egenhofer** is the director of the National Center for Geographic Information and Analysis (NCGIA) at the University of Maine, the Libra Professor of the College of Engineering, an associate professor in Spatial Information Science and Engineering, and a cooperative associate professor in Computer Science. His research interests include reasoning in geographic space and formalizations of spatial relations; user interface design for geographic information systems including spatial query languages; and the design of database systems for GISs. He was the General Chair of the Fourth International Symposium on Large Spatial Databases, the Program Chair of the 2nd ACM Workshop on Advances in Geographic Information Systems, and program co-chair of Interop '97, and currently chairs the panel on

Computational Implementations of Geographic Concepts under Varenius, NCGIA's project to advance geographic information science. He is a U.S. Delegate to the Technical Committee of the International Organization for Standards on Geographic Information/Geomatics (ISO/TC211), is a co-editor of *GeoInformatica*, and serves on the editorial boards of *the International Journal of Geographical Information Science*, *Transactions in GIS*, *Spatial Cognition and Computation*, and the *International Journal of Applied Software Technology*.

**Dr. Robin G. Fegeas** is a geographer with the National Mapping Division (NMD) of the U.S. Geological Survey (USGS) in Reston, Virginia. He has received degrees in geography from the University of Maryland and the State University of New York at Buffalo. Robin has been working with geographic information systems and spatial data models at the USGS since 1973. In the 1970s he designed and developed GIRAS, an early GIS used to build the USGS national 1:250,000 and 1:100,000-scale Land Use/Land Cover data base. During the 1980s Robin served with various geographic and cartographic data standards groups including the National Committee for Digital Cartographic Data Standards and the Federal Interagency Coordinating Committee on Digital Cartography. The work of these two committees eventually resulted in the Spatial Data Transfer Standard (SDTS). Robin was also instrumental in the early development of NMD's Digital Line Graph-Feature based model. Currently Robin serves as co-chair of the Feature Special Interest Group of the Open GIS Consortium Technical Committee and as a U.S. Expert to Work Group 2 on geospatial data models and operators of the International Organization for Standardization (ISO) Technical Committee 211 on Geographic Information/Geomatics.

**Dr. Clifford A. Kottman** is Vice President and Chief Scientist of the Open GIS Consortium. Prior to OGC, Cliff has served as Director of Geospatial Programs for The MITRE Corp., as Executive Manager for Intergraph Corp., as Senior Scientist for Lockheed, as Mathematician, Cartographer, Physical Scientist, and Program Manager for the Defense Mapping Agency, and as Associate Professor of Mathematics for Oregon State University. Dr. Kottman specializes in standards of all types that support automated cartography and geographic information systems. He has published numerous papers of original research in topology and functional analysis, and on the evolution of open GISs.

# 1 INTRODUCTION: THEORY AND CONCEPTS

Max J. Egenhofer

The integration of software components is often times seen as an engineering task, much like designing a converter from 220 to 110 volts so that appliances that were designed for 220 volts can be used in another environment. However, limiting the discussion of interoperating GIS to such an approach would miss a significant aspect: the semantics of geospatial information. We cannot talk about how to approach interoperating GIS without knowing what the software integration process should address.

The goal of interoperating GISs is to achieve an automated process that will allow us to use data and software services across the boundaries that their collectors and designers envisioned. An *ad hoc* integration may work in a few specialized cases, but it is often difficult or impossible to generalize such an approach. The deficiencies of *ad hoc* interoperability become apparent when interoperating subparts are supposed to be extended by integrating new components or adding functionalities or replacing subparts with new software pieces. The difficulties are primarily in the semantics of the diverse implementations. Compatible semantics of geospatial information are a key characteristic of interoperating GISs, and powerful methods to capture and describe geospatial semantics are critical.

As in many other fields, a solid theoretical foundation can assist in the development of interoperating geographic information systems by providing rationale and frameworks and giving guidance for implementations. This is not the first time that *theory* has been discussed within the realm of GIS. In the early days of geographic information science, Frank (1987) called for a unifying spatial theory. Also, the NCGIA solicitation (Abler 1987) included a statement on the need for research towards a theory of spatial relations, a particular subset of a spatial theory. With the advent of the conference series on spatial information theory (COSIT), researchers found a forum to discuss various theory components (Frank and Campari 1993; Frank and Kuhn 1995; Hirtle and Frank 1997). Of particular interest within that community is the linkage to cognitive aspects—how people think about geographic space. The concept of Naive Geography (Egenhofer and Mark 1995), a core of this perspective on spatial information theory, was defined as a set of theories "that provide the basis for designing future geographic information systems that follow human intuition and are, therefore, easily accessible to a large range of users." With interoperability, a new light is shed on spatial information theory, with a focus on capturing semantics and making software services work together.

An important consideration for research is that interoperating GISs need to go beyond providing interoperability for different geometries. Geometry is an integral part of GIS interoperability; however, the formation of particular geometries through shapes, orientations, and connectivities is driven in the first instance by different ways to conceptualize the world. If one considers the differences in spatial data sets that users want to integrate, one quickly realizes that they stem from different conceptualizations more often than from the choice of different geometric data models. These conceptual issues are a significant part of the semantics of spatial information. We need better models that capture how people express meaning. Computerized information systems contain symbols that need people's interpretations. Successful communication means that sender (often the data collector) and recipient (often the user) have the same mental models. In the realm of interoperability, there are two (or more) senders and the recipient is charged with the task of understanding the intended meanings of the senders *and* to integrate that information.

To address these issues, we have collected in this section nine chapters. To start out, we need to acknowledge that certain aspects of the discussion about GIS interoperability are not entirely new. Interoperability has been discussed within the broader information systems community for years and certain aspects have been implemented outside of GIS for years. It is important to be aware of these successes (and failures); therefore, this section on "Theory and Concepts" starts with lessons learned from interoperability research outside of GIS. Amit Sheth, who been involved in interoperability research both in academia and industry, reflects on how the focus has changed over the years. He organizes the developments into three generations: the early stages of data interoperability based on the concepts of federated databases; the second generation enabled by the availability of Internet technologies; and the third generation being concerned with capturing and exchanging semantics of information. Andrej Vckovski argues in the second chapter in support of building interoperability on top of a solid theory of spatial information. Such a spatial information theory need not be designed completely from scratch as there are many parts that have been developed and used over the past decades, but the important task is to bring them together and integrate these results. Some parts of this integration process are currently going on under the umbrella of the Open GIS Consortium (OGC). Cliff Kottman describes in the third chapter the OGC process and provides a vision for the not-so-distant future of how interoperating GISs may quite different from today's commercial products. His futuristic scenario goes beyond a dataset generated with one GIS to be accessed by another GIS, and includes software services that collaborate and negotiate with the goal to combine and integrate data from various sources, often through their geospatial footprints, locations, or names. In a "roadmap to interoperability," Kottman outlines some of the components that he believes are needed in order to preserve the meaning of the information stored across the different software services.

The three following chapters deal with semantics. Yaser Bishr, Hardy Pundit, Werner Kuhn, and M. Radian discuss issues of semantics within the context of information communities. They define a geospatial information community as a group of spatial data producers and users who share an ontology of real-world phenomena. Within such an information community there is more homogeneity with respect to the semantics of the geospatial information than there is across different information communities. To achieve a meaningful exchange of information among information communities, they advocate the use of *semantic translators* that map the a provider's information into a form that the recipient understands. Mark Gahegan provides a complementary model for the exchange of semantics, focusing on the often subtle differences between data models of different GISs. His model addresses four aspects—abstract properties, the geographic models, physical data structures, and system details—and he provides a language in the form of transformations that capture which of these four characteristics are preserved, and which ones are changed, from one system to another. Like in the case of information communities, this approach recognizes that often only a few aspects are so different that they need special attention and negotiation, while many other characteristics of data sets and services may show enough similarity that their differences may be neglected. Semantic commonalities and differences are also the topic of the chapter by Francis Harvey; however, he couches his discussions within a societal context. Therefore, he advocates the resolution of semantic differences as a participative task, in which so-called boundary objects play an important role. Boundary objects are seen as facilitators for achieving agreement, as they are the targets for mediation. A key concept in Harvey's perspective of semantic interoperabilty is the construction of portals to enable negotiations at boundary objects with differences.

The remaining three chapters of this section focus on different aspects of implementing interoperating GISs. Andrej Vckovski's second chapter builds on the lessons learned from the design of printer drivers and the languages developed to overcome incompatibilities between different devices. Of particular interest is the successful Postscript language, a programming language that enabled compatibility across different types of (laser) printers for the task of printing text and graphics. Vckovski discusses the pros and cons of this method, and also compares the Postscript approach with Java, the popular programming language for the Web, which has a similar design with the transmission of executable code that runs at the host's site. Particularly Java may play a much bigger role for future interoperating GISs in the form of an implementation framework. Silvia Nittel and Richard Muntz study the implications of differences in data types if different GIS services provide the same operations, but the operations have different signatures. Such operations are known in programming languages for simple data types (e.g., integers and floating point numbers) and overloading or coercion are typical techniques used there (Cardelli and Wegner 1985). For geospatial data types, however, such conversions would typically come with a loss of information, comparable with the rounding of floating point numbers to integers. Nittel and Muntz show how a negotiation may be achieved if an operation has at least one common argument. The last chapter of this section demonstrates how specification languages, particularly functional specification languages, can be used to describe semantics. The authors, Andrew Frank and Werner Kuhn, advocate the use of formal methods, beyond the description of scenarios in natural language, to overcome inevitable ambiguities. Since the proposed method separates conceptual issues from implementation details, it addresses interoperability at a higher level than many other approaches.

The nine chapters in the section provide a good overview of the current discussions about conceptual and theoretical aspects of interoperating geographic information systems. They demonstrate that much work on capturing semantics is still in its infancy, but it is very encouraging to see that the concerns in the GIS community have moved beyond the translation of data file formats and beyond discussions of differences in geometric data models. Advancements in the foundations for interoperating GISs require a close collaboration between GIS experts and computer scientists. This gives the term *interoperating GISs* a new meaning, requiring the interoperation of scientists with diverse backgrounds to design this next generation of information systems.

## References

Abler R 1987 The National Science Foundation National Center for Geographic Information and Analysis. *International Journal of Geographical Information Systems* 1: 303–326

Cardelli L, Wegner P 1985 On understanding type, data abstraction, and polymorphism. *ACM Computing Surveys* 17: 471–552

Egenhofer M, Mark D 1995 Naive geography. In Frank A, Kuhn W (eds) *Spatial Information Theory—A Theoretical Basis for GIS, International Conference COSIT '95, Semmering, Austria.* Lecture Notes in Computer Science 988. Springer-Verlag, Berlin: 1–15

Frank A 1987 Towards a spatial theory. Proceedings of International Geographic Information Systems (IGIS) Symposium: The Research Agenda: 215–227

Frank A, Campari I 1993 *Spatial Information Theory, European Conference, COSIT '93.* Springer-Verlag, New York

Frank A, Kuhn W 1995 *Spatial Information Theory—A Theoretical Basis for GIS, International Conference COSIT '95, Semmering, Austria.* Springer-Verlag, Berlin

Hirtle S, Frank A 1997 *Spatial Information Theory—A Theoretical Basis for GIS, International Conference COSIT '97, Laurel Highlands, Pennsylvania.* Springer-Verlag, Berlin

# 2 CHANGING FOCUS ON INTEROPERABILITY IN INFORMATION SYSTEMS: FROM SYSTEM, SYNTAX, STRUCTURE TO SEMANTICS

Amit P. Sheth

## 2.1. Introduction

Interoperability has been a basic requirement for the modern information systems environment for over two decades. How have key requirements for interoperability changed over that time? How can we understand the full scope of interoperability issues? What has shaped research on information system interoperability? What key progress has been made? This chapter provides some of the answers to these questions. In particular, it looks at different levels of information system interoperability, while reviewing the changing focus of interoperability research themes, past achievements and new challenges in the emerging global information infrastructure (GII). It divides the research into three generations, and discusses some of achievements of the past. Finally, as we move from managing data to information, and in future knowledge, the need for achieving semantic interoperability is discussed and key components of solutions are introduced.

Data and information interoperability has gained increasing attention for several reasons, including:

- excellent progress in interconnection afforded by the Internet, Web and distributed computing infrastructures, leading to easy access to a large number of independently created and managed information sources of broad variety;
- increasing specialization of work, but increasing need to reuse and analyze data, leading to creation of information and knowledge, and their subsequent reuse and sharing.

Any attempt to give a broad survey of information systems interoperability is difficult and complex for several reasons, including different levels of requirements, the variety of approaches, the number of technical areas involved, the large literature, and the large number of example systems. For summative and pedagogical reasons, we make some broad observations and avoid paying attention to many exceptions.

We propose to view information system evolution in the context of interoperability as occurring in three generations: Generation I which covers the period to roughly 1985, Generation II which covers the period of a decade through 1995, and Generation III which covers a period yet to be bounded since 1996. Our discussion primarily draws from database and information system research, while noting that relevant work can also be found in information retrieval, knowledge-based systems and AI, multimedia, and other disciplines. Also very little attention is given in this chapter to several application domains, for example, geographic information systems (Goodchild et al. 1997), in which interoperability has received significant attention.

One of the enduring approaches to studying the key interoperability issues in multidatabases and federated databases of the first generation has been to use the fundamental dimensions of distribution, heterogeneity, and autonomy (Sheth and Larson 1990). The same dimensions also provide a good starting point for studying interoperability issues in the subsequent generations. After a brief discussion of the distribution and autonomy dimensions, we will focus on the heterogeneity dimension as we study interoperability issues in the three generations.

### *2.1.1. Distribution*

The scope of interoperability during the first generation was primarily departmental and almost always within a company. Usually, the multidatabase systems involved just a few databases and computer nodes, either connected point-to-point or in a local area network. With the significant impact of the Internet and advent of the Web, the scope of interoperability during the second generation has been enterprise-wide as well as inter-enterprise. It was not unusual to find tens of computers and data repositories involved in a second-generation system. In the third generation, with significant improvements in communication technology, global information infrastructure, and distributed computing infrastructure, the dimension of distribution of data has achieved a very broad scope—from a single system to global. As the distributed nature of data and information is often hidden from the end users, the system developers face several new challenges. A few of the noteworthy challenges involve increasing use of large amounts of data and information sources—particularly involving visual data, use of a wide variety of communication modes with a variety of bandwidths, and a larger optimization space involving varying capabilities of the component systems. Compared to the first generation systems, the issue of optimization has received less attention in the second generation.

### *2.1.2. Autonomy*

The organizational entities that manage different information sources (including database systems) are often autonomous. Those who control an information source are often willing to let others share the data only if they retain control. Thus it is important to understand the aspects of autonomy and how they can be addressed when a database system participates in a federation or shares its data with new users or applications.

Let us look at a classification of autonomy issues in the context of federated database systems (adapted from Sheth and Larson 1990), noting that these can be adapted to other architectures by considering the various types of information sources and information system components involved. A component participating in a federation may exhibit several types of autonomy, including design, communication, association and execution.

*Design autonomy* refers to the ability of a component to choose its own design with respect to any matter, including

- the data or information being managed (i.e., the Universe of Discourse or domain),
- the representation (data model, query language) and the naming of the data elements (or the ontology used),
- the conceptualization or semantic interpretation of the data (or the context),
- constraints used to manage the data,
- the functionality of the system,
- association and sharing with other systems (see association autonomy below), and
- the implementation (e.g., record and file structures, concurrency control algorithms).

*Communication autonomy* refers to the ability of a component to decide whether to communicate with other components. A component with communication autonomy is able to decide when and how it responds to a request from another component. *Execution autonomy* refers to the ability of a component to execute local operations without interference from an external entity and to decide the order in which to execute external operations. Thus, an external system cannot enforce an order of execution of the commands on a component with execution autonomy. Execution autonomy implies that a component can abort any operation that does not meet its local constraints and that its local operations are logically unaffected by its participation in a federation. Furthermore, the component does not need to inform an external or federated system of the order in which external operations are executed and the order of an external operation with respect to local operations. Operationally, a component exercises its execution autonomy by treating external operations in the same way as local operations.

*Association autonomy* implies that a component has the ability to decide whether and how much to share its functionality (i.e., the operations it supports) and resources (i.e., the data it manages) with others. This includes the ability to associate or disassociate itself from the federation and the ability of a component to participate in one or more federations. Several first-generation systems in the database area paid significant attention to the autonomy issue because they also attempted to support updates. In comparison, few second-generation systems have considered update issues. Although autonomy of components has been assumed in almost all systems, there has been virtually no attention paid to the challenges posed by the various aspects of the autonomy dimension.

### *2.1.3. Heterogeneity*

Many types of heterogeneity are due to technological differences; for example, differences in hardware, system software (e.g., operating system), and communication systems. Researchers and developers have been working on resolving such heterogeneity for many years. Figure 2.1 shows one perspective on heterogeneity. Focusing on the crucial dimension of heterogeneity and corresponding solutions leads us to discuss different levels of interoperability—*system, syntax, structure*, and *semantic*. In this classification, we consider differences in machine-readable aspects of data representation, also referred to as formatting, to be relevant to syntactic heterogeneity. We consider representational heterogeneity that involves data modeling constructs to be relevant to structural interoperability. Schematic heterogeneity that particularly appears in structured databases is also an aspect of structural heterogeneity.

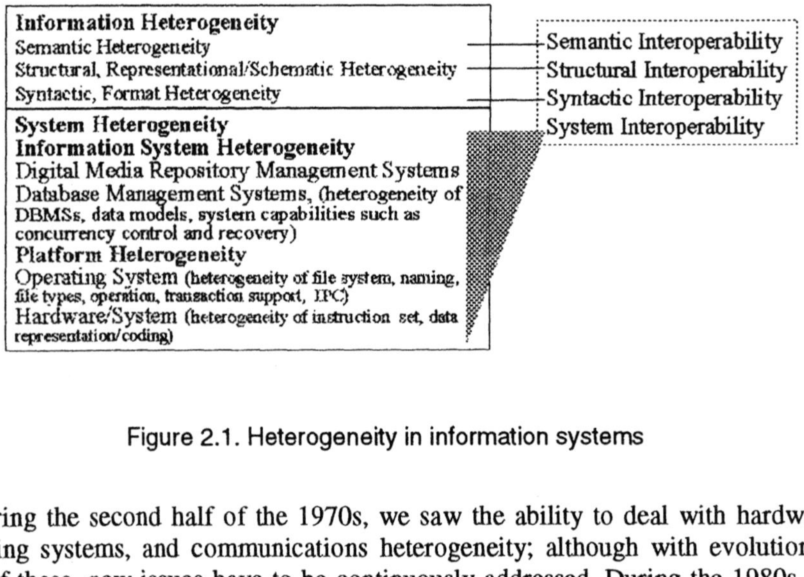


Figure 2.1. Heterogeneity in information systems

During the second half of the 1970s, we saw the ability to deal with hardware, operating systems, and communications heterogeneity; although with evolution in each of these, new issues have to be continuously addressed. During the 1980s, we saw significant progress in managing heterogeneity and support interoperability or integration in environments with structured databases and traditional database management systems (DBMSs). There is a large body of work during the first generation in dealing with heterogeneity associated with data models or schematic issues, DBMSs including query languages, concurrency control, commit and recovery, etc.

During the 1990s, the emergence of distributed computing, middleware technology, and standards has allowed us to increase focus on the heterogeneity that is intrinsic to data (or media). This has particularly supported syntactic and structural interoperability, and allowed us to address issues at the information level. As the future information system increasingly addresses the information and knowledge level issues, it will increasingly require semantic interoperability. Semantic interoperability requires that the information system understands the semantics of the user's information request and those of information sources, and uses mediation or information brokering to satisfy the information request as well as it can.

The remainder of this chapter provides an overview of the three generations of systems, with emphasis on the heterogeneity dimension of support for different levels of interoperability. Table 2.1 provides an overview of the three generations in terms of a variety of criteria.

| | Generation I | Generation II | Generation III |
|---|---|---|---|
| **Level of interoperation concern (new emphasis underlined)** | system, data | system, data, information | System, data, information, knowledge (incl. social), process |
| **Types of interoperability emphasized** | system (computer system and communication); limited aspects of syntax and structure (data model); transparency of location, distribution, replication, data models | syntax (data types and formats), structure (schematic, query languages and interfaces) | semantic (increasingly domain-specific) |
| **Dominant interoperability architecture** | multidatabases or federated databases | federated information systems, mediator | mediator, information brokering |
| **Scope of system interoperability** | handful of interconnected computers and databases | tens of systems on a LAN, databases and text repositories | enterprise-wide and global scope |
| **Software and information system architecture** | terminal access, point-to-point; also mainframes and minicomputers with remote access, client-server (two-tier); | client-server (three tier); | network, distributed, and mobile |
| **Communication infrastructure on which system interoperability solutions are built** | proprietary (IBM domination), TCP/IP | TCP/IP, http, CORBA | Internet/Web/Java, distributed object management, component, but increasingly higher-level such as multi-agent, mobile |
| **Types of data** | structured databases and files | structured databases, text repositories, semi-structured and structured and data in generic (e.g., SGML, HTML) and domain specific formats | all forms of digital media with increasing support for visual/spatio-temporal/ scientific/engineering data; |

| | | | |
|---|---|---|---|
| **Dominant information source/system model** | Relational and E-R | object-oriented | component-based; multi-modal |
| **Data/ information interoperability approaches** | structural and data model, data representation | understanding of a variety of metadata, comprehensive understanding of schematic heterogeneity | comprehensive use of metadata, increasing emphasis on semantics and ontology supported approaches |
| **Interoperability techniques (representative samples)** | data-level relationships, common/canonical data models, mappings, database exchanges, remote database interfaces, query transformations, schema translation, schema integration | schematic and metadata-level relationships, wrappers, extractors, single ontology, metadatabase, schematic heterogeneity, multidatabase consistency, mediators | multiple ontologies, information or semantic level relationships, context, media-independent information correlation, inter-ontological relationships, metadata consistency |
| **Key human roles in supporting interoperability** | data(base) administrators or experienced users, knowledgeable data structures and models, software developer written access programs | software developers to generate wrappers and mediators (with some toolkits) involving data level issues | domain experts for ontologies and for generating information correlations |
| **Access options** | database query language (SQL) for structured databases, keyword accesses for textual data/files | keyword-based attribute and (limited) content-based access, (limited) ontology-based access, | multimedia views; visual interfaces; information requests that are media-independent, multi-ontology based, context-sensitive and domain-specific |
| **A few representative applications** | integration of business databases or public databases | digital library, integrated access to heterogeneous data for a software team | digital earth, environmental phenomena, multi-step and multi-modal intelligence analysis |
| **One representative complex query** | Find a four star restaurant with less than $25 average cost that serves Mediterranean food in Richmond (a multidatabase query on distributed structured databases) | Find flowers suitable for winter gardens that look like *this* <image> with a soft smell (a keyword-, attribute-, and content-based query on text and image data repositories) | Find a block of land with urban land cover and moderate relief and population greater than 5000 and area greater than 1000 sq ft suitable for a strip mall (a query with terms whose meanings are understood by the system, and may involve multi-step processing against multi-modal data) |

| | | | |
|---|---|---|---|
| **Research prototypes** | ADDS, DDTS, Interbase, Mermaid, MRDSM, Multibase, Omnibase, see Sheth and Larson (1990) for more examples | GARLIC, Harvest, HERMES, InfoHarness/Visual-Harness, Information Manifold, InfoSleuth, RUFUS, SIMS, TSIMMIS, ... | |
| **Products (Companies)** | UniSQL/M (UniSQL), Mermaid (Data Integration), DataJoiner (IBM), OmniConnect (Sybase) | AdaptX/Harness (Bellcore), (Junglee), TIE (Tesserae), (Excalibur) | |

Table 2.1: An overview of three generations of interoperability R&D

## 2.2. First generation

By the 1980s, corporations had amassed large amounts of data in different departments to serve different applications, and on computers with different hardware and software (including DBMSs). The mantra of "data is a corporate resource" drove needs for exchanging and sharing the data between departments and within enterprises. Perhaps the most representative work in this generation occurred in the context of multidatabases (Litwin et al. 1982), or federated database systems (Heimbigner and McLeod 1985; Sheth and Larson 1990). We focus on some of the key understandings gained during this period (Drew et al. 1993; Elmagarmid 1992; Elmagarmid and Pu 1990; Elmagarmid et al. 1998; Hsiao et al. 1993; Kambayashi et al. 1991; Kim 1995; Ram 1991; Schek et al. 1993; Sheth 1987).

Much of the emphasis during this generation was on achieving system interoperability, in particular by addressing the heterogeneity due to differences in DBMSs, with some work on syntactic heterogeneity and schematic heterogeneity as appropriate to the structure of databases. Correspondingly, the emphasis was on data management and data (as opposed to information or knowledge).

Each DBMS has an underlying data model used to define data structures and constraints. Both representation (syntactic as well as structural issues including constraints) and language aspects can lead to heterogeneity. Following are some of the issues that received significant attention from the database and information systems research community.

**Differences in structure**: Different data models provide different structural primitives (e.g., the information modeled using a relation or table in the relational model may be modeled as a record type in the network database model). If the two representations have the same information content, it is easier to deal with the differences in structures. For example, address can be represented as an entity in one schema and as a composite attribute in other schema. If the information content is not the same, it may be very difficult to deal with the difference. As another example, some data models (notably semantic and object-oriented models) support generalization (and property inheritance) while others do not.

**Differences in constraints**: Two data models may support different constraints. For example, the set type in a schema based on a network database model may be partially modeled as a referential integrity constraint in a relational schema. However, a network database model supports insertion and retention constraints that are not captured by the referential integrity constraint alone. Triggers (or some other mechanism) must be used in relational systems to capture such semantics.
**Differences in query languages**: Different languages are used to manipulate data represented in different data models. Even when two DBMSs support the same data model, differences in their query languages (e.g., QUEL and SQL, or different versions of SQL supported by two relational DBMSs) could contribute to heterogeneity.
**Differences in the system aspects of the DBMSs**: Examples of system-level heterogeneity include differences in transaction management primitives and techniques (including concurrency control, commit protocols, and recovery), hardware and system software requirements, and communication capabilities.

A number of prototypes, including some of the well-known systems such as Multibase, Mermaid, DDTS, ADDS, or MRDSM addressed a variety of technical issues such those listed above. However there has been very little commercialization resulting from these efforts, and the success of commercial systems has been limited. Reasons are both technical and business (see Sheth 1995 for a discussion). Elmagarmid et al. (1998) provide a more recent review of technical work during this generation.

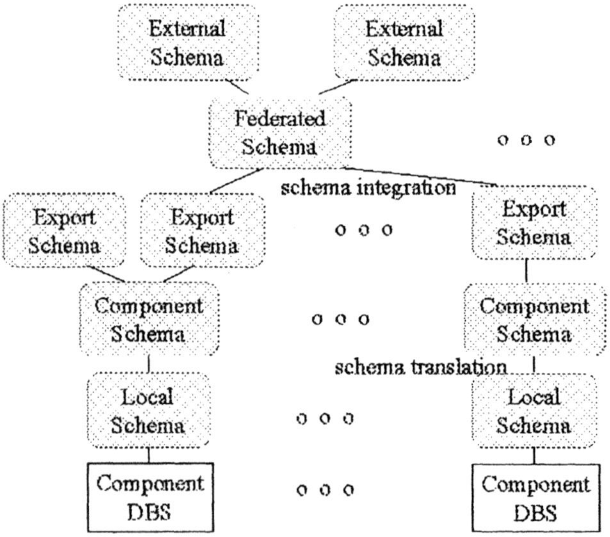


Figure 2.2. Schema architecture for a federated database system

Figure 2.2 shows how some of the distribution, autonomy, and heterogeneity issues for integrating component databases can be handled (see Sheth and Larson 1990 for more details). Briefly, the local schemas can represent data in the data model of respective DBMSs. To be able to compare the data objects modeled in different data models, one has either to perform direct and pairwise comparison—something like comparing apples and oranges—or to convert the schemas to a common or canonical model, preferably with an expressive power exceeding that of models for component databases, and then compare objects. Defining export schemas allows handling of one aspect of autonomy. Integration of export schemas into federated schemas allows for integrated or uniform access to objects managed by multiple component databases. Defining external schemas allows for handling additional types of heterogeneity.

DBMS-level heterogeneity covers only a small set of heterogeneity related to structured databases. Figure 2.3 shows one classification of a variety of conflicts related to achieving interoperability among or integration of multiple databases managed by traditional DBMSs (Kim et al. 1993; Sheth and Kashyap 1993).

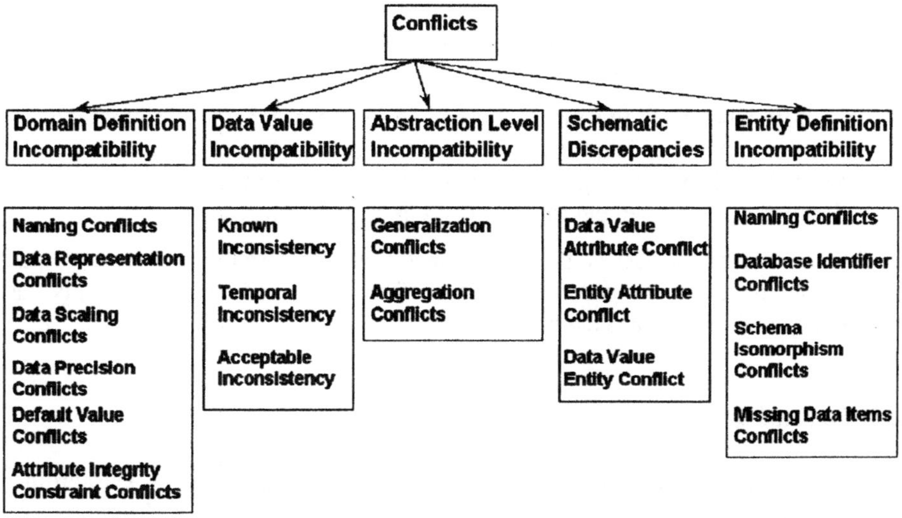


Figure 2.3. A classification of conflicts in structured databases

We believe that some of the lasting contributions of research in this generation are:

- understanding of the data models for structured databases, and the techniques to deal with data modeled with different models (schema translation);
- techniques for schema integration (Batini et al. 1986) and understanding the complexity (and at times the futility) of performing integration involving real-world schemas (Sheth 1995);
- understanding of and finding ways to deal with schematic or representational heterogeneity, and the early realization of the distinction between schematic/representational issues and semantic issues in dealing with relationships and possible integration of data managed by different sources (or

DBMSs; Drew et al. 1993; Meersman and Mark 1997; Sheth 1991; Sheth and Kashyap 1993);

- issues of managing consistency among data managed by different databases, and corresponding research in advanced transaction models (Elmagarmid 1992) and multidatabase transactions execution (Georgakopolous and Rusinkiewicz 1994).

The predominant architectural framework during the first generation systems was that of the federated databases (Heimbigner and McLeod 1985; Sheth and Larson 1990). A federated database system (FDBS) architecture consists of a schema component, such as the one shown in Figure 2.2, interspersed with processors (such as transforming processors to help with heterogeneity, filtering processors to help with autonomy, and constructing processors to help with integration of distributed data). Two broad subcategories of FDBS architectures emerged:

- a loosely-coupled architecture that provided for a more dynamic or flexible federation (where it is easier for a component to join or leave), usually with no support for updates, but with reliance on more end user involvement and less transparency of resources; and
- a tightly-coupled architecture that provided for more stable federation, with ability to support updates and distributed transactions as well as better integration of data managed by components, but requiring more systems administrator involvement for such tasks as schema integration.

One of the major issues that this generation addressed but that subsequent generations have yet to consider is that of updates, and correspondingly data consistency and use of transaction mechanisms.

## 2.3. Second generation

During the second generation two very important trends brought extraordinary opportunities for interoperability and exploitation of data: (1) proliferation of a variety of data—from structured databases, and semi-structured data, to digital media, including visual media (Gupta and Jain 1997), and (2) spread of the Internet and emergence of the Web. Applications such as digital libraries (Paepcke et al. 1998) and electronic commerce provided the context of interoperability.

Some of the key trends and achievements of this generation are (Papazoglou and Schlageter 1998; Sheth and Klas 1998):

- technology for dealing with heterogeneity of systems, data, and representational levels;
- support for a broader variety of data—not just structured databases, but also text, semi-structured, and unstructured (including image and video) data;
- use of a broad variety of metadata to support interoperability and integration; and
- use of knowledge representation and reasoning, especially for handling terminological differences.

### 2.3.1. Metadata

Metadata are usually defined as data about data. Often there is more than that, involving information about data as they are stored or managed, and revealing partial semantics such as intended use (i.e., application) of data. This information can be of broad variety, meeting if not surpassing the variety in the data themselves. Metadata can be regarded as an extension (albeit a significant one) of the concept of the schema in structured databases. They may describe or be a summary of the information content of the individual databases in an intentional manner. They typically represent constraints between the individual media objects that are implicit and not necessarily represented in the databases themselves. Some metadata may also capture content-independent information like location and time of creation. Examples of what we consider media types are structured data (data in relational or object-oriented databases), textual data (of different formats, such as Word files, source code, etc.), images (of possibly different modalities such as X-Ray, MRI scan), audio (of possibly different modalities such as monaural, stereophonic), and video. Sheth and Klas (1998) give an extensive discussion on types of metadata and their applications in managing and exploiting various digital media.

The criterion we use to classify metadata (Kashyap et al. 1995) is the extent to which they are successful in capturing the (data and information) content of the information asset (also called artifact or document in different contexts) represented in various media types. The level of abstraction at which the content of the assets is captured is very important. We believe that to capture the semantic content (i.e., at a level of abstraction closer to that of humans), it is important for the metadata to model application domain-specific information. Metadata descriptions present two advantages:

1. They enable the abstraction of representational details such as the format and organization of data, and capture the information content of the underlying data independent of representational details. This represents the first step in reduction of information overload as intentional metadata descriptions are in general an order of magnitude smaller than the underlying data.
2. They enable representation of domain knowledge describing the information domain to which the underlying data belong. This knowledge may then be used to make inferences about the underlying data. This helps in reducing information overload as the inferences may be used to determine the relevance of the underlying data without accessing the data.

One of several classifications of metadata (Kashyap et al. 1995) is as follows (also see Boll et al. 1998; Lagoze et al. 1996):

**Content-independent metadata**: This type of metadata captures information that does not depend on the content of the asset with which it is associated. Examples of this type of metadata are location, modification date of a document and type of sensor used to record a photographic image. There is no information content captured by these metadata but these might still be useful for retrieval of assets from their actual physical locations and for checking whether the information is current or not.

**Content-dependent metadata**: This type of metadata depends on the content of the asset it is associated with. Examples of content-dependent metadata are size of a document, maximum number of colors, number of rows, number of columns of an image. Content-dependent metadata can be further subdivided as follows:

**Direct content-based metadata**: This type of metadata is based directly on the contents of an asset. A popular example of this is full-text indices based on the text of the documents. Inverted tree and document vectors are examples of this type of metadata.

**Content-descriptive metadata**: This type of metadata describes the contents of an asset without their direct utilization. It often involves use of knowledge or human perception or cognition. An example is denoting the fragrance of an image containing a flower. Another example of this type of metadata is textual annotations describing the contents of an image.

**Domain-independent metadata**: This type of metadata captures information present in the document independent of the application or subject domain of the information. Examples of these are the C/C++ parse trees and HTML/SGML document type definitions.

**Domain-specific metadata**: Metadata of this type are described in a manner specific to the application or subject domain of information. Issues of vocabulary become very important in this case as the terms have to be chosen in a domain-specific manner. Examples of such metadata are relief, land cover from the GIS domain, and area and population from the Census domain. In the case of structured data, the database schema is an example of such metadata. Another interesting example is domain-specific ontologies, terms from which may be used as vocabulary to construct metadata specific to that domain.

Metadata may be precomputed (and possibly stored in a database) or they may be computed when needed (at query-processing time), in which case they may be represented by a computation, procedure, or method (e.g., an image processing routine giving values for land-cover metadata of a satellite image, executed when needed; Kashyap 1997).

Use of different types of metadata leads to different query or information access options. For example, three forms of information access, as supported in the VisualHarness system (Mudumbai 1997; Shah et al. 1997), are keyword-based access, attribute-based access, and content-based access. Table 2.2 shows relationships between various aspects of metadata for image data as exploited in the VisualHarness system. Furthermore, information access may involve iterative and customized (through use of different relative weights) use of one or more of these, as shown in Figure 2.4. Other types of information access options are possible, including standard, mediated, and immersive browsing (Grosky et al. 1998), semantic associative search (Kiyoki et al. 1998), and mixed media (Chen et al. 1998). Understanding the role of metadata and its use is likely to be one of the most enduring legacies of this generation of systems.

Some of the systems in this generation adapted the FDBS architecture to federated information systems architecture, where the DBMSs as managers of component information sources (databases) were replaced by a broader variety of information systems, including simple access protocols to access data, a broad variety of DBMSs (from network DBMSs to relational DBMSs to object-relational and object-oriented DBMSs), DBMSs specialized to manage specific digital media types (predominantly images), the Web itself as a manager of semi-structured data, and even expert systems. However, the mediator architectures (Wiederhold 1992) were clearly the dominant ones, involving wrappers for encapsulating heterogeneous information sources to provide more uniform interface to the rest of the world, and mediators to provide a broad variety of value-added services (Wiederhold 1997).

| Metadata | Content proc. used | Semantic (domain-specific) info. used | Typical format of metadata | Primary access option | Example for the metadata of this type from image data |
|---|---|---|---|---|---|
| **Content-indep.** | No | No | Attribute values | Attribute-based | Image height, width, size, date of creation, etc. |
| **Content-dep.** | Yes | No | Feature values | Content-based | Color, composition, texture, structure etc. (domain-independent) |
| **Content-based** | No | Yes | Attribute values | Attribute-based | Color (based on human interpretation), height (of the real-world object depicted by the image), fragrance, hybridizer of a flower; model, category etc. of an aircraft (domain-specific) |
| **Content-descr.** | May be | Either | Text and indices on text | Keyword-based | Descriptions of a flower, an x-ray, an aircraft (domain-specific); general descriptions of the image itself (domain-independent) |

Table 2.2: Image metadata classification and uses

During this generation, we saw increasing standardization or adoption of ad hoc standards, resulting in significant progress towards achieving system, syntactic, and structural interoperability. Acceptance of the Internet as a standard for interconnections between the systems, and evolution of infrastructures and middleware that support distributed computing (RMI, CORBA, and DCOM), and database connectivity (e.g., ODBC, JDBC etc. for relational databases), have had significant positive impact on achieving system interoperability. Syntactic interoperability includes the ability to deal with formatting and data exchange as supported by standards such as HTML for most current web-accessible documents, Z39.50 for bibliographic data, and MPEG-1 for pixel-level representation of image data. At the structural level, standardization for data modeling such as ANSI SQL, and object modeling standards and methodologies such as ODMG and UML have helped. Structural interoperability is further promoted by the use of IDL for distributed objects. Structural and a limited form of semantic interoperability are achieved by adoption of general purpose metadata standards, such as the Dublin Core (OCLC 1997), as well as metadata standards in various domains such as bibliography (e.g., LCC and DDC; Beard and Smith 1998), space and astronomy, and geography and environment (e.g., FGDC and UDK; Günther and Voisard 1998).

Among a large number of systems representing this generation, three classes of systems stand out: systems focusing on information integration or uniform access to heterogeneous repositories, systems providing more dynamic architecture or query processing mechanisms to process a user service or information request on demand, and systems that address domain-specific or semantic-level issues. A brief review of these three classes of systems follows.

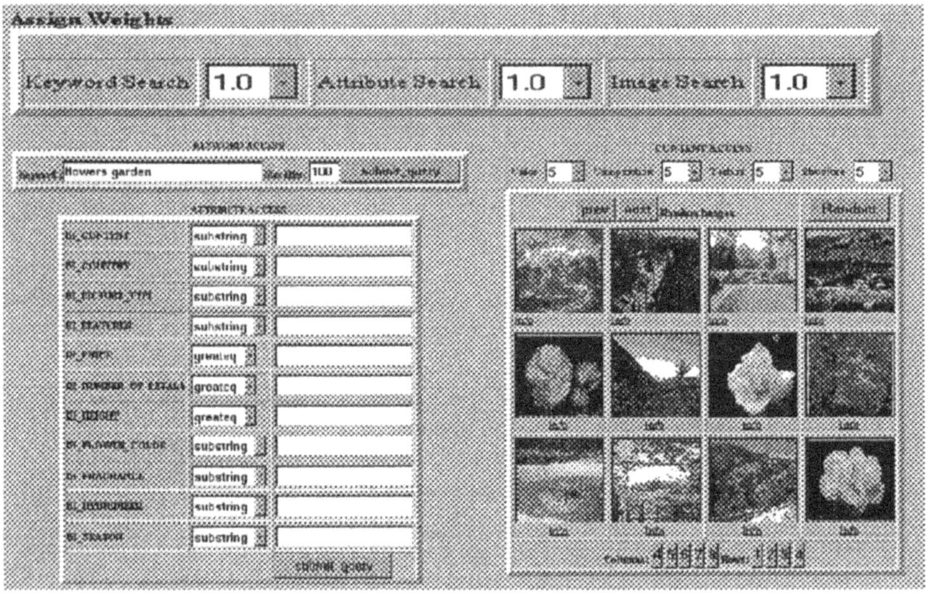


Figure 2.4. Keyword-, attribute-, and content-based access in VisualHarness

A small representative set of systems that support access to heterogeneous and distributed information sources includes TSIMMIS (*http://www-db.stanford.edu/tsimmis*), Information Manifold (*http://www.research.att.com/~levy /imhome.html*), GARLIC (*http://www.almaden.ibm.com/cs/showtell/garlic*), InfoHarness/VisualHarness (*http://lsdis.cs.uga.edu/infoharness*), SIMS (*http:// www.isi.edu/sims*), HERMES (*http://www.cs.umd.edu/projects/hermes*), and InfoSleuth (*http://www.mcc.com:80/projects/infosleuth*). SIMS, HERMES, and InfoSleuth also share capabilities with the third class as they support domain modeling or ontology, coupled with knowledge-based reasoning. WebSQL is an example of the second class of systems. There are also a number of systems that focus on specific media types or application domains, for example QBISM (Arya et al. 1994) and CoBase (*http://burton.cs.ucla.edu/*) for image data management, and THETIS (Houstis et al. 1997) and Drew and Ying (1998) for environmental data and geographic information systems, etc. For brevity, we only discuss some of these systems.

TSIMMIS (Garcia-Molina et al. 1995) uses a mediator approach to combine information from several sources containing textual and semi-structured data. Data sources are encapsulated using wrappers or translators that logically convert the data to a common information model by translating information requests and results to this common model. The mediator layer above the wrappers is responsible for routing queries to appropriate sources and for post-processing the results. An important focus of the system is to generate wrappers and mediators automatically for a set of specified rules. Thus, TSIMMIS provides a framework for users to specify information integration, which may be done manually or in a semi-automated manner.

The Information Manifold (Levy et al. 1995) is a system for retrieval and organization of information from disparate (structured and unstructured) information sources. The architecture of Information Manifold is based on a knowledge base containing a rich domain model that enables describing the properties of the information sources. The user can interact with the system by browsing the information space (which includes both the knowledge base and the information sources). The presence of descriptions of the information sources also enables the user to pose high-level queries based on the content of the information sources. One area of focus in the project has been to optimize the execution of a user query expressed in a high-level language, which might potentially require access to and combination of content from several information sources.

The InfoHarness system (Shklar et al. 1995) and its commercial counterpart AdaptX/Harness (Sheth 1996) provide browsing as well as keyword- and attribute-based access (involving typed attributes and logical operators) to a broad variety of Web-accessible heterogeneous documents (e.g., unstructured text, semistructured text such as AP news and emails, word processor and source code files, data accessible through well-defined interfaces such as NNTP news group server), and relational databases. Some of the key features are extensibility of metadata and corresponding dynamically-created query interface with typed attributes, logical remodeling of information space (such as browsing or searching source code files by function signatures), and support for multiple, third-party indexing strategies. The VisualHarness system (Mudumbai 1997; Shah and Sheth 1998) further extended it to support images using third-party visual image retrieval engines, and customizable access involving keyword-, attribute-, and content-based access.

WebSQL (Mendelzon et al. 1997a) is a high-level declarative query language for extracting information from the Web. WebSQL takes advantage of multiple index servers without requiring users to know about them, and integrates full text with topology-based queries. This enables definition of the content of domain-specific text indexes. WebSQL is used to define logical views on the unstructured global repository of Web-accessible documents. A level above Web SQL is the Web Semantics Query Language (WSQL) (Mendelzon et al. 1997b). This system has different layers of abstraction and provides mechanisms for describing the data that are available, for discovering the existence of data relevant to a problem, and for accessing discovered relevant data. WSQL has constructs for source discovery via controlled Web navigation, source registration in domain-specific catalogs, associative selection of sources from existing catalogs, and uniform access to data stored in heterogeneous sources.

There are several systems that employ the metadata-based semantic view of the world and employ an ontological layer above it. In the SIMS project (Arens et al. 1996) a model of the application domain is created using a knowledge representation system to establish a fixed vocabulary describing objects in the domain, their attributes, and relationships among them. For each information source a model is constructed that indicates the data model used, query language, network location, size estimates, etc., and describes the contents of its fields in relation to the domain model. Queries to SIMS are written in the high-level uniform language of the domain model. SIMS determines the relevant information sources by using the knowledge encoded in the domain model and the models of the information sources. These information sources are determined at run time based on their availability at that time.

The HERMES (Adali and Subrahmanian 1994) system follows the mediator architecture for semantically integrating different and possibly heterogeneous information sources (including those containing visual data) and reasoning systems. This integration is done using mediators that are very similar to the ones in the TSIMMIS system described above. Mediators, in HERMES, are logical guidelines of how information from different sources will be combined and integrated. In this framework, external information sources are abstracted as domains which execute certain functions with pre-specified input and output types. These domains are accessed in mediators using a logic-based declarative language. The system also provides a uniform environment for adding new external sources to existing mediators.

The InfoSleuth system (Bayardo et al. 1997) views an information source at the level of its relevant semantic concepts, thus preserving the autonomy of its data. Information requests to InfoSleuth are specified generically, independent of the structure, location, or even existence of the requested information. InfoSleuth filters these requests, specified at the semantic level, flexibly matching them to the information resources that are relevant at the time the request is processed. The InfoSleuth approach is to specify a common ontology for a domain, and local mappings from individual database schemas to the common ontology. These mappings can be thought of as views of the data that simplify query specification for selecting information. Given an appropriate set of mappings for a particular knowledge discovery task, the InfoSleuth system provides query support for selecting relevant information. It also pre-processes and transforms the underlying database data into records whose attributes consist of concepts from the ontology. Early emphasis in the InfoSleuth system was on harnessing structured databases and support for text data is also reported.

## 2.4. Third generation

One thing that has not changed for the third generation is that we are once again faced with more distribution, more autonomy, and more heterogeneity among the accessible information, information sources, and users. With the progress in global interconnectivity, we now need to deal with more heterogeneous information consisting not only of a broader variety of digital data, but also operations and computations (such as simulations) that can create new data and information. The scale of the problem has changed from a few databases to millions of information resources, and the new resources are added independently to the accessible set of resources, as other resources change rapidly or disappear. Currently favorite strategies that depend on keyword-based access or involve only representational or structural components of data are usually found to provide a poor quality of result, and their lack of precision leads to increasing information overload. We fully expect increasing standardization and interoperability at system, syntactic, and structural levels to address many issues—for example, see Paepcke et al. (1998) for relevant work in the domain of digital libraries. However, the key challenges to be faced are at the semantic level, where people would increasingly expect the information systems to help them not at the data level, but at the information, and increasingly knowledge levels.

Even a casual user of the Web is aware of the rapid increase in the amount and diversity of information available online. However, what is creating an even bigger challenge is the increased expectations of the user in terms of understanding of the context of the user's information need, increasing availability of semantically rich visual and new media, and a corresponding need to support semantic-level interoperability. The problem of information overload has turned the challenge of "So far (schematically) yet so near (semantically)" (Sheth and Kashyap 1993) faced by the previous generations into "So near (syntactically and structurally) yet so far (semantically)".

Although there are several uses and interpretations of *semantics* in information systems, our view is that future information systems will need to support a more general notion that involves relating the content and representation of information resources to entities and concepts in the real world (Beech 1997; Meersman 1997; Sheth 1997). That is, the limited forms of operational and axiomatic semantics of a particular representational or language framework are not sufficient (see Paepcke et al. 1998 for a relevant discussion on syntax and some types of semantics). Semantic interoperability will then support high-level (hence easier-to-use), context-sensitive information requests over heterogeneous information resources, hiding system, syntax, and structural heterogeneity. In essence, we need an approach that reduces the problem of knowing the contents and structure of many information resources to the problem of knowing the contents of easily-understood, domain-specific ontologies, which a user familiar with the domain is likely to know or understand easily.

Foundational research leading to building the third generation of information systems has been carried out in several umbrella projects and initiatives, including Knowledge Sharing Effort (*http://www-ksl.stanford.edu/knowledge-sharing*), Intelligent Integration of Information (*http://mole.dc.isx.com/I3*), and the Digital Library Initiative (*http://www.cise.nsf.gov/iis/dli_home.html*). Systems belonging to the third class of the second generation have also made contributions that the third generation systems can build on. Increasing standardization at different levels of information systems architecture for corresponding type of interoperability also plays an important role. Some of the examples are as follows.

- *System*: IIOP for interactions between distributed objects and components, KQML for interaction between agents;

- *Syntactic*: XML for all forms of Web-accessible data;
- *Structural*: RDF for general purpose description of information sources, various object models for web-based information exchange (Manola 1998), MPEG-4 for structural or object-level description video, MHEF-5 for multimedia and hypermedia, KIF for knowledge representation, OKBC for distributed knowledge bases;
- *Semantic*: MPEG-7 (still in progress) with likely support for limited forms of semantics with identification of context, objectives requirements, and applications.

We now focus our attention on a discussion of possible enablers of semantic interoperability. In particular, we identify three enablers and capabilities:
**Terminology (and language) transparency**: This will allow a user to choose an ontology of his or her choice (e.g., one based on LCC for querying bibliographic data or FGDC for geospatial data), while allowing the information source to subscribe to a related but different ontology (e.g., an ontology based on DDC or UDK, respectively. The latter recognizes some overlap between geospatial data sets and environmental data sets, and their respective modeling).
**Context-sensitive information processing**: The information system will recognize or understand the context of an information need and use it to limit information overload, both by formulating more precise queries used for searching information sources and by filtering and transforming the information before presenting it to the user.
**Semantic correlation**: This will allow the representation of semantically-related information regardless of distribution and heterogeneity (including various forms of media) by the user or the third party, and their use for obtaining all forms of relevant information anywhere.

Three key components of a possible solution are metadata (especially domain-specific and content-based metadata), contexts, and ontologies (Kashyap and Sheth 1998). We briefly discuss their role in developing semantic interoperability solutions. One key aspect of the third generation (operation or process interoperation) will not be discussed for brevity.

### *2.4.1. Ontologies and terminology transparency*

An ontology can be defined as a specific vocabulary and relationships used to describe certain aspects of reality, and a set of explicit assumptions regarding the intended meaning of the vocabulary of words (Gruber 1991; Guarino 1998). Among various other classification schemes and structures, including keywords, thesauri, and taxonomies, ontologies are often viewed as allowing more complete and precise domain models (Huhns and Singh 1997). Support and use of multiple, independently-developed ontologies is important for developing scalable information systems with multiple information producers and consumers (e.g., Arens et al. 1996; Dao and Perry 1996; Genesereth and King 1995; Kashyap and Sheth 1998; Khang and McLeod 1998 for need and use of multiple ontologies). One challenging issue in supporting semantic interoperability is how to allow both users and providers to subscribe to existing ontologies of their choice or create a new one (Kashyap and Sheth 1998). Processing an information request represented in terms of one ontology in an environment with information resources that subscribe to different (but related and relevant) ontologies may involve using inter-ontological relationships, such as synonym, hypernym, homonym, and other possibly domain-specific relationships. This work also requires understanding of and containing loss of information in multi-ontology query processing (Mena et al. 1998). One early example of research along these lines is the OBSERVER (sub)system (*http://siul02.si.ehu.es/~jirgbdat /OBSERVER*), which is a component of the InfoQuilt system (*http://lsdis.cs.uga.edu/infoquilt*).

### 2.4.2. Context

In characterizing the similarity between objects based on the semantics associated with them we have to consider the real-world semantics (RWS) of an object. It is not possible to define completely what an object denotes or means in the model world. We propose the *context* of an object as the primary vehicle to capture the RWS of the object. Understanding of the context of the information request can help the system to distinguish between whether the term *cricket* refers to an insect or a sports game.

Adapting from research in AI and Knowledge-Based systems (e.g., Shoham 1991), linguistics and other fields, modeling and representing context can lead to several benefits in dealing with information overload in a global information infrastructure (GII; see Kashyap and Sheth 1998 for more details):

- *Economy of representation*: In a manner akin to database views, contexts can act as a focusing mechanism when accessing the component databases or information sources on the GII.
- *Economy of reasoning*: Instead of reasoning with the information present in the database as a whole, reasoning can be performed with the context associated with an information source.
- *Managing inconsistent information*: In the GII, where information sources are designed and developed independently, it is not uncommon to have information in one source be inconsistent with information in another. As long as information is consistent within the context of the query of the user, inconsistency in information from different databases may be allowed.
- *Flexible semantics*: An important consequence of associating abstractions or mappings with context is that the same two objects can be related to each other differently in two different contexts. Two objects might be semantically closer to each other in one context as compared to the other.

There are several proposals for representing context. We believe that an effective approach needs to bring together metadata, user profiles, information modeling abstractions, and ontologies, as well as to allow their dynamic construction to model application domain and user needs. Besides their modeling and representation, a key challenge includes the ability to reason about or compare contexts (e.g., Kashyap and Sheth 1996; Lee et al. 1996; Ouksel and Naiman 1993). While there are many representations and associated reasoning techniques, practical application of context in GII is expected to be a key research challenge for achieving semantic interoperability in information systems.

### 2.4.3. Information correlations

One of the key applications of semantics in GII is to represent or specify information requests and semantic level information correlations regardless of the media (and other heterogeneity) and locations of information sources. These can involve queries over heterogeneous media assets represented at a higher level of abstraction in media-independent manner, using metadata and ontologies.

Two approaches to *representing* information correlations between independently-managed networked resources are Metadata Reference Links (MREFs; Shah and Sheth 1998; Sheth and Kashyap 1996) and Distributed Active Relationships (DARs; Daniel et al. 1998). They provide an initial step in specifying information correlation between heterogeneous digital media. Specifically, MREFs allow subscription to one or more ontologies in their specification, and the meta-information used in specifying an MREF is mapped to views involving keyword-based, attribute-based, and content-based specifications for various types of metadata of heterogeneous digital media. Specification and processing based on information correlations can be easily integrated with the Web technology. For example, MREF could be used anywhere a hypermedia link (HREF) is used, and its specification and processing can be supported using an RDF and XML-based infrastructure. However, many challenges remain in extending the current proposals to include non-standard resources such as datasets and procedures, integrating information correlation representation and processing with context and context mediation, and processing them efficiently in a very large information space.

### 2.4.4. Information-brokering architecture

It is hard to predict the architecture of the third generation of information systems. One proposed architecture, information brokering (Kashyap 1997; Kashyap and Sheth 1994), adapts and extends the concepts of (1) federated environments (Heimbigner and McLeod 1985; Sheth and Larson 1990) in which resources, metadata, and ontologies are created, administered, and enhanced independently; and (2) mediator architectures (Wiederhold 1992) which involve decoupling information creators and providers from information users and better semantic-level services and interoperability. However, we believe that the key to this generation of systems is their support for semantic interoperability, through exploitation of various forms of metadata, multiple ontologies, and contexts. Furthermore, we believe that before very general architectures that can support various domains can be developed, support for semantic interoperability demands that we focus on a specific domain first, such as GIS (Goodchild et al. 1997), and then extend what we learn to general-purpose and multi-domain environments. Figure 2.5 shows a schematic of a system for supporting semantic interoperability as described above in a geographical domain. It is too early to give representative examples of the third generation, but a few early efforts are described by Wiederhold (1996), and Papozoglou and Schlageter (1998); and see InfoQuilt (*http://lsdis.cs.uga.edu/infoquilt*).

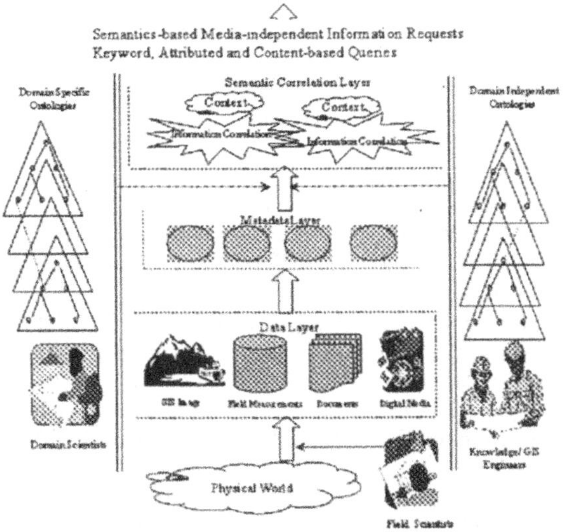


Figure 2.5. An architecture to support geographic information brokering

In closing, we believe that semantic interoperability is the key to progress towards a vision of Infocosm, a society whose members will have information anywhere, any time, and in many forms, for knowledge creation and use, effective decision-making, better learning, and more fun.

## Acknowledgments

Members of the InfoQuilt project, especially Vipul Kashyap, Kshitij Shah, Clemens Bertram, and Krishnan Parsuraman, have contributed to some of the ideas presented in this paper.

## References

Adali S, Subrahmanian V S 1994 Amalgamating knowledge bases, II: Distributed mediators. *International Journal of Intelligent and Cooperative Information Systems* 3(4): 349–383

Arens Y, Knoblock C A, Shen W 1996 Query reformulation for dynamic information integration. In Wiederhold G (ed) *Intelligent Integration of Information*. Kluwer Academic Publishers: 11–42

Arya M, Cody W, Faloutsos C, Richardson J, Toga A 1994 QBISM: extending a DBMS to support 3D medical images. *Proceedings of the 10th International IEEE Conference on Data Engineering, Houston, Texas*: 314–325

Batini C, Lenzerini M, Navathe S 1986 A comparative analysis of methodologies for database schema integration. *ACM Computing Surveys* 18(4): 323–64

Bayardo R, Bohrer W, Brice R, Cichocki A, Fowler G, Helal A, Kashyap V, Ksiezyk T, Martin G, Nodine M, Rashid M, Rusinkiewicz M, Shea R, Unnikrishnan C, Unruh A, Woelk D 1997 Semantic integration of information in open and dynamic environments. *Proceedings of the 1997 ACM International Conference on the Management of Data (SIGMOD), Tucson*: 195–206

Beard K, Smith T 1998 A framework for meta-information in digital libraries. In Sheth A, Klas W (eds) *Multimedia Data Management: Using Metadata to Integrate and Apply Digital Media*. McGraw Hill: 341–365

Beech D 1997 Data semantics on the information superhighway. In Meersman R, Mark L (eds) *Database Application Semantics*. Chapman and Hall

Boll S, Klas W, Sheth A 1998 Overview on using metadata to manage multimedia data. In Sheth A, Klas W (eds) *Multimedia Data Management: Using Metadata to Integrate and Apply Digital Media*. McGraw-Hill Publishers: 1–23

Chen F, Hearst M, Kimber D, Kupiec J, Pedersen J, Wilcox L 1998 Metadata for mixed-media access. In Sheth A, Klas W (eds) *Multimedia Data Management: Using Metadata to Integrate and Apply Digital Media*. McGraw Hill: 319–340

Daniel R, Lagoze C, Payette S 1998 A metadata architecture for digital libraries. *Proceedings of the Forum on Research and Technology Advances in Digital Libraries (ADL'98), Santa Barbara*: 276–288

Dao S, Perry B 1996 Information mediation in cyberspace: scalable methods for declarative information networks. In Wiederhold G (ed) *Intelligent Integration of Information*. Kluwer Academic Publishers: 43–62

Drew P, Ying J 1998 Metadata management for geographic information discovery and exchange. In Sheth A, Klas W (eds) 1998 *Multimedia Data Management: Using Metadata to Integrate and Apply Digital Media*. McGraw Hill: 89–121

Drew P, King R, McLeod D, Rusinkiewicz M, Silberschatz A 1993 Report of the workshop on semantic heterogeneity and interoperation in multidatabase systems. *SIGMOD Record* 22(3): 47–56

Elmagarmid A (ed) 1992 *Database Transaction Models for Advanced Applications*. Morgan-Kaufmann

Elmagarmid A, Pu C (eds) 1990 Heterogeneous databases. Special Issue. *ACM Computing Surveys* 22 (3)

Elmagarmid A, Sheth A, Rusinkiewicz M (eds) 1998 *Heterogeneous Distributed Databases*. Morgan Kaufmann Publishers

Garcia-Molina H, Hammer J, Ireland K, Papakonstantinou Y, Ullman J, Widom J 1995 Integrating and accessing heterogeneous information sources in TSIMMIS. *Proceedings of the AAAI Symposium on Information Gathering, Stanford, California*: 61–64

Genesereth M, King R (eds) 1995 *Reference Architecture, Intelligent Integration of Information*. Stanford University and University of Colorado. *http://logic.stanford.edu /achitecture/reference.html*

Georgakopoulos D, Rusinkiewicz M 1994 Using tickets to enforce the serializability of multidatabase transactions. *IEEE Transactions On Knowledge and Data Engineering* 6(1): 166–180

Goodchild M, Egenhofer M, Fegeas R 1997 *Interoperating GISs: Report of a Specialist Meeting Held under the Auspices of the Varenius Project, Panel on Computational Implementations of Geographic Concepts*. National Center for Geographic Information and Analysis, Santa Barbara

Grosky W, Fotouhi F, Jiang Z 1998 Using metadata for the intelligent browsing of structured media objects. In Sheth A, Klas W (eds) *Multimedia Data Management: Using Metadata to Integrate and Apply Digital Media*. McGraw Hill: 123–148

Gruber T 1991 The role of a common ontology in achieving sharable, reusable knowledge bases. *Proceedings of the 2nd International Conference on Principles of Knowledge Representation and Reasoning, Cambridge*: 601–602

Guarino N 1998 Formal ontology and information systems. *Proceedings of the 1st International Conference on Formal Ontology in Information Systems [FOIS'98], Torino*: 3–15

Günther O, Voisard A 1998 Metadata in geographic and environmental data management. In Sheth A, Klas W (eds) *Multimedia Data Management: Using Metadata to Integrate and Apply Digital Media*. McGraw Hill: 57–87

Gupta A, Jain R 1997 Visual information retrieval. *Communications of the ACM* 40(5): 70–79

Heimbigner D, McLeod D 1985 A federated architecture for information management. *ACM Transactions on Office Information Systems* 3(3): 253–278

Houstis C, Nikolaou C, Marazakis M, Patrikalakis N, Sairamesh J, Thomasic A 1997 THETIS: Design of a data repositories collection and data visualization system for coastal zone management of the Mediterranean Sea. *D-Lib Magazine, The Magazine for Digital Library Research*, November. *http://www.dlib.org/dlib/november97/thetis/11thetis.html*

Hsiao D, Neuhold E, Sacks-Davis R (eds) 1993 *Interoperable Database Systems (DS-5)*. North-Holland

Huhns M, Singh M 1997 Ontologies for agents. *Internet Computing* 1(6): 81–83

Kambayashi Y, Rusinkiewicz M, Sheth A (eds) 1991 *Proceedings of the RIDE-IMS'91: International Workshop on Interoperability in Multidatabase Systems*. IEEE Computer Society

Kashyap V 1997 *Information Brokering over Heterogeneous Digital Data: A Metadata Based Approach*. PhD Thesis, Rutgers University

Kashyap V, Sheth A 1994 Semantics based information brokering. *Proceedings of the 3rd International Conference on Information and Knowledge Systems*: 363–370

Kashyap V, Sheth A 1996 Schematic and semantic similarities between database objects: a context-based approach. *The Very Large Databases Journal* 5(4): 276–304

Kashyap V, Sheth A 1998 Semantic heterogeneity in global information systems: the role of metadata, context and ontologies. In Papazoglou M, Schlageter G (eds) *Cooperative Information Systems: Current Trends and Directions*. Academic Press: 139–178

Kashyap V, Shah K, Sheth A 1995 Metadata for building the MultiMedia Patch Quilt. In Jajodia S, Subrahmanian V S (eds) *Multimedia Database Systems: Issues and Research Directions*. Springer-Verlag: 297–319

Khang J, McLeod D 1998 Dynamic classificational ontologies: mediation of information sharing in cooperative federated database systems. In Papazoglou M, Schlageter G (eds) *Cooperative Information Systems: Current Trends and Directions*. Academic Press: 179–203

Kim W (ed) 1995 *Modern Database Systems: The Object Model, Interoperability and Beyond*. Addison Wesley

Kim W, Choi I, Gala S, Scheevel M 1993 On resolving schematic heterogeneity in multidatabase systems. *Distributed Parallel Databases International Journal* 1: 251–279

Kiyoki Y, Kitagawa T, Hayama T 1998 A metadatabase system for semantic image search by a mathematical model of meaning. In Sheth A, Klas W (eds) *Multimedia Data Management: Using Metadata to Integrate and Apply Digital Media*. McGraw Hill: 191–222

Lagoze C, Lynch C, Daniel R 1996 The Warwick Framework: A container architecture for aggregating sets of metadata. Technical Report TR96-1593. Cornell University, Department of Computer Science. *http://cs-tr.cs.cornell.edu:80/Dienst/UI/2.0/Describe/ncstrl.cornell/TR96-1593*

Lee J, Madnick S, Siegel M 1996 Conceptualizing semantic interoperability: a perspective from the knowledge level. *International Journal of Cooperative Information Systems* 5(4): 367–393

Levy A Y, Srivastava D, Kirk T 1995 Data model and query evaluation in global information systems. *Intelligent Information Systems* 5(2): 121–143

Litwin W, Boudenant J, Esculier C, Ferrier A, Glorieux A, La Chimia, J, Kabbaj K, Moulinoux C, Rolin P, Stangret C 1982 SIRIUS: systems for distributed data management. In Schneider H-J (ed) *Distributed Data Bases*. North-Holland, Netherlands: 311–66

Manola F 1998 *Towards a Web Object Model*. Object Services and Consulting, Inc. *http://www.objs.com/OSA/wom.htm*

Meersman R 1997 An essay on the role and evolution of data(base) semantics. In Meersman R, Mark L (eds) *Database Application Semantics*. Chapman and Hall

Meersman R, Mark L (eds) 1997 *Database Application Semantics*. Chapman and Hall

Mena E, Kashyap V, Illarramendi A, Sheth A 1998 Domain specific ontologies for semantic information brokering on the global information infrastructure. *Proceedings, International Conference on Formal Ontology in Information Systems (FOIS'98), Torino*: 269–283

Mendelzon A, Mihaila G, Milo T 1997a Querying the World Wide Web. *Journal of Digital Libraries* 1(1): 68–88

Mendelzon A, Mihaila G, Raschid L, Tomasic A 1997b Locating and accessing heterogeneous data sources. *Proceedings of CASCON'97*

Mudumbai S 1997 *ZEBRA: Customizable, Extensible Metadata-based Access to Federated Image Repositories*. M.S. Thesis, Department of Computer Science, University of Georgia

Online Computer Library Center, Inc 1997 *Dublin Core Metadata Element Set: Reference Description*. Office of Research and Special Projects, Dublin, Ohio. *http://www.oclc.org:5046/research/dublin_core/*

Ouksel A, Naiman C 1993 Coordinating context building in heterogeneous information systems. *Journal of Intelligent Information Systems* 3: 151–183

Paepcke A, Chang C, Garcia-Molina H, Winograd T 1998 Interoperability for digital libraries worldwide. *Communications of the ACM* 41(4): 33–43

Papazoglou M, Schlageter G (eds) 1998 *Cooperative Information Systems: Current Trends and Directions*. Academic Press

Ram S (ed) 1991 Heterogeneous distributed database systems. Special Issue. *IEEE Computer* 24(12)

Schek H–J, Sheth A, Czjedo B (eds) 1993 *Proceedings of the RIDE-IMS'93: International Workshop on Interoperability in Multidatabase Systems*. IEEE Computer Society

Shah K, Sheth A 1998 Logical information modeling of Web-accessible heterogeneous digital assets. *Proceedings of the Forum on Research and Technology Advances in Digital Libraries (ADL'98), Santa Barbara*: 266–275

Shah K, Sheth A, Mudumbai S 1997 Black box approach to image feature manipulation used by visual information retrieval engines. *Second IEEE Metadata Conference*

Sheth A 1987 *Heterogeneous Distributed Databases: Issues in Integration*. Tutorial Notes, 3rd International Conference on Data Engineering

Sheth A (ed) 1991 Semantic issues in multidatabase systems. Special Issue. *SIGMOD Record*

Sheth A 1995 *Multidatabase Interoperation: Perspective of Researchers and Practitioners*. Tutorial Notes, 11th International Conference on Data Engineering, Taiwan

Sheth A 1996 Bellcore's ADAPT/XHarness system for managing information on Internet and intranets. *Proceedings of the 22nd International Conference on Very Large Data Bases, Bombay, India*: 585

Sheth A 1997 Data semantics: What, where, and how? In Meersman R, Mark L (eds) *Database Application Semantics*. Chapman and Hall: 601–610

Sheth A, Kashyap V 1993 So far (schematically) yet so near (semantically). In Hsiao D, Neuhold E, Sacks-Davis R (eds) 1993 *Interoperable Database Systems (DS-5)*. North-Holland: 283–312

Sheth A, Kashyap V 1996 Media-independent correlation of information: what? how? Proceedings of First IEEE Metadata Conference

Sheth A, Klas W (eds) 1998 *Multimedia Data Management: Using Metadata to Integrate and Apply Digital Media*. McGraw Hill

Sheth A, Larson 1990 Federated database systems for managing distributed, heterogeneous, and autonomous databases. *ACM Computing Surveys* 22(3): 183–236

Shklar L, Sheth A, Kashyap V, Shah K 1995 InfoHarness: Use of automatically generated metadata for search and retrieval of heterogeneous information. *Proceedings of CAiSE-95*: 217–330

Shoham Y 1991 Varieties of context. In Lifschitz V (ed) *Artificial Intelligence and Mathematical Theory of Computation: Papers in Honor of John McCarthy*. Academic Press, Boston: 393–408

Wiederhold G 1992 Mediators in the architecture of future information systems. *IEEE Computer* 25(3): 38–49

Wiederhold G (ed) 1996 *Intelligent Integration of Information*. Kluwer Academic Publishers

Wiederhold G 1997 Value-added mediation in large-scale information systems. In Meersman R, Mark L (eds) *Database Application Semantics*. Chapman and Hall

# 3 INTEROPERABILITY AND SPATIAL INFORMATION THEORY

Andrej Vckovski

## 3.1. Introduction

The development and deployment of successful interoperability strategies requires standardization that covers many application areas, many different views and conceptualizations of the world. Interoperability between computing infrastructures needs—much like every information exchange—a set of common rules and concepts that define a common understanding of the information and operations available in every cooperating system. Standardization processes and interoperability initiatives such as those of the Open GIS Consortium try to provide an agreed-on set of such rules and concepts. Standardization processes are often driven by market forces and vendors trying to position their particular technology and product as a common concept. However, in spatial information systems (and also other areas) the common concepts include not only technical aspects but also fundamental questions on modeling spatial, real-world features, that is, problems which are beyond a specific technical approach and implementation.

The process of understanding a real-world phenomenon and providing a *code* for the communication of their states and relationships is the development of a theory. In that sense, a theory which is agreed-on by information-exchanging communities provides the previously mentioned set of common rules and concepts. Therefore, interoperability of geographic information systems needs such a theory, that is, a spatial information theory as the theoretical framework for interoperability.

This chapter discusses some of the required properties of a theoretical framework to be useful as an essential model of reality and thus, as a basis for standardization approaches. First, a definition of interoperability is given. Then, some requirements for the specification methodology are discussed. These requirements are apparently the same as those for a useful spatial information theory, that is, the specification methodology is the theory. Finally, a few comments on the existence of such a theory are discussed as a conclusion.

## 3.2. Interoperability

### 3.2.1. Why interoperability?

Interoperability has received significant research attention in recent years. Overcoming hardware limitations is not the main driving factor in the software industry anymore, and systems are increasingly designed according to what needs to be done and not what can be done. These systems do not live in splendid isolation but are typically embedded in larger information technology infrastructures with a growing need to exchange information and services. Geographic information systems (GIS) are becoming commodity products and data, hardware, and software are less expensive. In the 1970s and 1980s GIS projects typically collected their own data sets from analog sources (maps, field measurements, surveys), making the data collection and maintenance the main cost factor in a project. This has been replaced in many cases by exchanging, selling, and purchasing existing data sets, leading to the requirements for open and interoperable systems: systems that can be seamlessly embedded in larger infrastructures, systems that can be provided with external data and probably also external functionality, and systems that provide a certain level of interoperability.

Even though the research community tends to have a common understanding of what interoperability means, there is still some confusion regarding its definition. Here, a few definitions are given to surround the concept and provide a basis for the subsequent discussion.

### 3.2.2. Some definitions

Yaser Bishr (Bishr 1998) defines interoperability as "the ability of a system, or components of a system, to provide information portability and inter-application cooperative process control", based on Bordie (1992). A somewhat simpler definition is given by Vckovski (1998) as the "ability to exchange *and* integrate information", that is, two interoperating systems need to be able to exchange *meaningful* information. Therefore, interoperability always implies that there is some common concept on which information exchange and integration is based. In a computing infrastructure, interoperability exists on many levels. For example, two applications can be interoperable on the level of the targeted computing platform, that is, both applications run on the same operating system and hardware. The common ground is the operating system that provides a set of services and structures that are common to both applications. A communication protocol or a standardized data format implemented by two operating systems or applications is another level of interoperability. The higher the abstraction level on which interoperability needs to exist, the greater are the impediments because the concepts that need to be agreed on start to be more complicated. A physical layer of a network is more easily standardized than a spatial data exchange format because it needs to cover only a well-defined range of conceptualizations. The ISO/OSI reference model (Stevens 1990) is often used in the context of information exchange (communication) to organize such interoperability aspects into seven more or less well-defined layers. In ISO/OSI terminology, the interoperability issues discussed here relate to the

Application Layer 7. The layers below are usually taken care of by the underlying operating system and communication protocols.

Common to all aspects of interoperability is the necessity to have an agreed-on set of rules on which the information exchange and integration is based. Such an *agreement* is the basis of interoperability and the goal of every standardization process. The next section discusses some of the requirements that an agreement needs to fulfill in order to be useful for spatial information processing.

## 3.3. Standards for interoperability

### *3.3.1. Overview*

An agreement or standardization provides the *lingua franca* that is needed for the exchange and integration of information. The mere exchange of information is covered in most cases by well-known distributed computing infrastructures such as inter-process communication protocols, standardized storage media, or advanced environments such as CORBA. These mechanisms provide a well-defined way for some information to travel between applications, systems, or sites. The integration of information from different systems (e.g., integration of an external data set with some internal data) creates many problems that are not yet covered by the previously-mentioned methods. The integration is basically an identification of *relations* between elements of two sets of information. This identification, however, is often impeded by many heterogeneities on various levels, such as:

- Syntactic heterogeneities;
- Semantic heterogeneities.

Especially the latter are a major impediment for the interoperability of spatial information systems as they are not easily resolved using automatic procedures. However, syntactic and semantic heterogeneities can be minimized by providing an agreement or standardization on a high level of abstraction, on the level that maps reality to some abstract concepts being used in subsequent standardization of lower levels. To be useful, this agreement, which will be called spatial information theory later on, needs to fulfill a set of requirements discussed in the following. The overall objective is to have a semantically-rich way to conceptualize the real world. Being semantically rich minimizes potential semantic heterogeneities.

### *3.3.2. Requirements*

**Expressivity and simplicity**. An agreement needs to be expressive in the sense that it is able to express unambiguously as much as possible of what needs to be agreed on, that is, of the model of reality. When using a formal system or mathematical formalism to encode an agreement, this means that the set of symbols and rules in the formal system needs to be as small as possible. Having a high expressive power enables many real-world things to be expressed within the same formal system. An expressive system is usually also simpler in the sense that there are fewer concepts needed in the entire system. Simplicity also means that agreements should resemble common sense. Simple agreements are easier to understand and to deploy correctly. In the context of computing infrastructure simplicity can *never* get too much attention.

**Unambiguity and completeness.** An agreement needs to be unambiguous in order to allow precise identification of elements in the information set. To be well-defined means also a certain completeness regarding the real world elements being specified or modeled (Brassel et al. 1996). Avoiding an ambiguous and incomplete agreement needs explicit definition of all information elements. There are always (usually convincing) concepts within an agreement, standardization, or formal system that cannot be defined in terms of other well-defined entities. But everything else must be explicitly defined to avoid interpretation that pertains to implicit knowledge available in a specific scientific domain or application area.

**Extensibility and scalability.** No agreement or standardization will ever cover all possible application areas. Being complete as mentioned above always relates to a certain subset containing the concepts that are represented in the agreement. However, in order to achieve a stability and continued application, a standardization or agreement needs to be extensible and scalable. For example, many standardization approaches contain rules that define procedures for the change of the standardization over time. This includes versioning, integrity rules (can the set of rules be extended without introducing contradictions?). Scalability is another aspect of extensibility and completeness in the sense that a standardization remains applicable over various regions of system parameters such as "how many components are interoperating?"

**Acceptance.** A standardization approach does not make sense if it is not accepted by the software industry and users. There are many factors leading to a standard's acceptance, much like the acceptance of a product or service. The important notion here is that acceptance *matters*. Of course, the requirements mentioned above all are meant to lead to acceptance. However, there are still many other issues relevant such as *marketing* of the concepts.

### *3.3.3. Significance for spatial information processing*

In the context of spatial information processing the requirements discussed above lead to approaches such as the Open GIS initiative. The approach taken by OGC (the Open GIS Consortium) is to provide expressivity, simplicity, and so on both by defining specifications for the mapping from real world things to *abstract* entities and providing *implementation* specifications giving concrete implementations on a specific computing infrastructure (e.g., CORBA; CORBA 1995). Other approaches use data definition languages (e.g., INTERLIS; INTERLIS 1990) or specification languages to describe a model, which itself *implicitly* defines the mapping from reality to abstract entities.

All approaches need to address a methodological and a thematic question:

- How is the mapping formulated? That is, using graphical symbols, mathematical formalisms, computer languages, natural language, drawings.
- What is the content of the mapping, for example, the *story* written in the selected language?

These two questions seem to be very asymmetrical in importance. It seems much more important to define what the mapping is than to define how to express this. And, there are probably many equally-justified methods to express the same story, while there may not be so many stories to express if there are only few realities, after all. However, I believe that these two questions are not so asymmetrical as it seems at first glance. The methodology chosen is an important epistemic step and is more than a mere rule to encode real-world states. For example, a mathematical formalism needs only very few written (defined) statements to express infinitely many states. By stating that $a + b = b + a$ for all numbers $a$ and $b$, an infinite set of statements is implied such as for example $91 + 22 = 22 + 91$. If these statements are the reality to be described, it is definitely easier to encode this using a few formal statements than by encoding every single statement. Physics used the mathematical language not only to provide a simpler method to encode real world states or theories but also to get new insights. If the states are mapped into a formal system with a set of known rules (e.g., arithmetics), then the rules can be used to derive new states that have, for example, never been measured (what is the weight of a 1km cube of water?). These concepts are so basic to science that they tend to be forgotten because they are implied always.

In this context this means that the choice of an encoding methodology is not only a matter of taste but also can influence the understanding of the mapping itself. The next section extends the discussion of agreements and standards into something that might be called *spatial information theory.*

## 3.4. Spatial information theory

The intensified need to share data and systems made interoperability, that is, data exchange and integration, an important issue in the last decade. The underlying theoretical framework has not yet gained too much interest (e.g., Mark 1993). However, clear definitions of, for example, how spatial features are defined are a prerequisite for every standardization process. Currently most standardization processes do rely on such definitions and models, which usually are part of the standard. SDTS (Spatial Data Transfer Standard) includes, for example, exhaustive lists of known entity types and their attributes (Fegeas et al. 1992). What has this to do with spatial information theory? Andrew Frank and Irene Campari define spatial information theory as a theory that "attempts to discover the universally valid principles and to understand the differences of the particular solution" when relating any information to space (Frank and Campari 1993). Spatial information theory is seen as a theoretical base for the development of GIS.

The lack of such an agreed-on base for the development of GIS has produced many different concepts for dealing with spatial information, as the field–object duality clearly shows (Couclelis 1992). As long as systems are operated in isolation as a closed system without any data exchange and embedding into larger computing infrastructures this poses no problems. Whatever model of reality fits the particular objective best is chosen. Interoperating systems, however, need to agree on a common model of reality (at least for the data exchange and integration). This model serves as basis for the identification of relationships that are necessary for integration. Two systems exchanging information about hydrology need to share common concepts of the information entities relating to hydrology, that is, the agreement mentioned above. Spatial information theory is the theoretical base for any standardization that addresses interoperability of spatial information processing. These approaches (e.g., data transfer standards, OpenGIS™) always were based on a spatial information theory and sometimes even explicitly contain concise definition of the reality model used. However, it seems to me that a spatial information theory could be defined on a level that is even more general than a standardization approach. Most of the ideas that are part of the spatial information theory implicitly used today when developing GIS or standardization documents are a result of the research of the past decades and there are quite a few concepts which are more or less accepted by the research community. However, there is no unified way of communicating these concepts, and also no clear terminology. This leads to a situation often seen in the area of GIS where concepts are reinvented with a new name. It seems as if it might be sensible to provide a unification of the real-world models used, and the terminology and notation of spatial information, that is, to write down the spatial information theory already present in many GIS developers' heads, vendors' white papers, research articles, and books. There have been many approaches in the research area to provide such foundations for specific subsets of a spatial information theory (e.g., Goodchild 1992; Herring et al. 1990; Samet 1990; Worboys 1992), yet an exhaustive integration (and also simplification) of these results is missing.

### 3.5. Conclusions

The exchange and integration of information is a definition of interoperability that is simple, yet covers the core issues. The exchange and integration of information needs an agreement on how real-world features are represented, that is, an agreement on the mapping of reality onto information entities. A spatial information theory provides part of this agreement and serves as a foundation for interoperability approaches and GIS design. The development of a unified spatial information theory has been discussed in the research community for several years, but it is not at all clear whether there is such a thing. Spatial information may be too complex to be covered by a single, exhaustive, and useful theory. However, it still is an important research objective to search for an unified formula for spatial information. Focusing on such representational issues can lead to some mosaic stones of spatial information theory and probably in the long term simplify the concepts that are in use today. Fighting complexity is one of the major goals within information technologies. Research towards a probably-nonexistent single, simple, expressive, and useful spatial information theory might help to reduce complexity by providing models that are less specific, less proprietary, more interoperable, and more useful.

Interoperability will happen when there is an abstraction of reality that all interoperating components agree on. In that sense, interoperability research eventually will propose a spatial information theory, even if it will look like a standardization document. The research community can support industrial standardization efforts by supplying such a theory or components thereof. In many cases, such components already exist and need only to be summarized with a unified notation and terminology. This contribution does not provide any of these components but rather attempted to motivate research and industry to continue on the fruitful work being done by organizations such as the Open GIS Consortium.

## References


Bishr Y 1998 Overcoming the semantic and other barriers to GIS interoperability. *International Journal of Geographic Information Science* 12(4): 299–314

Bordie M L 1992 The promise of distributed computing and the challenge of legacy information systems. *Proceedings of the IFIP WG2.6 Database Semantics Conference on Interoperable Database Systems, Lorne, Australia, 16-20 November*: 1–31

Brassel K, Bucher F, Stephan E M, Vckovski A 1995 Completeness. In Guptill S C, Morrison J L (eds) *Elements of Spatial Data Quality*. Elsevier Science, Amsterdam: 81–108

CORBA 1995 *The Common Object Request Broker: Architecture and Specification*. Version 2.0. Object Management Group, Framingham

Couclelis H 1992 People manipulate objects (but cultivate fields): beyond the raster-vector debate in GIS. In Frank A U, Campari I, Formentini U (eds) *Theories and Methods of Spatio-Temporal Reasoning in Geographic Space*. Lecture Notes in Computer Science No. 639. Springer Verlag, Berlin: 65–77

Fegeas R G, Casio J L, Lazar R A 1992 An overview of FIPS 173, the spatial data transfer standard. *Cartography and Geographic Information Systems* 19(5): 278–293

Frank A U, Campari I (eds) 1993 *Spatial Information Theory, A Theoretical Basis for GIS*. Lecture Notes in Computer Science No. 716. Springer Verlag, Berlin

Goodchild M F 1992 Geographical data modeling. *Computers and Geosciences* 4(18): 401–408

Herring J, Egenhofer M J, Frank A U 1990 Using category theory to model GIS applications. *Proceedings of the 4th International Symposium on Spatial Data Handling, Zurich*: 820–829

INTERLIS 1990 *INTERLIS: A data exchange mechanism for land information systems*.Version 1.0e

Mark D M 1993 Toward a theoretical framework for geographic entity types. In Frank A U, Campari I (eds) *Spatial Information Theory, A Theoretical Basis for GIS*. Lecture Notes in Computer Science No. 716. Springer Verlag, Berlin: 270–283

Samet H 1990 *Design and Analysis of Spatial Data Structures*. Addison-Wesley, Reading

Stevens W T 1990 *UNIX Network Programming*. Prentice Hall, Englewood Cliffs, New Jersey

Vckovski A 1998 *Interoperable and distributed GIS*. Taylor & Francis, London (in press)

Worboys M F 1992 A generic model for planar geographic objects. *International Journal of Geographical Iinformation Systems* 6(5): 353–372

# 4 THE OPEN GIS CONSORTIUM AND PROGRESS TOWARD INTEROPERABILITY IN GIS

Clifford A. Kottman

## 4.1. The OGC perspective

The Open GIS Consortium (OGC) is a consensus-based association of public and private sector organizations dedicated to the creation and management of an industry-wide architecture for interoperable geoprocessing. It has a different perspective on interoperability from that of most technology developers and users because its focus is on user-centered business and purchasing models based on interoperable geoprocessing. OGC is driven by business needs and opportunities, and provides a forum for conducting rational industry planning processes.

The OGC vision and goal is the full integration of trusted geospatial data and geoprocessing resources into the mainstream of information technology. This means widespread use of interoperable, commercial, commodity geoprocessing software throughout the information infrastructure. Such a vision can be achieved only through the cooperation of many market sectors: vendors, integrators, academia, government agencies, and standards organizations. The vision also requires careful and continuing attention to the existing and evolving mainstream of geoprocessing, and to the currents that may shape the future mainstream. Especially important are the infrastructure technologies (of which geoprocessing is one) such as computing, communication, storage and access, distributed environments, trusted environments, and visualization.

Its vision and perspective put OGC in a unique position. It is at once a technology development organization, with attention to interface specifications and conformance testing, and at the same time a trade association, with a careful eye on developments in the marketplace that might shape the priorities and values of particular goods and services.A fuller treatment of the OGC perspective may be obtained in *The OpenGIS Guide* which is available on the Internet at ***http://www.opengis.org/techno/specs.htm.***

## 4.2. A review of OGC activities

OGC's technical activities fall in three categories: the development of the Abstract Specification, the development of the Implementation Specification, and the Specification revision process.

### *4.2.1. The Abstract Specification*

The Abstract Specification itself has two parts. The first is called the Essential Model, adopting the terminology of *Designing Object Systems: Object-Oriented Modelling with Syntropy* (Cook and Daniels 1994). It is a human-language (English) description of how the world ought to work. It identifies the components and behaviors of solutions to an enterprise challenge. The authoring of an Essential Model is probably the hardest part of the entire technology development process. This came as a surprise to most participants, probably because members assumed broad consensus pre-existed defining the current non-interoperable situation, as well as consensus on what to do about it. By abstracting away from implementations, and away from proprietary concerns, it is possible to achieve consensus on the actors and the behaviors of a yet-to-be-specified software environment. The Consortium has always striven for unanimous agreement on the Essential Model; such agreement here makes the rest of the consensus process relatively straightforward.

The second part of the Abstract Specification is called the Abstract Model, again following *Designing Object Systems*. The abstract model is a formal language expression of the Essential Model. The Consortium started with modeling language Syntropy, experimented with Object Modeling Technique (OMT), and is now moving toward Universal Modeling Language (UML). Several modern modeling languages are rich enough to hold the objects, relations, attributes, interfaces, and behaviors needed for the Abstract Model, and the final choice will probably be driven by the availability of software engineering tools.

The Abstract Specification is now organized into 14 Topic volumes. Each volume treats a portion of the whole GIS technology space. Each volume contains the Essential Model (in Section 4.2) and the Abstract Model (in Section 4.3) pertinent to its topic. An additional overview volume provides the logic for drawing the 14 volumes into a coherent whole. The Topic volumes currently under development are listed in Table 4.1. A recent version of the Abstract Specification may be downloaded from *http://www.opengis.org/techno/specs.htm.*

The Abstract Specification is constantly growing. All OGC members are able to contribute additions or other proposed changes, and these are brought to discussion and an acceptance vote at Technical Committee meetings held bimonthly. The Abstract Specification (both its Essential Model and its Abstract Model) form the basis for the next step in the OGC technology development process: the development of Implementation Specifications.

### *4.2.2. Implementation Specifications*

An Implementation Specification may defined to be a software specification that is so tightly specified that, if two software engineers write implementations, these implementations will plug and play for each other. In a very real way, Implementation Specifications are the ultimate technical deliverable from the Open GIS Consortium. It is through the technology of Implementation Specifications that OpenGIS brings rationality and interoperability to the GIS market place.

| |
|---|
| 1. Feature Geometry |
| 2. Spatial Reference Systems |
| 3. Locational Geometry |
| 4. Stored Functions/Interpolation |
| 5. The OpenGIS Feature and Feature Collections |
| 6. The Coverage |
| 7. The Earth Image |
| 8. Feature Relationships |
| 9. Quality |
| 10. Transfer Technology |
| 11. Metadata |
| 12. Services Architecture |
| 13. Catalog Service |
| 14. Semantics and Information Communities |

Table 4.1. The 14 OGC Topic volumes

Currently, OGC hosts three Implementation Specifications. All three specifications are in support of Simple Feature Access interfaces. They differ in that they each target a different Distributed Computing Platform (DCP). The three Implementation Specifications support the three dominant DCPs in use today: the Common Object Request Broker Architecture (CORBA) forwarded by the Object Management Group (OMG), Object Linking and Embedding/Component Object Model (OLE/COM) from Microsoft, and SQL including spatial extensions in work at ISO. All three are publicly available at the web site cited earlier.

It is a fact that the Distributed Computing Platforms in use today are not fully interoperable. Each has syntax and behavior patterns that preclude fully transparent distributed processing that requires access to more than one DCP in a single session.

### 4.2.3. The RFP process

OGC normally comes to consensus on Implementation Specifications through a Request for Proposal (RFP) process. Sometimes an RFP is preceded by a Request for Information (RFI). For example, the Consortium has issued two RFIs, one for catalog service information, and one for Earth imagery information. In both cases, responses have led to a more mature Abstract Specification, and one that is more closely aligned with the commercial mainstream.

In the RFP process, the Abstract Specification serves as a kind of requirements document. The RFP asks that Implementation Specifications be submitted that address specified parts of the Abstract Specification. The Consortium's first RFP was issued in late 1996, and the three Simple Feature Access Implementation Specifications were accepted by OGC as its baseline technology in August, 1997.

OGC's RFP process uses a double submission technique to assist the formation of consensus. It works like this. When an RFP is issued, two due dates are listed: one (about 90 days after RFP release) for a first submission and one (about 120 days later) for a second submission. The first submission proposals are open to the entire OGC membership, and the membership is invited to comment on them. Comments address the approaches, technology, implementability, style, and other facets of the proposals. These comments are also open to the entire OGC membership. In the span of time until the second submission date, vigorous teaming happens among the submitters. All submitters realize their strengths and weaknesses, and strive to find partners who will accept their most treasured strengths, while providing technology to patch their weaknesses. It has been the experience of the Consortium to date that only one team submits to each RFP item. In this way, the Consortium never has had a losing submitter, in the sense that another team's technology was accepted over its submission.

There is another avenue to OGC Implementation Specifications. This is the Request for Comment (RFC) process. At any time, a member may bring Implementation Specifications to the Consortium as an RFC. If no negative comments result, the Implementation Specifications may be added to the Consortium's baseline directly. This avenue is appropriate for ad hoc standards and technology found to be occupying the commercial mainstream.

## 4.3. OGC business structure and activities

The OGC organization is composed of four layers. A Board of Directors is composed of senior figures in business, government, and academe who volunteer their time and energy to give OGC a strategic plan for corporate development. The Executive Staff provides corporate management. My job there is to manage the technology development process. The real authority in the Consortium is the Management Committee, which is composed of Strategic and Principal Members. This committee is responsible for the OGC Business Plan and has the authority to accept or reject proposals and recommendations from the Technical Committee. The Technical Committee, in turn, is composed of representatives of all the member organizations, and its major function is the development of technology specifications. The fee structure, and additional member benefits may be found at the OGC web site: *http://www.opengis.org*.

The Consortium now consists of approximately 120 members, more than 30 of them from outside the US, and more than 30 of them university or research institutions. On the order of 100 individuals attend the Technical Committee and Management Committee meetings every two months. The Technical Committee is organized into three Task Forces: a Core Task Force that deals with technology important to nearly all GIS projects; a Domain Task Force that deals with special topics perhaps of interest only to particular domains or disciplines; and a Revision Task Force that maintains the Implementation Specifications by revising them when necessary to achieve or sustain interoperability in the ever-changing information technology environment.

### *4.3.1. SIGs and WGs*

The OGC Core and Domain Task Forces are organized into a number of Special Interest Groups (SIGs) and Working Groups (WGs). The difference between them is that Working Groups are coupled one-to-one with Requests for Proposals. The mission of a WG is to craft the RFP, ensure the Abstract Specification sections referenced are mature, to comment on the first submissions, to perform analysis and form consensus recommendations relative to proposals, and so forth. Currently, there are seven SIGs and WGs in the Core Task Force, as shown in Table 4.2:

| |
|---|
| The Service Architecture SIG |
| The WWW Mapping SIG |
| The Metadata SIG |
| The Feature SIG |
| The Coverage WG |
| The Catalog Services WG |
| The Transfer Technology WG |

Table 4.2. The seven SIGs and WGs in the Core Task Force

From the three Working Groups, you can tell that three RFPs are in progress. Two RFPs (Coverage and Catalog) were issued at the beginning of March 1998. The third (Transfer Technology) is currently scheduled for release in Summer 1998. The leadership of these SIGs and WGs comes from volunteers in the membership, vendors and users alike. The ultimate success of the Consortium hinges on the continued strong participation by the membership in these groups.

Three additional RFPs are anticipated later in 1998. The first centers on technology that supports additional behaviors of geospatial features, including a richer collection of consensus geometry types, the modeling of explicit feature-to-feature relationships, and additional support to feature identities. The second extends the family of consensus coverage types, and will probably include Triangulated Irregular Networks (TINs), Thiessen polygons, segmented lines, and others. The third is aimed at opening the technology for geospatial feature portrayal (also known as *symbolization* and *presentation*) in a way that allows a variety of alternatives relative to the allocation of portrayal services to computing components in a network environment. More RFPs are likely for 1999, including requests for interfaces that support the interplay between GIS, Earth imagery, and photogrammetry; transfer technology, such as XML; metadata; and quality.

### *4.3.2. ISO TC/211 coordination*

OGC holds a Class A Liaison relationship with the International Organization for Standards (ISO) Technical Committee 211 (TC/211). This committee, titled *Geographic Information/Geomatics*, is at the forefront of global standardization issues related to GIS. Two resolutions have been unanimously passed by the ISO

TC/211 relative to the work of OGC. The first, ISO TC/211 Resolution 47, which was passed at the Oxford, UK, meetings in September 1997, established a task force consisting of the chairman and TC/211 Work Group convenors to address specific opportunities for harmonization of ISO with OGC. Two technology areas were singled out for special attention: grid and raster structures, and metadata. Since then, OGC has adopted the ISO metadata standard, and a structure is in place to coordinate standardization efforts in grid and raster structures, which is a new work item in TC/211 as of March 1998.

The second is ISO TC/211 Resolution 54, passed at the March 1998 meetings in Victoria, BC. This resolution supports a draft of a much more far-reaching resolution that would establish procedural machinery to couple ISO and OGC closely and permanently. Essentially, the resolution would provide *Publicly Available Standard* or *International Standard Profile* or similar status upon commercial implementations that are determined to be OGC-conformant. For its part, OGC would agree to reference the ISO standards in its RFPs and evaluate all proposals for conformance to them in its evaluation of RFCs and responses to RFPs. The draft is now (early 1998) under review by the ISO Central Secretariat to determine the proper methods for ISO to institute the processes needed to support such a relationship.

## 4.4. Looking to an interoperable future: a story

The paragraphs above have addressed mostly institutional issues, not the technical issues closer to the heart of the OGC mission and vision. Let us now turn to the technical agenda of the Consortium and assess the progress toward interoperability from that perspective. Let's begin with a story that illuminates a possible future for GIS.

Imagine a future where you can say to your personal appliance (maybe you wear it on your wrist) "I'm hungry, show me the nearest pizza place." Imagine that the appliance performs speech recognition (or wirelessly connects to a service that does). Imagine that by wireless networks and protocols, a trader in restaurant information is found and queried, and through it a database of pizza places is accessed. The personal appliance signals your GPS position to the database, and the nearest three pizza places are located. These locations, and your location (and perhaps other information, such as your favorite projection and scale and symbol library) are handed to a map generation service (somewhere on the web), which creates a digital map showing three routes from your location to the three restaurants. This digital map is sent to your appliance where it is displayed for you. Perhaps also displayed is information about today's specials, and the approximate waiting time at each place.

This story is crafted to explain that geospatial data in tomorrow's business marketplace is different from today's map sheets and one degree cells of geodata. The whole marketplace of the future will be different from today's, especially when it comes to commerce in geodata. That is because today's geospatial information is in a pre-market era.

Before Edison invented the light bulb in 1879, electricity was in a pre-market era. Electricity was available in laboratories, but households had no use for it. There was no retail business based on electricity. It is more or less the same today with geospatial data. Hunks of geospatial data are available or for sale at various vendor or producer sites. They are not really for sale to the general public, because the general public has no tools to access and apply such data, just as the pre-Edison shopper (not owning a laboratory) would not buy electricity because there were few useful electric appliances. Now, there are many electric appliances, and electricity is a commodity in the wholesale and retail worlds. The analogy with geospatial information is clear. Today there are few geospatial appliances, and most of them are not interoperable.

The general public will buy commodities that are genuinely useful, however. Electric distribution to households began in the 1880s. Note that the households were motivated by the usefulness of electricity, not the commodity itself. Similarly, future users of geospatial information will be motivated by the need for information, not data. That is, users will want answers to specific questions, as in the pizza story, not raw data, as in a Spatial Data Transfer Format file.

The goal of OpenGIS is to enable visions such as the pizza story. Among other things, OGC is striving to enable interoperability between different specialty products, or appliances. The interoperability would allow the same map service used in the pizza story to be used for navigation, for emergency response, for *miss utility* applications, and so on. These additional applications would require additional specialty databases (like the restaurant database), and OGC is providing technology that allows all these databases to interoperate. Additionally, OGC is clarifying the relationship between geospatial commerce and the related components of the larger marketplace: the wireless services, the database services, the trader services, the security services, the fee collection services, the personal appliance services, and so on.

The really hard questions are left to the marketplace to answer: What is the role of today's GIS companies in the future scenarios? What partnerships should they form now to hold that market niche open for them? OpenGIS will not directly answer these questions. But, in the Consortium, members (especially at the Management Committee level) will tackle these questions, and hear how other interests (their potential partners) are answering them.

## 4.5. A road map to interoperability: follow the linguists' trail

The remainder of this paper serves to point at a possible development path for OpenGIS. The reader is cautioned that portions of the path outlined here have not yet gathered the consensus of the Consortium. The path should be understood as speculative, and as carrying no more authority than the imagination of the author.

The focus here will be on a special aspect of interoperability: the preservation of meaning. After all, GIS is not so much a scientific discipline in its own right, as it is a language for the representation of geospatial information (or a geography-related process) that may originate in any of the many geographically-related disciplines. GIS projects, therefore, always reflect the specific discipline or disciplines of their owners. Examples of geospatial disciplines in which GIS abstractions are frequently found include forest management, soil mapping, cadastre management, base cartography, surface hydrology, wetlands, transportation modeling, and so forth. Each of these disciplines has many sub-disciplines, and a GIS project can involve any combination of these.

It is exactly the multiplicity of languages at the project level that leads to the most difficult problems in interoperability between GIS information stores. Multiple languages are the source of much of the splintering that artificially divides the geospatial world. The situation is manageable, however, if sufficient care is taken in the formalization of the language constructs. In a way, we must become linguists: scientists of meaning and semiotics.

If geographic information is held in a language, then sharing geospatial information is like speech, and geographic processes are like reasoning. In Leviathan (Hobbes 1651: IV), Thomas Hobbes noted that there are four main uses of speech:

- To register what, by cogitation, we find to be the cause of anything;
- To show to others what knowledge we have obtained;
- To make known to others our wills and purposes;
- To please and delight ourselves, or others, by playing with our words, for pleasure...

In modern GIS terms, the first three of these uses might correspond to geographic data gathering, GIS report generation and publishing, and geospatial querying. The fourth use is clear if we may change a word or two: To please and delight ourselves, or others, by (virtually) playing with our *worlds*, for pleasure... Is this not the ultimate goal of geographic information: to enable the understanding of alternative future worlds, so that we can make wise choices now?

### 4.5.1. A language-based approach to interoperability

In the third book of the *Essay concerning Human Understanding*, John Locke observed that:

> "...since Sounds are voluntary and indifferent signs of any Ideas, a Man may use what Words he pleases, to signify his own Ideas to himself: and there will be no imperfection in them, if he constantly uses the same Word for the same Idea...
>
> The chief End of Language in Communication being to be understood, Words serve not well for that end [...] when any Word does not excite in the Hearer, the same Idea which it stands for in the Mind of the Speaker." (Locke 1731: III,IX, 2,4)

### 4.5.2. Information communities

Locke's observation is at the heart of interoperability. There is nothing special about one word or another relative to its symbolic use to represent a given idea. What is important is the consistent and uniform use of it for that idea. If more than one person is involved, there must be consensus on that usage. This leads to the definition: An *information community* is a collection of people and systems (a government agency or group of agencies, a profession, a group of researchers in the same discipline, corporate partners cooperating on a project, etc.) that, at least part of the time, share a common digital geographic information language and share common spatial feature definitions. This implies a common world view as well as common abstractions, feature representations, and metadata. We will also say that feature collections and geoprocesses that conform to the information community's standard language, definitions, behaviors, and representations belong to that information community. In a linguist's terms, an information community is a set of people and systems that speak the same language.

### *4.5.3. The parts of speech in a geographic information language.*

The next several paragraphs list the necessary parts of speech in a language for the expression of geographic information. Specifically, we impose fourteen formal and rigorous structures: *feature instances, feature types, geometry, corners, geometry schema, spatial reference system, property-value pairs, attribute schema, feature schema, project schema, point models, geometry models, feature models,* and *feature collection models.* All of these are to be understood as real world facts, not to be modeled in software at this level of abstraction. We forbid all language that lies outside these structures. Because all fourteen structures are essential for information sharing, we will discuss each of them, one at a time.

**4.5.3.1. Feature instances.** The primary structure we impose on the information community is that of the feature. The feature is the atom of geospatial information. The information community is completely defined by the phenomenon instances it recognizes: features. The next eight structures are, for the most part, substructures of the feature structure. We insist that the notion of a feature be completely definitive in an information community. The features of the information community carry far less, but more structured, information than reality, through the process of abstraction and simplification.

**4.5.3.2. Feature types.** The categories of features captured in an information community are feature types. Each feature type can be thought of as a template to be populated at each occasion of a feature instance that corresponds to that type. The template is populated with information specific to the feature instance.

**4.5.3.3. Geometry.** The feature of the information community is a cartoon simplification and abstraction of the information carried in the real world. We insist that features with spatial extent recognized in the information community be mentally drawn using simple primitive geometric shapes. We insist that the primitive shapes be instances of the Well-Known Types (WKTs) or Well-Known Structures (WKSs) of OpenGIS geometry, such as polygons, line strings, polyhedrons, and other OpenGIS shapes. We further demand that rules for representing each feature type with WKTs be explicit (for example, a rule may specify that a brick house is seen as a polyhedron). Finally, we demand that the mental model of the WKTs carries sufficient information to enable the reconstruction of the extents of the features to which they contribute. That is, we insist that the geometry components know how they contribute to complex geometries (for example, the highway segment geometries must know the sequence in which they concatenate to become an entire highway feature.)

**4.5.3.4. Corners**. The primitive spatial concept in a geographic Information Community, from which all phenomena with spatial extent are spatially conceptualized, is corner. A corner is a place that (directly or indirectly through a geometric construct, such as a spline) defines (in whole or in part) the spatial extent of one or more of the phenomena recognized within the information community. Each OpenGIS WKS can be constructed from a finite set of points that are its corners. We insist that the corners be explicit in the information community. That is, we insist that each abstraction recognized as a feature with extent in the information community be mentally drawn with geometry elements of WKTs, each of which carries additional abstractions, called corners, from which that geometry element can be constructed. We further insist that the mental model of the corners carries sufficient information to allow for the construction of the WKTs. For example, we insist that the sequencing of corners along a polyline is explicitly available within the information community.

**4.5.3.5. Geometry schema**. The collection of rules implicit in the previous two paragraphs are called the geometry schema of the information community. Examples include:

- For each feature type, the geometry WKTs that are to be used in its representation.
- For each WKT, the feature instances to which it contributes.
- For each feature instance, the WKTs that contribute to its extent.
- For each WKT, the structured list of corners which specify it.
- For each corner, the WKTs to which it contributes.

**4.5.3.6. Spatial reference system**. We treat a spatial reference system as a real-world fact. A good example is a well-monumented point-of-beginning used by a surveyor. The point-of-beginning, together with the tools and rules for surveying, enable one to attach coordinates to every point of interest in a project. The points of interest are precisely the corners needed for the geometric construction of the geospatial extent of any feature recognized by the information community. A commonly used spatial reference system is a datum, including a horizontal datum and a vertical datum, but there are others, such as linear reference systems.

**4.5.3.7. Property-value pairs**. An information community replaces the rich and complex natural language descriptions of features with an abstracted stick-figure language of property-value pairs. So far, each information community feature is associated with two structures: the feature type of which it is an instance, and the geometry consisting of an OpenGIS WKS. The geometry is expressed relative to a spatial reference system, and includes all corners needed for its construction. We now additionally insist that each feature instance be described using a strict grammar and vocabulary. The purpose of the grammar and vocabulary is to describe the attributes of interest unambiguously and repeatably for each feature recognized by the information community. In the information community, each feature type is associated with a list of properties. Our grammar demands that each property in the list have a formal property name, and that each property take a value of a type specific for that property. The vocabulary of an information community, therefore, is quite limited. We allow only the following terms:

- the names of the feature types recognized by the information community;
- property names;
- property values.

Of course, each of these terms can be used only where our grammar permits.

**4.5.3.8. Attribute schema and feature schema**. The collection of all feature types recognized by an information community (including the capture criterion for that feature type) together with the proscribed list of property/value pairs for each one, is the attribute schema of that project world. The attribute schema, together with the geometry schema, constitute the feature schema of the information community.

The list of property-value pairs recognized by an information community, and more importantly, the semantics behind these terms, reflects the interests of the scientists who conceptualize and formalize the community. We insist that the schema be extremely formal, so that all the individuals participating in an information community observe the same phenomena, and record them in exactly the same way. This rigor mandates that the attribute schema must be completely specific about the names and meanings of its properties, as well as the meaning (to include types and units) of its values. This implies the existence of a dictionary in each information community where the semantics of the terms in the vocabulary are made clear.

**4.5.3.9. Project schema**. Each distinct and unique feature collection needs to carry enough metadata to allow the feature information it carries to be understood and exploited. We hinted at some of these metadata when we spoke of a dictionary for the information community vocabulary, but there is more. The project schema is the formalization of this metadata, and it is best defined in the information community context. The project schema consists of at least four parts: *feature collection identification*, *lexical semantics*, *project semantics*, and *use semantics*. These notions are developed more fully in other OGC documents.

**4.5.3.10. Point models**. We assume that each information community project has a project-wide coordinate system. With the coordinate system, we are able to interface to the next level of abstraction: the point models. If the information community is concerned only with point features, this level of abstraction holds all of the spatial information of interest to that information community. Point models are perhaps best imagined in terms of a Cartesian space dotted with a finite collection of special points. Starting from any corner of any feature recognized by the information community, we assume the existence of a method named *survey* that abstracts the coordinates of the corner. Conversely, for each special point recognized by the information community, we assume a method named *locate* that locates that point as a corner of a feature. Of course, the survey and locate methods work for points other than the feature corners, but to recognize these additional points, one must enlarge the feature schema and allow new point phenomena to be recognized.

**4.5.3.11. Geometry models**. How do we know which points to connect, what geometry to build, and when and how to combine simple geometries into more complex ones? To answer these questions is exactly why the geometry schema was formalized as an earlier abstraction. The function of the geometry model structure is to recover from earlier abstractions the information needed to construct the geometry WKTs that model the feature, and to expose those WKTs in a manner recognized by the information community.

**4.5.3.12. Feature models**. Each feature type recognized by an information community is specified by its property set. A property set is a list of property/value pairs that is applied to all features of the corresponding type. The role of the feature model is to expose the property set of a specified feature.

**4.5.3.13. Feature collection models**. OGC is really about enabling commerce in geospatial information and process. We make the basic assumption that a basic unit of trade is a feature collection. The role of the feature collection model is to expose the information held in a feature collection.

### *4.5.4. The nouns and verbs of OpenGIS*

In the context of the parts of speech explained above, the current work of the SIGs and WGs in the Consortium can be understood as defining the nouns and verbs of OpenGIS. The high-level nouns in current development include *coverage*, *grid coverage*, *image*, *evaluator*, *trader*, *segment*, and *parameter*. Sample verbs in development include *query*, *evaluate*, *parameterize*, *portray*, *transform*, *analyze*, and *create catalog*. It is through the gradual introduction of these nouns and verbs into the Consortium, and through consensus on their meanings and behaviors, that interoperability in GIS will be achieved.

### *4.5.5. Information communities and their sub-communities*

As we have defined them, each information community has a fixed vocabulary. If we change a word in one information community vocabulary, say from *house* to *dwelling*, the result is a different information community. We need a tool that, when we have two implementations with such trivial differences, enables each information community to be able to share information with the other as if it were native. The situation should be the same if we make a minor change to the geometry schema, say by modeling a curve with a spline instead of an arc. The result must a different information community, but there should be a tool that bridges this one to the original information community.

It should be possible to move information easily and without semantic loss from information communities having naive schema into information communities with more sophisticated and inclusive schema. (Moving information the other way requires the truncation and loss of information.) An information community that is more naive than another is called a sub-community of the other. Here is a more detailed definition of sub-community:

Let $S_1$ and $S_2$ be two information communities. We say $S_1$ is a *sub-community* of $S_2$ if there are two functions, $R_1$ and $R_2$, that behave as follows:

I. $R_1$ is a mapping from the feature types of $S_1$ into those of $S_2$

II. $R_2$ is a mapping from the set of property/value pairs of all features in $S_1$ into the set of property/value pairs of all features of $S_2$ that preserves semantics, and is canonical with $R_1$.

The first condition guarantees that every feature in $S_1$ is paired with a feature of $S_2$. $R_1$ is the pairing function. The second condition states two things. First there is a (perhaps one-to-many) mapping $R_2$ from the property/value pairs of $S_1$ to those of $S_2$ with the following condition:

**The semantic condition**: if P/V is a property/value pair in $S_1$, then the semantics of P/V are a subset of the semantics of $R_2$(P/V). This is just a formal way to state that the information community $S_2$, is more sophisticated than $S_1$. It is, of course, a human function to affirm whether the semantic condition is met or not.

The second part of the second condition states that the first mapping ($R_1$, defined on features) and the second mapping ($R_2$, defined on property/value pairs) cooperate with each other. Specifically: if a feature $F_1$ in $S_1$ is paired by $R_2$ with a feature $F_2$ in $S_2$, then the semantics of any property/value pair of $F_1$ are a subset of the semantics of property/values of $F_2$. This means that within every category of information captured for feature $F_1$, the information held within that category is a subset of (or can be derived from) the information captured about $F_2$. The canonical mapping is shown symbolically in Figure 4.1.

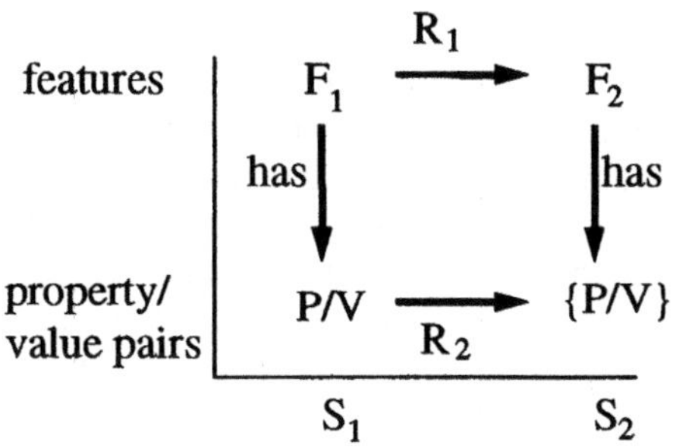


Figure 4.1. Symbolic representation of canonical mapping

The second condition in the definition of sub-community is designed to keep words (e.g., the use of Latin names, or the use of the jargon of a particular discipline) from being a barrier that divides information communities. It is intended that $R_2$ preserve semantics, but not necessarily the actual names of properties nor the units or domains of their values. In other contexts, the mapping $R_2$, would be an instance of a thesaurus.

If $S_1$ is a sub-community of $S_2$, then geospatial information can move from $S_1$ to $S_2$ without information loss. Moreover, the transfer can be automated with the help of formalization and crosswalks between the schema involved.

### *4.5.6. Isomorphic information communities*

**Definition**: Two information communities, $S_1$ and $S_2$, in *A* are *isomorphic* if each is a sub-community of the other.

The isomorphic relation clearly satisfies the symmetry and reflexivity conditions. For now we will assume it satisfies the transitivity condition as well. Therefore, the isomorphic relation is an equivalence relation. The immediate implication is that if two information communities are in the same equivalence class (that is, are isomorphic), then geographic information in either one can be successfully transferred without loss to the other.

It is important to note that isomorphisms between project worlds depend on the compatibility of their feature schema, and the lexical semantics of their project schema. In particular, the feature collection identification, the project semantics, and the use semantics do not play a role. They play a role, perhaps, in a user's decision to desire a feature collection, but no role in the user's technical ability to ingest it into the user's system, or into an existing feature collection there, with understanding.

In general, two information communities will not be isomorphic. However, each community may have a significant sub-community where the two sub-communities *are* isomorphic. This approach can be quite powerful, and leads to an interesting approach to translator technology in support of interoperability.

### *4.5.7. Translator technology*

Based on the ideas presented above, an automated translator can be conceived as a sequence of nine more fundamental mappings:

1. **Semantic mapping**. We begin by comparing the lexical semantics of the source and target information communities, and building a semantic mapping from the source to the target. (In this mapping, as in the mappings that follow, one may pass to a sub-community, if necessary, on either end). The semantic mapping is like a thesaurus in that it assigns valid target feature names, property names, and property values to the feature names, property names, and property values found in the source.

2. **Changing property name/value type pairs**. Using the semantic mapping, we build a second mapping. This mapping takes property name/value type pairs in the source schema, and assigns them to property name/value type pairs found in the target schema. We call the new mapping the *change-of-property name/value type-pairs mapping*.

3. **Changing the feature types**. Using the change-of-property name/value type-pairs mapping, we build a third mapping. This one has feature types in the source for its domain, and feature types in the target as the range. We call the new mapping the *change-of-feature-type mapping*.

4. **Changing the attribute schema**. A fourth mapping is constructed using the change-of-feature-type mapping. This one is called the *change-of-attribute-schema mapping*, and it maps the attribute schema of the source into the attribute schema of the target. The attribute reference systems of the source and target schemas will necessarily play a role in the definition of this mapping.

5. **Changing the geometry schema**. The geometry schema of the source must also be mapped into the geometry schema of the target. This is formalized with the

*change-of-geometry-schema mapping*. This mapping allows features to be modeled by geometry in different ways in the source and target environments.

6. **Changing the spatial reference system.** In order to allow the source and target to have different spatial reference systems, we need to construct a *change-of-spatial-reference-system mapping*.
7. **Changing the property name/value pairs.** The change-of-attribute-schema mapping that was constructed in Step 4 is now applied to the property name/value pairs found in the feature instances of the source feature collection. The result is a new collection of property name/value pairs conforming to the schema of the target.
8. **Changing the geometry.** The change-of-geometry-schema mapping that was constructed in Step 5 is now applied to the instances of geometry in the source. The result is valid geometry in the schema of the target (but still in the spatial reference system of the source).
9. **Changing the corners.** The last step is easy. Using the change-of-spatial-reference mapping that was constructed in Step 6, we transform the corners of the geometry that was created in Step 8 into the corners of the target. The feature collection is now complete in the schema of the target information community.

## 4.6. Conclusion

The road to interoperability in geographic information and geoprocessing is neither short nor smooth. Yet it is a path to which the OGC is committed. Along the way, there are two kinds of challenges: technology development and consensus gathering. The technology is not particularly daunting in its depth, but it has breadth as wide as geography itself.

This chapter has presented an outline of the kind of technology that might be developed in order to implement an environment of geospatial interoperability that spans two or more information communities. Alternate outlines are certainly possible, and, as the development path is actually followed, the judgments and decisions made in the marketplace will determine the technical twigs and branches that are destined to become part of the main trunk of the geospatial technology tree. It is because GIS is broad, but not too deep, that the parallel development of many technologies, such as those listed in this paper, is both necessary and possible.

An important role of the OGC is to provide mechanisms (especially consensus interfaces to software components) that enable parallel developments to interoperate, so that they have a chance of becoming a part of the commercial mainstream. Another role is to shed light upon and to influence the process by which the geospatial marketplace of the future is itself formed and shaped. Through the consensus process of OGC, many of the risks of parallel technology development can be managed, and the market development process itself can be rationalized and accelerated. These are the keys to progress toward interoperability at the Open GIS Consortium.

## References

Cook S, Daniels J 1994 *Designing Object Systems: Object-Oriented Modelling with Syntropy.* Prentice Hall

Hobbes T 1651 *Leviathan, or, the Matter, Form, and Power of a Common-Wealth Ecclesiatical and Civil.* Andrew Crooke, London

Locke J 1731 *Essay Concerning Human Understanding.* 10th Edition. Edward Symon, London

# 5 PROBING THE CONCEPT OF INFORMATION COMMUNITIES—A FIRST STEP TOWARD SEMANTIC INTEROPERABILITY

Yaser A. Bishr, H. Pundt, W. Kuhn, and M. Radwan

## 5.1. Introduction

Interoperability between different data sources and software systems is difficult to achieve due to the complexity of geodata. This complexity is caused by various factors, such as the underlying digital formats imposed by a particular software or acquisition method and the complexity of higher level descriptions, conventions, and rules imposed by individuals, organizations, and disciplines using the software (Buehler and McKee 1996).

The main motivation of this chapter is to provide a first step toward interoperability between different geospatial applications. Prior to seeking solutions to achieve such a high level of interoperability, a proper analysis of the problem is required. This chapter attempts to:

- Define the notion of an information community.
- Show differing facets of information communities.
- Have some hands-on case studies to exemplify these differences.
- Briefly introduce some direction to resolve these differences.

The notion of interoperability may imply different things depending on whether it is used by network designers, operating systems designers, or application software engineers. Today's desktop and enterprise environments lack interoperability between geographic information systems as well as on the application level. Section 5.2 presents a layered view of interoperability. In Section 5.3, we will define the concept of a geospatial information community and show its relationship to the notion of a project world. To identify the different types of heterogeneity between information communities, the abstraction process is analyzed in Section 5.4. Section 5.5 introduces the types of heterogeneity between information communities. To illustrate the various definitions of heterogeneity, Section 5.6 provides several case studies. In Section 5.7, we briefly discuss the concept and components of semantic translators, which act as a middleware between two heterogeneous project worlds. The chapter concludes with Section 5.8.

## 5.2. Levels of interoperability

As shown in Figure 5.1, interoperability can be viewed at six different levels; interoperability of network protocols is at the lowest level and interoperability of information communities is at the highest level. The emphasis in this chapter is on interoperability at the application level. We consider application interoperability and semantic interoperability as synonyms.

At the lowest level, a computer network consists of both hardware and software (Figure 5.1). The hardware includes network interface cards and cables that link them together. The software consists of network protocols (e.g., TCP/IP, SPX/IPX, etc.). Network protocols are rules and procedures used on a network scheme to communicate among the connected systems. At this level, the cabling scheme, hardware and software needed to accomplish communication between the underlying systems are defined.

File systems interoperability lies at the second level. This level was introduced because multi-protocol strategy only provides links at the network level. The ability of a UNIX workstation to communicate with a PC using TCP/IP protocol does not imply that it can run applications or access files on the PC. File system interoperability allows users to open files on other systems and display them in their native formats. This is not just a matter of file transfer and access; it includes files and directories naming, access control, access methods, and file management.

The network file system does not allow the execution of programs that reside on the machines. This occurs on the next level of interoperability by means of remote procedure call (RPC). RPC is a set of operations that executes procedures on remote systems. It standardizes the way programs run under the control of another operating system without having to adhere to the processing call of their underlying operating system.

The three levels of interoperability mentioned above are not sufficient to provide users with the ability to search and manipulate distributed databases. The fourth level is intended to provide technologies which allow users to do just this.

The lower four levels of interoperability provide a distributed computing platform on which interoperable GIS can be built. Within this context, GIS interoperability means that users can transparently access and share remote geospatial databases and services regardless of their underlying GIS platform. In response to the need for GIS interoperability, the Open GIS Consortium (OGC) works to provide a framework for this issue. This three part framework includes the open geodata model, services model, and the information community model (Buehler and McKee 1996).

Application interoperability is the highest level of interoperability (Figure 5.1). Different applications have different worldviews, representations, and, hence, different semantics. Heterogeneity at the application level is a semantic problem and is due to the differences in the way each individual views the real world, as well as the interpretation of the spatial data encoded in the database. These differences stem from the differences in the role and relevance of real world objects in each application.

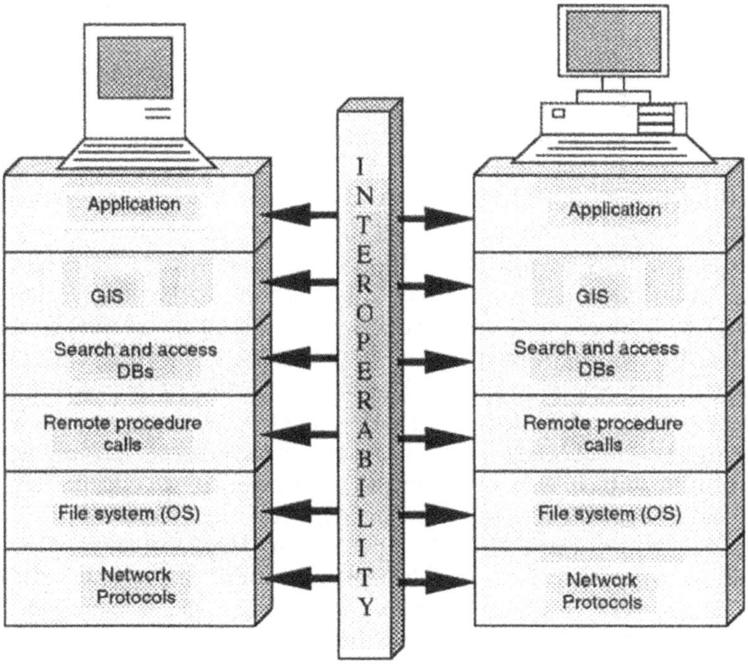


Figure 5.1. Levels of interoperability

## 5.3. What is a geospatial information community?

This section attempts to define the notion of a geospatial information community (GIC) and its relation to the notion of a project world. Three definitions of GIC have been provided by OGC:

> "GIC is a collection of people (a government agency or group of agencies, a profession, a group of researchers in the same discipline, corporate partners cooperating on a project, etc.) who, at least part of the time, share a common digital geographic information language and share common spatial feature definitions."

> "GIC is a collection of systems or individuals who can successfully share digital geospatial information, that is, features."

"GIC is formally defined in terms of the metadata of its members with which they are compatible."

The first definition does not distinguish between two distinct worlds, namely, the project world and the real world. A project world is the computer representation of a particular geospatial application domain. Information communities are formed, as mentioned in the definition, by people, organizations, etc., clearly outside the project world. Sharing geospatial information between two communities is still possible, although they don't share a common digital information language or common spatial feature definitions.

The second definition is problematic, since it assumes that geospatial information can only be shared within information communities. This restriction does not always apply; for example, a cadastre GIC and land use GIC can share their information, although they are different.

A definition of GIC should not be based on features, which are represented in the computer, but rather on the way a group of people, who belong to a particular domain, conceptualizes the real world. This domain-specific conceptualization is the universe of discourse of that particular GIC. In other words, the set of features in the real world, which are of interest to a particular GIC, forms its universe of discourse (UoD). UoD here is not restricted to a spatial extent, rather it refers to class of entities that is implied or explicitly delineated and deemed relevant for a GIC. Hence, people within a GIC may differ in the way they abstract concepts identified in their universe of discourse. This, in turn, creates different project worlds. A project world mainly consists of a fixed vocabulary, a fixed collection of schema elements, and an unambiguous set of feature instances.

The third definition states that a GIC is formed by geospatial applications having compatible metadata. The term metadata refers to information which describes a database, rather than a GIC. Conceptual modeling and metadata restrict the interpretation of database objects. A definition of information communities should not have this restriction, as we are not interested in providing a single standard data model for a particular community. It is then required to have a higher-level language, i.e. above the conceptual model, which formally defines the UoD of the underlying GIC without restricting its way of representation. We propose ontology as a higher level language.

Ontology is a term adopted from artificial intelligence (AI) to denote specification of conceptualization. It should capture and constrain a set of conceptualizations (Gruber 1995; Guarino 1995; Kashyap and Sheth 1995). The provision of ontology for a certain GIC means offering a way to specify the intended meaning of its lexicon. To answer the question "What is a geospatial information community?", we suggest the following definition:

"A geospatial information community is a group of spatial data producers and users who share an ontology of real-world phenomena".

Ontology can be considered a meta-language situated above data models. Our model requires that each GIC commits to a specific ontology in order to minimize the semantic loss between applications within and across information communities. It is also possible for two information communities to share their data if they, at least partly, share the same ontology. For example, if each of the information communities has its own ontology, the two information communities must share a common ontology to be able to share certain information.

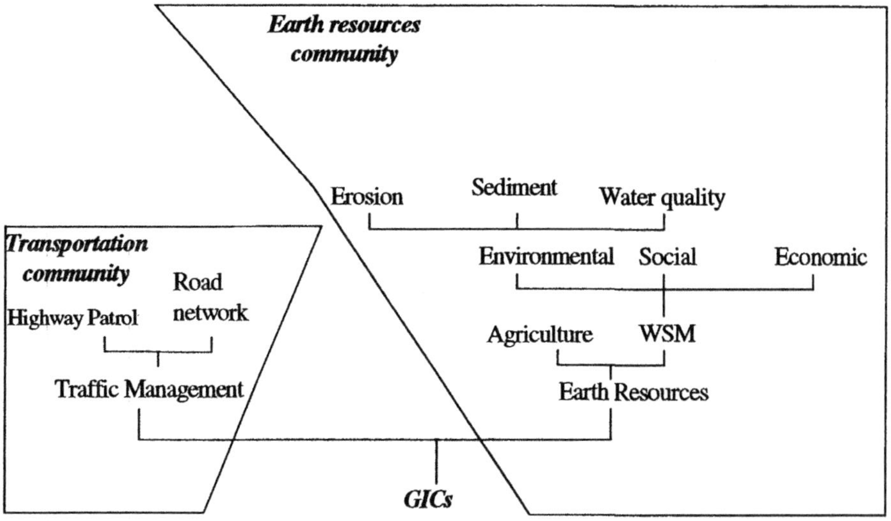


Figure 5.2. User groups are grouped together to form information communities.

As shown in Figure 5.2, GICs can be sorted in a tree that shows the relationship between them. Information sharing between project worlds within a GIC is possible with minimum information loss. On the other hand, information loss is more severe if the information is to be shared between project worlds that belong to different GICs. For example, information loss between project worlds of an agriculture GIC and a water quality GIC is less than the loss when it is shared between an Earth resources GIC and a transportation GIC.

## 5.4. The process of abstraction of real-world objects

For the provision of interoperability at the application level, it is essential to identify precisely the types of discrepancies that might exist between spatial databases. This can be achieved by a closer look at the abstraction process from the real world to the project world.

In Section 5.3, we have indicated that people within a GIC may differ in the way they abstract concepts identified in their UoD. These different abstraction processes define contexts. Within a certain context, the process of database design proceeds from the real world to the project world. The context world view is the specific conceptual model of the underlying UoD, thus an intensional definition of the UoD. It acts as a filter between the UoD and the project world, as shown in Figure 5.3. The context world view's definition includes three types of abstraction: the categories definition, class intension definition, and geometric description. These and other terms referring to Figure 5.3 are defined as follows:

*Categories* are collections of real-world features with similar characteristics.

*Class intension definition* is the process of defining rules by which real-world features are identified and associated with categories.

*Geometric description* is the process of outlining specifications to assign geometric types to features as well as specifications to represent them in the computer. The specifications depend on the objective and the type of analysis required by the underlying context.

*Context* is the domain where the process of abstracting the UoD to the project world occurs.

The *context world view* is the conceptual model of the underlying GIC's UoD. It includes categories definition, class intension definition, and geometric description.

The *project world* consists of an intensional and an extensional representation of the UoD.

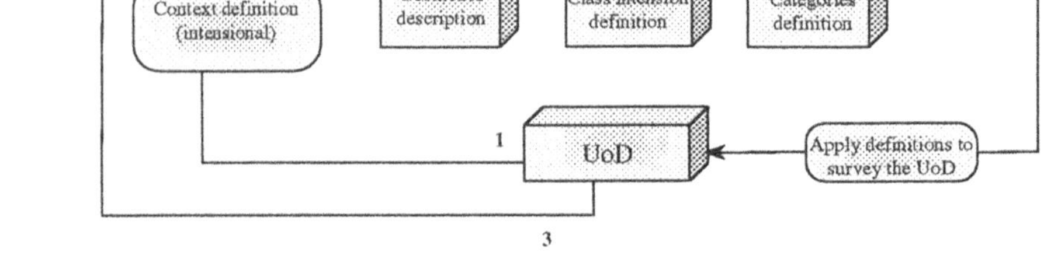


Figure 5.3. Abstraction of real world features: 1) define classes, rules for object/class membership, and geometric description; 2) apply these rules to survey the real world; 3) representation in the GIS domain.

These definitions are used to survey the UoD and represent it in the database. Therefore, the project world is an intensional and extensional representation of the underlying context's UoD. Mapping the context world view into the project world is representation, while the reverse process is interpretation.

## 5.5. Types of heterogeneity

Several types of abstraction were presented in the previous section. Based on these, the types of heterogeneity among database schemes are defined in this section. The most fundamental principle on which modeling rests is a one-to-one correspondence between real world objects and their symbolic representations in the database environment. Kent mentions that this assumption is strongest in object-based systems and weaker in value-based systems (e.g., relational databases; Kent 1991). The assumption starts to break down when dealing with several databases. We might have different objects or attributes in independent databases representing the same real world feature. The fundamental problem of sharing databases is to maintain consistency when objects move among databases.

```
Let x and y be two objects in two independent databases.
We say that x = y if they are the same.
We say that x ( y if they refer to the same real world feature.
We can assert that the first implies the second, but not vice
versa.
```

When we speak of interoperable geographic information systems, we mean that systems should be independent and yet be able to communicate transparently at a high level of semantics (Bishr et al. 1996). The following sections introduce three types of heterogeneity: semantic, schematic, and syntactic.

### *5.5.1. Semantic heterogeneity*

Semantic heterogeneity is usually the source of most information sharing problems. Individuals from the same or different GICs who would like to share their data are likely, at least partly, to share interest in a common UoD. It is important to recall that we use the notion of UoD as defined in Section 5.3. For example, a soil mapping and conservation group is interested in defining a detailed classification of terrain mapping units, their profiles, horizon, and rate of erosion. This differs from the way a land use group recognizes features. It identifies natural vegetation areas, types of land use at farm and plot levels, soil suitability for agricultural purposes, and irrigation schemes. The land use group regularly requires soil information, which can be retrieved from the soil conservation group. Thus, the soil information can be considered as the common UoD.

It may be concluded that semantic heterogeneity occurs mainly due to differences in the context world view, shown in Figure 5.3. Hence, it occurs in any or all of the following three forms:

- Differences in the class hierarchies and the attribute structure, i.e., categories definition.
- Differences in the rules that assign objects to classes, i.e., class intension definition.
- Differences in the geometric representation of the spatial objects.

These forms of semantic heterogeneity mainly occur due to the fact that features play different roles in different project worlds. In our opinion, differences in semantics among contexts lead to syntactic and schematic differences.

### *5.5.2. Schematic heterogeneity*

Schematic heterogeneity is the difference in the class hierarchies and attribute structure of two independent contexts. It occurs mainly due to semantic differences between applications. The schemata, that is, the classes, attributes, and their relationships, may vary within or across context. For example, several classification methods exist for ecological land classification (Sims et al. 1996). There are also several methods to classify soil units, such as the ASTM and the SOTER classification methods (ASTM 1985; Engelen and Ting-Tian 1995). Moreover, two contexts might adopt the same classification method and have different class and attribute structures, which results in the schematic heterogeneity.

### *5.5.3. Syntactic heterogeneity*

The syntactic heterogeneity is related to the difference in the thematic and the geometric representations, as well as the topologic relationships of spatial objects. For example, one GIS can have nodes, arcs, and polygons as primitive geometric types, while another can have nodes, lines, polygons, rings, polyhedral surfaces, and solids as its primitive geometric types. The Open Geodata specifications developed by the Open GIS Consortium, Inc. attempt to provide a common formal syntax for representing real world features in spatial databases.

## 5.6. Case studies for semantic heterogeneity

Semantics can vary within information communities and across their boundaries. However, semantic heterogeneity within a particular GIC is less severe than it is among GICs. Section 5.5.1 stated that semantic heterogeneity is the source of most information sharing problems and leads to schematic and syntactic heterogeneity. The following case studies should highlight the three forms of semantic heterogeneity mentioned in Section 5.5.1. This will help to understand the types of information needed to map between contexts.

It is important to mention that mapping between two geospatial databases is not only a schematic, but also a syntactic problem. However, the information needed to provide correct mapping is semantic. Essentially, one must resolve the semantic problem before one attends to the schematic and syntactic problems.

### *5.6.1. Semantic heterogeneity: categories definition*

We have investigated a case in which different applications require land use information. Human activities to gain benefits from the land result in practicing certain land use types. Land use types are descriptions about the functional role of the land in an economic context (Burrough 1986). Land use information is the collection of those descriptions. In general, land use information might be required by other GICs for taxation, valuation, mortgage, and spatial analysis and statistics.

Various criteria can be defined to partition space, which result in different categories of real world features. Three kinds of spatial boundaries can exist in land use:

- Plot: this is usually the basic feature in a land use database. It is a contiguous tract of land with uniform soil characteristics, under a specific land use.
- Parcel: this is usually the basic feature in a cadaster database. It is a contiguous piece of land with uniform tenure. This is usually the land use as viewed by a cadaster GIC.
- Farm (farm system): a decision-making unit needed by land management agencies. It consists of a farm household, crops, and livestock. A farm is an economic unit of agricultural production under single management and includes livestock and land of agricultural production purposes regardless of title, legal form, or size. The livestock production system comprises pastures and herds, as well as an auxiliary feed source transforming plant biomass into animal products.

In most cases, parcel boundaries can exceed or cross the boundary of plots. This is possible because the semantics of the classification are different. Farms are further aggregated entities which classify land use based on several socio-economic factors. Sharing information between these databases requires knowledge of the criteria used for such classifications.

### 5.6.2. Semantic heterogeneity: class intension definition

Class intension definition is the second form of semantic heterogeneity. Two land use databases from the same information community but different contexts may have the same classification schemas and yet different rules that assign objects to classes. Class intension definition provides rules to associate objects to classes. Consider the same case study, mentioned in Section 5.6.1, in which land use information is to be shared between two contexts within land use GICs. In Figure 5.4, we see that, although $Cont_1$ and $Cont_2$ have the same classification schema, they have different intensions. Differences in the attribute domains between the two contexts caused objects to be members of classes which bear different semantics. An object in $Cont_1$, which is a member of the class ***high suitability*** and has *suitability* = 2, is mapped to the class *low suitability* in $Cont_2$. Such mapping, for instance, can lead to wrong decisions for assigning crop types to a particular land type.

Another case study refers to land use classification. Fallow land (often denoted as *set-aside land*) represents a special kind of land use. The assignment of land parcels in certain areas to the class *fallow land* may vary between information communities. This variation depends on the rules by which parcels are assigned to the classes. Figure 5.5 shows some examples.

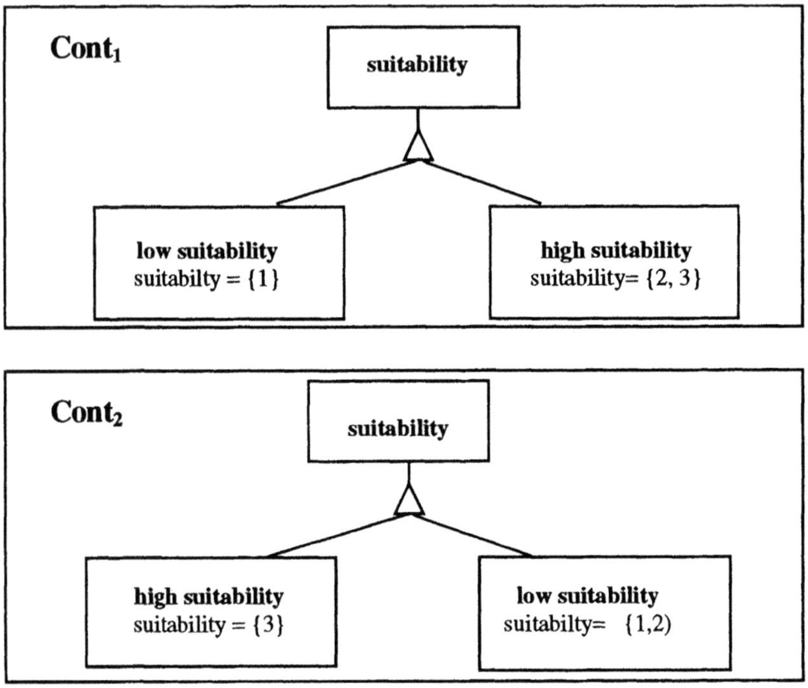


Figure 5.4. An object with *suitability* = 2 and member of class *high suitability* in *Cont*$_1$ is associated with class *low suitability* in *Cont*$_2$.

Basically, land use information is provided in the official topographic and cartographic information system of Germany (ATKIS). A clear definition of *fallow land* exists in the ATKIS (Object No. 4110 in the ATKIS object catalog). For the purpose of paying subsidies to farmers, the National Chamber of Agriculture is interested in information about *fallow land.* The rules for assigning parcels to the class based on the Chamber of Agriculture's definition are different from those based on ATKIS definitions. Other GICs, for example nature protection, are seriously interested in information about fallow land within the framework of specific projects. Many national nature protection laws define the term clearly, and yet differently. Consequently, the problem mentioned occurs repeatedly: it is neither possible for such nature protection projects to use ATKIS information, nor for the nature protection GIC to share data with the agriculture GIC. This case study represents a typical example of the limitations data semantics cause for interoperability between GICs.

In Nordrhein-Westphalia, political and financial reasons led to the merging of the Agriculture Agency and the Agency for Ecology, Landscape Development and Forestry (responsible for nature protection). This process gave rise to different types of semantic heterogeneity not only across GIC boundaries, but also within the same GIC.

| GIC Agriculture | GIC Nature Protection | Official topographic and cartographic information system |
|---|---|---|
| • Long term not intensively used parcels, where self greening and grass cutting is possible<br>• Parcels, where cultivation of biological raw materials is possible<br>• Land for biomass production | • Former agricultural land where now free succession of vegetation is possible<br>• Temporarily unused land with time-limited free succession of | • Unused areas, where the initial purpose was:<br>- settlement<br>- gardening<br>- traffic<br>- industrial |

Figure 5.5. Fallow land definitions in different GICs.

### *5.6.3. Semantic heterogeneity: geometric description*

Different contexts can have the same categories definition and class intension definition and yet have different geometric descriptions. The case we introduce in this section will show that the objects' role in the database can determine how they are represented geometrically. In other words, objects in different contexts can have different geometric representations, not only due to spatial resolution differences (i.e., scale), but also due to their role in their underlying context. For example, houses are point objects in a utility database, area features in a cadaster database, and further aggregated to form a block in a database for road management (Figure 5.6). Roads are defined by their boundaries in the utility and the cadaster databases, whereas they are represented by their centerline in the road management database.

Topological relationships, which are necessary to set up associations between spatial objects (for the purpose of spatial analysis), are important in determining an object's geometric representation. Road management requires road intersections to carry out analysis, thus the roads are represented by their centerline. For the utility context, the road boundaries are necessary to locate the underground utility accurately.

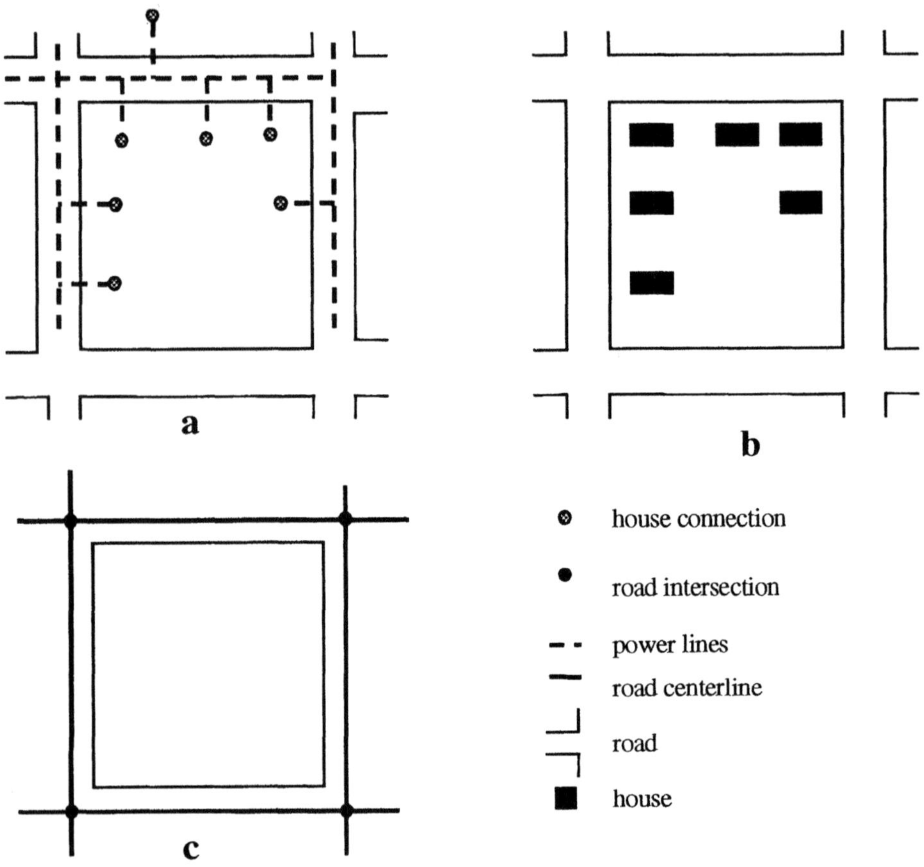


Figure 5.6. Difference in geometric representation due to semantic differences; a) utility context, b) cadaster context, c) road management context.

## 5.7. Semantic translators

A semantic translator is a middleware that provides project worlds within or across GICs with a mechanism to share information transparently, as shown in Figure 5.7. For two communities or subcommunities to share information, they should communicate via a semantic translator. Based on the concept of GIC hierarchy introduced in Section 5.3, semantic translators can be developed for each GIC; for example, a semantic translator to share land use information, another to share cadaster information. The functionality of semantic translators is analogous to that of a printer driver; the printer driver translates users' documents to a language which the printer understands. The semantic translator translates the shared information of a provider into a form which the receiving application understands. It consists of the following three main components:

1. A common ontology library, which describes the application domain, that is, cadaster, land use, etc. Ontology is used as a measure of the semantic similarity between the schema elements of the underlying project world and those of the semantic translator. Two schema elements are semantically similar if they map to the same ontological term.
2. A federated schema is a generic abstraction of the underlying GIC. Each element of the federated schema is readily mapped to the ontology of the underlying GIC.
3. Functions with interfaces, which are used to map between the schema of the underlying project world and that of the semantic translator.

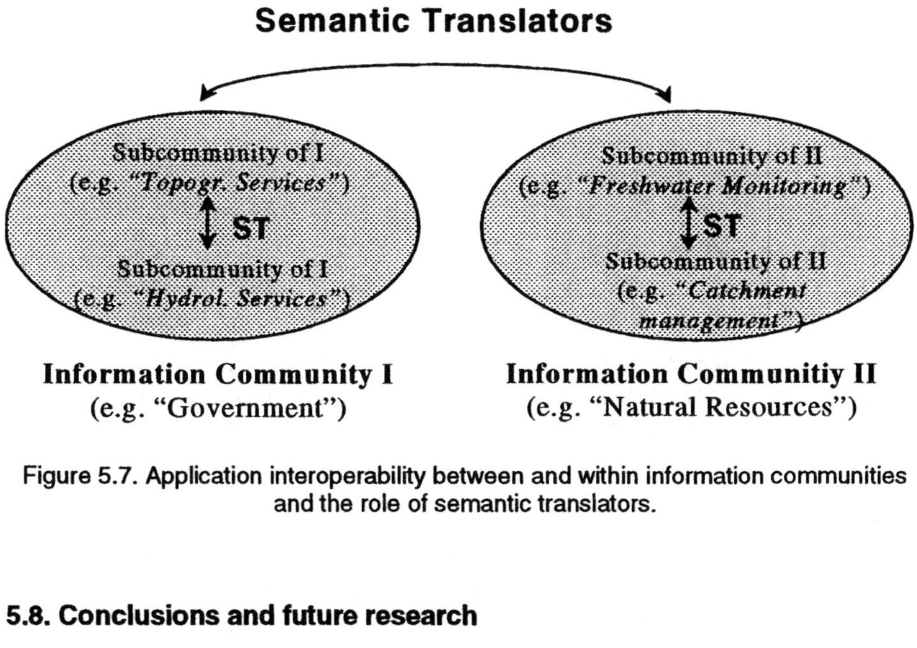


Figure 5.7. Application interoperability between and within information communities and the role of semantic translators.

## 5.8. Conclusions and future research

This chapter presented an informal introduction to the concept of geospatial information communities. It is essential to provide a structured way to arrange different geospatial information communities hierarchically and to define their relationships, as well as their differing facets. Instead of defining a GIC in terms of how it represents real-world features in databases, our definition is based on how each GIC conceptualizes real-world features. Ontology is presented as a high-level meta-language that can be used to construct formal description of GICs' conceptualization.

Three types of heterogeneity—semantic, schematic, and syntactic—were discussed. This chapter suggests that differences between GICs are mainly semantic. However, once the information of those GICs is stored in the database, the schematic and syntactic differences prevail. Semantic heterogeneity causes schematic and syntactic heterogeneity. Therefore, we can intuitively assert that solving semantic heterogeneity between applications will pave the road to true application interoperability.

Semantic heterogeneity is the most important and most difficult problem to solve. To gain more insight to the problem, we presented some case studies based on various semantic heterogeneity types. However, several research issues remain open:

1. *Types of semantic heterogeneity*: Three types of semantic heterogeneity were introduced: categories definition, class intension definition, and geometric description. This list is not necessarily complete. For example, two objects in two different databases can be similar in their categories definition, class intension definition, and geometric description, yet they might have different roles and methods in their respective databases.

2. *Case studies*: More case studies are necessary and further investigation is required in order to have a better understanding of the problem at hand. This will help to document possible semantic differences between geospatial applications within and across information communities.

3. *Identifying semantic similarities*: As introduced in this chapter, information sharing between geospatial applications is only possible if they have some semantic similarity. A formal methodology to identify the semantic similarities between geospatial applications is still required.

4. *A hierarchical structuring of GICs*: This is an essential step to study the semantic similarities between GICs.

5. *Resolving schematic and syntactic heterogeneity*: A formal method and specifications is required to map semantically similar and yet schematically and syntactically heterogeneous database features.

6. *Prototyping*: The next step is to implement the research issues discussed above in a working semantic translator prototype.

Since semantics describe and define a particular application, application interoperability and semantic interoperability are synonymous. And in spite of the progress we have made, we still have a long way to go before we can say we have achieved applications interoperability.

## Acknowledgment

We would like to thank Ann Hitchcock from the Institute for Geoinformatics, University of Muenster for her detailed proof reading and for pointing out some inconsistencies in the contents of this chapter.

## References

ASTM (American Society for Testing and Materials) 1985 *Annual Book of ASTM Standards.* (Section 4: *Construction*, Vol. 04.08, *Soil and Rock; building stones*). ASTM, Philadelphia

Bishr Y, Molenaar M, Radwan M 1996 Spatial heterogeneity of federated GIS in a client/server architecture. In *Proceedings, GIS'96, Vancouver, Canada*

Buehler K 1996 *The Open Geodata Interoperability Specification Part II: Abstract Specification.* Draft version. Open GIS Consortium, Wayland, Mass, USA

Buehler K, McKee L (eds) 1996 *The OpenGIS Guide: Introduction to Interoperable Geoprocessing.* Open GIS Consortium, Wayland, Mass, USA

Campbell J B 1983 *Mapping the Land: Aerial Imagery for Land Use Information.* Association of American Geographers, Washington DC

Gruber T R 1995 Toward principles for the design of ontologies used for knowledge sharing. *International Journal of Human and Computer Studies* 43(5/6): 907–928

Guarino N 1995 Formal ontology, conceptual analysis and knowledge representation. *International Journal of Human and Computer Studies* 43(5/6): 625–640

Kashyap V, Sheth A 1995 *Semantic and Schematic Similarities Between Database Objects: A Context-based Approach.* Technical Report TR-CS-95-001. Department of Computer Science, University of Georgia, Athens

Kent W 1991 The breakdown of the information model in multi-database systems. *SIGMOD Record* 20(4): 10–15

Sims R A, Corns I G W, Klinka K (eds) 1996 *Global to Local: Ecological Land Classification.* Kluwer, Dordrecht, The Netherlands

van Engelen V W P, Ting-Tian W (eds) 1995 *Global and National Soils and Terrain Digital Databases - SOTER. Procedures Manual.* Revised edition. International Soil Reference and Information Centre, Wageningen, The Netherlands

# 6 CHARACTERIZING THE SEMANTIC CONTENT OF GEOGRAPHIC DATA, MODELS, AND SYSTEMS

Mark N. Gahegan

## 6.1. Introduction

Geographic information systems (GIS) employ distinct conceptual models of geographic space (Goodchild 1992), often as a reflection of the origins of the software (e.g., computer-assisted design and image processing). Some of these models are radically different, such as the images employed by Idrisi compared to the object coverages used by ARC/INFO. Others are more subtly different, such as a topologically-oriented coverage compared to the *spaghetti* polygons used by many desktop GIS. The meaning of spatial data is not the same within these models (Morehouse 1990), and translation or interoperation that is based solely on the geometry can lead to logical inconsistencies within the translated data. Whilst a good deal of very useful progress has been made by the likes of ISO TC211 and the Open Geodata Interoperability Specification (and related models), as yet these standards fall somewhat short in addressing the semantics of the underlying geographic models. Efforts thus far have tended to concentrate on geometry, which can be rather ambiguous in defining the meaning of data in a geographic sense.

The work described here is motivated by research into interoperability and data translation in regard to three-dimensional geo-information systems, to support the needs of a wide range of geoscientists, including geographers, hydrographers, geologists, and geophysicists. The aim is to engineer a semantically rich environment by ensuring that objects are ascribed meaning based on their modeling role, as opposed to their properties (e.g., geometry). Interoperability issues are not restricted to the more standard GIS, but also include many of the available geological and exploration packages such as Surpac and Vulcan; these provide a wealth of further spatial primitives beyond the standard points, lines, regions, and surfaces, including volumes and profiles. Within the geosciences, there are a proliferation of data types, models, and application packages for which standardization in terms of data exchange and semantic content is both necessary and desirable.

### *6.1.1. Major research questions*

In order to set semantic interoperability in context, it is important to understand the major motivating research questions that underpin it. The motivating research questions pursued here are as follows:

1. What properties of geographic data need to be characterized to ensure that translation and interoperability can be achieved at the semantic level, in addition to the internal level of data structures and record formats?
2. How should these properties be described and categorized so that they can be understood and communicated?
3. How should the properties be represented and encoded within a metadata description?
4. How can the properties be used to improve data translation and interoperability?

No attempt is made in this chapter to cover all aspects of the problem, so the aim of this stage is not for completeness (something that should perhaps arise out of further discussion with interested parties), but instead for some illustrative examples, and a framework within which the details can be embellished later.

In earlier work (Gahegan 1996) a semantic notation was developed to describe the various transformations that occur as data are operated on or changed from one model of geographic space to another. It is based on a data communication protocol described by Pascoe and Penny (1995) which has been extended to encompass certain key geographic properties and both a conceptual and physical data model. The notation describes a *before* and *after* state for a given transformation and is useful for communicating the likely effects of a specific transformation in terms of the data properties that may change as a consequence. In turn this can highlight any changes in the underlying conceptual model that occur and furthermore can show where assumptions regarding the meaning of the data are invalid or need to be made explicit. More recently, further additions have allowed the specification of uncertainty characteristics within the data (Gahegan and Ehlers 1997).

## 6.2. The specification of an interoperable dataset

Interoperability, here defined as the sharing of data and functionality between systems, can occur at several levels of abstraction, ranging from the physical layer of file formats to the sharing of complex applications (Bishr 1997: Chapter 3). Discussion here is confined to the interoperability between GIS so that (explicitly) geographic datasets and functionality may be shared as a network resource (e.g., Albrecht 1994). This is the same level of abstraction addressed by the Open Geodata Interoperability Specification (OGIS; Buehler and McKee 1996).

In order that geographic data may be used and interpreted in appropriate ways it is often necessary to have a deep understanding of the data, extending beyond the impoverished descriptions employed in most commercial systems, where the description of geographic objects is limited to the structure of their geometry and associated aspatial attributes. Many semantic problems find their origins in this lack of richness for data and model description. Even in cases where a minimal description is sufficient, problems can still arise during translation and interoperation if the data are passed between systems where the geographic models employed are not semantically equivalent.

This chapter proposes a notation to help describe the (sometimes subtle) differences between the data models used by different GIS and thus to aid in the interoperability process by providing a concise and symbolic description of geographic data, specifying their semantic content as opposed to relying on the geometry to imply a meaning. This description can be used to represent both datasets and the operations and transformations conducted on them. It is represented by an expression of the form:

(⟨*abstract properties*⟩, ⟨*geographic model*⟩, ⟨*physical data structures*⟩, ⟨*system details*⟩)

*Abstract properties* describe the data as the user perceives them (equivalent to an external view); *geographic model* describes the implications and limitations of the geographic model of space under which the data exist; *internal data structure* describes how the geometric components of the data are logically represented (necessary since the choice of data structure can have an effect on other data properties); and *system details* describe the actual package and platform that the data reside in.

Notice that these components loosely follow the standard layered architecture approach taken in information systems design (Hadzilacos and Tryfona 1996; Tsichritzis and Klug 1978). In practice, each of these components is further broken down into a number of distinct parts. The following sections describe the important characteristics of these components. These are synthesized into a more formal notation in Section 6.3.

### *6.2.1. The abstract properties (external layer)*

The abstract properties describe the true geographic attributes of the data, independent of any data structure or system. For simplicity, discussion is restricted to the following:

$$a\ (d, s, t, u, v)$$

in which $d$ represents the data value(s), $s$ the spatial extents, and $t$ the temporal extents; $u$ represents some measure of uncertainty associated with $d$, $s$ and $t$; and $v$ represents the lineage or derivation history of $a$ in terms of the operations and transformations by which $a$ was formed. Together, $s$, $t$, $u$ and $v$ form the *context* over which $d$ is valid. Any geographic object, at any level of abstraction, may be specified in this abstract manner. For example, an entire dataset ($A$) may be described by:

$$A\ (D, S, T, U, V)$$

in which case the properties $D$, $S$, $T$, $U$, and $V$ represent some form of composite of the values for all $a_i \in A$ (Gahegan 1996).

### 6.2.2. The geographic model (semantic layer)

A geographic model is the conceptual or semantic basis under which the abstract data exist. One of the major impediments to translation and interoperability within the geosciences arises from the different models that form the basis of the systems that are currently in use. One reason for these differences is that the development of different systems (for utilities and infrastructure, for remote sensing, for environmental analysis, and for geological applications) have been largely divergent. The differences in geographic models reflect the types of data that the various packages use, and rightly so. None of these models should be regarded as correct or best since all have their place. For interoperability to be truly successful, it is necessary (in the author's opinion) to be able to recognize, describe, and communicate these models as part of the data translation and migration process; without this ability, it is not possible to state what the data might mean. Specifically, reliance on geometric structural equivalence is not sufficient to provide interoperability.

Following from this argument, the specification of an all-embracing interchange format such as OGIS does not help to improve the modeling capabilities or the semantic richness of existing systems. Unchecked, it offers the means to pass data into systems that are incapable of fully representing the intended content and of recognizing and reporting this fact.

**6.2.2.1. Geometric equivalence and semantic difference**. Some packages have radically different geographic models as their basis; for others, the differences are more subtle. As an example, compare the definition of a region (polygon) object as it occurs in MapInfo and ARC/INFO. Whilst it is a relatively simple operation to transform the defining geometry from the topologically-based ARC/INFO structure to the free-format boundary description used in MapInfo, this fails to account for the *subtle* differences in the geographic models used in these packages. A given polygon in ARC/INFO is defined as part of a coverage where topological closure is assured, neighboring objects will be boundary-coincident, but it is not possible for any other object in the coverage to be coincident with the object interior. The same polygon in MapInfo has no such constraints; objects may overlap and common boundaries are not recognized.

The model employed by ARC/INFO carries more constraints, so the associated data carries with it certain characteristics that cannot be expressed in MapInfo without extension of the underlying data model by customization. Data translation may therefore be *lossy* since the information content is being reduced. Other semantic differences may arise from whether boundary lines may cross each other and from the mechanism for dealing with holes or islands. Of course, in many cases such losses may be unimportant, but in others they may be quite necessary to correctness. When dealing with a data set imported from some other package, knowing its origin can help to ensure that it is used only in appropriate ways and is not ascribed any meaning that it does not possess. To reiterate, the conclusion is not that one model is superior, only that they are different.

As well as the differences between systems, users too may vary in the meaning they ascribe to data, even within the same system. Many systems support more than one geographic model concurrently, and some will allow semantically-inconsistent use of object descriptions, so the meaning of the data is somehow divided between the geographic model supported in software and the geographic model in the mind of the user. Showing semantic equivalence or providing semantic translation are difficult issues (e.g., Goh et al. 1994; Kuhn 1994), requiring abstract descriptions of models including the incorporated objects and methods. In contrast to GIS, traditional databases model such a limited and abstract set of object types (strings, numbers, dates, amounts) that issues of semantic differences between the data models of packages are less problematic (although semantic inconsistencies may still arise in applications built from these objects).

Geographic models can be categorized in a number of valid ways (e.g., Goodchild 1992; Peuquet 1988). Here we adopt categories defined by the semantic abstraction that the data have undergone (Smith et al. 1992). This attempts to show the relationship between primary measurable phenomena (field data and image data) and the higher geographic models that are often used for cartographic representations of objects and themes, and that also form the basis of most GIS. Some commonly used geographic models are summarized below in Table 6.1.

| | **Value Properties** | **Space Properties** |
|---|---|---|
| **Field Samples** | a vector of real values | individual point geometry |
| **Image Coverage** | a vector of quantised values | pixel image geometry |
| **Surface Coverage** | a height value | pixel / triangle / parametric |
| **Thematic Coverage** | a single class label | arbitrary shapes of classes |
| **Object Coverage** | a feature identifier (and other properties linked to this) | contiguous but arbitrary shape of objects |
| **Topological Object Coverage** | a feature identifier (and other properties linked to this) | topologically consistent shape of an object—usually defined in terms of relationships between lower order shape primitives |

Table 6.1. A summary of some different models of geographic space with their appropriate value (*d*) and space (*s*) attributes.

#### 6.2.3. The data structures (internal layer)

Spatial data structures are relevant only to the degree that the choice of data structure can affect various properties of the abstract data, such as their precision and the ability to represent linear objects or transition zones (Laurini and Thompson 1992). The most important distinction is, of course, between structures based on regular grids and those based on vectors.

#### 6.2.4. The systems and platforms (physical layer)

The choice of GIS package will affect the semantics in terms of the geographic model, expressiveness and constraints. Even where the system is chosen, it is possible for differences to occur between platforms (e.g., the PC product differs from the workstation product) and between versions, which may include additional modeling expressiveness and functionality.

### 6.3. A formal notation for geographic interoperation

Along the same lines as the high-level semantic descriptions proposed by Kuhn (1994), large-scale efforts have been undertaken amongst the geoscience community to establish a catalog of objects to which precise and unambiguous meanings and descriptions can be associated. This catalog may then become the basis for data transfer and interoperability at the semantic level (AMIRA P431 1995).

Whilst it is desirable to have perfect translation and interoperation, preserving all properties of the data, including geometry, attributes, and meaning, this is not always possible because of the constraints of working with existing (legacy) systems which are not readily modified. Where certain expressiveness is missing, the only short-term possibility is to *describe* the nature of the mismatch and not to fix it. A concise description of dataset properties and all functionality that transforms data can help uncover potential problems.

Data sets are represented by an expression that describes the properties outlined in the previous section. Transformations require expressions with both a left and a right side and show the changes imposed on the data:

$$(before\_state) \xrightarrow{transformation} (after\_state)$$

The *after* state contains a revised expression where any properties that have changed are flagged.

#### 6.3.1. A taxonomy of transformations

It is straightforward to build a taxonomy of transformation consequences in terms of the properties that change. A useful high level grouping is:

TYPE 1: transformations changing the abstract data properties. An example would be an overlay operation or region expansion.

TYPE 2: transformations causing the geographic model to change. An example would be the classification or interpolation of point data to form a thematic coverage, or the extraction of cartographic objects from a remotely-sensed image.

TYPE 3: transformations causing the physical encoding of the geometry to change. An example would be conversion of raster to vector representation or the creation of a triangulated surface from a series of points.

TYPE 4: transformations moving the data to another system. An example would be a data export or interoperation task.

It is quite possible that a transformation may fall in more than one of these categories— moving data between systems may cause the abstract properties and the geographic model to change (TYPES 1, 2, and 4) and performing semantic abstraction (e.g., extracting objects from remotely-sensed imagery) will usually involve a compound transformation of TYPES 1, 2, and 3.

### *6.3.2. Some example transformations*

As an example, using *A*, *G*, *P*, and *H* to represent the dataset properties (abstract data, A; geographic model, G; physical data structure, P; and host system, H, respectively), a transformation which moves data from one system to another, but using the same geographic model and data structures, is given by:

$$(A, G, P, H) \longrightarrow (A, G, P, H')$$

Obviously this expression represents the ideal situation where interoperability is completely unhindered and there are no implications or undesirable side-effects.

When considering interoperability, the transformation will often be made up of several stages: first moving the data to an interoperability format, such as the OGIS Open Geodata Model (Buehler 1994), then importing them into some new system, then operating upon them, and possibly moving them back again:

$$(A,G,P,H) \xrightarrow{export} (A,G',P',H) \xrightarrow{import} (A,G'',P'',H') \xrightarrow{operation} (A',G'',P'',H') \ldots$$

The *export* transformation moves the data into the interoperability format from the host system, changing their physical structure and (possibly) their geographic model. From there they are *imported* into the internal format of the new system, again changing their physical structure and (possibly) geographic model, as shown below in Figure 6.1. Next, some *operation* is carried out (here shown as only affecting the abstract data properties) after which the data may be passed back again to the original host. Clearly, if the models and data structures concerned with all stages cannot be mapped to each other exactly then there is a potential for conflicts to arise. For simplicity, only the highest-level properties are shown in the above example.

With the introduction of further properties, the transformation expressions can become quite specific in identifying exactly *what* has changed.

By a simple matching of the left and right side of expressions, a chain of transformations can be formed where the output of one becomes the input to another. In this way, data in some arbitrary form can be changed, step-by-step, to reach a desired outcome. It is also a simple matter for the user to become aware of the likely

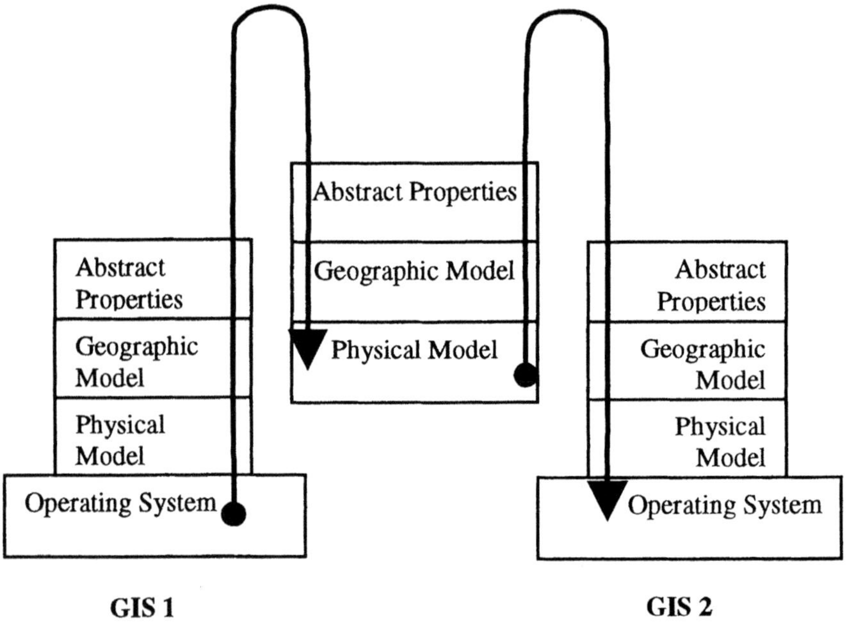


Figure 6.1. Schematic diagram showing the movement of data between systems via an interchange format

effects that various transformations may have on the data, by examining which properties of the data set are affected, that is, those properties that differ between the left and right side of expressions.

As an example, the following expression describes the input and output specification of a transformation which produces a thematic overlay from a collection of field points (all in vector format) via the following method.

points_to_theme():

$$(A(D,S,T,U,V), G_F, P_V, H_x) \rightarrow (A'(D',S',T,U',V'), G_T, P_R, H_x)$$

As can be seen, the method changes many of the abstract data properties and also transforms the data from a field model ($G_F$) to a thematic model ($G_T$). Before this method may be applied, all input data must first be mapped to a field representation in vector format, within the system that provides the functionality ($S_x$). So the required input format ($Z$) for the method can be described as:

$$Z_{\text{points\_to\_theme}} = \left(A, G_{\mathrm{F}}, P_{\mathrm{V}}, H_x\right)$$

This expression acts as a template to be matched against and achieves the following:

- It advertizes the fact that a transformation is available to change a set of points into a coverage.
- It describes the way in which the method will affect certain dataset properties.
- It shows where (in which system or at which remote site) the functionality resides.
- It gives an input specification for any dataset which is to be processed.

Any differences between the dataset expression and that of the method must be resolved via appropriate transformations before the method may be used. The movement of data to an interoperable format and the transfer of data in this format to other systems are examples of the types of transformations that might be required before a method can be applied.

## 6.4. Overseeing the transformation process

It is a relatively straightforward task to move from the kind of symbolic description outlined above to a set of rules and constraints that can determine if some interoperation is feasible or is likely to cause difficulties (Walker et al. 1992), by comparing a semantic description of a chosen operation in one GIS with a description of a chosen dataset within another. Any semantic differences between the description of the dataset and the left side of the transformation expression indicate a potential conflict in the meaning of the underlying data. Mismatches can be graded according to their severity, ranging from warnings to outright conflicts. In some cases, it may be possible to carry out any required conversion in an automated fashion; in others, some form of user intervention might be necessary. In either case, warnings can be issued and the mismatch reported.

Suppose now that we wish to overlay two datasets, both in thematic form. The first, $B_1$, is physically represented by a quadtree ($P_{\mathrm{Q}}$) in host system $H_1$ and contains 30 $\mathrm{m}^2$ pixels. The second, $B_2$, is represented by a grid or raster ($P_{\mathrm{R}}$) in host system $H_2$ and contains 20 $\mathrm{m}^2$ pixels. Before overlaying can take place both datasets must share a common system, grid and data structure as is shown by the expression below which represents functionality available in system $H_2$. We shall assume that both datasets already share a common temporal and spatial extent.

image_overlay():

$$\begin{pmatrix} A_1(D_1, S, \ldots, U_1, V_1), C_{\mathrm{I}}, P_{\mathrm{R}}, H_2 \\ A_2(D_2, S, \ldots, U_2, V_2), C_{\mathrm{I}}, P_{\mathrm{R}}, H_2 \end{pmatrix} \rightarrow \left(A_r(D_r, S, \ldots, U_r, V_r), C_{\mathrm{I}}, P_{\mathrm{R}}, H_2\right)$$

Suppose, also, that we have access to a set of conversion routines as specified below.

Quadtree_to_Raster():

$$(\ldots,\ldots P_Q, H_1) \rightarrow (\ldots,\ldots, P_R, H_1)$$

where … represents that no change will occur and the properties are irrelevant (i.e., 'don't care').

Image_Resample():

$$(A(D,S,T,U,V), C_I, P_R, H_2) \rightarrow (A'(D',S',T,U',V'), C_I, P_R, H_2)$$

Image_Interoperate():

$$(A(D,S,T,U,V), C_I, P_R, H_1) \rightarrow (A(D,S,T,U,V'), C_I, P_R, H_2)$$

Examining the desired operation, i_overlay(), reveals a target form, Z, for the dataset as:

$$Z_{\text{i\_overlay()}} = (A(\ldots,S,\ldots), C_I, P_R, H_2)$$

From the above description, dataset $B_1$ has the form:

$$B_1 = (A_1(D_1,S_1,T_1,U_1,V_1), C_I, P_Q, H_1)$$

whereas $B_2$ is described by:

$$B_2 = (A_2(D_2,S_2,T_2,U_2,V_2), C_I, P_R, H_2)$$

Three problems are immediately obvious. Firstly $B_1(S) \neq B_2(S)$ (the input datasets do not share the same spatial basis); secondly $B_1(P) \neq Z_{\text{i\_overlay()}}(P)$ (the input datasets do not share the same physical data structure); and finally $B_1(H) \neq B_2(H)$ (the datasets do not reside on the same host computer).

The physical schema of $B_1$ can be changed to the desired form (raster) by using quadtree_to_raster(), then, moved to host system $H_2$. The spatial basis of $B_1$ and $B_2$ can be made equivalent by resampling one (or both) of the datasets using image_resample(). Assuming that $B_1$ is resampled to 20m pixels then the pre-requisites are met and the overlay operation can be carried out. The above expressions allow the effects of the conversions to be reported; as well as having a different spatial extent and data structure, $B_1$ also has modified values and increased uncertainty. Following this conversion, $B_1$ and $B_2$ are now compatible and can be combined using image_overlay().

### *6.4.1. Evaluating alternative courses of action*

This example raises a tangential issue; there are sometimes many possible paths that could be taken in order to achieve some transformation. In the above example two choices must be made (i) the order in which to apply the conversion functions and (ii) the 'best' dataset to resample, giving rise to a number of possible scenarios.

In the event of more than one scenario resulting, all possible paths through conversion could be evaluated before a choice is made. Costing functions could be applied in terms of the number (and type) of the undesired changes made to a dataset, and the magnitude of these changes. In the above example, a reduction in grid resolution for a dataset might incur a high cost (say of $\left| resolution(A_1) - resolution(A_2) \right| * 2$) on the grounds that data resolution is being 'wasted'. By contrast, an increase in grid resolution might incur a reduced cost (say of $\left| resolution(A_1) - resolution(A_2) \right| * 0.5$) since the data are only resampled. Consequently, Scenarios (i) or (ii) would be favored.

Further work is currently being directed toward the use of automated reasoning for data interoperation. Specifically, the expert system shell CLIPS® is used to control the path through the various transformations, using a cost minimization search to establish the best method to integrate the data, and to report on potential problems or conflicts to the user.

## 6.5. Results and conclusions

A formalism was developed to represent both the abstract and concretized properties of data. The formalism was shown to be useful in characterizing the various transformations that are applied to geographic data, including moves from one model of geographic space to another. Transformation expressions form a sound basis by which to understand and describe the effects of the operations performed, in terms of the changed properties of the data. A taxonomy of transformations was presented that applies at both the semantic and physical layers, along with some relationships by which transformations may be categorized.

The research questions posed in Section 6.1.1 can now be revisited:

1. The properties that are required to characterize the data must be determined by the application. It is certainly possible to devise a minimum set (as in Section 6.2.1) and to build from there. The argument for including the geographic model as part of any interoperability specification (Section 6.2.2) is important because of subtle differences in the meaning of data that are not apparent when considering their geometry alone; so an exchange format based around geometry and associated attributes is not sufficiently rich to support informed use of the data.

2. Categorizing the various properties of data and underlying models provides a framework for the communication of meaning and may be useful to people and systems.

3. The description of such properties and their behavior can be achieved with a transformation notation which is easily encoded as part of a data description and may be manipulated as a series of clauses within an automated reasoning environment.

4. Once these properties are in place, the application and interpretation of data can be *guarded* by a sentinel process, that encapsulates some general rules and descriptions of transformations. Whilst this will not ensure that no misuse can occur, it will offer some protection, via warnings messages, to alert the user of potential changes and inconsistencies.

In short, as well as insights into the various transformations that interoperability gives rise to, there are some distinct benefits that might be gained by enhancing interoperability descriptions with semantics. This would also go a long way towards improving the handling of geographic semantics in existing systems. By including transformation descriptions for the functionality provided by each package some semantics can become, in effect, *reverse-engineered* into a system.

Further work is required to establish that the geographic properties modeled are necessary and sufficient. No such claim is made regarding the examples given in this chapter, which are provided as a starting point for discussions only. The cataloguing of the supported functionality for each commercial system and each dataset in the described format is also required. Finally, thorough examination of the implied geographic models used in different packages, and in different applications within the same package, is necessary before the semantic exchange of data and functionality becomes possible.

## References

Albrecht J 1994 Universal elementary GIS tasks—beyond low-level commands. In Waugh T C and Healey R G (eds) *Sixth International Symposium on Spatial Data Handling*: 209–22

AMIRA P431 1995 *A Data Dictionary of Geological Objects*. Australian Mineral Industries Research Association LTD, Melbourne

Bishr Y 1997 *Semantic Aspects of Interoperable GIS*. PhD Thesis, Publication Series Number 56. ITC, Enschede, The Netherlands

Buehler K 1994 *OGIS Geodata Model Overview*. OGIS project documentation 94-019. Open GIS Consortium

Buehler K, McKee L (eds) 1996 *The OpenGIS Guide: Introduction to Interoperable Geoprocessing*. Open GIS Consortium

Gahegan M N 1996 Specifying the transformations within and between geographic data models. *Transactions in GIS* 1(2): 137–52

Gahegan M N, Ehlers M 1997 A framework for the modeling of uncertainty in an integrated geographic information system. To appear in *ISPRS International Workshop on Dynamic and Multi-Dimensional GIS*, Hong Kong

Goh H, Madnick S, Siegel M 1994 Context interchange, overcoming the challenges of large-scale interoperable database systems. In *Third International Conference on Information and Knowledge Management*: 337–46

Goodchild M F 1992 Geographical data modeling. *Computers and Geosciences* 18(4): 401–8

Hadzilacos T, Tryfona N 1996 Logical data modeling for geographical applications. *International Journal of Geographical Information Systems* 10(2): 179–203

Kuhn W 1994 Defining semantics for spatial data transfers. In Waugh T C, Healey R G (eds) *Sixth International Symposium on Spatial Data Handling*: 973–87

Laurini R, Thompson D 1992 *Fundamentals of Spatial Information Systems*. Academic Press, London

Morehouse S 1990 The role of semantics in geographic data modeling. In Brassel K, Kishimoto H (eds) *Fourth International Symposium on Spatial Data Handling*. Department of Geography, University of Zurich, 2: 689–98

Pascoe R T, Penny J P 1995 Constructing interfaces between (and within) geographical information systems. *International Journal of Geographical Information Systems* 9(3): 275–91

Peuquet D J 1988 Representations of geographic space: towards a conceptual synthesis. *Annals of the Association of American Geographers* 78: 375–94

Smith T R, Ramakrishnan R, Voisard A 1992 Object-based data model and deductive language for spatio-temporal database applications. In Gambosi G, Scholl M, Six H-W (eds) *Geographic Database Management Systems*. Springer Verlag: 79–102

Tsichritzis D C, Klug A (eds) 1978 *The ANSI/SPARC DBMS framework—Report of the study group on data base management systems*. Volume 3. ANSI, USA

Walker D R F, Newman I A, Medyckyj-Scott D J, Ruggles C L N 1992 A system for identifying datasets for GIS users. *International Journal of Geographical Information Systems* 6(6): 511–27

# 7 DESIGNING FOR INTEROPERABILITY: OVERCOMING SEMANTIC DIFFERENCES

Francis Harvey

## 7.1. Semantic interoperability

Interoperability can be understood in a number of ways. In a minimal sense, even the capability to transfer data from one computer system to another without transformation loss can be identified as interoperability. In a broader sense, interoperability can be taken to suggest the ability of different applications to interact dynamically, facilitating the smooth interface of multiple information sources. This chapter examines interoperability in this second sense, specifically the role of semantics in facilitating the exchange of information.

OpenGIS documents lay out the basic technical infrastructure for interoperability. Standardized data formats, protocols, and transfer mechanisms permit a technically smooth exchange of data. The concept of information communities (Open GIS Consortium 1996) allows for the definition of consensual sharing of geographic information. This level of interoperability will change the way people interact with geographic information technologies. This technical foundation and framework is crucial for efforts to design a robust technical standard, but widespread application in heterogeneous environments will require broader consideration of semantic issues and the need to preserve geographic and attribute meaning in information exchange.

While many issues connected to the technical transfer of data have been resolved, the sharing of information still requires examination of the underlying semantic issues (Kuhn 1994). The exchange of information and its inherent meaning, although implicated in the discussions of data exchange mechanisms, does not find the same lucid technical solutions, nor discussion. In the OpenGIS abstract specification the authors acknowledge that information exchange between *World Views* requires application-specific knowledge beyond the existing technical description of interoperability and the geographic information community (Open GIS Consortium 1996).

Clearly, there is a need to develop mechanisms to facilitate the exchange of information between different information communities. Semantics plays a crucial role in linking the different conceptual worlds. The research I present here investigates an approach for considering semantic issues in the design of interoperable GIS. By articulating the semantics of each conceptual world in a rigorous manner, the technical prowess of interoperability can be harnessed to help resolve deeper data sharing issues. This approach relies on the construction of *portals* between the different realms. In particular, this work builds on a concept coming from ethnographic studies of science and technology, called boundary objects.

The chapter has two goals. Primarily, it describes a design approach intended to facilitate the exchange of information between GIS and to retain meaning. Second, it provides a conceptual framework for further work in geographic information science (GIScience) on geographic integration. The concept I focus on in this paper, semantic interoperability, elaborates a conceptual framework for considering such issues and others in GIScience.

## 7.2. Designing for semantic difference

This section moves directly to the theoretical framework needed to develop semantic interoperability.

### *7.2.1. Boundary object background*

The concept of boundary objects is the lynchpin in this theoretical framework. Coming from studies of science and technology, this concept and related work open new ways to understand the relationships between people, technology, and the environment. In this chapter, I will focus on their role in linking different conceptual worlds. This is one of the fundamental characteristics of boundary objects and is most important to semantic interoperability. Other characteristics will also be touched on below.

Boundary objects have broad applications in studying how multiple groups engage each other in the construction and use of technology. This builds on some reconceptualizations of the relations between humans and non-humans. Although there are many theoretical issues under debate, most importantly, people working in this field agree that technology is not separate from social endeavors. As a defining characteristic of civilization, technology is integral to all social, economic, and cultural processes. Therefore, to understand the role of technology we must reconceptualize our analyses to account for people and things both having active roles. Technology, like science, does not happen in isolation, but its practices extend into other social worlds. For example, Bruno Latour argues that Pasteur's success arose through the linking of multiple social worlds through his laboratory and the microbes it exposed. Pulling in the agenda of the hygienics movement, Pasteur allied their political power and translated their concerns into arrangements with other social groups in which the Laboratoire Pasteur plays a key role. The anthrax microbe isolated in the laboratory led to the extension of hygienic codes to farmers' barns and elsewhere, leading to fundamental changes in milk production, distribution, consumption, etc (Latour 1988). Pasteur's laboratory became not only a crucial link in solving health problems, but a focal point in a newly established network of humans and non-humans.

Latour's study bears a strong resemblance to studies utilizing boundary objects. Differences are many, but a discussion of them goes beyond this chapter. The key point is that boundary objects have never solidified as the work of a single group or author. The concept's roots are more varied and its conceptual development has led to significant differences to Latour's focus on certain artifacts, individuals, and translations between actants (Akrich and Latour 1992) in networks of relationships. Just as boundary objects themselves, writing about boundary objects involves multiple authors.

Fundamental to all descriptions of boundary objects is the central argument that the mediation of diverse social interests occurs in and during the construction of technology (Neumann and Star 1996; Star 1995; Star and Griesemer 1989). Negotiations may strive for consensus, or decay into coercion, but usually they lead to the stabilization of relations through the construction of artifacts and institutions that maintain some coherence. Stabilization is a key property of constructing technology, and although dynamic and therefore flexible, it provides the concepts and artifacts people refer to in constructing social relations. Boundary objects mediate the stabilization between different groups. Like geographic boundaries they have a dual function: at the same time they supply common points of reference, they also serve to distinguish differences.

Different interpretations of boundary objects still share the common understanding that boundary objects are crucial for stable translations between different perspectives. They form crucial intersections and translations between different social worlds. "Boundary objects facilitate the multiple transactions needed ... to engineer agreements among multiple social worlds" (Fujimura 1992: 172). In GIS technology there are already many examples. Data-sharing agreements usually involve the stipulation of technological requirements as the result of negotiation. They all may say wetland, but they mean different things, each called wetland. The layers in the multi-purpose cadaster (MPLIS) are also boundary objects: each layer usually remains in the jurisdiction of a particular agency, but is defined in a way that other agencies also have their needs served. Although each layer is the responsibility of an agency, the collaborative design of a MPLIS stabilizes certain attributes of this layer for other agencies. A boundary object, for example the wetland layer in a county GIS, connects multiple perspectives through its arbitration of differences, but, at the same time, maintains distinctions between the actors. Taking the example of a theoretical wetland layer in a county GIS, an attribute COUNTY_WET may be used for areas accepted by the county administrator as wetlands following the Cowardin schema, another attribute FEMA_WET may delineate wetlands adjudicated by the Corps of Engineers, and the attribute RIP_WET could be used by ecologists to designate environmentally sensitive wetlands protected by the county's riparian habitat preservation ordinance.

In such a case, the different mandates and legal frameworks of these three agencies necessitate the differentiation of unique wetland inventories. Perhaps COUNTY_WET overrides the others for general county purposes, but cases may well arise when it becomes necessary to reconcile the different inventories. At present, the only method we have for performing this activity is the cumbersome process of negotiation. Because of existing contentions and political wrangling, these processes frequently end up in *data wars* (King and Kraemer 1993). Since nearly one third of all information system projects are abandoned after large expenditures and commitment of resources (Ewusi-Mensah 1997), it seems much wiser to avoid struggle and waste and nip them in the bud. The same possibilities are faced in interoperability projects. Semantic issues are much more complicated, and the process of negotiation can readily turn into detente hearings resembling nuclear-disarmament treaty negotiations. Successful cooperation requires a focused process of design using negotiation to stabilize issues. Because the world, people, and institutions change, the ideal of consensus is not important; the coherence that lets people get to work and collaborate is.

Following these ideas, boundary objects provide the concept for articulating and conceiving semantic relations between different conceptual worlds and for finding coherence. "Boundary objects are objects which are plastic enough to adapt to local needs and the constraints of several parties employing them, yet robust enough to maintain a common identity across sites" (Star and Griesemer 1989: 393). Boundary objects moderate differences and establish a shared understanding. Semantic interoperability is the stabilized connections between participants that facilitate an exchange not only of data, but of information, and possibly knowledge.

In the process of design and implementation, this should also be understood in a broader sense as the construction or expansion of a network of social and technical elements. This network goes beyond the people and things directly involved and connects these participants with their institutions, disciplines, and colleagues to provide coherence for the construction of GIS technology and various other forms of relations. However, this means the inclusion of some groups and the exclusion of others. Coherency means that the position is acceptable for many, but not necessarily for all.

### *7.2.2. Think interoperability, act locally*

With the concept of boundary objects on board, it now is necessary to move on to the complicated issues of design and implementation. Both are very involved, political, social, and personal processes. Regardless of the size of the organization, design and implementation involve people with multiple agendas. Negotiations are unavoidable, but data wars are. Finding acceptable coherence in design involves compromising and face-to-face discussions. Participation is essential, and boundary objects help focus the discussions on substantive issues.

Basically, from the social constructionist perspective, all negotiations surrounding GIS technology involve boundary objects. Each particular GIS presents a unique local set of human and non-human elements that enable multiple social groups, with divergent or even contradictory values, to collaborate in some way. Of course, this is a dynamic process of social production, characterized by differences and contention, evolving chaotically. Stability is tenuous and subject to constant renegotiation. Because the coherence of boundary objects is as limited as the stability of involved social groups, involvement of these groups in all stages of design is fundamental to finding the dynamic stability necessary for chaotic systems.

The GIS design process is more open for heterogeneous environments. In a tightly closed system, for example, the requirements can be readily ascertained for a factory robot or a household appliance, turned into values, actions, and constraints and then modeled. This model is the basis for the design of a system that carries out the requisite actions in an assembly line, or cooking. In open systems (sometimes also called chaotic systems), in environments beyond control or accurate description with linear models, the need remains, to formulate robust solutions. The environment and human society are such open systems, that are further complicated due to wide-ranging, possibly mutually exclusive perspectives on phenomena.

In the design and use of GIS difficulties often arise when participating groups' understanding of phenomena diverge, or even contradict each other. These issues are often trenchant problems arising in different needs, mandates, purposes, perspectives, politics. A rational solution-finding process to the complex web of differences may lie beyond the reach of waterfall design procedures. Data wars, or costly and inefficient expenditures of resources (King and Kraemer 1993) may ensue. Consensus may not be possible, and utterly unsatisfactory coercion may be sought after. Conflict increases, cooperation decreases, and the organization's activities are crippled.

The design approach described below accounts for these issues and addresses the pragmatic problem at hand: reconciling different semantic interpretations of phenomena and finding the coherence necessary for organizational functions. Keeping the principles of interoperability and focussing on the distinct local issues and agendas provides a basis for designing and implementing semantically interoperable systems.

### *7.2.3. Design with a semantic difference*

Semantic differences are the crux of the problem in designing practical interoperable systems that account for the myriad conceptual worlds and the complexity of phenomena. This section presents a method for conceiving design as a participative task that involves all groups and focuses on resolving semantic differences. Because it is highly schematic, it is solely intended to outline a possible approach. Semantics are articulated in practice; a different arena from design. Participation requires unique approaches adapted to the local needs and relationships. The focus is on showing the role boundary objects play in linking actual practice to more abstract, conceptual, and formalized designs. In this sense, this section outlines steps to consider, but not actions to take.

Building on the concept of boundary objects, designing for semantic differences accounts for three different types, as shown in Figure 7.1. Differences are boundary objects that can be recognized a plain disagreements, or only in much more subtle issues. They are distinguished by the perceived semantic differences. For example, the Corps of Engineers delineates wetlands based on a combination of characteristics that is broader than the Cowardin system used by the EPA and other agencies. The boundary object may be at the first level of analysis the typology, but a more detailed study shows it to be the underlying semantics (Harvey and Chrisman, in press) resulting from different mandates.

The second type of boundary objects are commonalities. As the name implies, these are understood agreements. They can be formalized in standards, operating procedures, and so on, or they can be implicit. In local and highly hierarchical organizations, local knowledge and expertise can easily lead to informal arrangements being the basis for mission-critical operations. Some local organizations and cultures may even rest entirely on webs of informal arrangements.

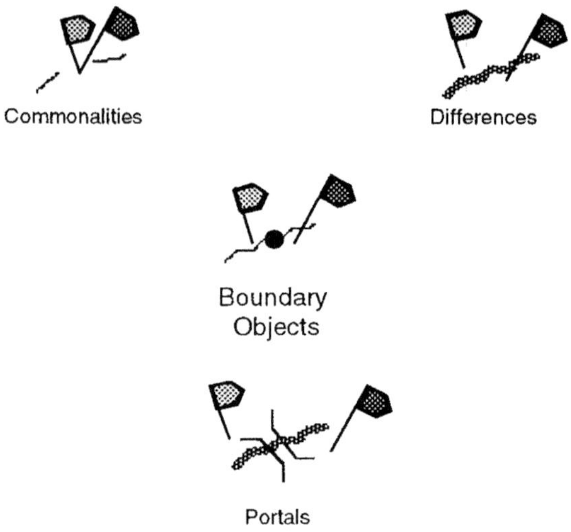


Figure 7.1. Three different types of boundary objects

The third type of boundary object is the most complicated and becomes the focus of design work. It is called a *portal*, and although it can emerge out of the woodwork of a project, most likely portals will need discussion, negotiation, and the long-term involvement of participants to uncover and understand collaboratively the semantic differences involved in such concepts as wetlands. The process of finding portals not only involves technical issues, but is crucial to organizational collaboration and addressing goals.

Portals are the result of collaboration to overcome differences. Instead of breaking down barriers of differences, portals work best as the result of collaborative efforts to find the thinnest places in the barriers and then the mechanisms to construct an opening that links the groups.

The key point about portals is collaboration. Institutional, social barriers and barricades are quickly built. And much quicker than portals. Portals are points of coherence, not coercion, between different groups or agencies. They permit a semantic exchange where previously opposition reigned. Obviously, collaboration and the construction of portals is not easy. It requires a commitment to participative design and a very open design process, that may not be distinguishable from implementation in the end.

The metaphor to visualize the design of semantically interoperable GIS is the iterative loop (see Figure 7.2). From some initial condition, for instance the mandate to build a Wetland GIS for a county, the groups involved meet together, find their shared points of agreement and gradually talk about major and minor differences and ways to work together. Commonalities are easily conceived of, for example land ownership as a base layer, and agreement found between the involved groups. Some differences may be readily solved through administrative procedure. Trenchant differences call for an analytical approach that aims to identify and refine possibilities for connecting differences. In a process of negotiation, these differences are laid out and usually only ways to circumvent them are discussed. In participative design, the design process needs at this point to probe the differences and seek possible points of agreement and also to ask the groups involved to inspect their positions and find possible ways to link with other groups. Admittedly, this involves a great deal of group dynamics and politics, meaning the process of participative design is much more involved and time-consuming than waterfall models of design. On the other hand, getting the people working with the information system involved in an early stage means they will have a much higher degree of identity with the system, and, in some ways even of greater importance, with their colleagues who before were on the other side of the barrier. The high initial investment will pay itself off later in more stable products, better inter-departmental relationships, and more contented staff.

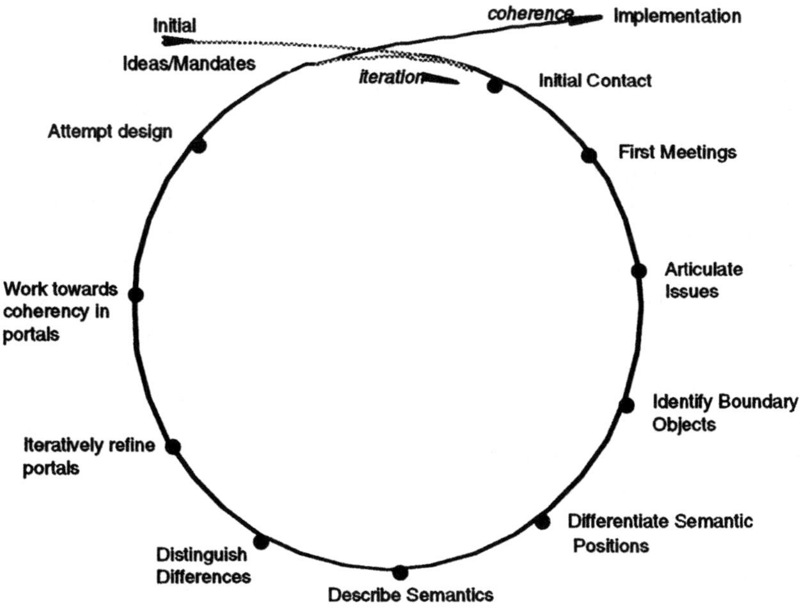


Figure 7.2. Designing for semantic interoperability

The other advantage of portals is the opening of a means to move towards a better formalization of semantic issues at hand. Boundary objects and portals are especially well suited for algebraic expressions of relationships and constraints (Herring 1990; Kuhn 1994). Morphisms described in an algebraic language preserve the semantics of the initial referents while linking them together through operations that change the results, not the inputs.

## 7.3. Example of a potential application

After this conceptual introduction to semantic design for interoperability, this section provides an illustrative example of the principles involved. Although this example comes for a case not explicitly involving interoperability and is furthermore a prospective study, it offers a lucid way to show what issues semantic interoperability involves.

Like many other state governments, the Swiss canton Vaud has several agencies with responsibilities for water resources. In the case in point here, two agencies had created separate, not spatially referenced, GIS data sets for the rivers and streams in the canton. In 1995 a conversion project merged their data sets in order to avoid redundancies, minimize data collection discrepancies, and provide a common data set basis for other cantonal activities. This is the starting point for this analysis, which goes beyond what has actually elapsed, and presumes the agencies are working on an interoperable GIS.

One agency is concerned with fisheries, the other with hydrological engineering. These different mandates mean that each data set has unique semantics for hydrography in the canton. One agency is solely concerned with hydrological branches. The second agency is concerned with hydrological segments, defined according to river characteristics.

In the terminology of semantic interoperability, the data sets shared some commonalities and a few important differences. The most important commonality was the use of a linear network for modeling. Second, there was almost complete agreement between hydrographic segment endpoints. Although these are crucial, still some differences were found. A very involved, hands-on design with the two agencies was used for the conversion project. The use of different scales was recognized as the crucial difference. Other differences recognized were questions of perspective; and if all necessary attributes could be related to the same linear network, they could remain in the domain of the individual agencies.

Due to the commonalities of network modeling, common attributes were joined and a shared data file created for these attributes. The unique attributes of each agency were connected to the hydrographic segments as independent files on a shared file server, allowing each agency to extract and use their own attributes. Now there is a single geometry linked with a set of common attributes and two sets of additional attributes related through external links.

The unresolved question of scale surfaces when considering updating procedure. Changes to the attributes of a segment, for example adding further detail to a section by one agency, may not be registered by another agency. Resultant attributal discrepancies may lead to unstable semantical relationships. Simply illustrated, recording the construction of a bridge in the more precise data set will not automatically ensure that the attributes in the other data file affected by this change are updated.

Clearly, these are issues that affect the quality of decision-making and should be considered in the future. Interoperative approaches will be faced frequently with such issues and will need to address them early on to provide viable information system solutions. Accounting for semantic issues is crucial.

Designing for semantic interoperability would require a series of activities to articulate and formalize the semantic issues. A first step to resolving these issues consists of recognizing the underlying semantics and identifying commonalities, portals, and differences as boundary objects. Where a review of the attributes updated reveals commonalities, rules and procedures for updating the second data set should be developed and implemented. Likewise, differences should be explored and the reasons elaborated as a basis for generating mappings and metaphors. Through an investigation of the first data set's attributes and semantics, together with a consideration of semantics and attributes from the second data set, rules and procedures to update the second data set partially should be developed. Part of this process involves participants elaborating common concepts to be used in describing the conceptual model for a data dictionary. Finally, the data structure for the implementation model would be elaborated.

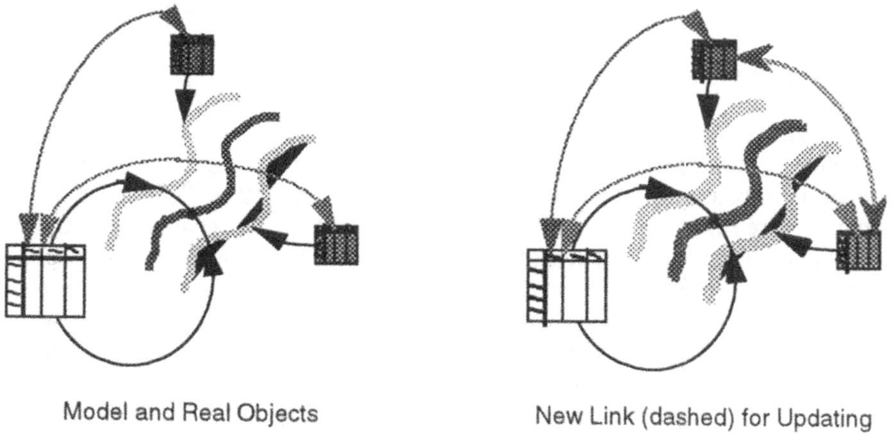


Figure 7.3. Updating in a semantic interoperative environment

In the example above of registering bridge construction or modification, a set of rules and procedures that transform the attribute changes in one database for the other would assure semantic interoperability. The process of negotiation and construction in participatory design makes it possible to identify commonalities between the different types of updates and construct portals to link the different semantics of the agency-specific data sets. Depending on the case and mandates, fundamental differences may remain, some changes may still not be accounted for, but it would be possible through this process to develop a data model that allows a partial updating of the data sets.

## 7.4. A conceptual model for semantic interoperability

Basically the concepts and example above provide a useful introduction to the issues and considerations involved in semantic interoperability. Because semantic interoperability is embedded in local situations, a universal set of rules or procedures for its design and formalization seem scarcely possible. Certainly though, the issues above should be presented in a more conceptual model to facilitate use in actual projects.

A key point in carrying out conceptual modeling is laying out an intermediate step between the group processes of identifying semantical differences, commonalities, and portals, and formalization of operations in the system. To do this, fundamental steps in design must be broadened to consider semantical issues and follow participatory design principles (Harvey 1997). This will lead to more robust data transfer mechanisms that preserve the semantic integrity of the transferred information. This modeling approach is suitable for formalization and integration with existing technical frameworks, applying specifications and outlining behavioral concepts for interoperability. Chains, actions, threads, and entities are behavioral concepts for the formalization of the requisite morphisms. Instead of wearisome negotiations, this approach opens ways of rigorously describing and formalizing interoperability operations, thereby enriching the exchange of geographic information.

Expanding on the basic OGC abstract specification is an important issue to consider in semantic interoperability. Beyond the conceptual model of interoperability as a recycling bag (Open GIS Consortium 1996: Figure 2-19), whose contents were separated and cleaned to ease later sorting and separation for the right processing plant and generation of new products, semantic interoperability needs to expand on the semantic issues of the world view, project schema, and feature schema (Open GIS Consortium 1996: Figure 2-21).

Semantic interoperability goes beyond the technical and builds its representations in relationships of meanings. Figure 7.4 points to some of the key differences between entity–relationship models and semantic views. What is indicated with a question in the top part of the figure is the focus of relationships in the semantic view. This is a simplified rendering of a semantic view whose purpose is to illustrate fundamental concepts. In the semantic view, different project views have different representations of a phenomenon, but are also related explicitly through chains (formalized connections) to the modeled entities and threads (ill-defined links) to the same entities. The focus of designing semantic interoperability lies in elaborating and expanding threads and putting them in relationship to chains, representations, and different project views. The entire collection in this figure should be called a universe of discourse, fully acknowledging at the same time that this is just one of many possible universes.

In the next stage of the design process, these elements are explicitly brought into the analysis and negotiations and articulated in terms of commonalities, differences, and portals. The two stages are interlocked and may occur iteratively as well. Participation is central to the design process and the foundation for successful semantic interoperability. The situatedness of participation restricts attempts to develop a universal method.

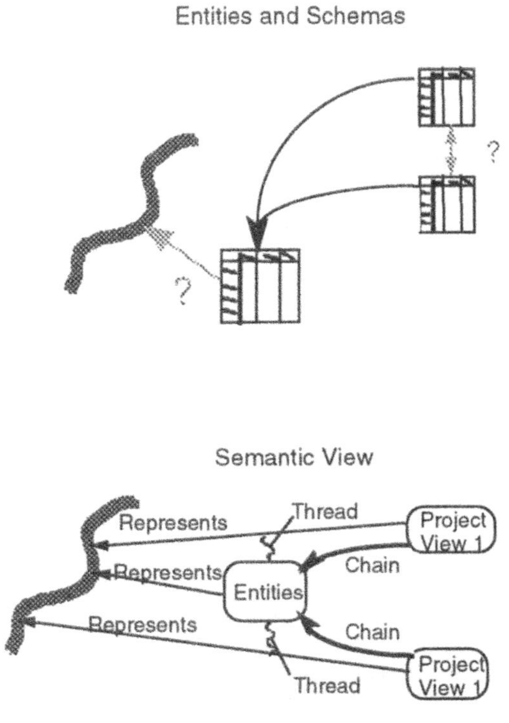


Figure 7.4. Modeling semantic interoperability

After this conceptual model is refined, the next step will be to move from the patterns to the processes involved in the arrows. This involves, on the one hand, describing processes in the figure, but, more importantly, identifying and formalizing different processes between project views. For example, updating requires a procedural link between the project views to develop processes and rules for dealing with semantic differences.

Because of the complex issues involved in linking the technical foundation laid down in the OGC documents and semantic concerns, this paper can only touch on this issue and call on future work in this domain to consider behavioral concepts as fundamental operators in the transformation of semantics between world concepts, project world, feature schema, and so forth. Further, the multiplicity of semantic concepts at each level and the practical relations should be considered.

## 7.5. The practice of interoperability

In a technical sense, interoperability is the linking of different simplified sets abstracted from a world we experience and change through our actions. In accounting for the complexity of different world views and the dynamics of society and its institutions, semantic interoperability is ultimately a question of practice; that is, for design purposes and interoperability, the complex interwoven web of relationships between our abstractions and our activities in the places and ways they occur.

Through abstractions and simplifications we make aspects of a complex reality malleable for GIS applications, but the focus on purposes and the implicitness of perspective only includes a certain range of interests at the cost of excluding others. Technical interoperability is in many ways limited to the closed settings of administrative or corporate environments with a distinct number of actors and distinct purposes. Open distributed processing, such as the provision and exchange of global change data for research, with highly variable uses, will require more advanced solutions that can build on this technical foundation and provide for semantic exchange as research.

As recently described in an article on conceptual modeling for environmental studies, semantics provide the means for scientists to distinguish between semantics and models (Park 1997). In this sense, semantic interoperability is a wide-ranging conceptual basis for geographic information integration. Beyond the static integration of layers, semantic interoperability opens ways to carry out dynamic integration of objects and features. This dynamicism can be extended as well to include geographic process.

In summary, the approach I describe here suggests a pragmatic, situated approach to enriching interoperability by bringing fundamental semantic concerns into the design process. Beyond the technical differences that previous work on interoperability has addressed so successfully, the broadening of considerations to include the semantics of interoperability is a crucial next step. Standards, like treaties, provide solid frameworks, but implementation remains a matter of being at the right place at the right time. Putting semantics in the context of design and a conceptual framework, is very useful in making the right step between technical abstractions and the complexity of practical applications.

## References

Akrich M, Latour B 1992 A summary of convenient vocabulary for the semiotics of human and nonhuman assemblies. In Bijker, W, Law J (eds) *Shaping Technology/Building Society*. MIT Press, Cambridge, Mass.: 259–264

Ewusi-Mensah K 1997 Critical issues in abandoned information systems development projects. *Communications of the ACM* 40(9): 74–80

Fujimura J H 1992 Crafting science: standardized packages, boundary objects, and 'translation'. In Pickering A (eds) *Science as Practice and Culture*. University of Chicago Press, Chicago: 168–211

Harvey F 1997 Improving multi-purpose GIS design: participative design. In Hirtle S, Frank A U (eds) *Spatial Information Theory: A Theoretical Basis for GIS*. Springer Verlag, Berlin: 313–328

Harvey F, Chrisman N R in press Boundary objects and the social construction of GIS technology. *Environment and Planning A*

Herring J R 1990 Using category theory to model GIS applications. *International Symposium on Spatial Data Handling (SDH '90)*. Zürich 2: 820–829

King J L, Kraemer K L 1993 Models, facts, and the policy process: The political ecology of estimated truth. In Goodchild M, Parks B O, Stayaert L T (eds) *Environmental Modeling with GIS*. Oxford University Press, New York: 353–360

Kuhn W 1994 Defining semantics for spatial data transfers. *Sixth International Symposium on Spatial Data Handling (SDH '94)*. Edinburgh 2: 973–987

Latour B 1988 *The Pasteurization of France*. Harvard University Press, Cambridge, Mass.

Neumann L J, Star S L 1996 Making infrastructure: the dream of common language. *PDC'96, Participatory Design Conference*. Cambridge, Mass. 1: 231–240

Open GIS Consortium 1996 *The OpenGIS Abstract Specification: An Object Model for Interoperable Geoprocessing, Revision 1*. Project Document No. 96-015R1. Open GIS Consortium

Park J 1997 Geographic information systems and problem solving environment. *Crossroads* 4(1): 3–8

Star S L (ed) 1995 *The Cultures of Computing*. Blackwell Publishers/The Sociological Review, Oxford

Star S L, Griesemer J R 1989 Institutional ecology, 'translations', and boundary objects: amateurs and professionals in Berkeley's Museum of Vertebrate Zoology. *Social Studies of Science* 19: 387–420

# 8 INTEROPERABILITY BY EXCHANGING EXECUTABLE CONTENT

Andrej Vckovski

## 8.1. Introduction

In the area of geographical information systems (GIS), the requirements posed upon a set of ***heterogeneous but interoperable systems*** are not easily met. The particular systems most often possess a certain degree of autonomy and are designed to meet specific requirements for their respective application domains. In order to be more than mere data exchange, interoperability needs to be based on a framework which is easy to understand for all participating communities and yet expressive enough to meet those application-specific requirements. One of the most difficult problems when designing such a framework is to find an appropriate depth or thoroughness of the items specified, that is, the question of what should be specified and what should be left to particular implementations. An interoperability specification that leaves many things open to the implementation is usually easy applicable but maybe not very useful if important things have been left out. In contrast, a specification that includes too many details and assumptions might avoid semantic heterogeneities and similar problems, but fail in acceptance by the software industry because it leaves no room for innovation and competitive differences. The decision on how much to specify is basically a question of the level of abstraction and can be compared to various abstraction levels that are used, for example, in layered software architecture (e.g., the ISO/OSI communication model; Stevens 1990). Such architectures are a result of vertical (in terms of abstraction) and horizontal (in terms of purpose) decomposition of complex computing systems, where the borders represent the interfaces defining the decomposition.

The history of information technology during the second half of this century contains several examples that show some of the lessons that have been learned when designing systems that have to provide interoperability in some sense. This contribution reviews two slightly related examples in the first two sections. The first example shows how the method to address and use peripheral devices in general and printing devices in particular within operating systems and application programs has changed to provide interoperability amongst many different printer models. The second example discusses the concept used by the programming language PostScript to overcome device dependency, again in the context of printing devices. The first example gives (one more) justification to interface-based approaches as taken, for example, by the OpenGIS Consortium (OGC; Buehler and McKee 1996). The second example shows that a distributed object system as is used as underlying architecture in most interfaced-based interoperability approaches does not need to mean *only* remote method invocation as it is used for example in CORBA (Common Object Request Broker; CORBA 1995), DCOM (Distributed Component Object Model; Stafford and Powell 1995) or Java RMI (Remote Method Invocation; Java 1996). Rather, methods can be also transferred to the caller and executed locally. This is sometimes called a transfer of executable content and plays an important role both in PostScript and Java.

The third section discusses how Java differs from PostScript with respect to the transfer of executable content and why this might be very useful in the context of spatial data interoperability.

In summary, there are three major points made here:

- It is sensible to specify interfaces instead of, for example, data structures;
- it is useful, for some purposes, to transfer the code behind the interfaces as well; and
- Java is a useful environment to implement interfaced-based systems and executable content.

## 8.2. Printing devices and printer drivers

Many of the problems discussed in the context of interoperability appear in various areas of computing. For example, the interfacing of peripheral devices such as storage systems, printers, scanners, keyboards, and so on in an operating system poses many similar problems to the interoperability of geoprocessing. Many different devices with different characteristics need to interoperate in a system, and most operating systems are designed in a way that lets the manufacturer of peripheral devices provide software components to access these devices and hide their particularities behind a well-defined interface.

The discussion here follows mostly the presentation given by Vckovski (1998) and is focused on printing devices as an example, but applies to other types of subsystems as well. The choice of printing devices was made because they are usually visible and their purpose (i.e., to produce output on paper) is easier to understand than for other types such as network hardware or similar components.

In the first phase of computing technology starting in the second half of this century, single-vendor solutions were predominant. There were no compatibility problems because the subsystems were designed to be part of a whole. The growth of the computing industry then led to many hardware manufacturers who sold not entire systems but specialized products such as printers and other peripherals. These products often had superior quality and lower prices compared to the products available from the manufacturer selling the main units. For printing devices, this meant that people started to buy printers from other vendors, because these printers most often had better performance, in terms of features, cost, or both. Therefore, the software had to become more versatile to support the new, probably slightly different printers and their features. The spread of varying features was naturally broader than within the well-controlled (and maybe no so innovative) product range of the large manufacturer of the main system. Even if the so-called third-party products were designed to be as compatible as possible (i.e., as close to the original as possible) in order to provide seamless integration, there was still some differences (mostly improvements) to motivate the clients to buy third-party products.

Initially, the support of peripheral devices was solved by hard-coding various peripheral types in the software as there was no abstraction layer available in the (if even existing) operating system.

In a second phase, application programmers started to provide configurable methods for access to different printers. This happened mostly by static descriptions of how certain device features could be enabled and disabled (e.g., *escape sequences*). Thus, it was possible to use a new printer that was not known at the time of application development, simply by configuring its characteristics into the application's printer description. Of course, this required the new printer to use the same underlying model of the way to address and control printers (e.g., "the printer gets data in a stream, and special features are controlled by intermixing control codes in that stream"). A new printer, which used a totally different model (such as the PostScript printers discussed later on), could not be embedded in that scheme.

At the same time the computing industry became aware of the importance of a generic layer over the hardware that provides basic services to application programs without bothering about hardware details such as writing a byte of data to a magnetic tape. This abstraction layer, that is, the operating system, initially rarely did provide support for different printer devices. It merely controlled basic input and output to and from these devices, but did not provide abstractions to access a printer's built-in features. This was to some degree due to the relatively fast change cycles of the printer hardware as opposed to the operating systems, which were large monolithic pieces of software in many cases. It was unwise to be very dependent on printer characteristics in these software systems.

The rise of very powerful printing devices (e.g., laser printers) with a large set of features required new methodologies to provide interoperability between different brands. Many conceptually-different ways to uses these features emerged, because it was not sufficient anymore to have a few codes that controlled paper feed and typeface selection. Applications started to embed driver modules that implemented a printer's special features and provided a standardized interface towards the application. The driver modules weren't static descriptions of control codes but executable code. This architecture was flexible enough to support various conceptually-different methods to control printers, while providing a standardized interface towards the application. Also, communication with a printer (or the driver, respectively) started to become bi-directional since the applications had to know some (abstract) properties of the printer such as for example the spacing of typefaces with proportional fonts.

Eventually, the operating systems took over the responsibility for the drivers. Today, the operating system in most desktop environments does provide a standardized interface to applications for printing support. The individual application does not need to provide drivers itself. In most cases, the interface is almost identical to similar interfaces available for other output devices such as the screen, and is called the graphic subsystem of an operating system. Today, the destination device of output is (almost) transparent to an application.

This short story on the development of the method to access printers (and it is similar for other peripheral devices) shows that it was necessary to introduce dynamic components (i.e., the drivers) to cope with the complexity and diversity of various printer devices. Today, printer drivers are software modules with well-defined interfaces. A printer manufacturer only needs to implement such a driver conforming to the interface to enable the device to be used with many applications—applications which were absolutely unaware of the specific device at the time they were written.

Spatial data sets are usually also developed (i.e., sampled, collected, maintained) without having every possible application in mind. Also, geoprocessing applications are developed without having precise knowledge of all the various data that might be used in that application. The example shows that it might be a good idea (as proposed by the OGC) to provide well-defined interfaces for applications and spatial data sets. Interfaces should be more than a data format and allow both applications (of course) and data sets to be seen as software modules in a computing environment. The approach decouples application and data complexity and provides much more interoperability. Actually, it probably even starts to blur the distinction between applications and data: being well in line with object-oriented terminology, both applications and data become objects with states and behaviors.

However, there is a small drawback in this setup. It requires applications (or consumer objects) and data (or producer objects) to be connected in some way so that the consumer can actually reach the producer's interface. That is, there needs to be something like a distributed computing infrastructure to be in place (where *distributed* does not necessarily mean physically distributed, it could also simply be a well-decoupled component within a software system). In the environment of printing devices, there has been and is a solution which provides the expressiveness and flexibility of a software module (or procedural description) together with remoteness. The page description or programming language PostScript is such an example and it is discussed in the next section as it demonstrates some of the advantages and disadvantages of the concept and because it relates well to the ongoing discussion on standardized data formats versus distributed computing infrastructures.

## 8.3. The PostScript approach

The programming language PostScript was released in 1985 by Adobe Systems Incorporated (Adobe 1990). PostScript is an interpretive programming language with powerful graphics capabilities. It was developed mainly to describe and distribute text and graphical shapes on printed pages in an environment of different printing devices. Various *raster output devices* such as laser, dot matrix, or ink-jet printers, and digital phototypesetters offered many sophisticated features such as high-quality typefaces and powerful graphic capabilities. From a user's perspective, however, the use of raster output devices caused major problems, especially because they were not only powerful but also incompatible with each other. Thus, one of the design goals for PostScript was a standardized access to such devices. Applications should be enabled to create output for various target devices without having to know all details of the specific device. Moreover, PostScript also addressed the aspect of physical distribution in the sense that the output device is often not physically available at the time of rendering output by the application. This is especially true for expensive high-quality and non-commodity devices such as phototypesetters. This means that output-producing and output-consuming happen at different times and locations. It might even be that the type of output device consuming the data is not known in beforehand. This is—by the way—a common situation in the context of spatial data as well: the possible use of a data set is sometimes not known at the time of sampling.

The approach taken by PostScript is a data exchange format that is based on a programming language. The entity exchanged between producer and consumer is a PostScript program. The consumer (e.g., a printing device, page viewer or even a Display-PostScript-based windowing system) interprets the program and creates—as a result—the content. This approach different from other encoding techniques (the PostScript program is, eventually, also just an encoding) as the expressiveness of the system is very high. Such an encoding scheme is sometimes called dynamic (Adobe 1990) because its results can only be seen by actually performing the decoding step, that is, running the PostScript program. Applications can create platform-independent output (e.g., graphical shapes and text) which is transformed into device-specific information by the device itself. This approach has many advantages contributing to the success of PostScript:

- *Platform- and vendor-independence*: The approach allows a high degree of interoperability between various producer and consumer platforms because platform-specific details are resolved within the PostScript interpreter.

- *Flexibility and expressiveness*: PostScript is a full-featured programming language allowing complex algorithms to be implemented. There are virtually no limits to what can be expressed in PostScript.

There is, however, a tradeoff between flexibility and ease-of-use because PostScript does not offer various views of a data set at different abstraction levels, for example, metadata. Generally, there is no other way to understand what a PostScript program does (i.e., what a PostScript data set contains) than running the interpreter on the program. For that purpose, a convention was specified (the so-called Document Structuring Convention), which defines a set of standardized comments within a PostScript program. These comments can be easily identified without understanding the PostScript code, that is, by simple pattern matching. The DCS grew more and more since there were many more metadata needed than were initially specified (e.g., more than just title of the document, who created it, and a time stamp). DCS starts to be an orthogonal data format (orthogonal in the sense that DCS appears as a comment for PostScript and has, therefore, no meaning to PostScript) within PostScript programs. One reason for the need of DCS was certainly also the procedural approach of PostScript, which does contain mechanisms to hide implementation (that is, the data) from its interface (that is—stretching the analogy a bit—the metadata).

The complexity of PostScript interpreter software (which mostly runs on embedded systems, e.g., a printer's microprocessor) and also rising user requirements led to cases where interoperability is missing even if both data producer and data consumer claim to be a PostScript driver or PostScript interpreter, respectively. The incompatibilities happen usually where the PostScript language cannot support new features of a device.

Today, most PostScript code is not written by human programmers, but by application programs or an operating system's device drivers. That is, the printer drivers are code generators for PostScript. In these cases, the full functionality of PostScript cannot be exploited. But nonetheless, the option for arbitrarily complex structures in a PostScript program provides almost an unlimited framework for special cases.

The following section will discuss another example of transferring executable code, the Java software system.

## 8.4. Java and executable content

Java is an object-oriented programming language and distributed computing environment that has become very popular in the past few years. One of the core concepts of Java is the same as in PostScript and other interpreted languages: hardware and operating system heterogeneities are resolved by interpreting a well-defined format on the target platform. However, unlike PostScript Java uses an intermediary format (*byte code*) which makes this stage of execution somewhat independent of the language itself (i.e., Java byte code need not necessarily be produced by a Java compiler). Java offers also other features, which make it even more suitable for the exchange of information in an heterogeneous world:

- Java-objects can be platform independent and dynamically loaded into a running program. A GIS could load Java objects describing some data into its running image and execute its content.
- Java has been partly designed to be used in distributed environments and supports therefore many corresponding primitives (e.g., class libraries for network communication).
- Java supports the notion of interfaces and interface inheritance, which is a flexible means to provide common ground (specification) within interoperable systems.

As has been said before, Java is both an object-oriented programming language and an environment for distributed computing. Java had its origin in 1990 in a research project at Sun Microsystems, which aimed at developing new techniques for programming embedded systems in the domain of consumer electronics, for example, video recorders, espresso machines, toasters, and so on. Software for such embedded systems have specific requirements (Flanagan 1996) such as portability (choice of hardware platform is cost-driven), reliability, and concurrency (many parallel threads). These requirements led to a design of a framework that is based on an architecture-neutral byte code, which is interpreted on the target platform. Java as a language has many constructs which support reliable programming (e.g., no pointers), multi-threading (e.g., synchronization constructs), and so on. However, the break-through of Java happened when it was used to build a World Wide Web browser. This browser, called *HotJava*, was entirely written in Java and able to change its functionality by dynamically adding executable code. For example, it could load an extension over the network, which would allow the system to handle a new, previously unknown protocol type. Or, it could load an extension within an HTML page that was responsible for rendering a rectangular area in that HTML page, a construct which later became famous as so-called *applets*. In other words, HotJava allowed executable content to be transferred and immediately activated, on various platforms and without (many) security risks. In that sense, the concept of Java is very similar to PostScript and other interpreted systems, except that the transferred code is not source code but a compiler-generated byte code. This allows the data (or code) producer to verify correctness at an early stage and still make use of the platform-independent byte code.

In the meantime, Java has evolved to a mainstream programming language and runtime environment and lost some of the other, initially very appealing aspects such as simplicity. In the context of the transfer of executable content, however, it still offers many advantages; advantages that weren't available in comparable systems such as for example the UCSD p family of languages (Wirth 1996).

Unlike PostScript, Java provides consistent methods to inspect byte code, that is, data sets. These methods can, for example, provide information on the type of the data contained, the interfaces implemented, or the metadata. These methods for class file introspection are part of an overall infrastructure within the Java runtime environment and class libraries which promotes distributed and component-based systems. From that point of view, Java already incorporates many of the requirements for an implementation environment for OGC specifications.

The last section draws some conclusions from these two existing frameworks for the exchange of executable content and relates them to the context of spatial data interoperability.

## 8.5. Conclusion

Spatial data interoperability needs basically two things:

- An agreed model of spatial data which describes our conceptualization of space, that is, a mapping of real-world entities onto a formal system.
- A method to exchange data (eventually, a sequences of bits) and provide their semantics with respect to the model.

Current interoperability approaches such as OGC approach the former by a set of interface specifications, a comprehensive collection of the data types and their operations. The latter is addressed via various environments for distributed computing such as CORBA or DCOM. This means that it is expected that the services and data sets (or data objects) are available on-line, that is, through some sort of network connection. The methods to access services and data sets are implemented in servers that can be accessed by client applications (however, there is not a strict client–server model as a server can always be a client for some other requests). The implementations of the methods do not migrate. There are, however, often situations where it might be desirable to distribute the method implementations as well. Motivations could be security aspects (no access to data sets from the outside within a secured network), volume (large data sets, with large chunks returned for every request such as satellite images), or context-dependent operations. Context-dependent operations might include, for example, graphical symbolization of some data sets that can only be determined at time of rendering. The symbolization for a city might be dependent on the current display scale. A map viewer needs to query the symbol for the object at time of rendering, parameterized by the display scale. In this situation it might not be wise to query a remote data service for the graphical shape to display for every entity. Rather, a local execution of the corresponding logic might be more sensible.

In these cases, a transfer of executable content can provide the same level of encapsulation (the container application does not need to know the details) even if there is no explicit distributed computing infrastructure in place. Transfer of executable content enables the same level of modularization that is available with separate services on various computers or within various process spaces, and also within a single process space. In the case of Java, it could be made transparent to the application developer whether a certain implementation of an interface resides on a remote server (accessible via CORBA or RMI), or whether the implementation has been loaded into the current process space (e.g., from a local file system, a database management system, or a remote server). A large data set on a CD-ROM could be accessed almost in the same way as an on-line data set. In that sense, transfer of executable content somehow bridges the gap between standardized data formats and distributed computing environments.

## References


Adobe 1990 *Postscript Language Reference Manual*. Second Edn. Addison-Wesley, Reading

Buehler K, McKee L 1996 *The OpenGIS Guide*. OGIS TC Document 96-001. OpenGIS Consortium, Wayland

CORBA 1995 *The Common Object Request Broker: Architecture and Specification*. Version 2.0. Object Management Group, Framingham

Flanagan D 1996 *Java in a Nutshell*. O'Reilly & Associates, Sebastopol

Java 1996 *JavaBeans—Component APIs for Java*. *http://www.javasoft.com*

Stafford P, Powell J 1995 COM: A model problem solver. *Microsoft Developer Network News* 4(2): 1–12

Stevens W T 1990 *UNIX Network Programming*. Prentice Hall, Englewood Cliffs

Vckovski A 1998 *Interoperable and Distributed GIS*. Taylor & Francis, London (in press)

Wirth N 1996 Recollections about the development of Pascal. In Bergin T J, Gibson R G (eds) *History of Programming Languages II*. ACM Press, New York

# 9 SUPPORTING INTEROPERATION OF GIS OBJECTS

Silvia Nittel and Richard R. Muntz

## 9.1. Motivation

Today, a promising approach to large-scale interoperability in geographic information systems (GIS) is a GIS implemented via a distributed object management system (DOMS; e.g., CORBA, Java, OLE/COM) based on a standardized spatial object model (e.g., OGIS; Buehler 1996). Geographic entities are available as location- and platform-independent objects, and are accessible via their interface, that is, their attributes and a set of operations. Using a DOMS, much of the hardware, network, and communication heterogeneity between interoperating GIS objects on different platforms and machines is handled automatically by the DOMS software, and is of no concern to the user. Also, via the notion of strongly-typed object interfaces, type checking is also performed automatically at compile time to ensure that correct argument types are passed between interoperating GIS objects.

However, to resolve the correctness of a method argument passed from a GIS client to a GIS object, it is not sufficient to verify only the correctness of the type of the argument, but it is necessary also to ensure that the content of an argument is matched. By content, we mean constraints related to the actual properties of an argument that are not easily captured in the type hierarchy. For example, let us assume the OGIS Geometry type hierarchy (OGC 1998). In this type hierarchy, the object type `LineString` has an operation `intersect()` which has the following signature: `Geometry intersect(in Geometry other);`

`Intersect()` expects an input argument of type `Geometry`. To perform the intersection operation correctly, both geometries have to be in the same spatial reference system (SRS), that is, in the same coordinate system with the same metrics for the axes. However, the requirement of this argument restriction is not explicitly expressed in the operation's signature, and cannot be automatically checked at compile time. Also, the client cannot be sure in which spatial reference system the returned `Geometry` object will be defined, and cannot specify explicit rules for the kind of spatial reference system it would require to be returned (e.g., for displaying objects). Other examples include a GIS service that handles the type `Image`, and has to apply operations to images of different format, resolution, size, compression, etc..

Today's DOMS systems offer little help to resolve issues of interoperability of GIS objects regarding the values that are passed, since their interpretation is usually based on property values of the user-defined GIS object rather than supported through a type system. There are several reasons why this information is not easily captured in an extended type hierarchy. First, it can be cumbersome to model all this information explicitly within a type system, and it would lead the original type system to explode if all these options are explicitly represented. Second, there are advantages to supporting a (meta) type system explicitly as we will see later. Third, a need exists to describe a range of types that are acceptable to a service that may not fit nicely into a type hierarchy. Today, it is the server's task to check which input parameters can be handled, and which cannot, and either provide implementation for different kinds of parameter content or return an error message. On the other hand, a client might have to conform explicitly to *value* rules of a service (e.g., providing a geometry object within a specific SRS), and to match the return value from a GIS server to its requirements. This is error-prone, cumbersome, and makes server and client code bulky.

In this chapter, we examine the issue of additional functionality in the type system for modeling restrictions of method argument values passed from one GIS object to another in a general distributed GIS. We do not necessarily assume an open system in this context, but a controlled type system as might be available in a distributed GIS used within a company or an agency. Key issues for which we discuss solutions concern the problem of how to decide whether two given GIS objects are compatible based on the argument values passed, or how to determine if they can be made interoperable through the introduction of ancillary mechanisms such as *reconciliators*.

For the proposed type model, we also define *sub-typing rules* supporting the conformance of GIS interfaces regarding the content of arguments; for example, `SRS[Cart,2D,1,km]` describes a 2-dimensional spatial reference system with a Cartesian coordinate system, tick marks as multiples of 1, and kilometer as unit. In a natural sense, an SRS with the value `SRS[Cart,2D,1,m]` is a type that subsumes `SRS[Cart,2D,1,km]` and therefore, it should be possible to narrow the former to the latter. In some cases, a sub-typing relationship might be too restrictive. For interoperability, it is sufficient if two GIS object types support at least one common argument descriptor. For example, a client might export a geometry argument in a range of SRS, and a service might accept geometry objects in a different range of SRS. For interoperability, it is sufficient to find one common SRS. If they support more than one common argument descriptor, a GIS object *negotiation* may take place to choose the descriptor (and information) that is actually used.

If two argument values (the one sent, and the one expected by the GIS server) do not support a common requirement, they might still be interoperable, if it is possible to use a reconciliator such as a converter of a spatial reference system or a converter for image encoding. Using this argument descriptor type model provides for automatic reconciliation of the content of arguments passed between a GIS client and server. The mechanism is reusable for all kinds of different GIS objects. Relieving both services and clients from the task of checking arguments makes their implementation simpler and more lightweight. Since the GIS object property reconciliation mechanism is extensible, that is, new description types and conformance rules can be added, GIS services can profit from the extension without any code having to be added to their implementation.

The approach that we discuss in this chapter is inspired by related work done in the area of interoperating multimedia systems. Here, a type model is defined to describe the characteristics of continuous media flows and allows run-time negotiation of data sources and sinks (Eliassen and Nicol 1996). We adopt the framework of Eliassen and Nicol, and apply this framework for related problems in GIS. We believe that this framework is of great value to the domain of interoperable GIS.

The remainder of this chapter is organized as follows. Section 9.2 defines the type model requirements for the proposed approach. We define the type model formally in Section 9.3, and also discuss subtyping, negotiation-based relationships and the problem of missing attributes. Reconciliator-based relationships and conformance checking are presented in the following sections. Section 9.6 contains our conclusions.

## 9.2. Type model requirements

In this section, we discuss requirements of a type model that is used for checking the binding of method arguments based on predefined restrictions for argument values.

### *9.2.1. Argument descriptor type model*

In the proposed type model, we distinguish between *operational interfaces* and *interfaces* that describe *restrictions on method arguments* (to be referred to as arguments in the remainder of the chapter). *Operational interfaces* for GIS objects are defined by standards such as OGIS; for example, for spatial features, feature collections, coverages, Earth images, etc. Operational interfaces encapsulate data objects and their values, define methods on data objects, etc. On the other hand, we discuss interfaces that describe the restrictions or characteristics of method arguments, for both input and output parameters of operational interfaces.

In the context of interoperating GIS, we can assume (and are mainly interested in) arguments that are objects (or value structures) such as `Geometry`. We model an argument that is passed between GIS services as an *object*, and use arguments as a basis for reasoning about type checking between GIS services. Note that in this context the object (and its type) *is an abstract definition of a method argument with certain requirements* (metadata type information is an object). It is not the concrete object that is passed as an argument at method invocation time.

In an interoperating, distributed GIS, there are clients and services. Correspondingly, we distinguish between a GIS object that a client will pass to a service, and a GIS object that a service requires. A service defines restrictions on its argument values, while a client defines the characteristics of arguments that it will pass to a service object. To distinguish between both kinds of arguments, we use the notation of client and server GIS objects.

For example, the implementation of an OGIS `Feature Collection` supports the standard OGIS interface for a `Feature Collection`, but it is implemented to deal with geometry objects in 2D and 3D Cartesian coordinate systems. Therefore, the object implementor defines the restrictions on the `Geometry` arguments of the `Feature Collection`'s methods. If a display client is implemented to call a *clip* on feature collections, the implementor knows that the display client is implemented for displaying 2D maps represented in Cartesian coordinate systems only. When using the *clip* method of the `Feature Collection` interface, the implementor describes the characteristics of its geometry arguments that will always be in a 2D Cartesian coordinate system. Therefore, it can be verified in advance which spatial services are able to interoperate (e.g., only feature collections with 2D objects).

We denote *GIS objects* as the overall abstraction for capturing restrictions on method arguments. In the proposed model, we use the method argument type of the operational interfaces as the *base entities* of our type model; for example, *Feature, Geometry, Coverage,* etc. Each base entity contains a *metadata element* that describes the constraints or characteristics defined for the base entity. The general notion is the following: `GISObject[MetaDataElement]`. For example, the GIS object `Geometry` has a *metadata element* `SRS` that contains the constraints for the geometry in relation to spatial reference systems (denoted as `Geometry[SRS]`).

In the proposed model, a GIS object is described by *one* metadata element type (note that a GIS object could also have two or more metadata elements such as `Geometry[SRS, LINESTRING]` , but for the remainder of the chapter we assume that each GIS object is described by one sufficiently expressive metadata element.

A metadata element type contains a description of a *generic element type* such as `IMAGE` or `SRS`, and a set of attributes specifying properties of the specific element type, the so-called *metadata descriptor type*: `GISObject[GenericElement MetaDataDescriptor]`.

For `IMAGE`, the set might include attributes such as `width`, `height` and `depth`, `encoding`, etc., and can be denoted via the following expression: `EarthImage[IMAGE<640,480,24>]` whereby `EarthImage` is the GIS object, and `IMAGE<640,480,24>` is the metadata element. In the following, we assume one *standardized* metadata descriptor type is defined for each metadata element (we assume its standardization comparable to operational interface in order to allow interoperability).

In the following, we introduce a set of definitions that are used for the description of our type model. To model the arguments of a GIS client and a service, we use the following notation:

`c:>>A` denotes an *available* GIS object `c` of type `A`; and

`d:<<B` declares a *required* GIS object `d` of type `B`.

With the symbol "|"we denote a functional pipelining, so that `c|d` declares that the GIS object `c` *has to be matched with* the GIS object `d`.

### 9.2.2. Requirements of a type model for defining restrictions on arguments

In general, a type model for the description of arguments should be capable of expressing the structure and semantics of commonly-used argument properties in the GIS domain such as geometry, image representation, etc. It also should be extensible such that new argument types can be accommodated.

The type model has to be highly structured so that base entities such as features, geometry, etc. as well as *collections* of such objects can be accommodated. Collections can be described via *homogeneous collections* (homogeneous with regard to the metadata of interest such as a common SRS) whereby one metadata element is attached to the collection or via *heterogeneous collections* in which each element contains its own metadata element.

As mentioned, we attach a *name* to a metadata element (such as IMAGE), and descriptive attributes to the metadata element (e.g., width, height, depth) in order to be able to distinguish between different *metadata element formats* at both the GIS client and server sides. Furthermore, the value of an attribute might be constrained or determine the value of other attributes within one metadata element; for example, the height and width of an image are dependent attribute variables. Therefore, it must be possible to define attribute value dependencies. The type model has to allow the representation of the *set of constraints* for an argument via one type definition.

Note that we distinguish between metadata element descriptor *types* and metadata element descriptor *values*. Therefore, a GIS object might support more than one metadata element descriptor (value) for its metadata element type. For example, a Geometry object might be defined in a Cartesian coordinate system, and elements can be defined in 2D and in 3D; therefore, definitions such as Geometry[SRS<Cart,{2D,3D}>] should be possible. This also allows the definition of a set of metadata elements in terms of possible combinations of attribute values. GIS-specific value sets such as time, units, etc. can modeled explicitly via sets of enumerated type (e.g., Unit:Set(Enum(Meter, Kilometer, Mile)) ).

### 9.2.3. Relationships between argument descriptors

The binding of operational interfaces in DOMS is based on the notion of subtyping, where subtypes obey the rule of *substitutability*. For GIS objects, the subtyping rules can additionally be based on the *metadata element type* and its *descriptive attributes*. For example, type S is a subtype of T if they are of the same metadata element type (e.g., IMAGE), and S specifies a more narrow range of its potential descriptive attributes than T. Accordingly we can, for example, reason that Image[640,480,24] is a subtype of Image[640,480] since the supertype allows any image depth.

In more general terms, we define the symbol < to denote a *subtype relationship* for metadata element types. Then we can reason that if A, B denote metadata element types such that A<B, then GISObject[A] < GISObject[B].

A *GIS object subtype relationship* as described is useful for checking the consistency and correctness of evolving distributed GIS, especially when upgrading existing objects or services. One has to make sure that a new version conforms in the defined substitutable sense to the old version to guarantee the continued servicing of existing clients. A proper definition of a subtype relationship has to make this kind of control possible.

However, there is a need for more flexibility regarding the subtyping relationship. One can assume that, as described above, a GIS object might support more than one metadata element descriptor for its metadata element type. For example, `EarthImage` contains `IMAGE` as its metadata element type, and supports different image encoding mechanisms to encode the image. Another GIS object might accept GIS objects with images in certain encodings; the two services are compatible if there is at least one common kind of image encoding that both support. If there are several, a *negotiation service* might choose the best image-encoding algorithm.

Therefore, we define the term *compatibility* in our type model. Two GIS objects are *compatible* if their metadata element types are compatible, denoted by the symbol <>. We might conclude that `GISObject[A] <> GISObject[B]` if `A<>B`. Thus, two GIS objects are compatible if they support at least one common metadata element value.

In some systems, subtyping is based on the existence of a type reconciliator (filter). A reconciliator-based type relationship cannot be supported within the framework of any of the type relationships mentioned above. If a GIS client and server do not support a common kind of metadata element descriptor value (e.g., the SRS value cannot be matched), an intelligent system would automatically determine if one or more type reconciliators exist that can be inserted as filters into the GIS object interaction.

## 9.3. Argument descriptor type model

We now present the details of a type specification notation for GIS objects and their properties.

### *9.3.1. Argument description type specifications*

The general form of a GIS object type specification is: `type S = GISObject [T e]`. The specification states that a metadata element has a type `T e` whereby `T` denotes the type of the generic metadata element, and `e` stands for its metadata element descriptor.

We assume a metadata element descriptor `e` is specified as a set of `{attribute_name, value}` pairs (e.g., `width:480`). An attribute value is generally interpreted as a set of *atomic* values; this means that a type system conforming to this model treats them as invisible, opaque data. From the user's point of view, however, they can be of any (complex) type. Furthermore, we assume that the *equality operator* and for ordinal types the *relational operators* such as *greater than, less than,* etc. are defined. We allow that an attribute value can be defined as a single *atomic value*, a *set of atomic values*, or a *subrange of an ordinal type*. Furthermore, we assume that for each attribute the function value(a) is defined.

A *metadata descriptor type* (e above, e.g., IMAGE[width,height,depth]) is defined as a tuple type denoted as $([a_1:T_1, \ldots a_n:T_n])$ where $a_1, \ldots a_n$ are attribute names, and $T_1, \ldots T_n$ are type expressions. Each type expression $T_i$ can be a union of the form S + Set(S) + (c..d), where S is an atomic type, and (c..d) is a range specification such that c and d are constants of type S. Ranges can be specified if S is an ordinal type. S is the *base type* of the attribute $a_i$. For example, a hypothetical metadata descriptor type IMAGE is defined as the following tuple type: IMAGE[width:int,height:(0..480),depth:(1,2,4,8)])

A *metadata descriptor value* of a metadata descriptor type $[a_1:T_1, \ldots a_n:T_n]$ is a tuple denoted $[a_1:v_1, \ldots a_n:v_n]$ such that the type of $v_i$ is $T_i$ for $i = 1, \ldots, n$. We assume that the function *attr* is defined for any descriptor $e = [a_1:v_1, \ldots a_n:v_n]$ and descriptor type $E = [a_1:T_1, \ldots a_n:T_n]$ such that $attr(e) = attr(E) = \{a_1 \ldots a_n\}$, and we get the attribute names of the descriptor. Subsequently, we will use the term descriptor when we refer to the metadata descriptor value.

A missing $\{a_i:v_i\}$ pair in a descriptor of a GIS object specification is interpreted as *don't care*, that is, any value is supported. For example, if a metadata element IMAGE does not require a specification for a depth value, than the object implementation can handle images of any depth.

### *9.3.2. Interpretation of metadata element descriptors*

We chose to specify a metadata element attribute value as a set of atomic values so that the GIS object type definition can be represented as a set of possible GIS object type properties. For example, if a GIS object type has a metadata element SRS, and defines for the SRS descriptor's *CoordinateSystem* attribute a set of possible values, it actually declares that it can handle GIS objects of any of the indicated coordinate systems. Therefore, we can *interpret* a GIS object type as a set of optional GIS object properties.

**Definition 1**: *interpretation of a descriptor attribute value*

The interpretation $I(v)$ of a metadata descriptor attribute value v is its value expressed as a set such that:

$I(v) = \{v\}$ if v is an atomic value

$I(v) = v$ if v is a set value

$I(v) = r$, if $\{v\}$ is a subrange of an ordinal type, and r is the set of values of the subrange.

**Definition 2**: *interpretation of a metadata element attribute descriptor*

The interpretation $I(e)$ of a metadata element attribute descriptor e such that $attr(e) = \{a_1, \ldots a_n\}$ is the Cartesian product of the interpretation of each of the attribute values of e; that is, $I(e) = \{ ([a_1:s_1, \ldots, a_n:s_n] : s_1 \varepsilon I(value)a_1), \ldots ([a_1:s_1, \ldots, a_n:s_n] : s_n \varepsilon I(value)a_n)\}$

Intuitively, the interpretation $I(e)$ of a metadata element attribute descriptor e produces n-tuples of all possible combinations of defined descriptor attribute values;

that is, a set of tuples where each tuple denotes a concrete metadata element descriptor. This interpretation presents the opportunity to apply relational algebra operators to the definition of GIS object types.

**Definition 3**: *descriptor attribute value constraint*

A descriptor attribute value constraint is an expression of the form a.v ⇒ b.w where a and b are attribute names, and v = $<v_1, \ldots v_n>$ is a list of type correct attribute values for a. The symbol w denotes a list of elements, w = $<w_1, \ldots w_n>$ where each $w_i$ either is a type correct attribute value for b or a computable function, w = f(a), such that f is a partial function from type(a) to type(b).

When w is a set of attribute values, the intuitive interpretation of the attribute value constraint is that whenever attribute a has value $v_i$ then attribute b must have value $w_i$. The case where $w_i$ is a function expresses a *computable* relationship between a and b values that must always hold.

**Definition 4**: *descriptor attribute value*

A metadata element descriptor value is a pair e c where e is a metadata element and c is a metadata element descriptor constraint. A metadata element descriptor constraint is a set of descriptor element value constraints defined over attr(e).

**Definition 5**: *interpretation of a metadata element descriptor*

The interpretation I(d) of a metadata element descriptor d = e c is the relation selection S of the interpretation I(e) of e satisfying the predicate P( c ), I(d) = S P( c ) (I(e)).

Assuming we have c = $(c_1, \ldots c_n)$, the predicate P( c ) is obtained from c by taking the conjunction of the set if predicates {p( $c_1$ ), ... p( $c_n$ )}, i.e. P( c ) = $\prod^{n}_{i=1}$ p( $c_i$ ), where each predicate p( $c_i$ ) is obtained from the value constraint $c_i$ = "a.v ⇒ b.w" as follows:

p( $c_i$ ) = $\prod^{n}_{j=1}$ (I.a ε I($v_j$)) ⇒ B), where

B = (I.b ε I($w_j$)), if $w_j$ is an attribute value

B = (I.b = f(I.a)), if $w_j$ is a formula f

B = q(I.a, I.b)), if $w_j$ is a predicate q

where v = $<v_1, \ldots v_n>$ and w = $<w_1, \ldots w_m>$ and I.a, I.b denote attribute names in the relational table I(e) corresponding to the attributes a and b in e. The value of an attribute in I(e) is always atomic.

The intuitive idea behind these definitions is that the interpretation of a metadata element descriptor is obtained by computing first a set of tuples representing all possible combinations of atomic values, and then filtering out tuples that do not satisfy the value constraints.

**Example 1**: Consider the expression e = [x:{A,B,C},

| x | y |
|---|---|
| A | 1 |
| B | 2 |
| C | 2 |
| C | 1 |

y:{1,2}]{a.<{A},{B}> ⇒ b.<{1}, {2}>}, which depicts a metadata element descriptor e with a value constraint, and the interpretation I(e). A tuple in the expression represents a possible combination of attribute values within a metadata element descriptor satisfying the specification.

In the following, we introduce a couple of example metadata element descriptors.

**Spatial reference descriptor:**

**type** SRS_D = [CoordinateSys:Set(Enum(Cartesian,Spherical,Projected)),
axes, units:Set(Enum(Meter,Mile,Kilometer))]

**Image descriptor:**

**type** Image_D = [height:(1..480), width:(1::640),
encoding:Set(Enum(jpeg_1, jpeg_2,.. jpeg_n, mpeg_i,..)),
colormodel: Set(Enum(RGB,CMYK,..)) ]

**Example 2**: This example includes a value constraint:

**type** EOImage=GISObject[IMAGE[height:(1..480),width:(1..640)]
{height.(1..480) => width.PosInt((4*height)/3)} ]

### *9.3.3. Subtyping rules for metadata element types*

In the following, we define a *subtyping rule* for GIS object types based on the definition of the interpretation of a metadata element descriptor that we presented above. This subtyping rule is based on set inclusion of the interpretations of metadata type descriptors. We use the symbol < to denote both the *subtyping relationship* and the *inclusion relationship*.

In the definitions below, we need to be able to refer to the *type* of the interpretation of a descriptor. The type of the interpretation of a descriptor e is denoted type(I(e)), such that $attr(e) = \{a_1, \ldots a_n\}$ is $Set([a_1:S_1, \ldots a_n:S_n])$ where $S_i$ is the base type of $a_i$. We also need to define the *projection* of a metadata element descriptor. Such a descriptor $e=[a_1:v_1, \ldots a_n:v_n]$ over a subset B of attr(e) is a metadata element descriptor, denoted $P_B(e)$ (projection B over e), such that $attr(P_B(e)) = B$, and $P_B(e)$ and e have identical values in attributes in B; that is, if $B=\{a_{i1}, \ldots a_{in}\}$ then $P_B(e)=\{a_{i1}:v_{i1}, \ldots a_{in}:v_{in}\}$.

**Definition 6**: *metadata element descriptor inclusion*

Given the two metadata element descriptor values $d_1 = e_1\ c_1$ and $d_2 = e_2\ c_2$, $d_1$ is included in $d_2$ (denoted as $d_1 < d_2$) if:

$type(I(e_1)) < type(I(e_2))$

$P\ attr(e_2)(I(d_1)) \subseteq I(d_2)$

**Definition 7**: metadata *element subtype*

Given the two metadata element type specifications:

**type** D = a:S s

**type** E = b:T t

where a, b are element type labels with generic element types S and T, and s, t are metadata element descriptors. Then *D is an element subtype of E* (denoted **D < E**) iff:

1. S = T (i.e., the generic element types are identical), and
2. s < t (i.e., the descriptor of D is a subtype of the descriptor of E)

### *9.3.4. Metadata element type compatibility*

The subtyping relationship might be too strict in the sense that a GIS client and service can still interoperate if they can at least support one common kind of GIS object type property. Therefore, we introduce a definition of a *compatibility relationship* for GIS objects. Intuitively, we have compatibility between two metadata element descriptors when all their possible metadata descriptor values have at least one metadata element in common, or when they do not have any attributes in common. More formally, we have compatibility between two metadata element descriptors if the interpretation of the descriptors projected to their common attributes has at least one metadata element descriptor in common, or when they do not have any attributes in common. The latter reflects the situation where the attributes specified in descriptor `A` correspond to the missing *don't care* attributes in descriptor `B`. This is defined as follows:

**Definition 8**: *metadata element descriptor compatibility*

Given the two metadata element descriptor values $d_1 = e_1 \; c_1$ and $d_2 = e_2 \; c_2$, $d_1$ is compatible with $d_2$ (denoted as $\mathbf{d_1} <> \mathbf{d_2}$ ) iff:

$P_{at(e1) \cap at(e2)} \; (I \; (d_1)) \cap P_{at(e1) \cap at(e2)} \; (I \; (d_2)) \neq \emptyset$, or

$at_{(e1)} \cap at_{(e2)} = \emptyset$

Based on this definition, we define that two metadata element types are compatible if their metadata element descriptors are compatible.

**Definition 9**: *metadata element type compatibility*

Given the two element type specifications:

**type** `D = a:S s`

**type** `E = b:T t`

where `a`, `b` are labels, `S`, `T` are generic element types and `s`, `t` are element descriptors. Then `D` and `E` are metadata element type compatible (denoted **D <> E**) iff.

1. `S = T` (i.e. identical generic element types)
2. `s <> t` (i.e. the descriptor of D is compatible with of the descriptor of E)

### *9.3.5. Dealing with missing attributes*

In terms of our typed GIS object model, the problem that we are trying to solve via the *compatibility relationship* can be formulated as follows: Consider the client object `out:<<M` and the server object `in:>>N`. The problem is to bind the GIS objects `in` and `out` (i.e., the method arguments). If `N` and `M` are compatible GIS objects, `in` and `out` can be bound directly. Otherwise, they cannot be bound without some additional means.

As mentioned, the type system should also accommodate missing attributes in a GIS object type specification. There are several approaches to handle this problem: one approach might be that a type is allowed to define a constraint that all its attribute types need values. Another approach would be to introduce labels to attributes that define whether an attribute is mandatory or optional. Neither of these approaches is desirable since it makes a negotiation of common policy necessary, before binding can be attempted. We choose an approach based on relational algebra that allows solution of the problem of missing attributes making complex protocols of negotiation unnecessary.

To compute the common set of concrete metadata elements, we assume we have two compatible metadata element descriptor types M and N:

**type** M = a:T $d_m$

**type** N = b:T $d_n$

where $d_m$ = $e_m$ $c_m$ and $d_n$ = $e_n$ $c_n$ are the metadata element descriptors (e.g., M = img:IMAGE [height,width,depth] {width_constr}). In this case, E is the standard descriptor type of the metadata element type e in M and N. Since attributes are allowed to be missing, attr($e_m$) and attr($e_n$) are subsets of attr(E).

The set **F** of concrete metadata element descriptors that $d_m$ and $d_n$ have in common is defined by the Cartesian product of the sets $F_1$ and $F_2$, whereby $\mathbf{F_1}$ is the set of *combined* concrete element descriptor types from M and N that have *compatible* attribute values in M and N, and the interpretation of $\mathbf{F_2}$ is the *don't care* attributes common to M and N (i.e., computation of all possible metadata element descriptors of the *don't care* attributes).

Let us first compute $F_1$. The set $F_1$ consists of all possible combinations from I($d_m$) and I($d_n$) that have the same values in their common attributes. More formally, this is the join (denoted by •) of the two element descriptor interpretations, that is,

$F_1$ = I($d_m$) $\bullet_c$ I($d_n$)

where c = attr($e_m$) $\cap$ attr($e_n$). If $e_m$ and $e_n$ do not have common attributes, $F_1$ is computed as the Cartesian product of I($d_m$) $\bullet_c$ I($d_n$).

The set $F_2$ is obtained by the interpretation over the don't care element attributes of each M and N, that is, the set given by (attr(E) - (attr($e_m$) $\cap$ attr($e_n$))). For this, we need the definition of the extension of an attribute type. In this context, we assume that the *extension of an attribute type* a is defined by all possible values an attribute of that type may take, denoted by dom(a). Hence, we can define $F_2$ by the Cartesian product of the domains of its corresponding attributes, that is:

$F_2 = \bigotimes_{a \in D} \mathrm{dom}(a)$

where D = attr(E) - (attr($e_m$) $\cap$ attr($e_n$)).

## 9.4. Type reconciliation via filters

The introduced interoperability policies still require that a GIS *client* and *server* support a *common* metadata element to allow binding. However, an object such as a Geometry object in a certain spatial reference system could still be used as an input argument, if the geometry could be transformed into the SRS that is expected by a GIS server.

Various mechanisms for transforming GIS data, especially spatial referencing systems, are well-defined and standardized in the GIS domain. Today, the implementation and use of these mechanisms, however, is often restricted to conversion code that is hard-coded within GIS client and servers leaving little room for extensibility, flexibility, and reuse. Some changes, however, are on the way; for example, GIS packages such as ESRI's Projection Engine support transformation code that can be reused across products. OGC plans to define a standard service for the transformation of spatial referencing systems. Externalizing the conversion of geometry to a reusable service, however, does not yet solve the problem of automatic matching of arguments.

In this section, we describe an approach to allow otherwise incompatible GIS clients and servers to be bound through the use of a *reconciliator*. In our type model, a reconciliator (service) is a *filter* that transforms an argument that is exchanged between a GIS client and a service. A filter `f` with an available argument type `A` and required argument type `B` is declared using the notation

`f: A Ʒ B`

Through the notion of reconciliator-based relationships, we attempt to solve the problem of automatic GIS object transformation during the interaction of a GIS client and server. In terms of our type model, we might formulate the problem as follows. Consider the available GIS object `out` of type `M` (`out:>>M`), and the required GIS object `in` of type `N` (`in:<<N`), where `M` and `N` are incompatible GIS object types (since their metadata element descriptors are incompatible). They are, however, compatible if there exists a `filter p:M' Ʒ N'` such that `M` and `M'` as well as `N` and `N'` are compatible. This allows a binding of `out` and `in` via `p`, that is, we must construct the functional pipelining `out|p|in`.

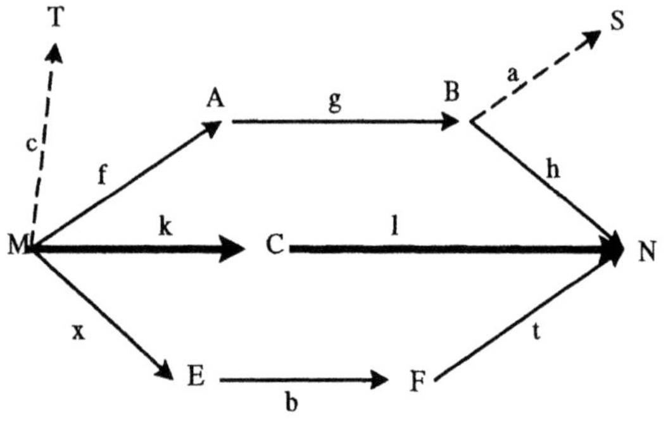


Figure 9.1. Determining a reconciliator-based relationship

For the type reconciliator system, we build a knowledge base that is graph-structured (collection of directed graphs) where the **nodes** represent *GIS object types*, and the **edges** represent *filters*. If two types `S` and `T` are part of the knowledge base, the interpretation of an edge from `S` to `T` is labeled `f`, that is, f is a filter that takes arguments of type `S` and produces output of type `T`. A possible approach to determine the available filters between two given incompatible GIS object types is the following: the type reconciliator first locates the two sets of nodes $S_M$ and $S_N$

from the knowledge base. The set $S_M$ consists of GIS object types compatible with M, while the set $S_N$ consists of GIS object types compatible with N. For each node in $S_M$, the type reconciliator system then identifies the optimal path to each node in $S_N$ resulting in a set of paths P. Finally, the best path (to some measure) in P would be selected. The edges of the chosen path represent the sequence of filters that has to be applied to perform the transformation. If no paths can be found, we assume that in the current implementation the GIS client and server cannot interact.

## 9.5. Conformance of GIS services

If a service S can be replaced transparently by another service T, and clients depending on S can be serviced by T without modification to their code, we say *that T conforms to S.* In this section, we will define conditions under which service T can substitute service S and conform to it. We already defined compatibility between client and service method arguments. To achieve conformance between two services S and T means that when a GIS service S is replaced with a conforming service T, the *compatibility* of each GIS object function must be preserved from S to T.

We base the definition of our conformance rules on the concept of conformance developed in the Emerald system (Black 1987) for distributed object-based systems, which are widely accepted today. For operations, it is based on the rule of *contravariance*. This means that an operation f conforms to the operation g if the argument type of f is a generalization (i.e., supertype) of the argument type of g, and the return type of f is a specialization of the result type of g (i.e., subtype). Therefore, we define *conformance* for operational interfaces (as introduced earlier) in the following sense:

An operational interface type S *conforms* to an interface type T, iff;

1. S provides at least the operations of T.
2. For each operation of T, the corresponding operation of S conforms according to the rule of contravariance.

If we apply this definition to GIS objects, we have a GIS client F that is replaced by the client G. Therefore, we have the available GIS objects f:<<M (the GIS object f of type M made available by F) and g:<<N (the GIS object g of type N made available by the new client G). According to the above rules, g may replace f anywhere where g is expected, if g does not produce metadata descriptors that cannot be produced by f. This requires that N is a subtype of M (N<M). On the other hand, consider the GIS services P, Q and their GIS objects p:>>M and q:>>N. Then, q might replace p anywhere where p is expected, if q is able to accept all metadata elements that are accepted by p. This requires that M is a subtype of N (M<N). Here, the problem is with this definition of conformance that the client GIS object g is allowed to produce *fewer* metadata descriptors that the client GIS object f that it replaces. This, however, leads to the problem that compatibility with other GIS services from f to g is not always guaranteed.

Assume that the (client) GIS objects are f, g, and h; here, g conforms to f (g can replace f, g<f), while f is compatible with h (f<>h). For f to be compatible with h, we require that f produces at least one metadata element descriptor that it has in common with h. If g replaces f and is allowed to produce fewer metadata element descriptors, we cannot guarantee that it will produce the one that makes it compatible with h. Therefore, we cannot guarantee g<>h.

We can guarantee compatibility, however, if the GIS object f's type M is a subtype of the GIS object g's type N. According to Definition 7, if N is a supertype of M, it does produce at least all the metadata element descriptor values that M produces (i.e., the inclusion relationship applies).

**Definition 10**: *GIS object interface conformance*

A GIS object interface type T conforms to a GIS object interface type S if and only if:

1. T produces at least the GIS objects and metadata elements of S; and
2. for each GIS object g of S, the corresponding GIS object of T is a supertype of g.

This definition of GIS object type conformance preserves the computational compatibility from S to T.

## 9.6. Conclusions

In this chapter we presented an approach to simplify interoperation in large-scale distributed GIS. We showed that the notion of strongly-typed object interfaces ensures that correct argument *types* are passed between GIS clients and servers, but cannot help to resolve the correctness of a method argument regarding the actual *properties* of an argument which are not easily captured in the type hierarchy. Based on the work done for the interoperation of continuous media objects in multimedia (Eliassen and Nicol 1996), we have applied the framework developed in this work to the domain of GIS.

In the proposed system we introduced an *argument descriptor type model* that allows method arguments to be described in a general way, and restrictions to be defined for them via a new type. We also define subtyping rules supporting the conformance of GIS interfaces regarding the content of arguments. For interoperability, subtyping is not always necessary to allow GIS client and service interaction; in our model, it is sufficient that two GIS objects support at least one common argument descriptor. If they support more than one common argument descriptor, a GIS object *negotiation* may take place to choose the descriptor (and information) that is actually used. Furthermore, if two arguments do not support a common requirement, they might still be interoperable, if it is possible to use a reconciliator such as a converter between the GIS client and server.

Overall, the model has the advantage that it relieves both GIS services and clients from the task of checking arguments, which makes their implementation simpler and more lightweight. For the applicability of the presented framework, it is relevant if the approach can be implemented easily in different, widely-used distributed object platforms such as CORBA, OLE/COM, Java, etc. Parameterized types might be an adequate concept for implementation. Other, less widely adapted environments such as functional programming languages already provide language concepts for the implementation of meta information about types.

## References

Black A, Hutchinson N, Jul E, Levy H, Carter L 1987 Distribution and abstract types in Emerald. *IEEE Transactions of Software Engineering* SE-13(1)

Buehler K, McKee L 1996 *Introduction to Interoperable Geoprocessing*. TC Document 96-001. Open GIS Consortium Inc.

Eliassen F, Nicol J R 1996 Supporting interoperation of continuous media objects. *Theory and Practice of Object Systems* Special Issue on Distributed Object Management Systems 2(2)

Mesrobian E, Muntz R, Shek E, Nittel S, LaRouche M, Kriguer M 1996 OASIS: An open architecture scientific information system. *Workshop on Research Issues in Data Engineering (RIDE 96), New Orleans*. IEEE Computer Society

Open GIS Consortium 1998 *The OpenGIS Abstract Specification Model*. TC Document 98-101. Open GIS Consortium Inc.

# 10 A SPECIFICATION LANGUAGE FOR INTEROPERABLE GIS

Andrew U. Frank and Werner Kuhn

## 10.1. Introduction

Specifications of software interfaces are essential for interoperability. If an interface is not clearly specified, software that exposes or uses the interface necessarily contains assumptions that can make it non-interoperable with other software based on different assumptions. Eliminating, or at least coordinating such assumptions requires a precise and complete specification language. If such a language also allows for prototyping of the specified components, it is a much more powerful tool in the specification process and can even support conformance testing. This chapter describes why and how functional languages serve all these purposes.

Specification, in the sense of software engineering, is the key term distinguishing industry-driven standardization processes from the more traditional committee standards defining desirable properties of data collections and their description. The practical problem is that, while the latter can use satisfactory languages [such as EXPRESS (ISO 1992) or UML (Eriksson and Penker 1998; Rational 1997)], interface specification processes lack appropriate tool support. In particular, there is a widely acknowledged lack of means to specify the semantics of operations and there is not even a standard way of defining the syntax.

Interface specifications must be expressed in a *formal* language in order to avoid ambiguity and achieve the necessary precision (Frank and Kuhn 1995). The formal specification language should ideally be tied into the compliance testing process. Otherwise, testing remains an activity that is decoupled from specification, and can lead to unpleasant surprises too late in the product development process.

Selecting a formal language to write specifications with is one of the most crucial and pressing methodological problems in interoperability research and practice. The problem has not yet found a satisfactory approach. This chapter argues that such an approach, that is, a foundation in some kind of language paradigm, must fulfill certain requirements that are currently only met by *functional languages*. It shows that interface specifications can already be written in an executable functional language with clearly defined semantics. It presents the semantic foundations of such a specification approach and discusses the implications for compliance testing.

## 10.2. State of the art

Most current interface specification approaches consist of some combination of formal and natural language descriptions. The formal parts generally cover the syntax of interfaces (describing names and parameters of operations) and in some cases parts of the semantics through first-order logic predicates.

### *10.2.1. Some examples*

A typical example is the specification for an equality operation on geometric objects of the Open GIS Consortium (*http://www.opengis.org/techno/specs.html*). The Abstract Model states the types of inputs (*Geometry*) and outputs (*Boolean*) and gives a natural language description of its semantics:

```
Boolean Equal(Geometry) – determines whether or not the passed
geometry is spatially equal to the geometry.

Note: A discussion of the use of tolerance for tests of equality is
needed.
```

The CORBA Implementation Model does essentially the same in CORBA syntax:

```
boolean equals (in Geometry other);
equals – returns true if this and other are equivalent geometries.
```

This example contains formal and informal parts. The formal parts are limited to syntax specifications. The informal (natural language) parts attempt to describe the semantics. For any programmer in charge of developing components with the specified behavior, they raise more questions than they answer:

- what do *spatially equal* and *equivalent geometries* mean?
- does the Note in the Abstract Model request a discussion of tolerances in the Implementation Model (which is lacking) or does it mean that the discussion will need to be held later (in either case, the programmer needs to decide)?
- what if users need several kinds of *equals* (based on coordinates, ID's, tolerances, etc.)?

An example where first order logic is used to specify semantics is the following curve definition from the same simple feature specifications of OGC:

```
∀c∈Curve, [a,b]=c.Domain, c.IsSimple⇔(∀x1,x2∈(a,b]x1≠x2⇒f(x1)≠f(x2)
)∧(∀x1,x2∈[a,b)x1≠x2⇒f(x1)≠f(x2))
```

While such predicates suggest impressive precision, they are difficult to write (for specifiers, particularly in committees) and read (for programmers). Who can *see* from the above statement that (if?) it correctly describes simple curves that do not self-intersect? The likelihood of specifying something inaccurate with high precision, or of misinterpreting something accurate, is quite high.

In current practice for the discussion of program semantics, the most complete and precise way of stating semantics formally is through pre- and post-conditions on operations expressed in first-order logic (Hoare 1969). While pre- and post-conditions have the big advantage of stating observable behavior, this advantage can only be fully exploited when the specified behavior is fully implemented. In practice, programmers interpret conditions and write code which they believe satisfies them.

They are not to be blamed for this, as it is often difficult and time-consuming to attempt a full validation based on logic predicates. By the time software is tested, the conditions have disappeared and test data are expected to reveal any non-conformant behavior.

### *10.2.2. Syntax specifications*

Formal specifications of interface syntax are not an issue any more. They can be written (and tested) in any language that provides user-defined types and procedures. In current practice, most specifications formally describe structures for data types and the signatures of operations. They use some kind of formal syntax from the theory of formal languages (regular grammars, parsers, signatures from algebra). As long as there are precise scoping rules for names of operations and types, there will be no danger of ambiguity or imprecision. Interface declarations in any object-oriented language (such as header files in C++) or some flavor of a specialized interface definition language (IDL) will do. Nevertheless, tools that check completeness and consistency at the syntax level are still not used early and often enough in the specification process.

### *10.2.3. Semantics specifications*

Formally defining the semantics of interfaces is a different matter. By nature, programming languages define the semantics of operations in program code. All languages currently used by software developers are far too low-level and verbose to achieve the required clarity and level of abstraction for stating *what* an operation does, without constraining *how*. Therefore, natural language or logic descriptions are being used instead, as shown previously.

Because natural language specifications of semantics require human interpretation and may thus lead to different understandings of what has been specified, they are not appropriate and even dangerous. The consequence of their use can be different implementations of components that do not interoperate or, worse, interoperate and produce uncontrollable or undetected errors.

To compensate for the shortcomings of natural language, logic-based formalisms are not satisfactory either. They can be as precise as a specifier wants, but not as clear and understandable as the programmer needs them to be. They are, consequently, difficult to interpret and almost impossible to check for correctness. Also, it is difficult to express processes and change in logic by pre- and post-conditions (Floyd 1985; Hoare 1969; Wirth 1976); Reiter has recently reviewed situation calculus and contributed some substantial clarifications, mostly useful for database design (Reiter, in preparation).

In summary, the current state of the art in specification languages for software interfaces is sufficient for the syntactic, but deficient for the semantic aspects. Any split between these two aspects (as it exists in practice) is undesirable. It not only adds to the multitude of language environments that play a role in software development, but it separates two concepts that are not separable in any natural language. And it removes us even further from the tighter coupling of specification, prototyping, and testing that is so badly needed.

## 10.3. Functional languages as specification tools

Functional languages are the approach proposed in this chapter to eliminate the shortcomings of current interface specification methods. They support the definition of an operation as a pure mathematical function (acting on something and yielding a result). This idea is very close to the needs of interface specifications. It allows for a functional view of the services specified, without involving procedural or state notions. Indeed, the syntax and semantics of modern functional languages is very close to that of algebraic specification languages (e.g., Larch; Guttag, Horning et al. 1985).

Pure functional languages *avoid side effects* (i.e., something else than the function's arguments being operated on at the same time) and make sure that the values of variables, once assigned, never change. This results in referential transparency (i.e., the same name always refers to the same thing) and allows for a *mathematical* reading of the code as a collection of equations.

Strictly typed functional languages come with a *type theory* (Cardelli and Wegner 1985; Milner 1978) which restricts the interpretation of formulae to properly typed models. This avoids some of the fundamental logical paradoxes and other puzzles of mathematics.

Class-based functional languages permit an *object-oriented* style with multiple inheritance. The classes in these languages are collections of data types sharing some behavior. The best known examples of class-based functional languages are *Haskell* (Hudak et al. 1992; Peterson et al. 1996, 1997) and notably *Gofer*, which allows classes with multiple parameters (Jones 1991, 1994, 1995). The underlying theory, however, is substantially simpler and more transparent than that of standard object-oriented languages (Abadi and Cardelli 1996), where the asymmetry between reference to *self* and to other objects causes theoretical and practical problems. The clean distinction between behavior inheritance and implementation inheritance is another important contribution to conceptual clarity.

## 10.4. A functional specification example

Suppose we want to deal with the issue, brought up in Section 10.2, of specifying equality among geometric objects. The functional language Gofer offers a built-in equality class that defines the general behavior of the equal function (==) as follows:

```
class Eq a where
    (==), (/=) :: a -> a -> Bool
    x /= y      = not (x == y)
```

The name of the class is *Eq*. It stands for the collection of all data types that can be compared for equality. The parameter *a* represents such a data type. For all of these data types, the equal function (==) and its negation (/=) have the signature (syntax) *a -> a -> Bool*, that is, they map two values of type *a* to a Boolean value (comparing them for equality or inequality, respectively).

The third line of the class definition states the property that inequality (/=) is the logical negation of equality (==). This is all one can specify about equality independently of the choice for data type *a*. This specification of the syntax (signature) and semantics (logical relationship between equality and inequality) of equality are both as general and as specific as they can be for all possible data types.

Now we can inherit the specified behavior from this class and specialize it for the case of geometric data types. In order to do this, we first have to define some geometric data types. For the sake of simplicity, we only show this for points. Using the same syntax for class definitions as in the Eq class, here is a specification for the class of all data types that are points:

```
class Points a where
   newPoint :: String -> Int -> Int -> a
   name :: a -> String
   x, y :: a -> Int
```

The class definition introduces the signatures of four operations through which Points can be characterized: a constructor function *newPoint* (taking a String for the name and two Integers for the coordinates of the point) and three observer functions (returning the name and the two coordinates). Note that at this stage, points cannot be compared for equality, and nothing has been said about the semantics of the operations.

Gofer lets us define data types to create exemplars to the abstractly defined classes. For example, the following is a definition of a data type called *PointN* (standing for points with names) and its declaration as an instance of the class *Points*:

```
data PointN = NewPointN String Int Int

instance Points PointN where
   newPoint n cx cy = NewPointN n cx cy
   name (NewPointN n cx cy) = n
   x (NewPointN n cx cy) = cx
   y (NewPointN n cx cy) = cy
```

With this instance declaration, the semantics of the point operations has been fully specified: the *name* function retrieves the String *n* that went into the constructor function, and similarly the coordinate observers retrieve the two Integers. Of course, there is nothing spectacular about these particular semantics, but the demonstrated kind of algebraic specification applies to more complex cases as well. Still, nothing has been said about equality of Points. This follows now, with the declaration of the data type *PointN* as an instance of class Eq.

```
instance Eq PointN where
   p == q = (name p) == (name q)
```

Based on the standard Eq class, this simple statement adds all there remains to be said about the semantics of point equality: two points p and q are considered equal if their names are equal. An operation not equal (/=) is automatically given (from line 3 in the class definition) with this instantiation. Note that this relies on a previous declaration for Strings as instances of Eq.

Obviously, we could have decided on a different semantics for point equality. For instance, the following code segment specifies a data type *PointC* whose values (points) are considered equal if their coordinates (rather than names) are the same:

```
data PointC = NewPointC String Int Int

instance Points PointC where
   newPoint n cx cy = NewPointC n cx cy
   name (NewPointC n cx cy) = n
```

```
    x (NewPointC n cx cy) = cx
    y (NewPointC n cx cy) = cy

instance Eq PointC where
    p == q = ((x p == x q) && (y p == y q))
```

Thus, the question from Section 10.2 about the meaning of *equal geometries* can be answered succinctly in any desirable way (or several of them concurrently). Nothing has been said about the implementation of points in actual production code. Points could be represented by any data structure. Nevertheless, the above code is a complete and executable Gofer program. Therefore, it can be used *during the specification process* to demonstrate and test the intended behavior of points.

## 10.5. The semantic foundations of functional specifications

### *10.5.1. General criteria*

When selecting a language to write specifications, one must assess its:

- semantics: how clearly and cleanly are the language constructs defined;
- expressive power: what can or cannot be stated in the language;
- ease of expression: how easy it is to state something.

The semantic foundations of the language itself (as opposed to the semantics of the interfaces described) are essential. Any ambiguity in the semantics of the language will automatically be carried forward as an ambiguity into the specifications. Natural language specifications of interfaces are a good example of this propagation of ambiguity.

The expressive power of current interface specification languages is seriously lacking with respect to the semantics of the specified operations, as exemplified above. There are either no means to express semantics (as in an Interface Description Language) or one has to use several different languages to express everything that needs to be stated about the behavior of an interface (but not more). In practice, interface specifications often use three languages: a formal IDL, a logical calculus, and natural language to fill the gaps and link the other two.

The ease of expression is not just a matter of taste or training. It measures and compares the complexity of a certain statement formulated in different languages. For example, it is straightforward to express certain topological relationships (such as containment) in natural language (by prepositions), but more complex to state the exact same relationship in logic or diagrams.

We have found Haskell (and its interactive version Gofer) to be currently the language with the least semantic ambiguities in its typing systems and execution procedures (Hudak et al. 1992; Jones 1995). Given that they are computationally complete programming languages, their expressive power is better than any specialized interface definition language. And their ease of expression for the functional aspects relevant to interfaces is far better than that of any procedural or logic-based language.

Haskell and Gofer are *pure* functional languages. This means in practice that they have minimal ambiguity in their semantic foundations. These are directly related to *denotational semantics* (Scott 1977), and more readable (Stoy 1977). Evaluation of programs is done by *substitution*, the expression on the left-hand side being replaced by the expression on the right-hand side of an equation. No procedural notion is involved here that would remove the semantics from their purely mathematical interpretation.

Built into these languages is only the Boolean data type with its operations (and, or, not). It can be exhaustively and easily tested for accuracy. The finiteness of computers limits the representable integers to a finite subset. For simplicity, the languages typically use the integers and floating point numbers of the underlying hardware. This means effectively that the ANSI specifications for numbers and their implementation are imported into any specification written.

### *10.5.2. Type theory*

The key strength of functional languages in the Haskell tradition is their elaborate and sound type theory. In contrast to traditional procedural object-oriented languages, Haskell makes a clean and understandable distinction between data types (as collections of values responding to the same operations) and type classes (as collections of data types sharing some common behavior). This distinction provides the basis for a unified treatment of single and multiple inheritance, and thereby for object-oriented specifications without compromises.

Object-oriented design in general is motivated by the dominance of objects in human thinking (Frank 1996). It is mathematically based on universal (or abstract) algebra (Birkhoff 1945; Birkhoff and Lipson 1970; MacLane and Birkhoff 1967). An algebra consists of an (abstract) carrier sort (implemented as a data type) and a set of operations with their signatures. Axioms define the behavior of the operations and group algebras with common operations together.

Classes in Haskell define classes of (polymorphous) algebras and instances define particular models. Multiple instantiations of a class can exist in the same program, as exemplified by the data types PointN and PointC above. For a particular data type, the behavior of multiple classes can apply. This is the mechanism to achieve multiple inheritance. It was demonstrated above with both PointC and PointN inheriting the operations of the class Eq (== and /=).

Classes can be built on top of other classes, which is a conceptually different kind of inheritance. A new class LocatedObjects contains the operation *distance*. The operation should only be applicable to objects of the previously defined class Points and it can be defined for all such objects (independent of their implementation, provided they have the operations of the class Points).

```
class Points p => LocatedObjects p where
   distance :: p -> p -> Float
   distance p1 p2 = sqrt (fromInteger (sqr (x p1 - x p2)
                    + sqr (y p1 - y p2)))
```

## 10.6. Testing functional specifications

Functional specifications written in a functional programming language can be executed. This determines if the specification contains the intended ideas. The formal description is brought to life and the effects can be observed. Human beings are much better in spotting deviant behavior than errors in formalism.

Haskell is a functional *programming* language and must therefore restrict expressions to executable ones. Mathematically, this means the language must be constructive: stating the existence of something, without providing a method to construct it, is not possible. This coerces the formal statements of (spatial) theories into computational tractability and thereby helps to make specifications implementable. At the same time, it provides a prototyping mechanism to the specifiers—something that is very badly missed in practice today.

Gofer separates the construction of a theory from that of a model for this theory. *Classes* define theories. *Data types* are the actual carriers of data from which *instances* build models for the theories. Thus, models are implementation examples and can be used for prototyping.

Theories are restricted to constructive axioms. For a constructive definition of square root, it is not sufficient to give the well-known equality:

```
a = sqrt x <=> a * a = x;
```

it is necessary to provide a method to compute the square root (e.g., using Newton's rule). Properties that cannot be explained as constructive axioms can be stated as testable theorems.

## 10.7. Limitations of functional specifications

### *10.7.1. Lack of side effects*

Purely functional programming languages do not use side effects. This interferes with the programming of databases and user interfaces—which appear as a side effect to the execution of the program. From a programmer's point of view, it is a drawback of Gofer that its widely available, public-domain version does not offer user-interface building tools. The language also has no mechanism for persistence and databases. This makes the use of the language for rapid prototyping somewhat more difficult, but both these limitations are not an issue in specification writing.

There are several very powerful support packages available for interface programming (Hallgren and Carlsson 1995). A design of a database for storing and accessing Gofer code has been completed at TU Vienna and is currently being tested. The use of monads allows for a solution to both the user-interface and database connection in a theoretically clean way (King and Wadler 1993; Peyton Jones and Wadler 1993).

### *10.7.2. Scalability*

Gofer (in its current implementation) has some space limitations. However, the limitations of human thinking appear to be more serious: we simply cannot comprehend complex interaction in a specification unless it is well structured.

Our experience shows that *clean* parts of a specification can be combined without restrictions on depth (i.e., layer can be put onto layer). Any *unclean* detail will surface several layers higher up and confuse issues profoundly. This is most likely not linked to language specifics, but to a general problem of specification writing; a sharp tool like Gofer only makes it painfully visible.

## 10.8. Conclusions

The use of a pure functional language with a class-based type theory is highly recommended for writing specifications of open GIS interfaces. The semantics that come with such a language are mathematically sound and provide a good foundation for building specifications. The constructive nature and executability of the code ensures no-nonsense specifications and offers prototyping capability.

Gofer is public-domain software and implementations for UNIX, Macintosh, DOS and Windows are available over the Internet.

## Acknowledgments

Christoph Rüther contributed examples and ideas to the arguments in the paper. Roswitha Markwart and Damir Medak helped with the final preparation of the manuscript. The research reported here is partly supported by the Open GIS Consortium and a grant by the European Commission (DG III, GIPSIE Project). Attendance by the authors at the Interop Conference was partially funded by NSF through NCGIA's Varenius Project. We thank all contributors for their support.

## References

Abadi M, Cardelli L 1996 *A Theory of Objects*. Springer-Verlag, New York

Birkhoff G 1945 Universal algebra. *Proceedings of First Canadian Mathematical Congress, Toronto University*: 310–326

Birkhoff G, Lipson J D 1970 Heterogeneous algebras. *Journal of Combinatorial Theory* 8: 115–133

Cardelli L, Wegner P 1985 On understanding types, data abstraction, and polymorphism. *ACM Computing Surveys* 17(4): 471–522

Eriksson H-P, Penker M 1998 *UML Toolkit*. John Wiley, New York

Floyd C 1985 On the relevance of formal methods to software development. In Ehrig C F H, Nivat M, Thatcher J (eds) *Formal Methods and Software Development*. Springer-Verlag 2: 1–11

Frank A U 1996 The prevalence of objects with sharp boundaries in GIS. In Burrough P A, Frank A U (eds) *Geographic Objects with Indeterminate Boundaries*. Taylor & Francis, London: 29–40

Frank A U, Kuhn W 1995 Specifying open GIS with functional languages. In Egenhofer M, Herring J (eds) *Advances in Spatial Databases: 4th International Symposium on Large Spatial Databases, SSD'95, Portland*. Springer-Verlag, New York: 184–195

Guttag J V, Horning J J, Wing J M 1985 *Larch in Five Easy Pieces*. Digital Equipment Corporation, Systems Research Center

Hallgren T, Carlsson M 1995 Programming with Fudgets. In Jeuring J, Meijer E (eds) *Advanced Functional Programming*. Lecture Notes in Computer Science 925. Springer-Verlag, Berlin: 137–182

Hoare C A R 1969 An axiomatic basis for computer programming. *Communications of the ACM* 12(10): 576–580

Hudak P, Peyton Jones S L, Wadler P L and others 1992 Report on the functional programming language Haskell, Version 1.2. *SIGPLAN Notices 27*

ISO 1992 *The EXPRESS Language Reference Manual*. ISO TC 184. International Standards Organization

Jones M P 1991 *An Introduction to Gofer*. Yale University

Jones M P 1994 *The Implementation of the Gofer Functional Programming System*. Report No. YALEU/DCS/RR-1030. Yale University

Jones M P 1995 Functional programming with overloading and higher-order polymorphism. In Jeuring J, Mejer E (eds) *Advanced Functional Programming*. Lercture Notes in Computer Science 925. Springer-Verlag, Berlin: 97–136

King D J, Wadler P 1993 Combining Monads. In Ayr S, Launchbury J, Sansom P M (eds) *Proceedings of the Functional Programming Workshop Glasgow 1992*. Springer-Verlag, Berlin

MacLane S, Birkhoff G 1967 *Algebra*. Macmillan, New York

Milner R 1978 A theory of type polymorphism in programming. *Journal of Computer and System Sciences* 17: 348–375

Peterson J, Hammond K, Augustsson L, and others 1996 *Report on the Functional Programming Language Haskell, Version 1.3*. Yale University Research Report YALEU / DCS/ RR-1106. *http://haskell.cs.yale.edu/haskell-report/haskell-report.html*

Peterson J, Hammond K, Augustsson L, and others 1997 *The Haskell 1.4 Report*. *http://haskell.org/report/index.html*

Peyton Jones S L, Wadler P 1993 Imperative functional programming. *Proceedings of the ACM Symposium on Principles of Programming Languages (POPL), Charleston*. ACM: 71–84

Rational 1998 *UML Notation Guide*. Rational Software Corp

Reiter R (in preparation) *Knowledge in Action: Logical Foundations for Describing and Implementing Dynamical Systems*

Scott D S 1997 Logic and programming languages. *Communications of the ACM* 20(9)

Stoy J 1977 *Denotational Semantics*. MIT, Cambridge

Wirth N 1976 *Algorithms + Data Structures = Programs*. Prentice Hall, Englewood Cliffs, NJ

# 11 INTRODUCTION: MODELING IN DISTRIBUTED ENVIRONMENTS

Michael F. Goodchild

Much of the motivation for the drive towards interoperability in the past few years has come from the enormous power of the Internet and its applications, particularly the World Wide Web (WWW). Five years ago the Internet was an interesting way of connecting computers; today, it is the basis of an industry that is registering explosive growth and intruding into many aspects of our lives. It is commonplace to compare the WWW to traditional libraries: instead of books that take years to publish, and are deposited relatively inaccessibly in massive physical structures, the WWW allows everyone with a simple electronic connection to be an instant publisher and a reader of up-to-date information. With the WWW, we appear to be drowning in information.

Geographic data offer what is arguably one of the most compelling areas of application of the WWW. Geographic data are of relatively limited interest—a street map of Paris is of interest mostly to people who are there, and of much less interest to people in California. Geographic data change through time, so the lengthy process of publishing on paper can be too slow. They are difficult to handle in traditional paper and photographic media, and difficult for libraries to catalog for easy retrieval—but the WWW makes physical media irrelevant, and readily supports multidimensional indexing.

Not surprisingly, then, there has been massive interest in the use of the WWW and similar technologies for disseminating geographic data. But the potential of distributed environments like the Internet goes much further, because users have options about where to store, where to process, and where to obtain results. The traditional model of a GIS project has all software and data being assembled on a single workstation, probably on the user's desk. This process can take months, especially when differences of format, structure, and semantics have to be dealt with. Paradoxically, then, the electronic speed offered by computer-based decision-making is often overwhelmed by the glacial speed of data reformatting and checking. Instead, the Internet and WWW make it possible to avoid shipping data, by instead requesting services from remote systems, and shipping results.

In this section we present six chapters on various aspects of distributed environments. In the first, Pedro Pereira Gonçalves and his colleagues discuss interoperability between distributed software components, within the framework of ActiveX. Rather than a single, monolithic solution that must be coupled to other solutions when its functionality is inadequate, component-based solutions use a single interface to access a variety of smaller software components for specific needs. In their solution the interface is a spreadsheet application, through which it is possible to access GIS and dynamic model functions interoperably. They also show how it is possible to build the interface as a virtual environment.

The second chapter, by David Bennett and his colleagues, takes this one step further by building a modeling environment across heterogeneous computers. Intelligent agents are used to search for and retrieve suitably-*wrapped* data across a network, and dynamic models are built using a simple geographic modeling language. Next, Agnès Voisard and Marcus Jürgens discuss two alternative models of data dissemination in distributed networks: the *eager* approach, in which data are first collected in a data warehouse; and the *lazy* approach, characterized by a set of distributed data repositories and a mediator. They then propose a hybrid solution, and illustrate it with a geospatial example.

David Pullar and Kristin Stock are concerned with how agencies model their data and system requirements, and how interoperability can be achieved across multiple agencies. They propose a metamodel described in the *Unified Modeling Language* (UML), and discuss its application to geospatial applications in Queensland.

Finally, we include two chapters on the important task of data search. Libraries assist their users through a complex process of information abstraction known as *cataloging*, using three sets of keys: author, title, and subject. The WWW has no catalog, but it has been possible to develop intelligent agents that can extract and build keyword-based indexes, and by some clever guesswork it is possible to ensure that the indexes capture the most important keywords of each WWW page or site. In effect, search engines build catalogs, but lack the sophistication of professional catalogers. Moreover, today's search engines are not able to solve the problem libraries have always faced in cataloging geospatial material. Library catalog specialists using traditional methods are unable to deal with the fact that the primary key, geographic location, is multidimensional; search engines are unable to identify the same key because it is not possible to express it in keywords.

Ian Finch and Eleanor Small describe an approach to this problem that allows interoperability between different forms of geospatial metadata. They describe a prototype system, and although the current version makes use of metadata built by hand, they look forward to more intelligent agents that will take on the metadata construction task automatically. Finally, Brandon Plewe and Steven Johnson discuss the problems of metadata interpretation by general GIS users. Metadata defined by geospatial data producers can be dauntingly complex; they propose a layer of metadata interpreters and processors that take over some of the task of metadata use.

# 12 INTEROPERABILITY OF GEOGRAPHIC INFORMATION: FROM THE SPREADSHEET TO VIRTUAL ENVIRONMENTS

Pedro Pereira Gonçalves, Nelson Neves, João Pedro Silva, Joaquim Muchaxo and António Câmara

## 12.1. Introduction

The extended use of GIS varies from simple map making to environmental modeling. In each, different users have specific data needs, formats, and most important of all they possess unique problems. As a response to this, the majority of GIS vendors supply a normalized *GIS-in-a-box* package, containing the data structure, formats, and geographic functions that will ultimately impose the vendor flavor in the design and operability of the problem. The user becomes limited in the design of his or her own problem to the functions supported by the vendors. On the other hand, the need to increase market share imposes a constant need to upgrade the spatial functions and data structures available. This creates an exponential growth in software complexity. In the long term the user depends on data and software standards and needs constant upgrades to training.

Environmental modeling, statistical data analysis, and GIS are three examples of computational activities that are usually relegated to separate, large, and sophisticated computer systems. Usually each of them has more functionality than is required by the user, while the user's needs exceed the range of a single one (e.g., both GIS and statistical data analysis). In the first two, the data organisation benefits from a GIS perspective and some very simple issues like the quality of spatial coverage of the data can be resolved by a simple geographical display. Likewise, the quality of the time coverage over a geographical region can be combined with a geographical display by enhanced visualisation. For these two examples (spatial and temporal coverage of data), for interfacing models over a large range of possible input parameters, and for many other circumstances, it is beneficial to have a level of integration which is much greater than a simple file transfer mode.

Depending on the hardware and software used, the integration of spatial models (including data and processes) and GIS are usually divided in four main categories with increasing level of coupling, namely isolated applications, loose coupling, tight coupling, and full integration (Burrough 1997; Wegener 1996). These start from a simple data transfer between data models using ASCII files where the user is the interface and running probably in different hardware environments, to a full integration where the data exchange is based on a common data model and database management system.

These approaches focus on the transference of the needed functions to one of the systems. Their goal is to develop one application that has all the spatial and modeling tools needed by the user. In the end, this interaction can be very efficient, but it is first necessary to include all the model functions and data formats in the GIS. There is an obvious duplication of software, and the approach does not take in account the evolving nature of software.

It is not possible to design an all-in-one solution that integrates spatial models and GIS. To achieve interoperability, with the sharing of data and functionality between the systems, one must proceed with the decomposition of the GIS into small components that are available to other applications. If the user just needs to visualize the data, or query the database, or even overlay maps, it is necessary to make available only those procedures, and not the overall application. It is necessary to provide the building blocks of spatial models, that combined with the use of macro-languages (similar to BASIC) or Internet-based languages (like Java or VBScript) will further increase the power available to the end-user.

But, even the classic GIS graphic interface must be reconsidered. Does it regard the map as a tool for communication or as simple store of information? Does the interface adapt itself to the user or the inverse? It should be possible to adapt the GIS interface to different uses, working in different places with the same data.

Transforming the GIS into components will also enable the design of customer-specific interfaces, adapted to knowledge about the user. If the user is adapted to work with Internet browsers, or word processors, or even spreadsheets, then the GIS should be embedded in that application. From fields like simulation, multimedia, and the Internet, the GIS should take advantage of the possibilities already available in established applications. To achieve this, the GIS must be deconstructed. In the end the GIS application should be invisible to the end user; only the geographic concepts should be visible, working within other applications.

In this chapter a closer look is taken at how GIS technologies can be implemented as components, enabling the exchange and sharing of information and functions between applications for the design of environmental models. A component-based structure to the GIS, called *iNovaGIS*, is proposed and implemented as ActiveX objects. These objects allow the separate use of visualization and query or map algebra procedures within other Windows-based applications through the use of Microsoft macro-languages (Visual Basic) or Internet-based languages (Java and VBScript). A cellular automata forest fire model was implemented to illustrate the potential applications of this approach. The interface is embedded in a spreadsheet application with two main interactions: the GIS and the cellular automata model functions. The spatial model results are analyzed in the spreadsheet application taking advantage of the already available functions. Additionally, the objects were changed with a Virtual Environment (VE) application to allow an enhanced spatial visualization of model outputs.

## 12.2. Background

In this section, a brief introduction is given to three main issues for geographic information interoperability. These are the integration of different data sources and formats, the accessibility to spatial models within other software applications, and the enhanced visualization of geographic data.

### *12.2.1. Integration of different data sources*

To achieve geographic information interoperability, the development of a generic GIS conceptual model is important. It must possess the structural aspects of the data, including data types, file formats, and relationships. This will provide a coherent ensemble of constructors representing abstract objects in a structural manner and encapsulating the properties and methods of these objects.

The advent of the Open Geodata Interoperability Specification (OGIS) opens for the first time a real opportunity to develop data-structure-independent GIS applications (Buehler 1994). The OGIS documents lay out the basic technical infrastructure for interoperability of geographic data and methods. The implementation of the GIS conceptual model as an object-oriented data model allows easier description of the user's understanding of the data structure than does a relational model (Worboys et al. 1990). A relational model requires the user to consider all data in tabular form, whereas the object-oriented model allows for the consideration of complex objects as individual units that may relate more directly to the user's concepts. Like records, objects are structured data types that contain fields that hold values. Objects, however, also contain methods. Methods are procedures and functions that act on the data in the object's fields. These GIS objects will serve as the building blocks of spatial models since they integrate the necessary abstract structures. This will also minimize program complexity by splitting functions into separate components and structuring spatial models as a set of distinct modules.

### *12.2.2. Accessibility to spatial models: design and use interface*

In the past few years the need for spatial models has grown in response to awareness of the complexity of environmental problems, and to the realization that spatial models should represent real world processes (Heywood et al. 1995). An integrated model combining atmospheric, hydrologic, land-surface-subsurface or ecological process modeling, is an important issue for environmental modeling. Being geographic information a main component of these models it is important to guarantee the access to a set of tools that facilitates the wider use of GIS.

However, conventional GIS data models emphasize static representations of reality (Yuan 1996). Geographic information for a given area is decomposed into a set of single-theme layers as regular (raster) or irregular (vector) tesselation models (Frank and Mark 1991). These layers constrain the ability of the GIS to represent dynamic information, such as transitions and motion. Raster cells encode attribute values at every given location with no consideration of the spatial characteristics of the theme they represent. Geometrically-indexed vector objects, on the other hand, "force a segmentation of the entities being represented into separate layers whenever they interact in time or space: adopting this representational method forces compromises on most environmental modeling" (Raper and Livingstone 1995). GIS needs a complete and rigorous framework for geographical data modeling to overcome the difficulty in handling geographic complexity, scale differences, generalization, and accuracy. The lack of data representation schema to integrate GIS data with models for spatial-temporal processes appears to be a major shortcoming in current GIS.

The availability of a generic toolbox may facilitate the wider use of spatial information and models. However, since the models can be easily applied at the press of a button, the misuse of data may occur. Without an appreciation of the underlying concepts that govern the use of spatial information, the models cannot be correctly implemented. To prevent this, the tools must include the clear identification of model components and assumptions.

Normally environmental models are accessed following a *black-box* approach. The user inserts the necessary data as input and gets a result. Apart from the change of parameters, no other interaction is possible between the user and the spatial model. However, it is accepted that the user should have access to tools of model calibration, validation, and investigation (Burrough 1997), such as the study of the effects of scale in spatial models. Those tools should be expressed in terms of direct access to the properties and methods of GIS primitives. This access will allow the development of spatial models that do not follow the rigid structure imposed by any particular GIS application. On the contrary, it is the user that will choose the methods and properties for the spatial model.

If it is necessary to visualize an ongoing process of the spatial model, or if the data inputs for the spatial models require query to the database, or even a map overlay, it is necessary to make available those procedures. The use of macro-languages, or Internet-based languages (like Java or VBScript), will further increase the potential for spatial model building.

### *12.2.3. Visualization of environmental processes*

One of the most frequent criticisms of GIS is that it has failed to give adequate attention to principles of cartographic design, or for regarding the map as a simple store of information rather than a tool for communication (Goodchild 1990). Data visualization enhancement is often cited as both an advantage of and a justification for investment in GIS development. Visualization allows more data to be examined and assimilated in less time (Buttenfield and Ganter 1990). In the domain of analysis, visualization supports spatial and temporal processes through an understanding of structure and pattern. Both maps and charts may be used in the context of exploration and inference, to describe a data set, to generate hypotheses, or to check for errors.

The use of Virtual Environment (VE) techniques has the potential for spatial-temporal representation with great fidelity because users can interact directly and immersively with the elements which form the basis of their decisions (Neves et al. 1997). A VE provides much deeper knowledge of an environment than does a typical desktop environment because the user is actually in the scene. It facilitates human computer interaction, due to the use of a three-dimensional representation and direct manipulation of virtual objects. In VE systems the user is immersed in a 3D world generated by the computer using, for example, a head-mounted display and position and orientation sensors.

The assumption in providing visual analytical tools for the visualization of environmental processes is that immersion in large amounts of data, combined with greater interaction (scale change, graphic symbolization and classification, surface depiction, etc.) will produce new insight and innovation in the design of environmental models.

## 12.3. System implementation

The iNovaGIS is an ActiveX server that creates an interface to geographic data and information. It allows the use of GIS-like variables and functions in programs like Excel, Visual Basic, Delphi or any other Windows software. All these software products have the capability to run macros or even to compile the source code. The system functions are well separated, but to the end user the system works as one. This interconnection is achieved by means of ActiveX technology.

### *12.3.1. The iNovaGIS ActiveX server*

ActiveX technology extends the Component Object Model (COM) and Object Linking and Embedding (OLE) to support new features required for ActiveX controls. ActiveX controls are visual controls that run on in-process servers, and can be plugged into an OLE container application. They are not complete applications, but can be thought of as prefabricated OLE controls that are reusable in various applications. ActiveX controls make use of OLE automation to expose their properties, methods, and events. Features of ActiveX controls include the ability to fire events, bind to data sources, and support licensing.

A COM server is an application or a library that provides services to a client application or library. A COM server can be an in-process server, meaning a DLL running in the same process space as the client; or a local server, meaning an EXE running in a different process space but on the same machine as the client; or a remote server, meaning an application running on a different machine from that of the client. COM servers are the modules in which COM objects exist. A COM server that contains the code for OLE automation objects, ActiveX controls and Active Windows is an ActiveX server.

With iNovaGIS methodology, a generic data object (called *iData*) was created with two main descendents, *iValue* and *iGeodata*. These two objects are respectively non-geographic and geographic structures with properties and methods that communicate through their common ancestor. The iGeodata object has two descendants: *iVector* and *iRaster*, which implement the properties and methods for dealing with vector and raster data. The descendants of these objects implement the specific methods of different GIS file formats. From the user's point of view, data access is transparent and independent of the data format. This allows users to focus directly on abstract objects corresponding to concepts in their applications.

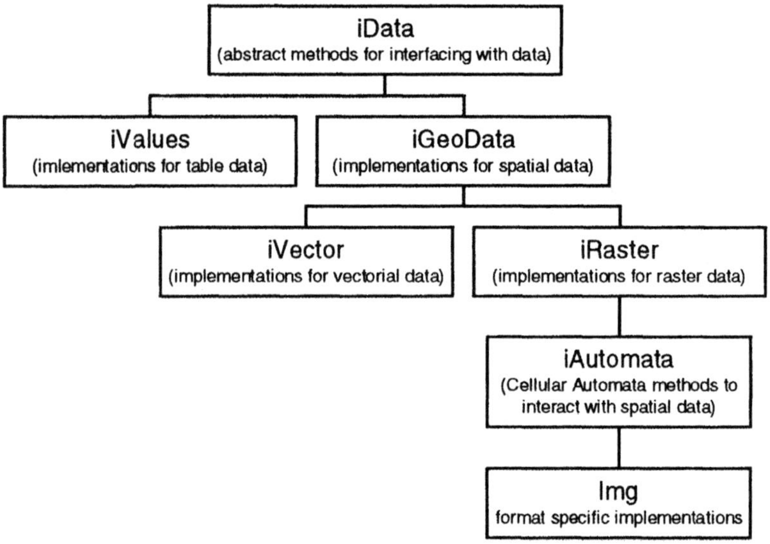


Figure 12.1. iNovaGIS object structure

As a descendant to the raster object, an object called *iAutomata* was implemented that works as a cellular automaton (CA) and enables a spatial connectivity function model. CA models have proven to be valuable modeling tools in the fields of the physics of growth, physical chemistry, and hydrodynamics, for they make it possible to follow the mechanisms of interaction between automata and to observe the state of the lattice at any given moment in the iteration. Several types of connectivities of varying ranges exist (Weisbusch 1991). The connectivity inherent to these mathematical models is ideal for systems in which many simple components act together to produce complicated patterns of behavior (Wolfram 1986). Each cell takes on $k$ possible values, and is updated in discrete time steps according to a rule $f$ that depends on the values of sites in some neighborhood around it. Figure 12.1 describes the overall structure of the implementation. Due to their common ancestor it is possible for these objects to exchange information and interact.

One of the main advantages of the use of this technology is the transformation of GIS to a simple common macro language using normal variables. Using these objects, traditional geo-processing services can be added to our software. Normally it is possible to define a variable as an *integer, byte* or *float*; with *iNovaGIS* it is possible to define one's own geographic data and to access a multitude of geographic functions. Figure 12.2 shows a source code example in Visual Basic Editor from Microsoft Excel to query a raster map in IDRISI format (Clark University 1998).

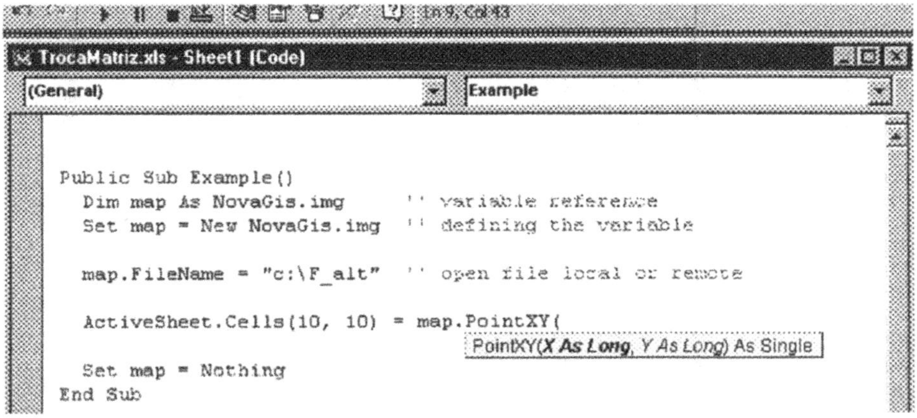


Figure 12.2. Source code example in Visual Basic editor from Microsoft Excel

Functions like defining map attributes, the study area, or a set of data become simpler due to the potential of many macro languages present in common office applications. There is no need to invent or rebuild new scripts or macros to work with these geographic information structures. The full potential of software dedicated to environmental modeling, statistical data analysis simulation, or even data visualization can enhanced by adding GIS concepts and functions. At the same time, the GIS application remains invisible to the user. When the variable is defined the user works directly in the data structure and functions. In Figure 12.3 an example of a simple implementation of visualizing raster data in Microsoft Excel is shown. Because the information is becoming more and more dynamic, it is necessary that the GIS becomes more flexible in dealing with its remote sources. The iData object implements methods that allow the remote access either by FTP or HTTP protocols.

## 12.4. Application

Forest fire modeling was chosen as an example to illustrate the potential of the proposed approach, using the model *FIREGIS* (Gonçalves and Diogo 1994). To prove the feasibility of the system, two intertwined applications were built. In the first, the forest fire model interface was built in a Microsoft Excel spreadsheet, with all the data input, model parameters, and map and chart visualizations. At the same time the spatial outputs of the model are seen in a Virtual Environment (VE) using the *VirtualGIS* system (Neves et al. 1997). However, it is first necessary to explain the forest fire model definition, followed by the two examples.

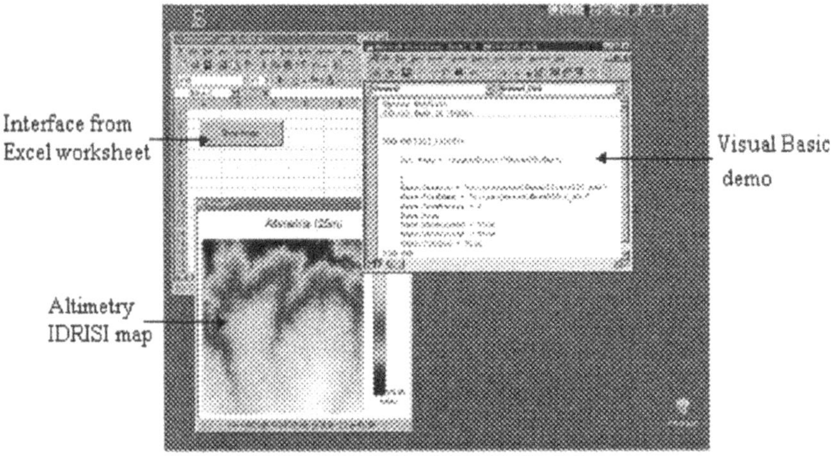


Figure 12.3. Calling geographic information from Microsoft Excel

### *12.4.1. Forest fire model*

The area devoted to forestry in Portugal is nearly 3 million hectares, being 45% of that area occupied by ***Pinus pinaster*** woodlands, employing 100,000 workers (2.5% of the national workforce) and representing almost 20% of national exports (Ventura 1992). Between 1974 and 1987 approximately 300,000 hectares of forest were destroyed by fire (10% of the total forestry area). This scenario repeats itself in other southern European countries, and the majority of these fires are caused by human-related activities (Moreno 1994), with important land resources being consumed each year by this phenomenon. Considerable research effort has been made in the prevention of forest fires, with the use of forest fire models being one of its major subjects. These models are used to simulate fire spread under given environmental conditions. Spatially-explicit environmental modeling provides an effective means of analyzing 'what-if' scenarios relevant to spatial land use problems. Such models serve as the basis for making decisions regarding the management of environmental resources.

The model used in this study is a cellular automaton named *FIREGIS* (Gonçalves and Diogo 1994). The rules defining the automaton are stochastic, with the probability distribution defined through the use of a deterministic mathematical model. The model used was the physical semi-empirical model developed by Rothermel (1972), based on vegetation characteristics, slope, and wind. It forecasts the rate of spread and the reaction intensity by knowing certain properties of the fuel matrix and the environmental conditions in which the fire will occur.

Three groups of parameters are requested by the model, in order to determine the Rate of Spread (RS):

- fuel particle properties (heat content, mineral content and particle density);

- fuel array arrangement (loading by size class, mean size, and fuel's depth); and
- environment-related values (wind velocity, fuel moisture content, and slope).

The proposed model is based on the three main assumptions. The Rothermel (1972, 1983) model is representative of the fire's speed of propagation within each cell (where the vegetation is homogeneous). In each time step, each cell can propagate to one of the eight adjacent cells (Figure 4) only if they are not already extinct or without vegetation. It assumed that the fire ignition always occurs in the center of the cell.

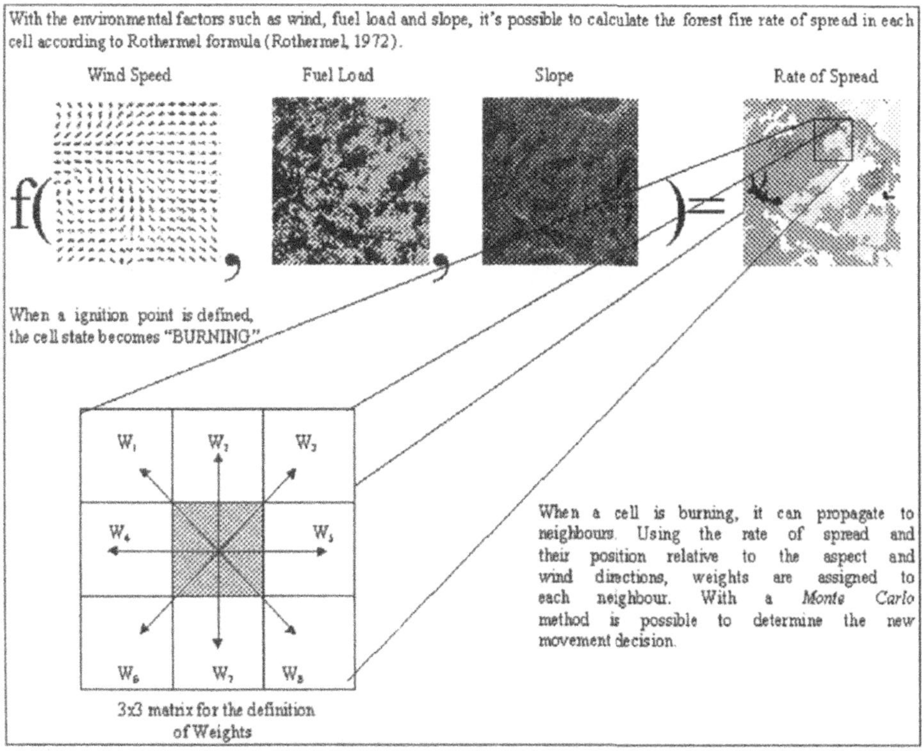


Figure 12.4. The FireGIS model

Fire ignition may start in one or more cells. Each cell may then have one of four possible states: unburned, ignition, combustion, and extinct. For each cell in the combustion state, there are eight adjacent cells where is possible to propagate. To determine the RS associated to each cell, one has to apply the Rothermel equations (1972, 1983).

### *12.4.2. Spreadsheet application*

To implement the FIREGIS model, it is necessary to have direct access to the data required. Since the data were in IDRISI format, a set of raster objects called *img* were used. These objects inherit all the properties of the iData, iGeodata, and iRaster objects, and implement the required procedure to read and write in IDRISI format. To achieve the spatial transport functions necessary to the model, the methods and properties implemented in the iAutomata ancestor were used. Within the application, and using the Visual Basic Editor, it was possible to access all the methods required to build the cellular automata model. Figure 12.5 show how the data input and visualization are done within the application.

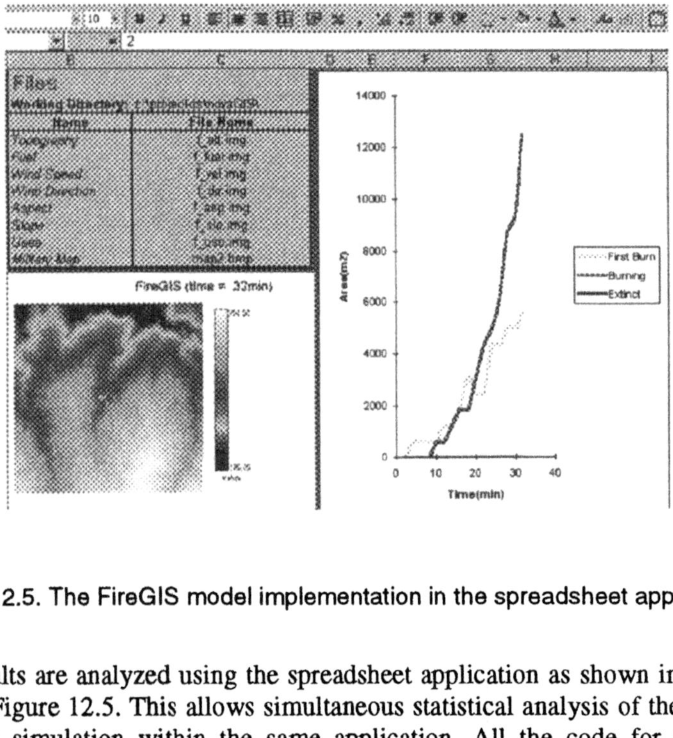


Figure 12.5. The FireGIS model implementation in the spreadsheet application

The results are analyzed using the spreadsheet application as shown in the graph at right in Figure 12.5. This allows simultaneous statistical analysis of the results of an ongoing simulation within the same application. All the code for the spatial analysis and functions is embedded within the iNovaGIS objects, leaving the spreadsheet application with the necessary computer resources.

### *12.4.3. Virtual environments (VE)*

The forest fire propagation model also outputs the results to a virtual environment (VE) using the *VirtualGIS Room* system (Neves et al. 1997). This system is an interface solution for GIS users that uses digitizing tablets and mouse along with position and orientation sensors and head mounted displays. The VE application and the iNovaGIS system run independently but exchange objects with each other.

The VirtualGIS Room application updates the Digital Terrain Model (DTM), where a correspondence is established between cell sizes in the virtual environment and the simulation environment when the model starts running. As new simulation steps are computed, the information about the cells that changed is sent to the VE, reporting the cell location (*x*,*y*) and the new cell value. At the same time the user can manipulate the data inputs and model results in a common spreadsheet program. Figure 12.6 shows a snapshot of the visualization of the outputs from the forest fire propagation model in the VE.

Figure 12.6. Visualization of the FireGIS model result in a VE

Exploration of this virtual world is enhanced by the ability to scale the world, and by the existence of points of view normally unobtainable in the real world. The use of a virtual environment to visualize and interact with spatial information and associated simulation models allows users to explore information recreating their usual interaction in a real GIS room.

## 12.5. Conclusions

With the use of this iNovaGIS, the definition of map attributes, study area, or set of data is only restricted in use by the potential of the macro languages used. The objects created can be used in different applications, allowing their use in a diverse range of applications, from word processors and spreadsheets to programming tools (like Visual Basic, Delphi, or C++). At the same time there is no need to invent or rebuild new scripts or macros to work with this structure. The full potential of software dedicated to environmental modeling, statistical data analysis, or even data visualization is enhanced by the addition of GIS concepts and functions. The interoperability of the system allows the design of different user interfaces according to different needs, as well as a more integrated analysis of the results.

So far all of the tests were done in a single computer. In the case of several computers the distributed version of COM will be used. Distributed COM (DCOM) is a technology that extends the local capabilities of COM to cross-network communications between objects on different computers. This will enable the interconnection of systems in different computers and even in completely different locations.

The expectation is that iNovaGIS will serve as a fast and inexpensive way to implement prototypes of spatial databases and environmental models. Its use in education is encouraged; the software is available as freeware from *http://gasa.dcea.fct.unl.pt/iNovaGIS/*.

## Acknowledgments

The work of Pedro Pereira Gonçalves and Jorge Nelson Neves is funded by Junta Nacional de Investigagação Ciêntifica e Tecnolόgica (JNICT) through PhD fellowships. This work was partially supported by JNICT under research contract PEAM/C/FGI/633/95.

## References

Buehler K 1994 *OGIS Geodata Model Overview*. OGIS Project Document 94-019. The Open GIS Foundation, Cambridge, Mass.

Burrough P A 1997 Environmental modeling with geographical information systems. In Kemp Z (ed.) *Innovations in GIS 4*. Taylor and Francis, London: 143–53

Buttenfield B P, Ganter J H 1990 Visualization and GIS: What should we see? What might we miss? *Proceedings of the 4th International Symposium on Spatial Data Handling, Zurich*: 307–316

Clark University 1998 *IDRISI Software Homepage*. *http://www.idrisi.clarku.edu/*

Frank A U, Mark D M 1991 Languages issues for GIS. In Maguire D J, Goodchild M F, Rhind D W (eds) *Geographic Information Systems: Principles and Applications*. Longman, London: 147–63

Gonçalves P, Diogo P 1994 Geographic information systems and cellular automata: A new approach to forest fire simulation. *Proceedings of the European Conference on Geographical Information Systems (EGIS 94), Paris*. EGIS Foundation: 702–712

Goodchild M F 1990 Keynote address: spatial information science. *Proceedings of the 4th International Symposium on Spatial Data Handling, Zurich*: 3–12

Heywood D I, Oliver J, Tomlinson S 1995 Building an exploratory multicriteria modeling environment for spatial decision support. In Fisher P (ed.) *Innovations in GIS 2*. Taylor & Francis, London: 127–36

Moreno J 1994 Fire in the Mediterranean region: implications of land-use changes in a global change context. In Viegas D (ed) *Proceedings of 2nd International Conference on Forest Fire* Research, Coimbra: 61–2

Neves N, Silva J P, Gonçalves P, Muchaxo J, Silva J, Câmara A 1997 Cognitive spaces and metaphors: A solution for interacting with spatial data. *Computers & Geosciences* 23(4): 483–8

Raper J, Livingstone D 1995 Development of a geomorphological spatial model using object-oriented design. *International Journal of Geographical Information Systems* 9(4): 359–84

Rothermel R 1972 *A Mathematical Model for Predicting Fire Spread in Wildland Fuels*. Res. Pap. INT-115, U.S. Department of Agriculture, Forest Service, Intermountain Forest and Range Experiment Station

Rothermel R 1983 *How to Predict the Spread and Intensity of Forest Fire and Range Fires*. Gen. Tech. Rep. INT-143, U.S. Department of Agriculture, Forest Service, Intermountain Forest and Range Experiment Station

Ventura J 1992 *Propagação de Fogos em áreas Florestais*. Instituto Superior Técnico, Lisboa

Wegener M 1996 Spatial models and GIS. Presentation for the ESF-GISDATA and NSF-NCGIA 1996 Summer Institute in Geographic Information held at Villa Borsig, Berlin, 24 July – 1 August 1996

Weisbuch G 1991 *Complex Systems Dynamics*. Lecture Notes Volume in the Santa Fe Institute Series in the Science of Complexity. Addison-Wesley

Wolfram S 1986 Cellular automaton fluids 1: Basic theory. *Journal of Statistical Physics* 45: 471–526

Worboys M F, Hearnshaw H M, Maguire D J 1990 Object-oriented data and query modelling for geographical information systems. In *Proceedings of the 4th International Symposium on Spatial Data Handling, Zurich*: 679–88

Yuan M 1996 Temporal GIS and spatio-temporal modeling. *Proceedings of Third International Conference/Workshop on Integrating GIS and Environmental Modeling, Santa Fe*. National Center for Geographic Information and Analysis, Santa Barbara (CD-ROM)

# 13 GEOGRAPHICAL MODELING IN HETEROGENEOUS COMPUTING ENVIRONMENTS

David A. Bennett, Greg A. Wade, and Raja Sengupta

## 13.1. Introduction

Solutions to spatial problems often impact a variety of stakeholders. Geographic information systems (GIS) and spatial decision support systems (SDSS) provide useful tools for the analysis of spatial problems and can be used to investigate the impact of alternative solutions on individual stakeholders. These technologies are, however, often expensive and difficult to implement. As such, some suggest that GIS and SDSS provide an unfair advantage to those with the economic means needed to implement the technology.

The Internet provides new opportunities for the sharing of information and technology and, thus, offers a partial solution to the problem of technological parity. The full potential of the Internet for geographical problem solving, however, remains untapped because many potential users lack the time or technological expertise required to integrate analytical tools, data, and software into a system capable of supporting problem-solving activities. Advances must be made on at least three fronts before distributed GIS technologies reach their full potential. First, new geoprocessing technologies are needed that bridge multiple vendor formats and heterogeneous computing environments. Second, institutional arrangements must evolve that facilitate the distribution of problem-specific data and analytical tools to applicable stakeholders. Finally, advances in the theory and application of public participatory GIS are needed to better understand how and to what extent geographical concepts and problem-solving techniques can be communicated to the public at large. If such advances can be made in the theory and application of geoprocessing technologies then decision-makers, analysts, and stakeholders will possess: 1) access to computing resources that may not otherwise be available; 2) an electronic forum for the exchange of ideas in written, graphic, and cartographic form; and 3) a more level technological playing-field on which to build consensus and compromise.

The work presented here represents the intersection of three related research initiatives that are being pursued by the authors. Each of these initiatives is focused on some element of distributed GIS technology and should be viewed as work in progress.

### *13.1.1. Geographical modeling across a network of heterogeneous computers*

To address social and environmental problems it is often necessary to understand the impact of human activity on a complex set of geographical processes that interact across space and through time. Spatially-explicit computer models can provide insight into spatio-temporal relations and, thus, contribute valuable information to the decision making process. The creation and management of spatially-explicit models, however, is often difficult and time-consuming because the real-world systems that they represent are large and complex. Furthermore, the use of such models often requires large repositories of data and computational power. The utility of distributed processing for the resolution of combinatorially-complex geographical problems is well documented (Armstrong and Marciano 1996). Here, distributed processing capabilities that provide seamless access to data, models, and computational power across a network can help reduce the organizational and technological overhead associated with geographical modeling. To take full advantage of such opportunities, however, new geoprocessing technologies are needed that facilitate the use of spatial models and data stored in multiple vendor formats and on multiple computing platforms. These technologies should provide the capability to: 1) search network-accessible repositories of geographical models, model components, and geographical data; 2) provide a user-friendly mechanism for integrating geographical data and model components; 3) transform geographical data stored in multiple formats into a common format and referencing system; and 4) execute models across a distributed modelbase and database.

The construction of links among disparate software products and data formats requires a common communication protocol. Such a protocol can be developed by providing a common import/export format or by providing software hooks into proprietary data structures. Such protocols are in development for geographical databases (e.g., SDTS, OGIS). However, there is little support for the distribution and sharing of geographical models and analytical tools. To accomplish this level of interoperability we have focused our research on the development of: 1) a framework for geographical modeling across a heterogeneous computing environment; 2) prototype software that implements and tests some of the more unique aspects of this approach; 3) a cooperative system of intelligent agents that use this GML to acquire and transform these data and models into a usable form; and 4) a geographical modeling language (GML) that is designed to facilitate the search for geographical data and models.

## 13.2. A framework for the representation of dynamic geographical systems on heterogeneous computing environments

Most existing frameworks for the representation of geographical data are built on data models that reflect the cartographic or image-processing heritage of GIS. Although these data models have proved useful for a wide variety of applications they have significant shortcomings with respect to the representation of dynamic geographical systems. Most importantly, they lack the spatio-temporal data structures needed to capture change through time and the modelbase management capabilities needed to store, manipulate, and execute geographical models. In the modeling framework presented by Bennett (1997a) modelbase management and geoprocessing technologies are integrated into a single knowledge-driven system that supports the digital representation of dynamic geographical systems. Within this framework the form and function (*CBehavior* and *CStructure* in Figure 13.1) of geographical features is encapsulated into a single object (*CFeature*). Geographical models are comprised of a set of geographical features or sub-models.

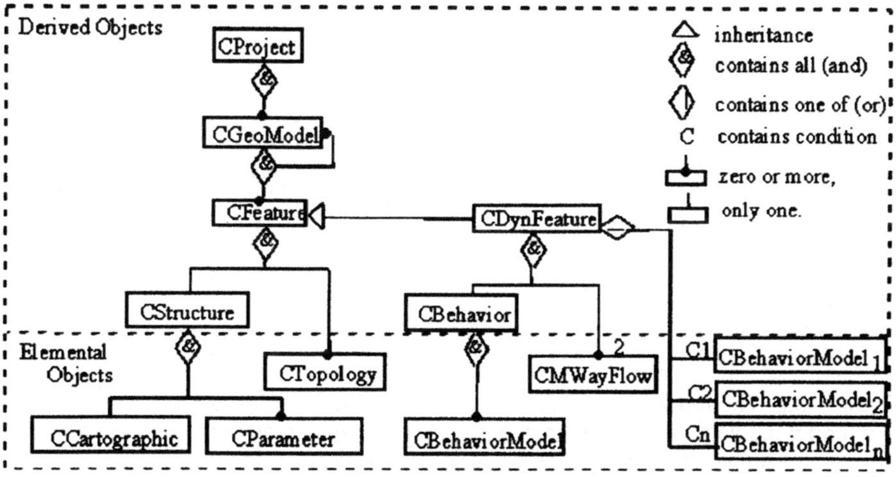


Figure 13.1. The geographical modeling system is an object-oriented prototype software package that supports the construction and use of dynamic geographical models (after Bennett 1997)

Model design is viewed as the capture of geographical knowledge and the organization of this knowledge into a model graph that emulates the spatial processes of a particular geographical system (Bennett 1997b). The representation of geographical systems as directed graphs provides a modeling topology well suited to a distributed implementation. Nodes in this graph represent geographical features and models. Graph edges represent *contains* relations (e.g., a watershed *contains* sub-watersheds) and *flow* relations (e.g., output from one sub-watershed *flows* to another sub-watershed). Because of the modular structure of this framework the digital representation of geographical objects can be distributed across a computer network without loss of semantic content.

This concept is illustrated by the following example. A user at computer "A" wants to execute a watershed model that supports the decomposition of a landscape into sub-watersheds (e.g., the SWAT model). The model is stored on computer "B" and the data needed to execute this model is distributed across a set of computers linked together via the Internet. A graph representation of this model is presented in Figure 13.2. Edges in the graph capture the flow of water from one sub-watershed to the next and the nodes represent computational units that calculate the state change associated with specific sub-watersheds through time. The user can take advantage of the computing power distributed across the network by mapping the computational units to specific computers.

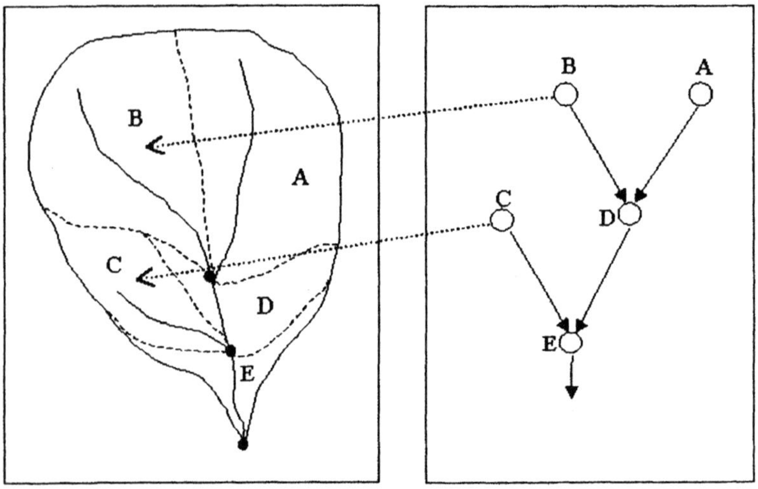


Figure 13.2. The modular structure of the geographical modeling framework provides a model topology well-suited to a distributed processing environment. Here, each sub-watershed can be modeled on a separate processing unit

Three distinctly different options exist for the implementation of this model graph. First, the data and model could be downloaded to a local machine, reformatted into a common framework and executed. This approach fails to take advantage of the distributed computational power that is available to the user across the network. A second approach would be to bring the data and model to a local machine, reformat the data and model into a common framework, recompile the model, redistribute model components across a network, and finally execute the model. Nodes in the model graph map to individual processing units while edges map to the communication pathways existing on the network. This computing approach allows the user to take advantage of data, models, and processing power that are distributed across the network, but requires a high degree of software and hardware compatibility. A third approach would be to allow data and models to reside on multiple machines and in multiple formats and to link these elements by constructing a communication protocol that facilitates interoperability. This approach has the potential to provide a high degree of flexibility and allows users to take advantage of a network of heterogeneous computers. Platform independence, however, requires the development of interoperability tools that facilitate the access, manipulation, and use of geographical data and models on a remote machine or host. Furthermore, if data and models reside on different machines and in different formats then geographical processes can no longer be encapsulated into a single object as described in the original GMS framework. To overcome this limitation of the GMS two new objects are defined, *virtual models* and *model components.*

### *13.2.1. Extensions to the GMS framework*

*Virtual models* reside on a host machine and coordinate the interactions among *model components* and data that are distributed across a network (Figure 13.3). Each model component represents an autonomous element within a simulation model consisting of static inputs, local variables, dynamic inputs, computation modules, and outputs. Static inputs to a component include any value that does not change through the course of a simulation. Local variables and dynamic inputs are allowed to change during a simulation; however, the source of their values varies and their purposes differ. Local variables are used by model components as state variables and their values are calculated within the computation modules. Dynamic inputs on the other hand get their values from the outputs of other virtual models and model components. Model components perform the role of the *CFeature* class in the original GMS implementation.

Data from the output of one model component are passed to the input of another through *communication channels.* These channels use the communication technologies built into the host machines, thus allowing components running on distributed systems to communicate. The distribution of independent components among processing units can be used to exploit the natural parallelism of some models. This form of parallelism follows from the Dataflow computer architecture Hurson et al. 1990).

## 13.3. A prototype system for distributed geographical modeling

A prototype distributed geographical modeling system was developed using the Java programming language. The extended GMS framework presented above was used for the representation of dynamic geographical systems and distributed model management capabilities were developed to support the instantiation and use of these model components across a network. This work was first presented by Wade et al. (1997) and will be reviewed briefly here.

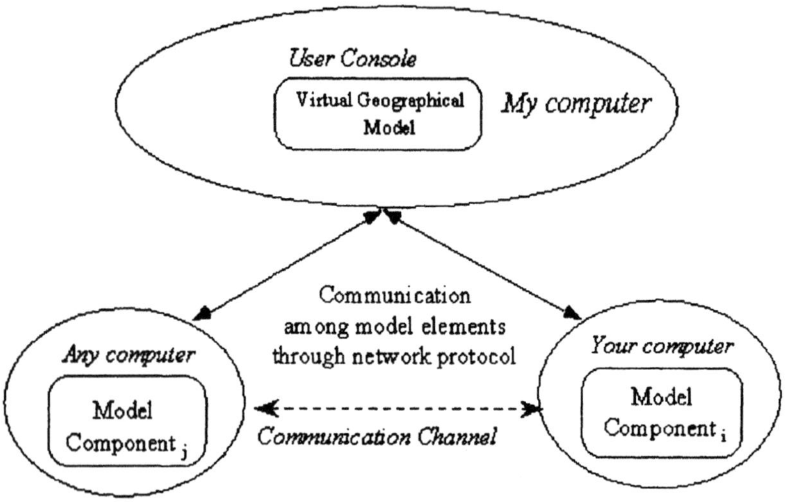


Figure 13.3. The geographical modeling system was extended to support distributed computing by adding two new classes that represent *virtual models* and *model components*

### *13.3.1. Software architecture*

The management and integration of data and models across a network is orchestrated by four model-management software components that together represent a virtual geographical modeling system. These components include a model repository, data server, object repository, and compute server (Figure 13.4). All elements within this system (virtual models, model components, model management components, and geographical data) are instantiated as Java classes. A user console provides a graphical user interface to the virtual model.

**13.3.1.1. Model repositories.** Each model repository consists of a database management system (DBMS) and a model server. Model servers store model components that simulate the dynamics of particular geographical features and processes. The DBMS stores metadata associated with all models stored in the model server. This database can be queried to locate model components that meet user-defined criteria. These metadata include a description of the functional capabilities of the model, and a list of keywords, a list of model inputs, and a list of model outputs. The input list can be used by the system to search for requisite data. Elements in the output list can, in turn, be made available as potential data sources.

**13.3.1.2. Data servers.** Data servers provide access to data stored using existing GIS software or generated by geographical models. Each data server consists of two elements, a DBMS and zero or more GIS software packages. The DBMS is used to locate spatial data based on metadata supplied by a user or stored as part of a model component. The metadata format used here follows the guidelines set forth by the spatial data transfer standards (NIST 1992). In order to support network communications, this format is extended to include the Internet address of the server that stores the data set, the type of Java object that the data can be loaded into (e.g., ARC/INFO chain can be loaded into a GMS chain), the individual who created the object definition, the creation date, and modification dates. A geographical name-server (GNS) designed to expedite the search for relevant data has been proposed and tested by Wade and Bennett (1997). As implemented here, data extracted from a data server are instantiated as GMS objects that can be shared across a network. The instantiation of geographical objects using data stored in vendor-specific formats provides a bridge between GIS and geographical models. Note that the choice of the GMS as a common format is implementation-specific and other data models could be used to support distributed geographical modeling.

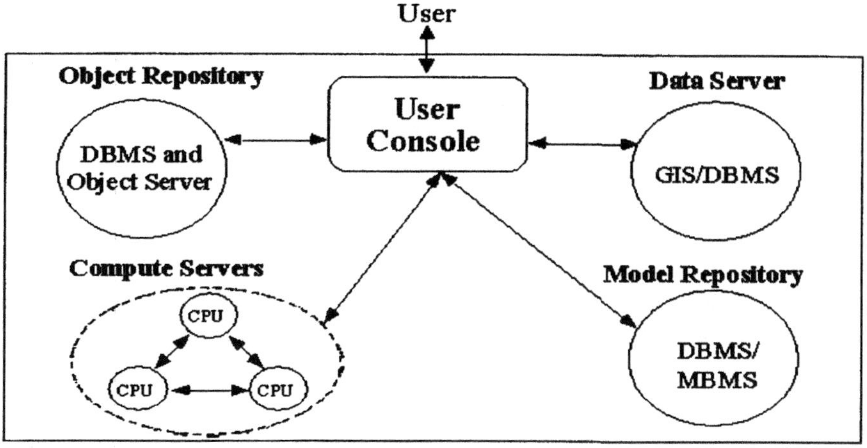


Figure 13.4. Distributed modelbase and database management capabilities are added to the GMS through the construction and instantiation of object repositories, model repositories, data servers, and model servers

The Java Database Connectivity (JDBC) API provides an interface between the Java programming language and databases. These drivers were used to link the prototype distributed modeling software to Microsoft's Access DBMS software (Figure 13.5). Database queries return the information needed to locate code that is dynamically loaded and instantiated as HORB objects. HORB extends Java to provide object request broker capabilities that allow clients to create and use instances of objects on remote machines (Hirano 1996). Using HORB, data and models reside on remote machines and are never transferred to the host running the user console. Communication among objects is accomplished through the use of proxy objects on a client (compute servers or user consoles) that link to a remote object (derived from data and object servers) instantiated on the host machine. In future work, HORB will be replaced with the more standard technique of Java Remote Method Invocation (RMI).

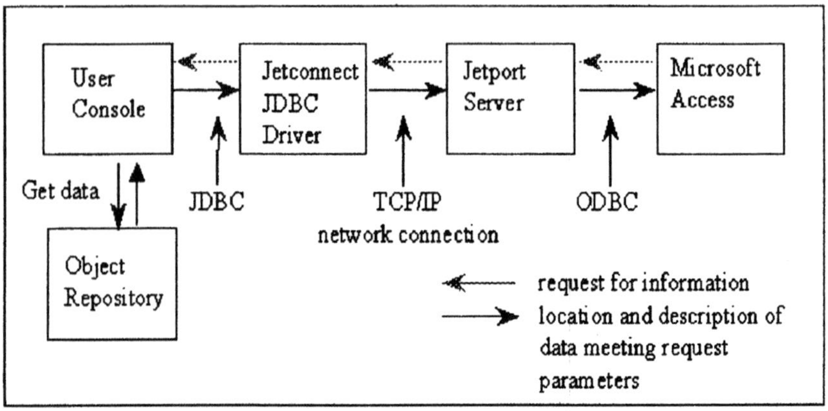


Figure 13.5. Access to geographical data distributed across a network is provided through the use of JDBC and ODBC connectivity tools

**13.3.1.3. Object repositories.** Object repositories store class definitions for the representation of geographic form. Objects stored in these repositories, therefore, represent distributed and more general implementations of the *CCartographic* and *CStructure* classes defined within the GMS framework (Bennett 1997a). These repositories are accessed and managed by a DBMS and an object server. Each record in this DBMS contains the name of a defined class and the Internet address of the host where the program code for the class is stored. These class structures can store geographical data directly or they can reference data created and stored in an established GIS using a data server. Access to geographical objects stored on remote machines was provided by HORB.

Users can expand an object repository to meet new modeling demands. Model and data searches, therefore, may return a reference to an object that is unknown to a local system. As such, a mechanism must exist to distribute knowledge about new classes across a network. Performing this task is the role of an object server. When an unknown object name is encountered a *request-for-information* is sent all known object servers. If an object server contains the class definition for that object it will return its binary representations to the requesting object. These classes can then be dynamically loaded into the user console or saved to disk for later use. A custom Java ClassLoader was created to dynamically load the compiled class files returned by object servers. The user console also implements a cache as part of its class loader to reduce the overhead of repetitively searching for and downloading classes.

**13.3.1.4. Compute servers.** Compute servers are used to execute model components built to the user's specifications. The only responsibility of these servers is to support the computing demands of geographical models. In performing this function multiple compute servers can be linked together and data can be passed through communication channels. Coordinating the flow of data among compute servers via these channels is the responsibility of the virtual model.

The user console begins model execution by instantiating model components on compute servers. A communication channel is created for each model component having outputs. This channel is created on the compute server that executes the associated model component. The model graph created by the user determines how other model components link to these channels and retrieve dynamic inputs (Figure 13.6).

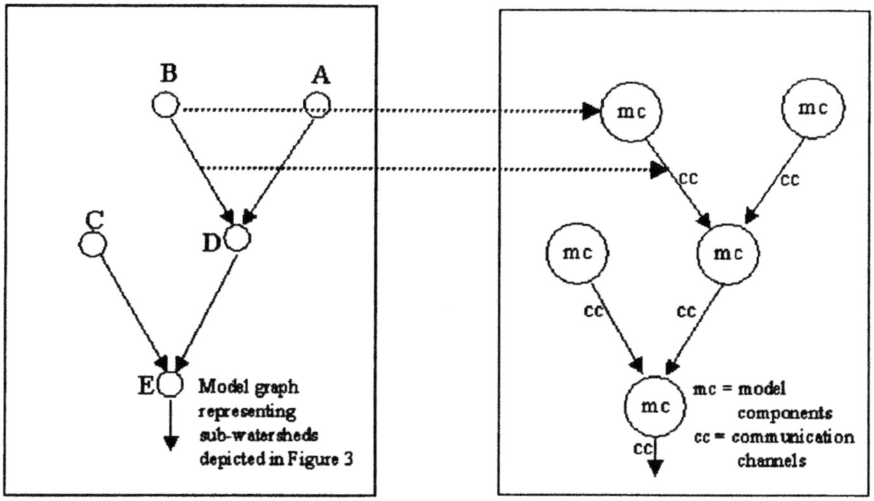


Figure 13.6. A model graph constructed by a user can be mapped to model components and communication channels. Here the model graph represented in Figure 13.3 is mapped to its corresponding extended GMS elements

## 13.4. Retrieving and transforming datasets using agents

The prototype system described above was designed to test the feasibility of distributed geographical modeling. While the system accomplishes this initial goal, it only supports a rudimentary representation of spatial data structures. To support real world problem solving, the system framework needs to accommodate a variety of spatial data structures in multiple vendor formats that reside on several machines. Towards this goal, work is in progress to develop a distributed geographical modeling system that utilizes intelligent "agents" collaborating with one another to identify and integrate useful model components and datasets.

An Agent as defined by Russell and Norvig (1995) is "(an entity) that can be viewed as perceiving its environment through sensors and acting upon that environment through effectors". This basic definition of an agent is extended by Shoham (1993) to include "collaborative computing". Shoham's Agent-oriented programming (AOP) paradigm "promotes a societal view of computation, in which multiple agents interact with one another".

In the proposed system, collaborating agents are utilized to retrieve model components and datasets, and to integrate model elements and match datasets to models through the use of "wrappers" and data transformations.

The retrieval of model components and datasets across a network requires the use of metadata and heuristics. Metadata files are used to identify the contents of the data servers, model repositories, and object repositories. The number of possible datasets available on the Internet/Intranet can be huge (Figure 13.7a). The selective and intelligent use of information stored as metadata can be used to facilitate the retrieval of relevant data (Figure 13.7b).

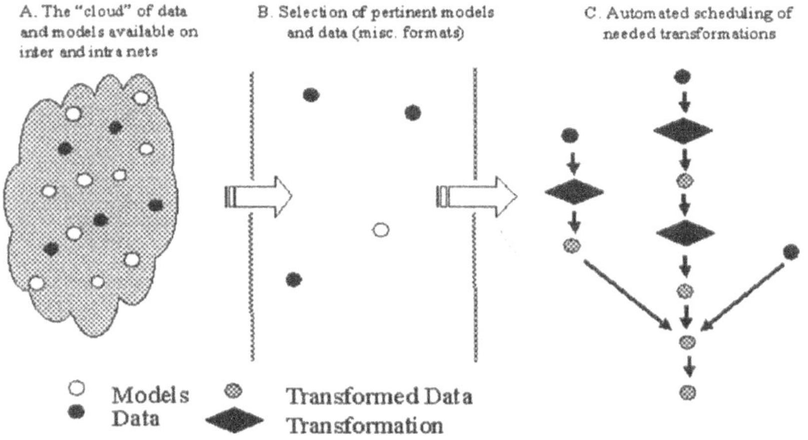


Figure 13.7. Intelligent search agents assist the user in the identification and selection of relevant data stored in accessible networks. A series of transformations may be required to make these data usable in a particular problem solution

The integration of model elements and the matching of datasets to models can be achieved through the use of "wrappers". A "wrapper" is a piece of software that creates a link between different components by providing a standard interface between these components (Figure 13.8). This interface creates a mapping of data and model requirements to a common format, such as the extended GMS. A variety of data transformations may also be required to convert all input data into a usable form (e.g., data structure transformations, projection transformations, and conflation). These transformations can be managed by scheduling agents (Figure 13.7c).

To perform data transformations and to link data and models into a common system, the scheduling agents must be able to communicate with one another using a common language that supports modeling, data transformations, and spatial query. A Geographical Modeling Language (GML) was developed to support these requirements (Sengupta et al. 1996). A GML provides an inter-agent communication protocol for spatial model and data handling. Through the use of the GML, agents can retrieve, manipulate, and store geographical databases and modelbases. In addition, an interoperability extension to the GML allows the agents to request data structure transformations (e.g., Raster to Vector) and transformations between different vendor formats (e.g., IDRISI to ESRI-GRID).

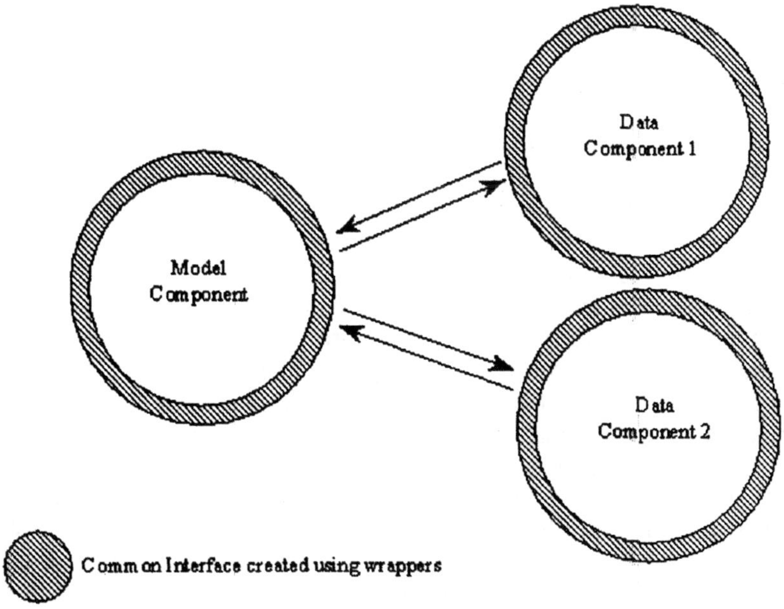


Figure 13.8. Data and model components can be integrated through the use of wrappers that provide a common interface

### *13.4.1. The retrieval of datasets using a knowledge-directed search*

The identification and retrieval of relevant data on a network can be facilitated if selection criteria are first placed into priority order. The priority of a particular criterion is based on the importance of that criterion to the problem solution and the computational intensity associated with the analysis of that criterion. This ordered list can be used as a heuristic for the retrieval of geographical data across a network.

This heuristic acts as a filtering mechanism that ensures that computationally intensive manipulations occur only on those datasets that are most likely to be useful. For example, an intersect operator would be applied only after it has been ascertained that a dataset contains needed information in a usable coordinate system. To implement an efficient search and retrieval operation the system enforces a kind of operator precedence onto the actions that an agent takes. For example, search agents will first check for thematic content. If the dataset contains the proper theme then the query process continues and the search agent will analyze metadata associated with specific attributes. If the dataset does not have the desired attribute information (e.g., Soil erodability factor K) it may still prove useful if a related non-spatial database can be located that contains the needed data. Accuracy issues are analyzed next. This set of criteria includes positional accuracy, attribute accuracy, and logical consistency. Query operators that require geometrical calculations (e.g., intersect) come last in the precedence order (i.e., implemented last) because they are the most computationally intensive.

### 13.4.2. Use of scheduling agents in transforming the datasets

A failure to meet a particular specification does not necessarily remove a dataset from consideration since a properly scheduled sequence of transformation may convert an otherwise unusable dataset into valuable information. These transformations include data structure transformations, vendor transformations, projection transformations, and database joins. Usually, human experts are required to determine the set of transformations required and the sequence in which to apply these operators. Human expertise is required because of the combinatorial complexity of the problem, since a potentially large number of transformation operators are available in the real world. A traditional algorithmic approach to determine an efficient sequence of transformations applies a brute force search method (every possible sequence of operators is applied to all elements of the dataset) to determine the solution. This proves to be an unsatisfactory solution, as the combinatorial complexity makes a real-time response unlikely. An alternative solution is the use of intelligent scheduling agents, directed by human expertise encoded into the system, that can be used to retrieve and manipulate only those datasets and models that are potentially useful to the user.

Scheduling agents reduce the search space through the use of heuristics and constraint mechanisms. To construct a useful transformation schedule, agents must recognize contiguous relationships among transformations. In some cases, the order in which these transformations are to be applied to a dataset must be recognized. For example, it may be necessary to bring a dataset into a particular vendor format before changing its data structure. In other cases, the sequence of transformations may be independent of each other. For example, the data structure and the projection of the dataset could be modified independently and, as such, the order in which these operators are applied is unimportant. Once the solution set has been queried and an efficient sequence of transformations identified, the system-identified transformations are converted into a consolidated GML statement (an example of a consolidated GML statement is given in Section 4.3). If this task can be decomposed into a number of sub-tasks, then it can be allocated to separate processing units.

Work is currently in progress to develop the structure of the scheduling agents. Within the proposed framework, the transformation operators will be implemented as daemons or threads running in the background that respond proactively to a transformation request by the scheduling agent.

### 13.4.3. A geographical modeling language (GML)

A GML was developed to facilitate communication about spatial operations between the agents (Sengupta et al. 1996). To support the query and manipulation of spatial data this language must implement applicable spatial operators and lend itself to parsing by agents. It has been extended in this paper to include operators that support transformations required for interoperating geographic datasets.

To identify an appropriate set of spatial operators, we looked to the GIS literature. From this search, we found that the work done by Egenhofer and Franzosa (1991) and Tomlin (1990) provided a suitable foundation upon which to base the GML. Note that others have used these research efforts as a foundation for similar research efforts (Clementini et al. 1993; Smith and Park 1992; Wesseling et al. 1996). Logic, control, and iteration operators were borrowed from traditional programming languages to support the development of problem-specific algorithms (Kernighan and Ritchie 1988). To provide support for interoperability, three new GML operators have been introduced in this paper: the operators *VendorTransformFrom*, *VendorTransformTo*, and *DataStructTransform*.

Using GML, the agents can retrieve and transform relevant datasets into a common format. As an example, consider the spatial query to find all cities within a 15,000 meter radius around St. Louis. Given a *USCities* layer in the UTM coordinate system and in ARC/INFO vector format, the following steps will construct the desired dataset:

1. Select the city of St. Louis
2. Draw a buffer around St. Louis of 15000 meters
3. Find which cities fall within this buffered region

The above steps can be represented in GML as follows:

*NewLayer = (Overlaps(Buffer(Select( USCities,Name = "St. Louis")),15000),USCities,2)*

Further, the layer generated above can be transformed into any format that could be required by the modeling software. As an example, modeling software may require the NewLayer generated above in a delimited ASCII text format as input. Since the original USCities layer was an ARC/INFO point file, then the required transformation operations would be:

*Temp1 = VendorTransformFrom(ESRI-VECTOR, NewLayer))*
*OutputLayer = VendorTransformTo(ASCII-DELIMITED, Temp1))*

## 13.5. Conclusions

The GMS framework has been extended to support distributed modeling through the use of model repositories, object repositories, data servers, and model servers written in Java. The utility of this approach was demonstrated through the development of prototype software. Presently, this software only supports a rudimentary representation of spatial datasets. Work is underway, however, to extend the capabilities of this software to support the construction and instantiation of distributed geographical models from data and models stored in multiple formats. The management of these data and models will be the responsibility of intelligent agents that automate the search, retrieval, and transformation of data and models that are distributed across a network and the integration of these elements into a virtual geographical model.

Two benefits are expected to accrue from the implementation of this system. First, distributed computing will allow the use of computing power of several machines. This will allow users who possess limited computing power to perform computationally-intensive analyses. Second, the automated search, retrieval, transformation, and integration of data and models will simplify the process of model development and facilitate the sharing of data and tools among stakeholders.

## References

Armstrong M P, Marciano R J 1996 Local interpolation using a distributed parallel supercomputer. *International Journal of Geographical Information Systems* 10(6): 713–730

Bennett D A 1997a Managing geographical models as repositories of scientific knowledge. *Geographical and Environmental Modelling* 1(2): 115–133

Bennett D A 1997b A framework for the integration of geographic information systems and modelbase management. *International Journal of Geographical Information Systems* 11(4): 337–357

Clementini E, Felice P D, Van Oosterom P 1993 A small set of formal topological relationships suitable for end user interaction. In Abel D, Ooi B C (eds) *Advances in Spatial Databases, Proceedings of the Third International Symposium on Spatial Data Handling*. Lecture Notes in Computer Science No. 692. Springer Verlag, New York: 277–295

Egenhofer M J, Franzosa R D 1991 Point set topological spatial relations. *International Journal of Geographic Information Systems* 5(2): 161–174

Hirano S 1996 What's HORB?. *http://ring.etl.go.jp/openlab/horb/doc/what.htm*

Hurson A R, Lee B, Shirazi R, Wang M 1990 A program allocation scheme for data flow computers. *Proceedings of the 1990 International Conference on Parallel Processing*: I415–I423

Kernighan B W, Ritchie D M 1988 *The C Programming Language*. Prentice Hall, Englewood Cliffs, NJ

NIST 1992 *Spatial Data Transfer Standard (SDTS)*. Federal Information Processing Standards Publication No.173. Computer Systems Laboratory, National Institute of Standards and Technology, Gaithersburg, MD

Russell S J, Norvig P 1995 *Artificial Intelligence: A Modern Approach*. Prentice Hall, Engleweed Cliffs, NJ

Sengupta R, Bennett D A, Wade G A 1996 Agent mediated links between GIS and spatial modeling software using a model definition language. *Proceedings of GIS/LIS '96*. American Congress on Surveying and Mapping, Bethesda, MD: 295–309

Shoham Y 1993 Agent-oriented programming. *Artificial Intelligence* 60: 51–104.

Smith T R, Park K K 1992 Algebraic approach to spatial reasoning. *International Journal of Geographical Information Systems* 6(3): 177–192

Tomlin C D 1990 *Geographic Information Systems and Cartographic Modeling*. Prentice Hall, Englewood Cliffs, NJ

Wade G A, Bennett D A, Sengupta R 1997 An interactive distributed architecture for geographical modeling. *Proceedings of Auto-Carto 13*. American Congress on Surveying and Mapping, Bethesda, MD: 307–316

Wade G A, Bennett D A 1997 GNS: A distributed hierarchical topical nomenclature for geographic data. *Proceedings of GIS/LIS '97*. American Congress on Surveying and Mapping Bethesda, MD

Wesseling C G, Van Deursen W P A, Burrough P A 1996 A spatial modeling language that unifies dynamic environmental models and GIS. *Proceedings, Third International Conference/Workshop on Integrating GIS and Environmental Modeling, Santa Fe, NM, January 21-26, 1996*. National Center for Geographic Information and Analysis, Santa Barbara, CA. *http://www.ncgia.ucsb.edu/conf/SANTA_FE_CD-ROM/main.html*

# 14 GEOSPATIAL INFORMATION EXTRACTION: QUERYING OR QUARRYING?

Agnès Voisard and Marcus Jürgens

## 14.1. Introduction

Interoperability within geographic information systems (GIS) is concerned with both data and operations. We focus here on the data aspect, more specifically on the access to multiple, distributed, heterogeneous, and autonomous information sources and on their integration. For this purpose, we need to take into account the special requirements of applications dealing with geospatial data, such as:

- the vast amounts of data considered;
- the existence of complex and highly-structured data;
- the co-existence of many different geographic formats;
- the increasing trend towards reuse of geographic data all over the world;
- the distributed nature of geographic data.

In addition, temporal aspects play an important role in many geospatial applications. For instance, in geomarketing applications, it is useful to keep many versions of data for analysis over time. Applications such as traffic control or intelligent transportation require fast access to highly dynamic data. GIS are concerned with many tasks, such as input and store, query and analyze, display and select geographic information. In this chapter we concentrate on the data management aspect. The participating information sources may vary widely, from legacy systems to geospatial repositories. The three main actions we focus on are the following: (1) collecting data from the information sources, (2) organizing data in a geospatial store or keeping track of data organization in different systems, and (3) querying data. In what follows we refer to these actions as 3-step data handling.

In the database area, common approaches to data integration are as follows (Widom 1995). In the *eager* approach, data is collected in a data warehouse, which allows direct querying and analysis. The query-driven or *lazy* approach, on the other hand, hinges on the concepts of a mediator together with wrappers built on top of the participating repositories. In practice, it turns out that a combination of both approaches is often more appropriate. In this chapter, we present a hybrid alternative to geospatial data integration. Geospatial data sets are either integrated through wrappers and mediators or part of a data warehouse. In this context, a warehouse plays not only the role of a participating system but also the role of an efficient storage for further reuse. This storage can be related to data themselves but also to application sessions. For instance, results of queries likely to be posed in the near future are stored in the warehouse for efficiency reasons.

This chapter is organized as follows. We first present the two common database approaches to data integration and compare them for different types of geospatial applications. We then present our hybrid solution, which we illustrate with a flood control application. We give a list of qualitative parameters for tailoring such general frameworks to particular application requirements. We conclude with the main research issues that need to be addressed in order to design such systems efficiently.

## 14.2. Common approaches to data integration

This section starts with a description of the two common database approaches to data integration, that is, the eager approach and the lazy approach. We then give examples drawn from geospatial applications in order to illustrate their use, before studying the main differences between the two solutions.

### *14.2.1. Eager approach (data warehouse)*

In the eager or in-advance approach, data is integrated in advance in a data warehouse, which is defined as a subject-oriented, integrated, non-volatile, and time-variant collection of data (Inmon 1996). It allows the multidimensional storage of data. A common integrator component extracts data sets from different sources and integrates them before sending them to the warehouse. Figure 14.1 illustrates this approach.

Following the 3-step data handling mentioned in the introduction, we now describe the major tasks necessary to run a data warehouse, namely data warehousing, warehouse organization. and query processing.

**Data warehousing (1).** The process of collecting the data is known as data warehousing. New data are added to the data warehouse from different sources, which generally handle data in different formats. Metadata describe how data have to be transformed between formats and how they are integrated. Before data are added to the warehouse, consistency checks are performed and a mechanism guarantees that all data are loaded exactly once. The load time becomes a critical issue when the number of sources or the size of sources increases. Attention should be paid to keep the load time as short as possible.

Regarding the moment when data are fetched to the warehouse, two options are possible. With the push mechanism the source decides when to send the data to the warehouse. In the pull mechanism the warehouse asks explicitly for the data at a certain time. A push or a pull is either event- or time-driven. To reduce the amount of data transferred to the warehouse, only the changed data could be shipped. Sending the delta between two data sets saves significant amount of time and space in applications with small changes from time to time. This turns out to be useful in applications handling large maps, especially in the case of raster maps with changes in a few cells.

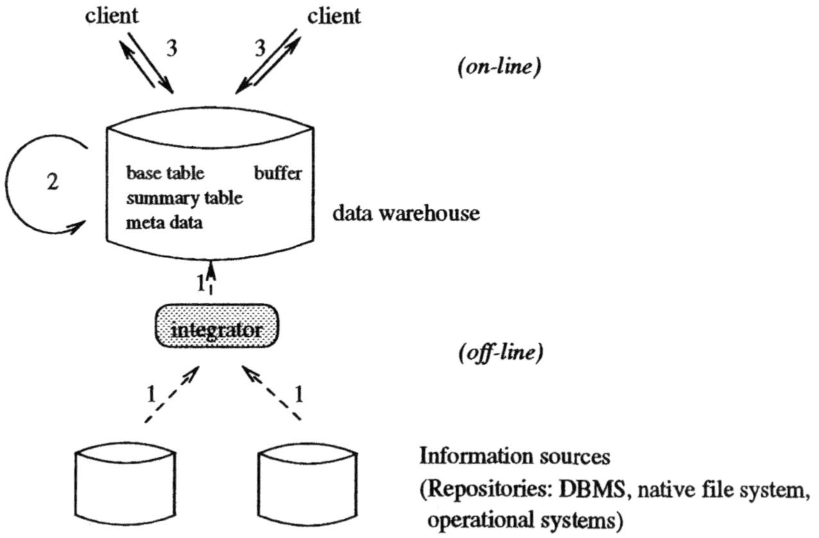


Figure 14.1. Eager approach to data integration

**Warehouse organization (2).** The resulting warehouse database is somewhat special as there is usually no delete or update operation possible. The problem of constant growth can be solved either by moving the data that are not used to less expensive media or by summarizing them further. The role of data warehouses goes beyond data storage. Raw data are (re)-structured, and aggregated in such systems and stored in base tables. In addition there are aggregates held in summary tables. Aggregation rules are defined in the metadata. Further aggregation is useful because most common queries analyze subsets of data. Users are sometimes not interested in particular objects but in aggregated values, such as the population of a state computed from a table gathering the population of all its counties. If these results have to be calculated on the fly, that is, when the query is processed in the system, the response time can be intolerably long. Results of aggregation are calculated in advance during times when the system is not used for analysis (e.g., during nights or weekends). They are stored in summary tables (Mumick et al. 1997).

The more data are aggregated, the faster can queries be processed, but the higher are update and storage costs. Wiener et al. (1996) present algorithms for selection of summary tables and indices. Aggregation is appropriate when clients require specific predictable portions of information. It is noteworthy that, even if this mechanism provides high performance when answering queries, the warehouse does not necessarily show either the most recent state of information sources or all details, because summarized data and not all raw data are stored there. However in many applications it makes sense to get a rough result fast rather than a precise value after long processing time.

**Query processing (3).** Users typically pose ad-hoc queries against the warehouse and expect answers quickly. To ensure that typical queries will be answered fast, queries are logged and users profiles are generated from the log-files. These profiles are used to adapt the organization of the warehouse to users' needs and to generate indices for faster access and better performance. Because profiles change over time, the reorganization of the warehouse might be necessary on a regular basis.

During the data warehousing and warehouse organization phases the warehouse is off-line and no queries can be processed. It is on-line only when no changes are made. This architecture is called a *read-mostly* environment. The clear distinction between on-line and off-line modes makes transaction monitors superfluous while being in on-line mode. This decreases response time.

### *14.2.2. Lazy approach (mediator)*

In the lazy or on-demand approach, when a query is issued, the appropriate information sources are selected and a mediator generates the subqueries or commands that will be addressed to the information sources via their wrapper (Figure 14.2). A wrapper converts data objects of specific data sources into objects of a common information model (Step 1). A catalog keeps meta-information related to the information sources (Step 2). Data conversion works both ways. Queries from the mediator are transformed to queries understood by the sources and data from the sources are converted to a format processable by the mediator (Garcia-Molina et al. 1995; Step 3).

In such systems, the results are collected from the information sources and the mediator performs translation, filtering, and merging of data before returning the answer to the client. A mediator is a software module that refines in some way information from one or more sources. The wrapper does the technical transformation (e.g., unit transformation) whereas the mediator's task concerns the semantical integration of data (e.g, generalization). The great advantages of this architecture are low storage costs and flexibility for integration of new sources. A disadvantage is that time analysis is only possible on data provided by the data sources.

One may wonder about the differences between a mediator/wrapper architecture and a federated DBMS. In the first architecture, the sources are often semi-structured, and no schemas are available. Source integration occurs at access time. The traditional method of first integrating the schemas and then the data does not apply in this context. Because the format of data sources often changes, one important research issue is the development of tools to support the efficient development of wrappers and mediators.

### *14.2.3. Summary: warehouse vs. mediator approach*

Following our 3-step view of data handling, Table 14.1 sums up the functionalities of each approach.

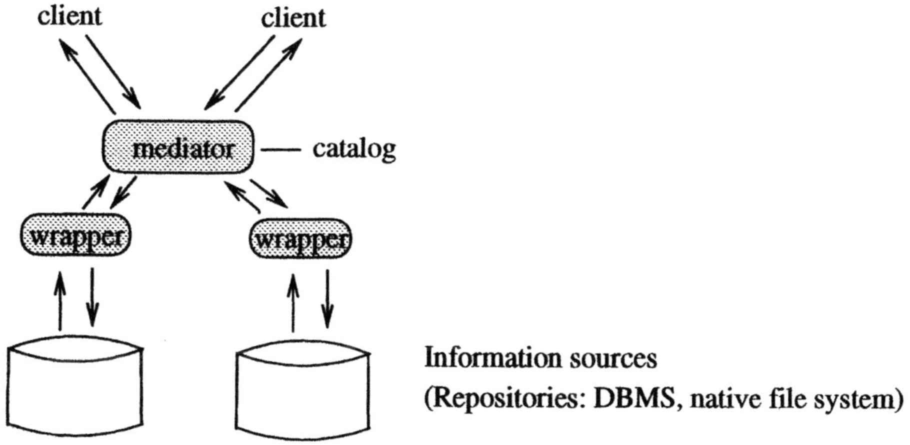


Figure 14.2. Lazy approach to data integration

| | Data Warehouse (eager appoarch) | Mediator (lazy approach) |
|---|---|---|
| Data Collecting | Load/Clean/Consistency Check Aggregation Rules for Base Data | Wrapper |
| Data Organization | Summary Tables/ Base Tables/ Aggregation Rules for Summary Data | Catalog |
| Query Processing | Warehouse Querying | Distributed Query Processing |

Table 14.1. 3-step data handling in eager and lazy approaches

**14.2.3.1. Using a warehouse in geospatial application.** A major interest in warehousing is to avoid converting large amounts of data in and out by having information at hand. This is an important issue in geographic information systems where map transformations are usually complex and time-consuming. From time to time, snapshots are taken from the data origins and loaded to the warehouse (cf. Section 14.2.1).

Clearly, this approach brings new potentials to data handling, with the fact that data can be further processed, for example, annotated, modified, structured, and aggregated. Annotating data is particularly useful in geographic applications: Sometimes, old data cannot be used because information on their origin or on their accuracy is missing. Moreover, historical information can be associated with the data. Techniques borrowed from related disciplines, such as statistical databases (e.g., aggregation or data cube metaphor), versioning, data mining, or OLAP (On-Line Analytical Processing) can greatly improve data handling in this context.

Data warehouses are often used for ad-hoc queries and what-if analysis. What-if analysis calculates different scenarios for the future. This functionality is well supported by data warehouse architectures. It allows new data to be calculated from the already stored data. This predicted data can be labeled differently from the data loaded from the real world. Such analysis would be difficult with the pure mediator approach. Geospatial applications dealing with analysis over time gain from using a data warehouse. In such systems the life-expectancy of data is 5–10 years compared to 60–90 days in operational systems (Inmon 1996). In some applications such as the ones handled by decision support systems (DSS), expected response time is usually relaxed and the order is minutes or hours. Hence the analogy with *quarrying* data.

**14.2.3.2. Using the wrapper/mediator approach in geospatial application.** In contrast, the mediator approach is in particular appropriate when data and data sources change rapidly. It is obviously well-suited to electronic marketplaces, in which prices for goods or services are assumed to alter constantly. However, the delay induced by significant processing needed for data transformation and shipping may make its use inefficient. Moreover, if the same queries are issued multiple times, numerous accesses to data are unnecessarily expensive. This approach is appropriate for near-real-time applications, such as intelligent transportation systems or environmental problems monitoring. Note that it cannot be used if ad hoc querying is not feasible. This approach is inefficient for applications where static data from different sources have to be expensively combined many times in the same way (for instance, joined for a map overlay).

With each object in the warehouse is associated a validity time, whereas in operational systems the data are true at access time. This is one fundamental distinction between OLAP and OLTP (On-Line Transaction Processing), where the snapshot view of data in data sources makes data analyses impossible.

## 14.3. Hybrid approach

Given these two general approaches to data integration, one may wonder about the most appropriate solution for geospatial data handling. Part of the answer is that it depends on the type of application and on the type of data. We place ourselves in a context where data are coming from extremely heterogeneous sources, from a semantical and physical viewpoint. For instance, they can be available freely on the Internet but also purchased from agencies (e.g., governmental institutions). Moreover, data can be either static in the sense that they are unlikely to change over time (e.g., state contour) or dynamic (e.g., traffic flow, population, weather, stock market, currency exchange rates, entertainment programs, or sales data). Note that there are many degrees in dynamicity. A challenge is to integrate these different types of data within one coherent framework. Hence in many geospatial applications a hybrid approach is needed.

This section starts with the basic concepts of such architectures. We then illustrate their use through an example drawn from an application that predicts water floods in a given geographic space. We conclude with a list of qualitative parameters to be considered when designing such flexible frameworks. This list is based on both the type of applications and on the types of data in the sources.

### 14.3.1. Basic concepts

In hybrid approaches, some of the data are collected, processed, and integrated in advance in the warehouse, while other data are fetched when users pose queries against the overall system. The data catalog knows where to find appropriate data (the size of this registry is negligible compared to the volume of the data involved). Figure 14.3 depicts a hybrid solution to data integration. This architecture is a generalization of the two architectures described in Sections 14.2.1 and 14.2.2, respectively. If everything is loaded from the data sources $S_1$ ... $S_n$ to the geodata warehouse, every query can be answered by using data from the warehouse. This approach is the pure first architecture described in Section 14.2.1. If the data warehouse is left empty, it cannot support answering queries. All data necessary for answering queries have to be shipped from the sources, wrapped, and processed by the mediator. This is the pure second architecture as explained in Section 14.2.2.

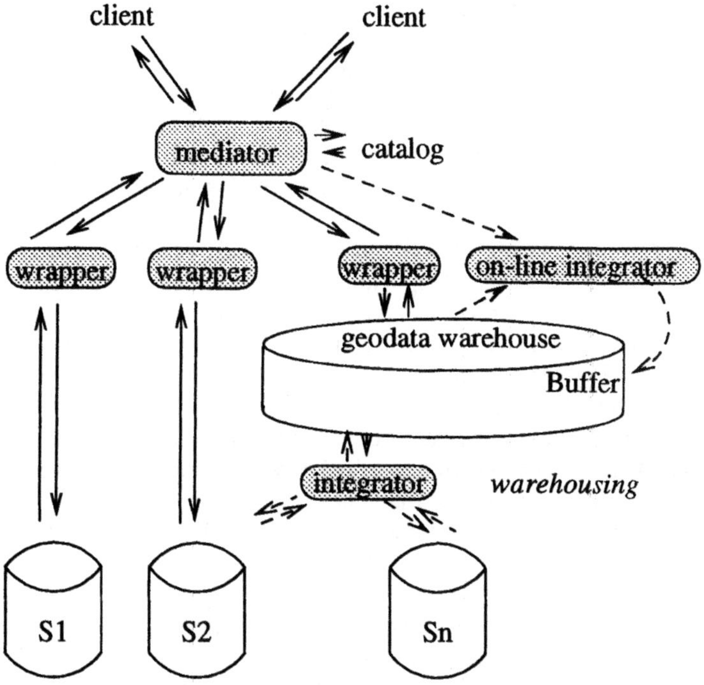


Figure 14.3. Hybrid approach to data integration

To our knowledge, even though they are more and more used in practice, such architectures have not been studied in detail by the scientific community. The problem is obviously more complicated than considering a data warehouse as a regular information source. The warehouse is a special kind of data storage. In contrast to the other information sources, which are fixed and cannot be changed by the system, the warehouse is freely configurable by the system itself. The fact that the system can precompute and store regularly the required data can improve the performance considerably. In addition, other data related to applications and sessions can be stored for further use, which opens new horizons to many geospatial applications.

Many degrees of hybridization should be considered from the pure data warehouse approach to the pure mediator approach, and a major design issue is to determine which parts of each should be considered. A first step is to consider the sources where data may come from. Once the sources with data stored in the data warehouse are determined, the question is whether raw data should be stored or if it is, for space and time reasons, more sensible to store (pre)aggregated data. This is a common issue in pure data warehousing. There are many possibilities regarding data aggregation. First of all, data of only one source can be aggregated, such as the population of counties in one state aggregated in the population of a state. Besides, because of the multiple repositories, data from different sources can be joined and generate aggregated data; for example, the population of cities joined with sales data from other sources to get sales data for cities with more than 100,000 inhabitants. Once data are collected in the data warehouse, the later plays the role of a participating system, hence the wrapper depicted in Figure 14.3 (right-hand side).

In this approach, the data warehouse can be used for more than storing data coming from different sources, which we refer to as application-driven contents. The selection of the data to be stored is done when the system is installed. In practice, such a data warehouse serves many purposes. Following are the three types of data it hosts:

**Application-driven contents (100% manual work).** A warehouse is useful for archiving data related to a whole application field. In order to use it in the long run data are stored in one part of the warehouse. This is useful in geomarketing applications where data, being socio-cultural behaviors or statistical data, are considered over long periods of time. A mechanism enforces the storage of data (even some dynamic data) in order to keep them, and make studies over years for a specific application. These application-dependent data sets are selected by users or system administrators, who decide in particular which tables have to be stored in the warehouse together with the update frequency.

**Usage-driven contents (programmed selection of data).** As mentioned in Section 14.2.1. (query processing in data warehouses), it is useful to monitor the user's queries and to generate user profiles. This is done during the data organization phase. This treatment can be applied to metadata as well, that is, to save user sessions (query graphs) over time. These metadata hold data about typical user behaviors and can be also exploited for analysis. They can also be referred to as session-driven contents.

**Data-driven contents (0% manual work).** First, one could think of abusing the data warehouse as a kind of cache. A cache is a small fast memory which stores the most recent accessed values from a large slow storage. It is based on the assumption that the user will require some of the most recent used data again. He or she can access the data much faster by using the cache than accessing the data from the large slow memory. Second, one could consider the integration of data as they come to the application and not necessarily in advance; after being processed, they are stored for further use within the same application (e.g., retrieval and analysis). Collecting data in this context is more than simply merging data sets. This is the role of the *on-line* *integrator* displayed on top of the geodata warehouse in Figure 14.3. One of its tasks is to check that data are not stored many times.

### *14.3.2. Illustration example*

Let us consider an application for predicting floods of major European rivers, and suppose that we concentrate on rivers that flow through more than one country, such as the Rhine. The overall system (the geospatial data manager) gets information from various repositories located in different countries. Some of these sources are static (e.g., topographic maps) while others are dynamic (e.g., water levels at some stations). To predict floods, we assume the existence of a model that runs on the following data sets: the topography of the landscape (altitude); the water network; the present water levels and speed of flow at specific reading points; and finally the present and predicted precipitation. All these sources are overlaid and the simulation model returns a set of points that are likely to be flooded.

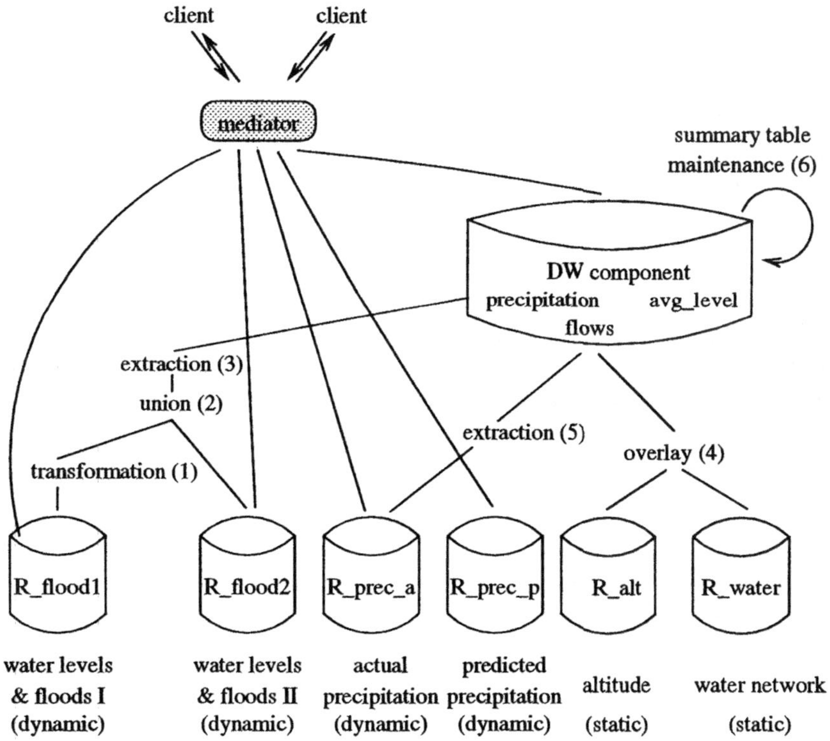


Figure 14.4. Example application

In the following, we start with a more complete description of the data sources. We then describe the major processing steps as well as the new relations to be created. Operations to be performed are given in a pseudo-code whose syntax is supposed to be intuitive. Database queries and schema definitions are given in our SQL-like language.

**Supply of data by different data sources.** As depicted in Figure 14.4, the data sets come from six data sources. A static map with altitude information is stored in *R-alt*. The static map *R-water* contains the water network where in particular the positions of all rivers and other waters bodies are stored. Two dynamic sources (containing the maps *R-flood1*, *R-flood2*) make available the water levels and the speed of flow of selected rivers for different regions, with updates every hour. *R-prec-a* keeps the actual precipitation and *R-prec-p* contains the predicted weather map.

We assume that these maps are stored in relations whose schema is given below. Regarding attribute types, we assume the existence of basic types and of geospatial types, such as point, line, and region (Güting 1988; Scholl and Voisard 1989).Timestamp is a combination of types *time* and *date*.

- R-alt (Location: point, Altitude: real)
- R-water (Rivername: string, Area: region)
- R-flood1 (Location: point, Level: integer, Speed: real) [level in cm, speed in km/h]
- R-flood2 (Location: point, Level: real, Speed: real) [level in m, speed in m/s]
- R-prec-a (Location: point, Precipitation: real)
- R-prec-p (Location: point, Precipitation: real)

We assume the existence of an operation **transform** (Relation-name: string, Attribute-name: string, Old-unit: string, New-unit: string), which, for all tuples of relation Relation-name, applies a unit transformation from Old-unit into New-unit for the value of attribute-name.

**Table creation.** In the data warehouse, two types of relations are considered: Base tables and summary tables. Base tables store the most detailed information handled in the warehouse. The summary tables are generated from the base data. The base tables are Topography (on which the model will be run), Flow, and Precipitation. They are defined as follows:

**create table** Topography (Rivername: string, Location: point, Altitude: real, Rivername: string, Area: region);
**create table** Precipitation (Location: point, Time: timestamp, Precipitation: real);
**create table** Flow (Location: point, Time: timestamp, Level: real, Speed: real);

**Sequence of operations.** The pseudo-code corresponding to the first sequence of operations is given below:

Step 1: Data transformation
R-flood1' := **transform** (R-flood1, Level, "cm", "m");
R-flood1" := **transform** (R-flood1', Speed, "km/h", "m/s")

Step 2: Union of the two dynamic sources R-flood1 and R-flood2
R-flood := **union** (R-flood1", R-flood2);

Step 3: Data extraction from R-flood

The following query is triggered every hour and extracts data from R-flood. The value of attribute *Time* corresponds to the access time.

**insert into** Flow (Location, Time, Level)
**select** Location, Access time, Level
**from** R-flood;

Step 4: Data extraction from the precipitation prec-a
The following query is triggered every 12 hours and extracts the present precipitation from the source prec-a.

**insert into** Precipitation (Location, Time, Precipitation)
**select** Location, Access time , Precipitation
**from** prec-a;

Because data about flows and precipitations are needed for analysis in the system, relations Flow and Precipitation are stored in the warehouse.

For the above **insert** operations the warehouse has to be *off-line*. To avoid switching the system on and off, new data could be stored in a buffer first.

Step 5: Map overlay
map := **overlay** (R-alt, R-water);

Maps R-alt and R-water are assumed to be static. Therefore it is efficient to do the **overlay** only once and to store the resulting map in the warehouse.

Step 6: Maintenance of summary tables
In addition, tables with aggregated values over the base tables are stored in the warehouse. These summary tables can be defined by views, e.g.:

**create view** avg-level(Location: point, Year: integer, Month: integer, Day: integer, Level: real)
**as**
**select** Location, Year (Time), Month (Time), Day (Time), **avg**(Level)
**from** prec-a
**group by** Year, Month, Day
**union**
**select** Location, Year (Time), Month (Time), NULL, **avg**(Level)
**from** prec-a
**group by** Year, Month
**union**
**select** Location, Year (Time), NULL, NULL, **avg**(Level)
**from** prec-a
**group by** Year
**union**
**select** Location, NULL, NULL, NULL, **avg**(Level)

**from** prec-a

**Remark on aggregated values.** If the view above is materialized, it creates tuples with the average level of water at different locations for different time periods. This implies the existence of a function that computes the year, the month, and the day from a given timestamp. It should be mentioned that in this example, there exist **group by** operations over calculated values that are not supported by standard SQL. The extension of SQL with a cube operator to overcome this problem is discussed by Gray et al. (1997). One way to avoid using a **group by** over calculated values is to consider a star schema. In a star schema the dimensions are not directly stored in the fact tables but in dimension tables. The fact tables themselves contain foreign keys to entries in the dimension tables. Fact tables are placed in the middle of dimension tables (hence the name star schema). To get the desired data, the fact table and the dimension table have to be joined and the **group by** is performed on the joined table. These operations can be defined in standard SQL although they are complex to write.

Instead of being aggregated over time, data could also be aggregated over space. Sets of locations (points) could be summarized to regions and, for example, the average precipitation per region could be calculated. It could also be aggregated over both time and space at each possible aggregation level. However, the type of data that can be summarized have to be carefully chosen. Lenz and Shoshani (1997) define a framework and a set of conditions for summarizability of data in statistical databases and OLAP.

### *14.3.3. Configuring parameters: a qualitative approach*

Within this mixed solution, the criteria for using either mechanism are based on infrastructure, that is, on the characteristics of the networks and the collection of data sources, as for instance:

**Nature of data**, that is, whether they are static or dynamic. Dynamic data are less likely to be integrated in the data warehouse because they increase the update cost. In our example, the altitude map and the weather map are static and are stored in the warehouse whereas the predicted precipitation is highly dynamic and is therefore not integrated.

**Cost for transforming (integrating) the data and existence of wrappers and protocols.** High transformation costs can be reduced by transforming the data from the sources to a neutral format in the data warehouse only once. This is an important criterion in geographic applications where transformation of map formats is a costly operation.

**Cost for accessing data.** Repeatedly paying high access cost can be circumvented by loading the data from the sources to the data warehouse, as above.

**Size of the data sets.** The bigger the data set the more it is unlikely to be integrated in the data warehouse unless it can be easily compressed, for example in the case of raster images.

**Speed for data shipping and networks used.** The trouble caused by slow connections to data sources can be alleviated by storing the data in the data warehouse if possible.

**Frequency of access to particular sources.** Some sources are needed more often than others. Sources rarely used should not be materialized in the warehouse.
**Sources availability.** There is no need to integrate sources in a data warehouse if they are always available. But in case of low availability integration should be considered.

On the application side, examples of parameters to be taken into account are:
**Possibility of predicting queries** (e.g., the number of times that queries are issued). If queries can be predicted and if they do not require dynamic (e.g., real-time) data, then it is advisable to precompute them and to store the results in the data warehouse.
**Needs for data analysis** (e.g., time). If data are not stored in the warehouse, data analysis can be tedious and even sometimes impossible.
**Possible joins over many sources.** Joins over different sources can be expensive and time can be saved by precomputing them (even though the results are large).
**Granularity of the fetched information.** If data granularity is high, the results can be precomputed easily and stored in the warehouse.
**Size of the query results.** Small answers in terms of size are more likely to be precomputed and stored in the data warehouse than large ones.
**Required response time.** The shorter the required response time the greater the need for precomputing data.
**Possible compromise on the accuracy of data.** If less accurate data can be used, data should be condensed to save space, which is sometimes a compromise in large geospatial applications.

Systems based on hybrid architectures should be tailored to accommodate the parameters above. A quantitative approach would consist in defining a cost model that takes these parameters into account.

## 14.4. Conclusion

The goal of this chapter was to study solutions to the integration of geospatial data coming from multiple, distributed, and autonomous sources. These sources are moreover heterogeneous, both physically and semantically, and their only common aspect is to store geospatial data. We assumed that the applications we deal with know the sources where the data come from, hence considering transparent access to these sources was out of the scope of this chapter. However, if users do not know where data are coming from, one requires a mechanism to allow transparent access (e.g., that finds appropriate resources and dispatches queries). We presented the two popular database alternatives to data integration, namely the eager approach (data warehouse) and the lazy approach (mediator). We showed that the concept of a data warehouse goes far beyond a simple storage for terrabytes of data. Data are not just copied from the sources to the data warehouse but are also integrated, cleaned, and aggregated. In the wrapper/mediator approach, on the other hand, data are collected on demand and not stored further, or only locally. However, both approaches have limits to handle various kinds of applications, as illustrated by many geospatial examples. We then presented a promising approach based on the two paradigms, and we illustrated its use through an example drawn from a flood control application. We finally gave a list of qualitative parameters for configuring a particular instance of such frameworks.

It is our belief that the future should bring adaptable, scalable frameworks based on the combination of the two common database approaches described above. However, even though these approaches are now well-understood by the community, lots of work remains to be done in the design of such hybrid systems. An important issue concerns the geospatial data warehouse organization itself, both at a logical and at a physical level. For instance, issues such as the adaptability of such systems to various applications and time-space tradeoffs still need to be addressed. Han (1997) shows a first step towards that direction with selective materialization, based on the fact that computing on the fly is expensive, saving all combinations has a huge space overhead and that rough approximations lead to accuracy problems.

Conceptual models to describe hierarchies in many dimensions (e.g., time and space) are also missing. These models encompass aggregation rules in addition to geospatial objects. Work from the statistical database community could be adapted to take the notion of space into account, as initiated by Scholl and Rigaux (1995). Another issue concerns the granularity and the partitioning of data. What level of detail has to be shown? How to break the data into separate physical units that are handled independently? Data can be partitioned by date, organizational units, geography, and clustered according to users' needs. Combinations should also be considered. Classical techniques (e.g., materialized views) are based on redundancy and cannot be used in the context of geographic information because of the huge amounts of data involved.

More related to software engineering issues is the design of the whole system. The ideal situation would be to have a list of quantitative parameters to tailor the system to particular application requirements and to adapt it automatically to users' needs. The input of the function that defines a suitable configuration would be a list of parameters together with possible compromises, as listed in Section 14.3. Its output would be a set of given configurations. This problem, which is related to cost models issues, is currently being addressed.

## Acknowledgments

Support from the German Research Society, Berlin-Brandenburg Graduate School in Distributed Information Systems (DFG grant no. GRK 316) is gratefully acknowledged. We also wish to thank Oliver Günther, Hans-J. Lenz, and Heinz Schweppe for stimulating discussions. Part of this work was carried out while the first author was visiting the International Computer Science Institute in Berkeley, California.

## References

Garcia-Molina H, Hammer J, Ireland K, Papakonstantinou Y, Ullman J, Widom J 1995 Integrating and accessing heterogeneous information sources in TSIMMIS. *Proceedings of the AAAI Symposium on Information Gathering*: 61–64

Gray H, Chaudhuri S, Bosworth A, Layman A, Reichart D, Venkatrao M 1997 Data cube: A relational aggregation operator generalizing group-by, cross-tab, and sub-totals. *Data Mining and Knowledge Discovery* 1(1): 29–53

Güting R H 1988 Geo-relational algebra: A model and query language for geometric database systems. In Schmidt J W, Ceri S, Missiko M (eds) *Advances in Database Technology. Proceedings of the International Conference on Extending Database Technology (EDBT' 88)*. Lecture Notes in Computer Science No. 951. Springer-Verlag, Berlin: 506–527

Han J 1997 Spatial data mining and spatial data warehousing. *Tutorial Notes of the International Symposium on Spatial Databases (SSD'97), Berlin*

Inmon W H 1996 *Building the Data Warehouse*. Wiley Computer Publishing, New York

Lenz H J, Shoshani A 1997 Summarizability in OLAP and statistical data bases. *Proceedings of 9th International Conference on Statistical and Scientific Database Management*. IEEE Computer Society Press

Mimick I S, Quass D, Mumick B S 1997 Maintenance of data cubes and summary tables in a warehouse. *SIGMOD Record* (ACM Special Interest Group on Management of Data), 26(2): 100–110

Scholl M, Rigaux P 1995 Multi-scale partitions: applications to spatial and statistical databases. In Egenhofer M J, Herring J (eds) *Proceedings of the 4th International Symposium on Spatial Databases (SSD'95)*. Lecture Notes in Computer Science No. 951. Springer-Verlag, Berlin: 170–183

Scholl M, Voisard A 1989 Thematic map modeling. In Buchman A, Günther O, Smith T R, Yang Y F (eds) *Design and Implementation of Large Spatial Databases, First Symposium SSD'89*. Lecture Notes in Computer Science No. 409. Springer-Verlag, Berlin: 167–192

Widom J 1995 Research problems in data warehousing. *Proceedings of the 4th International Conference on Information and Knowledge Management (CIKM'95)*

Wiener J, Gupta H, Labio W, Zhuge Y, Garcia-Molina H, Widom J 1996 System prototype for warehouse view maintenance. *Proceedings of the ACM Workshop on Materialized Views: Techniques and Applications*: 26–33

# 15 GEOSPATIAL MODELING: A CASE STUDY FOR A STATEWIDE LAND INFORMATION STRATEGY

David Pullar and Kristin Stock

## 15.1. Introduction

This chapter goes beyond the exchange of spatial information to look at how tools for manipulating that spatial information are commonly understood and transferred across application boundaries. In this respect interfaces for spatial information systems are examined at a conceptual level. We are more concerned with the expression of user requirements and system structure than the component level interfaces between system modules.

The chapter reviews data modeling methodologies for GIS-specific applications. In everyday terms we are interested in specifications, design artifacts, and database schemas for land information systems. Two forces motivate our work. Firstly, we believe appropriate attention has not been given to design and analysis methodologies within the spatial information industry. A great deal of attention within the GIS scientific community has focused on spatial query operators and spatial access methods. By comparison, research into application level design has been lacking. This is despite the fact that more people would be drawn into defining system requirements and design objectives for a GIS application than would be tasked with building the system. A second motivating force is that most user requirements are either expressed or understood to relate to a small set of concepts, for example maps, coverages, networks, etc. While new paradigms will emerge, we believe that only a few key concepts relating the geographical context are needed in conceptual modeling. These model elements define a high-level system with application-specific semantics.

This chapter limits its consideration of geospatial modeling to developing applications for a statewide land information system. We believe this to be an important example because of organizational and system requirements. The GIS applications discussed operate within an organizational structure where many people participate in the product design, that is, data custodians, resource managers, end users, etc. The system requirements are also demanding as most information products need to provide a continuous service that operates across a distributed and heterogeneous computing environment.

The chapter will:

- review the applicability of object-oriented modeling and design (OOMD) methods for GIS;
- describe and test the usability of a special geospatial modeling notation; and
- describe interoperability requirements for land information products.

The outline for the chapter is as follows. The next section describes the setting for the development of information products in a statewide land information system called the Queensland Spatial Information Infrastructure Strategy (QSIIS). Section 15.3 discusses object-oriented modeling and design (OOMD) methodologies in relation to modeling techniques. Section 15.4 reviews specific OOMD modeling languages for spatial classes. Section 15.5 describes how a general OOMD modeling language may be extended to include spatial requirements. Section 15.6 describes this in the context of information products being developed for QSIIS. The conclusion discusses the usability of our geomodel and its applicability to providing interoperable tools in a statewide networked information system.

## 15.2. QSIIS interoperability environment

The Queensland Spatial Information Infrastructure Strategy (QSIIS) is a plan for providing spatial information services to one Australian State. It is largely driven by government departments, and its aim is to provide 40 information products that are defined by the collection, maintenance, and use of spatial data sets by each department. This is based on the assumption that the departments involved (13 out of a total of 18) are representative of the main purchasers and providers of spatial information (Alexander-Tomlinson 1997).

Each of the 40 information products is developed to fulfil a particular set of requirements. For example, the property interests analysis product (currently under development) is concerned with land administration. Currently, to obtain full information about a property, users must search across multiple sites holding land information records, with related issues in timeliness; incomplete or non existent electronic records and difficulties in knowing where to search. The property interests analysis information product aims to allow single point access to all property based information. This is expected to provide benefits including:

- increase efficiency and accuracy in assessing land values and setting rents for government land;
- simplify bureaucratic processes for registration of interests in land (with implications for decision-making on land use);
- make information about encumbrances (i.e., easements, flooding, contamination, pest control, native title, etc.) available to determine constraints on land use (Alexander-Tomlinson 1997).

The immediate concern of system designers is to make information available across a number of different sources by providing interoperability. However, as indicated by the above list of requirements, for users the concern is more focused on the outcomes of the availability of information, in terms of what economic and administrative impact it might have.

## 15.3. Information modeling

We need to build a model for a GIS when it is too difficult to comprehend the system in its entirety. As the complexity of the system increases, so does the importance of good modeling techniques. Some essential factors for geographical application development are the use of a standard application design methodology (Hadzilacos and Tryfona 1996), and a model specification using a rigorous modeling language. Geographical applications make use of mathematical models of space and their representation, including geometry, map projections, cost surfaces, statistical rendering, etc. If modeling space from a mathematical perspective is considered to be an open problem then it is not surprising that geographic application methodologies are also open to various interpretations. This is evident in the constant tension between users who want semantically expressive models for their particular application domain and vendors who need to provide general-purpose tools. So given that it is a fallacy to believe in the existence of a *universal data model* for geographic reality (Morehouse 1990), it seems appropriate that we concentrate our energies on describing metamodels from which we can build different models.

So what are metamodels? If models are representations of some reality then metamodels are descriptions of models. Metamodels use a rule language to define all well-formed models that may be represented. Likewise to describe a metamodel flexibly, in theory, sometimes requires a meta-metamodel. Architecturally this is shown in Figure 15.1. A meta-metamodel defines an abstract language for specifying metamodels. A metamodel is an instance of a meta-metamodel, which is a language for specifying models. As with any modeling process, the model cannot be expected to provide complete knowledge of its subject, but a good model should provide a reasonable interpretation of the real situation.

The meaning of the abstract concepts in Figure 15.1 may be understood using some real-world illustrations. These examples are adapted from Crawley et al. (1997). A meta-metamodel would be a type system for CORBA IDL, then the metamodel would contain schemas of CORBA IDL types and relationships between them, and the entities in the model would be CORBA objects. Alternatively, if the meta-metamodel defined a language system for design notations, then the metamodel elements would be specific notations, and the entities in the model would be design diagrams.

We want users to construct object models for geographic applications. The metamodel language we chose to describe these models is the *Unified Modeling Language* (UML). It represents the best practices in the object technology industry derived from leading object-oriented methods, including OMT (Rumbaugh et al. 1991). UML (1997) is "a language for specifying, visualizing, constructing, and documenting the artifacts of software systems, as well as for business modeling and other non-software systems." UML's modeling language includes:

- Model elements—fundamental modeling concepts and semantics;
- Notation—visual rendering of model elements;
- Guidelines—idioms of usage within the trade.

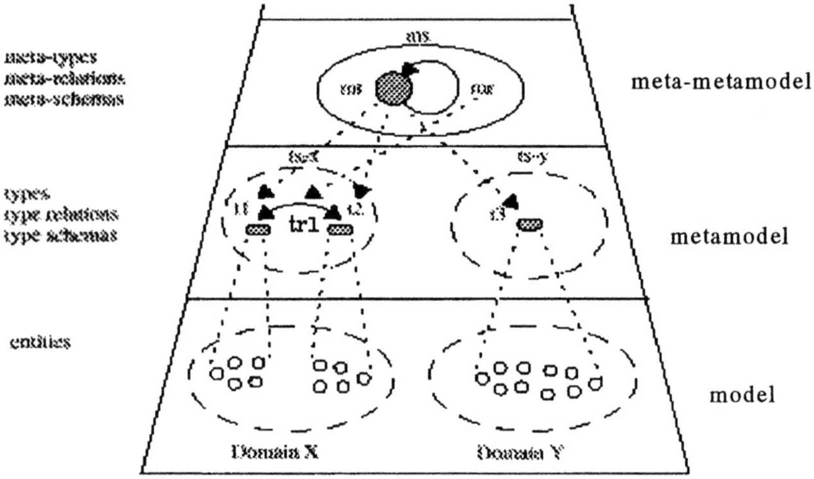


Figure 15.1. Meta-metamodel, metamodel, model architecture (Source: Crawley et al. 1997)

The most prominent aspect of UML is its visual notation. It includes a method to capture the dynamic aspects of a model, such as interactions, collaborations, and state histories of objects in a system. A static *class diagram* is used as an integrating framework for the system specification. The class diagram shows a collection of declarative (static) model elements, such as classes with their contents and relationships. UML has the advantage that it is relatively easy for professionals to understand and to facilitate their involvement in the design process. But it still includes formal definitions for the structural aspects of class diagrams and a set-theoretic language for expressing well-formedness constraints.

A brief description of some modeling elements for UML class diagrams is shown in Figure 15.2. *Class* symbol is a composition of a class name, its attributes and operations. A binary *association* is shown as a line connecting class symbols. The end of an association where it connects to a class is called an *association role*. Roles signify important aspects of associations including multiplicity, ordering, qualification, navigation, and aggregation relationships between classes. Association paths are adorned with an association name and constraints. A *constraint* is a semantic relationship among model elements that specifies conditions and propositions that must be maintained as true. It is shown as a text string in braces { }. All model elements that include text strings (such as class names, attributes, association names, constraints, etc.) represent information that has both syntax and interpretation. We have presented just a small subset of UML notations, a detailed description can be found in the UML documentation set (UML 1997).

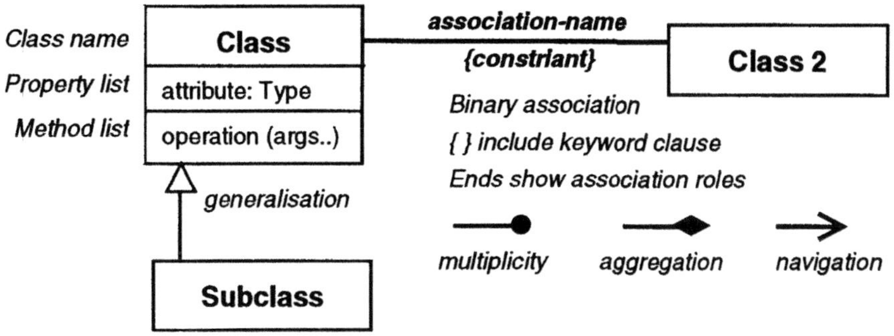


Figure 15.2. Notations for several model elements in UML

UML also includes language extensions for specifying process-specific models, for instance to define a special type of association between classes. These metamodel constructs are sufficient for a user to create a customized specification of an application-specific domain, without having to resort to a meta-metamodeling layer. The next section describes GIS-specific models that have been proposed by various researchers. Sections 15.5 and 15.6 illustrate how a geospatial data model can be defined using the UML extension mechanisms.

## 15.4. Literature review

A number of modeling methods have been advocated for GIS application development, most notably those using an OOMD model (De Oliveira et al. 1997). A recent criticism of object-oriented modeling is that there are no standard practices or commonly accepted conventions. Worboys (1994) examines the causes for the proliferation of competing models, all claiming to represent a true object-oriented modeling methodology. The main cause is the lack of general consensus on the key features and definitions that constitute an object-oriented model.

In recent times, the debate over object-oriented data models has been defused somewhat by distinguishing between object-oriented modeling and design (OOMD) methods, and object-oriented programming (OOP) languages and databases (Booch 1996). It is important to make the distinction between modeling concepts and implementation tools. OOMD provides the method for analyzing problems and framing them in terms of objects. The benefits of OOMD for an application can be realized without the need for an OOP implementation, but using OOP to build system applications has obvious advantages. Similarly OOMD can be implemented on a non-object oriented database management system.

Concepts for OOMD have been applied to GIS in a number of different ways. Three approaches are reviewed:

- GeoOOA, which is an extension of the object-oriented model (Kosters et al. 1997)
- GISER, which is an extension of the Enhanced Entity Relationship Model, itself an extension of the Entity Relationship Model to include some object-oriented concepts (Shekhar et al. 1997)
- GMOD, which is an extension of Camara et al.'s (1994) model, itself an extension of the object-oriented model (De Oliveira et al. 1997).

The remainder of this section reviews these three models in terms of four basic questions:

- What problems are the models intended to solve (that is, why were they developed)?
- What representational semantics do the models include?
- How do the models handle space and time?
- What metamodel characteristics do the models provide?

### *15.4.1. Problems addressed by the models*

The three modeling methods differ in the problems they are intended to solve, or the justification for their existence. GeoOOA defines the following problems with conventional object-oriented modeling methods:

- they do not support the user's view sufficiently;
- no easy way to distinguish between spatial and non-spatial classes;
- no obvious way to determine the type of a spatial class (that is, point, line, or region);
- no distinction between topological and conventional relationships;
- important topological constraints are hidden in textual specifications; and
- frequent topological constraints lead to redundant textual specifications (Kosters et al. 1997).

GISER's main contribution is to address the problem of the unification of field-based and object-based spatial data model approaches. It includes concepts for the discretization of field-based models using an interpolation technique, as well as procedure-valued attributes and support for the entire GIS process (input, modeling, manipulation, and presentation; Shekhar et al. 1997).

Finally, GMOD's contribution to the use of object-oriented modeling for geographic objects has some of its justification in common with GISER. It attempts to isolate users from implementation details, allowing the inclusion of semantics (De Oliveira et al. 1997).

### 15.4.2. Representational semantics provided by the models

The term *representational semantics* defines how well a model expresses static, dynamic, and behavioral aspects of an application. Models that describe the workings and use of application objects are more meaningful. GeoOOA, for instance, provides a graphical notation to express conventional OOMD-model representations (e.g., object class, generalization, aggregation). But it includes specialized class descriptions for spatial primitives and specialized association relationships for topological whole–part spatial structures. The model expresses relevant spatial information directly within the model, including the syntax and semantics for all GeoOOA primitives and their standard attributes, services, topological relationships, and constraints. It is visually explicit in that it ensures that the objects and their intent are visible (Kosters et al. 1997).

GISER extends the Entity Relationship (ER) model to include discrete and continuous representations of space. Conventional ER-model notations (e.g., entity and association) are enhanced to express geometry and topological properties. GISER's graphical notation provides a functional view of geographic phenomena as input, data modeling to extract spatial information content, manipulation, and result presentation (Shekhar et al. 1997). Apart from this methodology there are no mechanisms to express the behavioral aspects of a geospatial model.

GMOD uses an OO-model notation with special notations for geospatial information (Camara et al. 1994), temporal dimension and causal relationships (De Oliveira et al. 1997). The geospatial classes are not immediately obvious from the example diagrams other than by class name. Detailed class semantics are described for each geospatial class using formal specifications. GMOD allows progressively more detailed models to be developed in the typical style of OOMD (Rumbaugh et al. 1991).

### 15.4.3. Model concepts of space and time

GeoOOA (Kosters et al. 1997) defines geoclasses for point, line, region, and raster. These classes encapsulate specific geometric behavior and topological constraints. The model defines special composition notations for the network structure (including edge links and node junctions). In addition, it provides an association for whole–part composition, including covering, containment, and partition relationships. Spatio-temporal objects are also supported using a timestamping function.

GISER (Shekhar et al. 1997) represents the continuous field class using space–time extents, from which features (both names and unnamed) are identified in space. These may then be discretized to spatial objects in a coverage. The model also supports derived features and relationships.

GMOD (De Oliveira et al. 1997) divides its classes into conventional classes and geoclasses. Geoclasses can be either geoobjects (object-based view) or geofields (field, or continuous view; Camara et al. 1994). Both geoobjects and geofields have an attribute which describes their georegion (geometric location). Time is included through the definition of a time class, which any other object can have as a component (De Oliveira et al. 1997). GMOD includes two temporally-based relationships: versioning and causation.

### *15.4.4. Metamodeling characteristics provided by the models*

The questions examined above in relation to each model indicate that there are significant similarities between the models in terms of the problems they address, the solutions they pose, and their concepts of space and time. The differences in these three areas relate mostly to the representational semantics used to express geospatial classes and relationships. Some were based upon ER models and others on OOA models. But all advocated the need for special treatment of geospatial concepts in the modeling language.

Some of the modeling approaches included the development of a CASE tool, GeoOOA being the most notable and having an open software development environment (Kosters et al. 1996). Bot none of the papers discussed how geospatial models could be shared or communicated across software development environments for GIS. To express and communicate the semantics for special modeling concepts requires an abstract language, namely a metamodeling language.

None of the models reviewed provide metamodeling tools. The provision of metamodeling concepts and tools removes the dependence on proprietary systems, as the tools provided at the metamodel level support specific geospatial models. The next section shows how UML is used to define geographic models using metamodeling language extensions.

## 15.5. GIS application-specific extensions to UML

The UML (1997) metamodel includes built-in mechanisms to facilitate domain-specific extensions to its metamodel without needing to resort to a meta-metamodeling layer. These are essentially variants of the core OOA modeling elements (i.e., class and association) and their semantics (i.e., constraints) that can be tailored for specific application areas. Two important extension mechanisms supported by UML include *stereotypes* and *constraints.*

Stereotypes may extend the semantics, but not the structure, of pre-existing types and classes in the metamodel. Certain stereotypes are predefined in the UML; others may be user-defined. The general presentation of a stereotype is to place the name of the stereotype within matched angle brackets «*stereotype name*», or to depict it by a graphic icon appropriate for the model element being described.

Constraints are semantic conditions on the relationship between model elements expressed as a text string. Certain constraints are predefined in the UML; others may be user defined. UML does not prescribe the language in which the constraint is written, but ideally a process-specific constraint is described in a formal language with a specified syntax and interpretation. A constraint is shown as a text string in braces { }.

It is conceivable that the models described in Section 15.3 may be specified using the process-specific extensions from UML. Take GeoOOA for example. Stereotypes may be used to describe *geoclasses* with an identical graphical syntax. Topological *whole–part* structures, including specializations for *covering*, *containment*, and *partition,* may be defined as association stereotypes. This would appear as a text string (as yet there is no graphical syntax in UML for user-defined association stereotypes). The *abstract network* structure may be expressed as a constraint between association paths for *network-link* classes and *network-node* classes. The semantics for geometric standard services may be completely specified using behavioral model aspects of UML.

Therefore UML is capable of expressing concepts of space and time using special geomodeling constructs. It provides the same desirable features as GeoOOA, such as visible distinction of spatial classes and explicit topological constraints. Both syntax and semantics can be expressed in a mathematical language to enforce domain information. UML also considers a supplementary methodology (i.e., guidelines) as an essential part of building models for complex systems. The additional benefits of using UML are that it allows tailoring of the metamodel to allow for variant models, and provides a mechanism for model interchange and interoperability.

### *15.5.1. Example UML geomodel*

As an example of an UML metamodel we relate our own experience in developing geographical applications for statewide land products. At a conceptual level we noticed product requirements made few references to spatial concepts. Most users described their needs in terms of the geographical scope of an application that controlled sets of spatial features having common thematic properties. We distinguished three high-level generic concepts: (1) a *spatial realm* to define the context and interaction for geospatial information, (2) *spatial themes* to define passive states of features and coverages, (3) *temporal events* affecting the information. It was only in the later stages of analysis that details for features such as map projections, scale, dimension, topological relations, and accuracy were expressed. Figure 15.3 shows stereotypes developed for these modeling elements.

### *15.5.2. Usability of the UML geomodel*

The success of a metamodel can be measured by how well it communicates the meaning of a model to different groups. To measure how well our geomodel is understood we conducted usability tests (Rubin 1996). A small reference group of several people with knowledge of OOMD techniques and some GIS application development experience were surveyed. The geomodel was presented along with two case scenarios, and respondents were questioned on: (1) how well the model matched requirements, (2) how well they understood elements in the model, and (3) if they could proceed to an implementation. Clearly most respondents understood the significance and meaning of the geomodel elements, although responses did vary on the level of understanding. Also of interest, we found that most respondents felt overwhelmingly that an interaction diagram, which describes how groups of objects collaborate for a particular case scenario, significantly helped in interpreting the class diagram specifications. In other words, showing the behavior of objects as they are used gives users greater insight than class description or textual specifications. Interaction diagrams should be given greater consideration as a modeling technique.

The last set of questions of the survey judged the usefulness, effectiveness (ease of use), and learnability of a geomodeling technique. This received a very positive response, indicating that the geomodels helped to express and communicate GIS application requirements. Furthermore, respondents felt that the technique could be readily applied to domain applications in their own discipline.

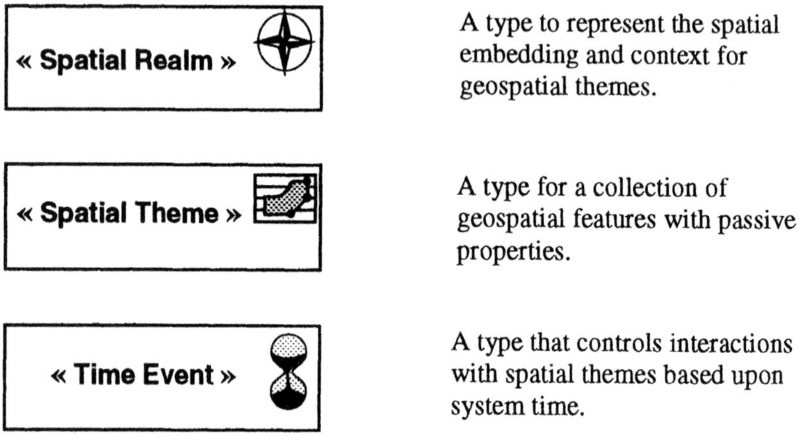


Figure 15.3. Class stereotypes for the geomodel

## 15.6. Example pest information application

We describe a sample application dealing with land pest infestations to demonstrate the use of these geomodeling elements. We have based this sample application on a real system called *PestInfo*, an information product owned by the Department of Natural Resources (DNR) of the Queensland Government in Australia. The requirements for the pest infestation information system are as follows. Plant and animal pest *infestations* for a land *property* are recorded by an *inspection*. A land officer may visit a site several times to record the state of an infestation, including its spatial extent measured using a GPS traverse. Land officers enter pest information into their field system, which is periodically synchronized with the central office system. A *registry* is used to coordinate this exchange. Figure 15.4 shows part of the class diagram. A complete requirements analysis (not shown here due to space limitations) includes attributes and operations in the class diagrams, and behavioral modeling diagrams for interactions and collaborations between objects.

Even this simple example embodies several complex interactions and interoperability considerations. Firstly an inspection acts as an *event* to record the current *state* of a pest infestation. A time-event-modeling stereotype based upon concepts for temporal systems (Langran 1992) is used. Synchronizing the field and central office systems requires a registry to connect and interchange information over a networked connection. The central office system could be viewed as a replicated database with the registry acting as a replication manager to control authorization and upload and download of data. The sample application described was developed using Arcview from ESRI. Our work suggests that the extended geomodel from the system design can be translated in a very mechanical fashion into a prototype application.

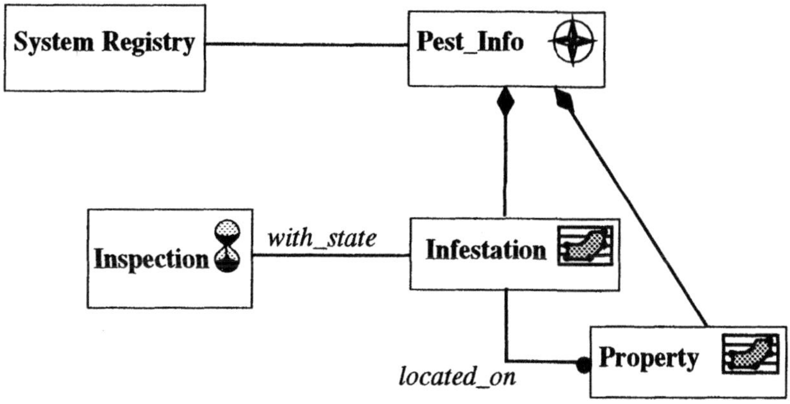


Figure 15.4. Geomodel for sample pest information system

### *15.6.1. Land information as a Yellow Page service*

UML compels a system level view of application problems, and defines a model as a semantically-closed abstraction of a system. Therefore system-specific terminology is introduced even during requirements analysis. This bias is well-suited to a requirements analysis for information products described in the Queensland Spatial Information Infrastructure Strategy (QSIIS). Unlike environmental planning and control applications where there is a high level of semantic information related to the real-world process (De Oliveira et al. 1997), most of the applications for QSIIS reflect information access, and fiscal and institutional considerations about land ownership and land assets.

Applications like the pest information system in the future will be part of an online directory. A trial technology architecture was explored in 1996 that linked directory access to several GIS databases. The objective of the trial linkage was to help identify standards and protocols to support interoperable access to spatial information systems (QLIS 1996). It is simple to connect two sites, but linking several sites with various client users and services adds significantly to the complexity. The trial architecture explored linking several sites through a *Yellow Page* directory. This is implemented as a broker that maintains a directory catalog of services and providers, which it can look-up in response to user requests (Buschmann et al. 1996). Subscribers, such as PestInfo, advertize and unadvertize their services (Figure 15.5). This is a very dynamic environment where the QSIIS broker adds and removes Yellow Pages listings as instructed, and also knows how to dispatch user requests to the appropriate service. All user requests are processed through the QSIIS directory and not directly to the service's application. The connected client broker and server broker manage service transactions. The transaction protocol begins by a user request. The QSIIS broker sets up a communication session to the listed service. The client broker will advise if it can answer the request giving the cost, in time and monetary units, to fulfil the request. The server broker relays this information back to the user; once accepted, the server broker using the same session identifier communicates again with the client broker to complete the request. The QSIIS directory records the details of the transaction for business accounting purposes.

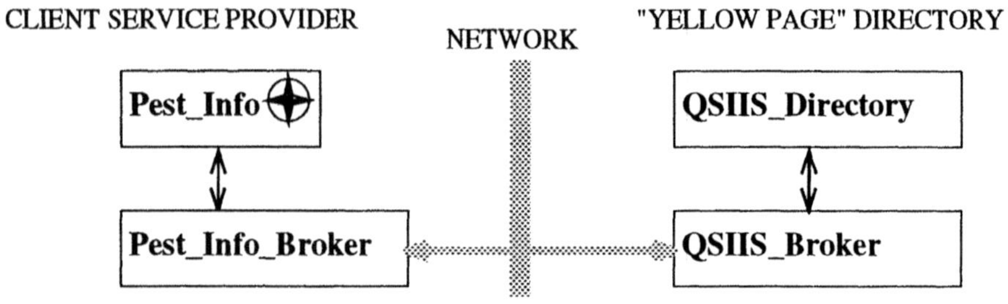


Figure 15.5. Client–server object classes to list PestInfo services with Yellow Page directory

A prototype system was implemented in C++ using Tuxedo middleware software on a UNIX Server to control multidatabase access. Arcview's Shapefile format was used for interchange of spatial information. It is understood that a proposed system would provide greater dynamics for providers to list and deploy services. A review of this trial confirmed the need for high-level interoperable components that adhere to standard protocols. In particular, the particular needs of spatial information, its access methods and interpretation, were seen as depending upon future standards efforts like OpenGIS (QLIS 1996).

It is expected that GIS vendors will soon support OpenGIS interface specifications, but these will be relatively low-level component interfaces and do not solve the need to share high-level definitions of geospatial information between user communities. We recognize the need for metamodeling to define geomodels. Interoperability between these geomodels will rely upon well-defined standards and protocols, just as interchange of data relies upon standard formats. A model interchange format should try to balance its formal specification with readability.

## 15.7. Conclusion

We have defined several metamodel elements used in the system analysis and system design. We have undertaken a usability study and end-user surveys to ascertain their utility, and received very positive results. It is our belief that application design operates in a collaborative environment between end users, experts, and software engineers. The usability tests indicated that intuitive and readable models are desirable. We are using three criteria to judge overall readability: usefulness, effectiveness, and learnability (Rubin 1996).

The future of QSIIS is to support an open system marketplace for access to spatial information and special information services. This will include an interoperable environment where providers list new and integrate existing information products. The machinery for building such a system is not available today, or at least the technology is in its infancy. The strategic development of QSIIS, and similar land information systems, will depend upon communicating meaningful data models between networked systems.

This chapter has discussed spatial data models with regard to application development methodologies. We have reviewed several methodologies, each addressing slightly different problems and user requirements. Each has advanced a geomodel they feel incorporates the necessary concepts for space and time, and representational semantics. But each is slightly different. Therefore it is apparent that there is not one geomodel for all information communities. We have explored how metamodels can be defined using an industry-accepted modeling language. We are currently testing the usability of a geomodel for system analysis and system design purposes.

We believe the standards effort for interoperability should progress at two ends of the software engineering spectrum. As is occurring now, the lower end will deliver common data types used to define spatial information classes. Specifications for such data types rely upon mathematical principles for spatial data types. They will provide the basic building blocks to describe the spatio-temporal characteristics of spatial information. The middle part of the spectrum includes defining spatial data models to suit real world situations. This part should retain a high level of user input. In other words, system designers should have the opportunity to define their own geomodels (or metamodels) for an information community. This challenges the standards effort to adopt meta-metamodeling facilities to allow designers to describe their own metamodels.

## References

Alexander-Tomlinson 1997 *Queensland Land Information Strategy (QLIS) Benefit Study. Volume II: Situation Report (Assessment of Government and Industry Spatial Information Requirements)*

Booch G 1996 *Object Solutions: Managing the Object-Oriented Project*. Addison-Wesley, Menlo Park, California

Bushmann F, Meunier R, Rohnert H, Sommerlad P, Stal M 1996 *A System of Patterns*. Wiley, England

Camara G, Freitas U M, Souza R, Casanova M, Hemerly A, Medeiros C 1994 A model to cultivate objects and manipulate fields. *Proceedings 2$^{nd}$ ACM Workshop on Advances in GIS*: 20–28

Crawley S, Davis S, Indulska J, McBride S, Raymond K 1997 Meta information management. *Formal Methods for Open Object-based Distributed Systems (FMOODS) Conference, Canterbury, UK, July 1997*

De Oliveira J L, Pires F, Medeiros C B 1997 An environment for modeling and design of geographic applications. *GeoInformatica* 1: 29–58

Hadzilacos T, Tryfona N 1996 Logical data modelling for geographical applications. *International Journal of Geographical Information Systems* 10(2): 179–203

Kosters G, Pagel B, Six H 1996 The GeoOOA tool and its interface to open software developments for GIS. *Proceedings of the 4th ACM Workshop on GIS, Rockville.* ACM Press: 165–173

Kosters G, Pagel B, Six H 1997 GIS-application development with GeoOOA. *International Journal of Geographical Information Science* 11(4): 307–335

Langran G 1992 *Time in Geographic Information Systems.* Taylor & Francis, London

Morehouse S D 1990 The role of semantics in geographic data modelling. *Proceedings 4th International Symposium on Spatial Data Handling, Zurich*: 689–698

QLIS 1996 *QLIS Project Report: Technology Architecture Research & Development Project.* Queensland Land Information Council, State of Queensland Department of Natural Resources, Australia

Rubin J 1996 *Handbook of Usability Testing: How to Plan, Design and Conduct Effective Tests.* John Wiley & Sons, New York

Rumbaugh J, Blaha M, Premerlani W, Eddy F, Lorensen W 1991 *Object-Oriented Modeling and Design.* Prentice-Hall

Shekhar S, Coyle M, Goyal B, Liu D, Sarkar S 1997 Data models in geographic information systems. *Communications of the ACM* 40(4): 103–111

UML 1977 *Universal Modeling Language (UML) Document Set.* Rational Software Corporation, Santa Clara. *www.rational.com/uml*

Worboys M F 1994 *GIS: A Computing Perspective.* Taylor & Francis, London

# 16 INFORMATION BROKERS FOR A WEB-BASED GEOGRAPHIC INFORMATION SYSTEM

Ian Finch and Eleanor Small

## 16.1. Introduction

This chapter presents a prototype system which allows the geographic information user to locate data efficiently. By adapting ideas from business systems research (particularly from the field of multi-agent systems), a distributed network of information brokers has been designed. These information brokers understand geographic metadata and use it to communicate with the user through a graphical user interface (GUI). This approach removes many of the disadvantages of general Web search engines (which have no understanding of metadata). Furthermore, it allows specialised information brokers to be created, which can be situated in the best part of the Internet for their user community. This overcomes a major problem with current geographic metadata servers, where a server in America becomes unusable from Europe at most times of the European working day.

Furthermore, by adopting a distributed approach, different information brokers can use different metadata standards allowing richer communication between users of a given information broker. However, access to information using other metadata standards is still possible, since a broker can act as a mediator, finding common metadata and converting from that common vocabulary to the specialised metadata standards.

The power of this approach arises from the fact that the information brokers (which have been coded in Perl in the prototype system) and the GUI (coded in Java) speak a common language (KQML) and can therefore easily exchange information. Furthermore, all the participants share a common ontology which ensures that when they refer to some entity (for example a map) during a conversation, they both have the same understanding of what that entity is and how it can be manipulated.

A method has also been developed for allowing the information brokers to gather their metadata from the Web and this too will be described here.

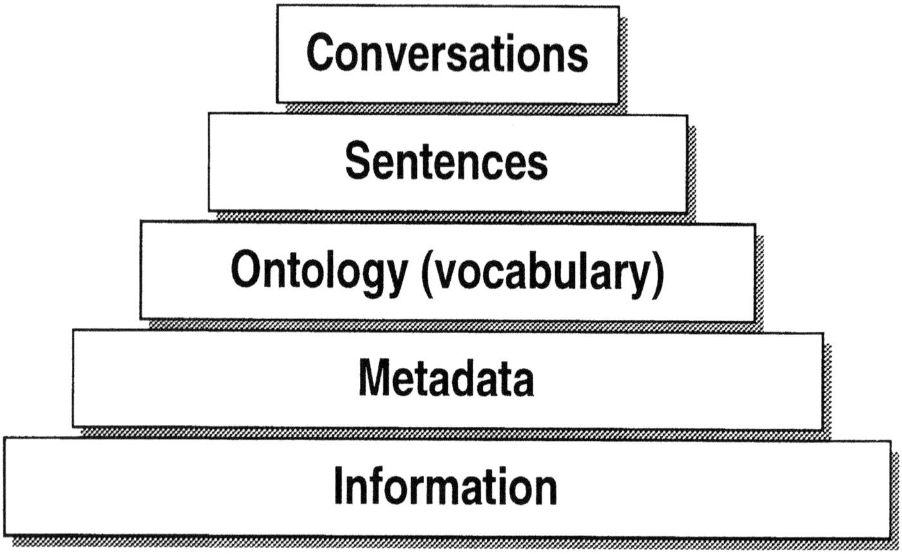


Figure 16.1. The five layers of information

### 16.2. From metadata to a common language

In order for efficient retrieval of geographic information, the search engine (or other agents being used) must understand the data (for example, by its associated metadata). Furthermore, a means must be defined by which exchanges of information can be conducted using the data and metadata. The system described here defines five such levels of information, each layered on top of the other. These are shown in Figure 16.1.

Information is, quite simply, the geographic data which may be of interest to a user. This could be, for example, population statistics or an image of a political map of some region. Metadata is the information about this geographic information. Continuing with the map example, this would include the location and extent of the area to which the map refers, together with information about the projection used for the map. The ontology enables the various agents (brokers and GUI) to communicate. Just as for human communication, it is essential that all participants mean the same thing by each word they use. Each word in the ontolgy will therefore be known by each agent and there will be a single definition of its meaning. The word *map* will thus refer to an image, together with metadata giving its location, extent, and projection. Note that both the image and its metadata must be present for some entity to be a map. Thus, if an image is a map to a human, but has no metadata associated with it, then it cannot be a map as far as the agents are concerned. Conversely, an image may not be a conventional map to a human, but as long as it has the correct metadata associated with it, then the agents will consider it to be a map.

Moving on from the ontology, it is necessary to create rules which specify how the words from the ontology can be combined to create valid requests and answers. These requests and answers—sentences—make it much easier for the agents to understand each other. The rules restrict the infinite number of combinations of words to a very small set. Finally, to allow exchanges between agents to take place easily, it is necessary to have rules governing how sentences can be combined. These rules allow standardisation of the conversations between agents. For example, one valid conversation may be a request for information from one agent followed by a rejection of the request by another agent.

To clarify these levels of information and their relationship with each other, the following section discusses how the types of information were represented in the prototype system, using the Knowledge Query and Manipulation Language.

## 16.3. The knowledge query and manipulation language

The Knowledge Query and Manipulation Language (KQML) was developed by DARPA to allow agents to exchange information in a standard manner (Finin et al. 1993; Labrou and Finin 1997). It has been used by many researchers in the fields of artificial intelligence and distributed systems, and has proved to be both powerful and versatile.

The basic idea behind KQML is that agents send each other *performatives*; instructions to perform some operation (such as evaluating an expression and sending back the result). These performatives are encoded as lisp expressions for easy parsing, and sent as simple ASCII strings. A performative consists of a keyword indicating the type of performative it is (an *ask-one* performative, for example, asks an agent to evaluate an expression and return one answer) and then a series of key/value pairs indicating the language (or protocol) which is to be used, the ontology to use and the actual content of the performative (which in the case of the ask-one performative is the expression to be evaluated). As an example, one agent might send the following performative to another agent asking it to perform an addition:

```
(ask-one :language arithmetic
         :ontology general
         :content "1 + 7")
```

To show how KQML is used in the prototype system, it is first necessary to see how the five layers of information discussed earlier can be represented in KQML. Since all data must be available on the Web, each piece of data can be uniquely described by its Web address (that is, its URL). Metadata must then be associated with these data. The actual method for implementing this would be up to each agent in the system. There is no need to standardise this across the system, provided that each agent has the appropriate metadata available for it to use or pass on.

In order that the agents can communicate effectively, they need to share an ontology, knowing, for example, the metadata that is associated with a satellite image. Thus, one agent (the user's Web browser, for example) can request a map from another agent (an information broker, maybe), confident of obtaining the correct response. As an example, consider the map. In the system discussed here, a map is an image which can be manipulated by the user, allowing points and areas to be specified. Thus, the user could call up a map of Europe, draw a box around the British Isles, and then zoom in on this area, perhaps automatically loading a new map with more detail (rather than simply scaling the area in question). It is obvious, therefore, that a map is more than a simple image; there must be associated metadata linking the pixels in the image to a geographic location reference system.

So, within the ontology, a map is an image, together with metadata which allows it to be linked to the real-world area it represents. A simple scan of a page from an atlas, with no associated metadata, is therefore not a map, but just an image. Conversely, a radar image which has the appropriate metadata associated with it is a map. The ontology as a whole is given a name (for example *gis*) and this can then be used as the value for the ontology key in the performative, to ensure that a common ontology is being used.

The next level above the ontology is the sentence level, combining concepts from the ontology into sentences. A sentence in English might be 'find all maps of Liverpool'. This would become a KQML performative in this system, for example:

```
(ask-one :language lisp
         :ontology gis
         :content (find :type map
                        :area liverpool))
```

The final level of information is the conversation, which consists of a sequence of sentences between two or more agents, according to a set of rules. In the system described here, one agent sends an ask-one sentence. This can only be responded to using one of four performatives: *sorry*, *error*, *later*, or *reply*. The *sorry* response means that this is a valid sentence, but the second agent is unable to fulfil the request. The *error* response means that there is something wrong with the performative. The *reply* response means that the evaluation has been successful and here is the result. The *later* response means that the evaluation can be performed, but the result will not be available immediately. This response will then cause the first agent to respond indicating whether it wishes to *callback* to retrieve the result at some later time, *subscribe* (let the second agent send the result when it evaluates it) or *discard* the result (in which case the second agent can abort its evaluation). In this way a set of rules can be constructed specifying whether a given conversation is legal or not.

Thus, using KQML, the five levels of information described in the previous section can be represented in a computer-understandable form.

## 16.4. The information brokers

Search engines typically consist of four parts: a crawler which finds new Web pages, an indexer which takes these Web pages and identifies words which can be stored in a database, a user interface into which the user can type queries, and a query engine which takes the user's query and interrogates the database to satisfy the query (Koch et al. 1996).

There are a number of disadvantages to this approach:

- The crawler can only locate pages in a haphazard fashion;
- The database needs to contain a large part of every Web page in order to allow effective searches;
- It is difficult to find relevant pages, since the exact words used in the Web page must be present in the query;
- It is easy to find irrelevant pages, because a desired term may be present in many Web pages in a different usage or context;
- It is difficult to keep the search engine up-to-date, since the database is only updated when the search engine has time, rather than when a given page is updated.

Note that many of these were identified by McCauley et al. (1996) as being specific problems for the geographic information community, and they are clearly not addressed by conventional search engines.

The information broker approach described here overcomes these problems in a number of ways. The system relies on the metadata and ontology layers discussed earlier. When an author creates a Web page containing geographic data, he or she embeds HTML META tags in it, containing metadata about the page and the ontology used. An information broker is then informed about the page and it visits the page, extracting and storing the metadata. Whenever the page is updated, the author informs the information broker and it updates its stored metadata.

When a user wishes to locate some geographic information on the Web, he or she contacts an information broker (using the Web-based GUI) and submits a query. Because the GUI understands the metadata and ontology for geographic systems, the query can be formulated using intuitive tools (for example selecting an area of a map) and passed unambiguously to the broker. Once the broker has been asked a query, it can look for an answer itself, but it can also ask other brokers to help if it cannot locate some resource. Finally, when appropriate information has been located, the broker can either retrieve it and pass it back to the user or it can simply inform the user where to locate it (and leave it to the user to fetch).

This approach solves the problems mentioned earlier in the following ways:

- The brokers know of all relevant pages, since they are informed about them by the authors;
- The database only needs to store the metadata for a page. Furthermore, because an ontology is associated with the metadata, the terms used can be very concise;

- Relevant pages can be found easily, since the ontology provides an explicit, isomorphic mapping from a concept to a metadata term;
- Irrelevant pages are not found for the same reason;
- The broker is kept up to date because the author notifies it of changes.

## 16.5. Using multiple metadata standards

One problem which arises when retrieving geographic information by its metadata is that different groups use different metadata standards. It is not possible to use a single metadata standard, since it would be impractical to organize and would require that a compromise be reached for nearly every type of metadata. Each of the specialized groups will lose some of the richness of their own metadata representation.

A similar problem arises in the business systems community (at the ontological level), and a number of projects have been initiated which seek to minimize it. For example, the KRAFT project (Gray et al. 1997) proposes a solution whereby related agents can communicate using their own, specialized ontology which is simplified into a cruder common ontology when other, more distantly related agents are involved. This is achieved by providing *wrapper* agents which can translate from one ontology to another. This approach is particularly well-suited to the information broker idea presented here. An information broker which has been created for a specific research group can use the metadata (and ontology) which best suits the group's needs. However, when the group needs to access information which has been marked up with a different metadata standard, a wrapper agent can be called upon to identify a common ontology and perform appropriate translations.

## 16.6. The prototype system

As discussed above, the prototype system uses a structured communication protocol, consisting of five levels of information, from the actual geographic information to the rules for conversation. The system consists of three types of agent: the user's GUI (implemented as a Java applet running in a Web browser), the information brokers (implemented as servers written in Perl) and the Web servers (not implemented as part of the prototype system).

An example interaction will illustrate the operation of the system. Suppose that the user wishes to obtain some radar imagery of Liverpool. The first step is for the user to open the Web page containing the Java GUI. This will automatically connect the user to his or her default information broker. The GUI begins by requesting from that broker the most general map it possesses (usually a map of the world). This is sent to the GUI, which then displays it for the user. The user can now use a *rubber rectangle* to select portions of the map and zoom in for more detail. Each time the zoom button is pressed, the GUI asks the broker for more detail. If the broker knows of a more detailed map, it retrieves it for the user and that is displayed. Otherwise, the GUI simply scales up the appropriate portion of the map which it already possesses.

When the user is happy with the area selected, he or she can fill in information about the imagery (for example the resolution or the band, in the case of Synthetic Aperture Radar). The GUI will ask the broker for this information. In this case, if the data is not immediately available, the user will be consulted (using a dialog box) to check whether it is acceptable to wait for the data. The user can then select an appropriate response, which will be converted to a KQML sentence by the GUI and sent back according to the defined rules for the conversation.

## 16.7. Conclusions

This approach offers many advantages over other ways of locating geographic information on the Web:

- The system understands about geographic information, due to the various layers of information, notably the metadata and ontological levels. It is therefore more suited to the needs of the GIS community than a general search engine.
- The system reduces traffic to the Web servers containing the actual information because the information broker only passes on the location of relevant data (as opposed to the greater amount of data sent when browsing the Web site).
- Information brokers can be tailored for specific groups of users. For example, brokers can be set up to know about the UK or maritime-related GIS or radar-related GIS.

Currently the system does not automatically obtain information from Web servers, but the information brokers use a hand-built database. This is obviously the next stage in the development of this system. The authors will also consider interaction between the classes of agents which do not currently communicate, for example, allowing two brokers to exchange information. Future work will consider mismatches between various metadata standards and will investigate how the use of wrapper agents can allow users to communicate using a rich metadata standard within their community, whilst still being able to exchange information with other communities (albeit with some loss of richness).

## References

Finin T, Weber J, Wiederhold G, Genesereth M, Fritzson R, McKay D, McGuire J, Pelavin R, Shapiro S, Beck C 1993 *Specification of the KQML Agent-Communication Language*. Working paper, ARPA Knowledge Sharing Initiative

Gray P M D, Preece A, Fiddian N J, Gray W A, Bench-Capon T J M, Shave M J R, Azarmi N, Wiegand M, Ashwell M, Beer M, Cui Z, Diaz B, Embury S M, Hui K, Jones A C, Jones D M, Kemp G J L, Lawson E W, Lunn K, Marti P, Shao J, Visser P R S 1997 KRAFT: knowledge fusion from distributed databases and knowledge bases. *Database and Expert System Applications (DEXA '97), Toulouse, France*

Koch T, Ardo A, Brummer A, Lundberg S 1996 *The Building and Maintenance of Robot-Based Internet Search Services*. Technical Report, Lund University

Labrou Y, Finin T 1997 *A Proposal for a New KQML Specification*. Technical Report TR CS-97-03, University of Maryland

McCauley J D, Navulur K C S, Engel B A, Srinivasan R 1996 Serving GIS data through the World Wide Web. Proceedings 3rd International Conference/Workshop on Integrating GIS and Environmental Modeling, Santa Fe, New Mexico. National Center for Geographic Information and Analysis, Santa Barbara

# 17 AUTOMATED METADATA INTERPRETATION TO ASSIST IN THE USE OF UNFAMILIAR GIS DATA SOURCES

Brandon S. Plewe and Steven R. Johnson

## 17.1. Introduction

Geography is by its nature interdisciplinary, and almost any geographic task involves the integration of a wide variety of information about the world. The primary advantage of geographic information systems (GIS) over other methods of analysis has been their ability to overlay various types of geographic information visually; however, the primary drawback of GIS has been the difficulty of collecting the necessary data before one can do anything interesting. Because obtaining data directly from the field usually requires considerable time and financial resources, one would prefer to obtain existing data from another source if possible.

However, in the past there have been considerable obstacles to the effective sharing of geographic information within and between organizations. These obstacles fall into three types, which have led to three different branches of GIS interoperability research:

*Data/Service Interoperability* (Sheth, this volume). These are obstacles to the physical communication between systems containing GIS data and systems containing user applications. Advances over the past few years in the Internet and other Distributed Computing Environments (DCE), and the field of Distributed Geographic Information (DGI), which includes GIS interoperability, have begun to solve the technical issues. Most notably, the Open GIS Consortium (OGC) has been actively involved in the development of technical means for allowing GIS software from different vendors to incorporate data directly from heterogeneous sources, both internal and external to an organization. Currently, several GIS vendors are developing and introducing software based on the OGC standards which begins to overcome this obstacle.

*Institutional/Societal Interoperability* (Meredith 1995). There is not a strong tradition of sharing detailed information between and within public and private organizations, and making information sharing possible requires changing the attitudes and infrastructures of the GIS community. The Federal Geographic Data Committee (FGDC) and similar state-level GIS councils have been rather successful at spreading the message of data sharing and its potential advantages to all involved. Governmental mandates such as the U.S. Executive Order 12906 (Clinton 1994) have been moderately effective in getting public agencies to share information.

*Semantic/Syntactic Interoperability* (Bishr et al., this volume). Often, the producers and users of a data set associate different meanings to various parts of the data set, including features, attributes, and relationships. There are also problems encountered when users do not understand the basic syntactic rules of the data set, such as its data structure and quality. This issue is the focus of this chapter.

## 17.2. Semantic interoperability

This last problem of semantic/syntactic interoperability is potentially much more difficult to address than the other two. It is one thing to be able to obtain the data set one needs; it is another thing entirely to be able to understand the data set and use it effectively. In traditional, closed-environment GIS installations, this was not a problem, because most of the data an organization used was information it had produced, and with which it was intimately familiar. As interoperability becomes increasingly possible technically and acceptable socially, people will more frequently be using data with which they are unfamiliar.

This lack of knowledge of and concern about the semantics and syntactics of data sources commonly results in faulty geography (Gersmehl 1985). People may use these data sets in applications for which the data are entirely inadequate, or they may use them in the correct situations, but with incorrect methods or expectations. For example, geographic data from the U.S. Census Bureau's TIGER database, which have a high degree of topological accuracy, but very poor positional accuracy, are commonly used for making large-scale street maps. In general, the types of problems caused by lack of semantic interoperability include at least the following:

1. Using a data set in an application for which it is poorly suited (in accuracy, semantics, etc.).

2. Using two data sets simultaneously (overlay, maps, etc.) which are not compatible.

3. Drawing incorrect conclusions from a procedure due to misinterpretation of a data source.

In this book, several solutions are presented for overcoming the obstacle of semantic interoperability. Most of these are theoretically elegant and have a great deal of promise for a real solution, but will require significant changes to GIS software and the way it is used. This chapter looks at the potential of a solution which is more immediate and pragmatic, if less rigorous and elegant.

### *17.2.1. The metadata solution*

*Metadata* have long been touted as the solution to the problem of semantic interoperability; in fact, long before the problem was given this name (e.g., Frank 1992; Goodchild 1995: 421). The purpose of metadata is for data producers to describe the semantics and syntactics of each data set fully, so that other users can have all the information they need to use it effectively. This includes such information as:

- Physical structure: file format, projection, units of measure, etc.
- Thematic data: subject matter, attribute tables, classification schema, etc.
- Lineage: sources, dates, processing history, etc.
- Quality: horizontal and vertical positional accuracy, attributes, completeness, etc.
- Usage guidelines: intended purposes, legal restrictions, etc.

Standards such as the FGDC Content Standard for Digital Geospatial Metadata (CSDGM) have defined exactly what pieces of information must or should be included in a metadata record, and to a lesser extent, the format for encoding that record. In the currently most common model for how metadata are used, the metadata record for a specific data set exists as a separate ASCII text file. The most common application for these metadata files to date has been *data discovery services*, such as the NSDI Geospatial Data Clearinghouse, the Alexandria Digital Library, and GeoLens (Behrens et al., this volume). The metadata records for an organization's data sets are collected in a catalog, which GIS users can search to locate any data sets which may be useful for a given purpose.

On the other hand, metadata could have a more profound impact on semantic interoperability if they are used in the GIS along with the data set. In a normal situation of this type, the metadata file is packaged with the data set when it is transferred, so the data user can read through the metadata to learn about the characteristics and appropriate usage of the data.

The main problem with this approach to using metadata is that the last step is based on two flawed assumptions: first, that users would actually be willing to read through the metadata file; and second, that once read, users would be able to use the raw facts to make judgments concerning appropriate use. In real-world situations, both of these assumptions rarely hold. Metadata records based on standards such as the CSDGM tend to be very long (there are over 300 fields), have a complex structure, and are difficult to read and understand for all but the creator. Also, not enough GIS users have a firm-enough grasp on concepts such as scale and resolution, accuracy, classification, lineage, and projection to make informed judgments.

The best solution to the latter obstacle is the one espoused by GIS faculty for many years: education. Although today's GIS software requires less training than before, a great deal of knowledge about geography and cartography is still necessary to use it effectively (in the same way that knowing how to use a word processor well is not the same thing as knowing how to write well; DiBiase 1996). This is necessary for judgments such as "if the stated horizontal accuracy of this data set is 100 feet, I should not use it to make a map at a scale of 1:5000." A firm knowledge of geographic information science (not just geographic information systems) can overcome all three of the potential problems discussed at the beginning of Section 17.2.

Another approach which should at least partially overcome the obstacles of reading and understanding metadata is to involve the GIS software itself in the interpretation of the metadata record. While the complex structure of metadata is difficult for comprehension by human readers, it is well-suited to automatic parsing. Also, many (but not all) of the judgments which can be made concerning appropriate use follow well-known rules, such as "do not use data derived from a source which is at a much smaller scale than the scale of your application." These rules lend themselves to automatic processing by the computer.

## 17.3. The metadata interpreter

Thus, it is conceivable that tools could be added to existing GIS software which would use existing metadata to aid people in using unfamiliar data. These tools could serve several purposes:

1. *Preprocessing.* Often data sets from several local and remote sources will be combined in a project, so the computer would use syntactic information in the metadata to perform any data preprocessing necessary to make them match automatically (e.g., format conversion, coordinate transformations), with little to no involvment by the user.

2. *Automatic usage judgments.* By interpreting the metadata for a data set in the context of a particular task, using a set of rules, the software would determine whether it is appropriate to use the data set. If it were deemed inappropriate, the software could then either prevent the task, warn the user, or educate him or her about the problem and its potential consequences.

3. *Assisting user unterpretation.* The software would present the metadata to the user for review, but would modify them to make them easier to use. This might include reformatting the structure and appearance; using charts, graphs, and maps to display some fields; and including extensive documentation to educate users on the meaning of various parts of the metadata.

4. *Assisting user tasks.* For some tasks, the question is not whether the task is appropriate, but exactly how it should be performed. The software would use the metadata to advise the user on how to perform a task, such as which variable to use in classification, or whether to worry about polygon topology.

Purposes 1, 2, and 4 all use decision rules, and are especially suited to the first and second type of semantic interoperability problem given in Section 17.2. However the third type of problem (misinterpretation) is much more difficult to formalize, so the third purpose helps overcome this problem by simply aiding the user in interpreting the semantics of the data set, rather than trying to automate the interpretation.

A prototype Metadata Interpreter (MI) system including tools for these purposes is currently under development at Brigham Young University. However, the implementation is still in its early stages, so this chapter will focus on the design choices which have been made. The MI design consists of several scripts running on ESRI Arcview GIS, which will augment many of the standard functions of Arcview. When a theme is added to a GIS project, the associated metadata record is also retrieved and parsed. This information can then be used throughout the project, whenever a task may be affected by a data set characteristic which is described by its metadata.

The Metadata Interpreter uses metadata conforming to the CSDGM standard mentioned above, which is the most common metadata standard to date. This standard enables encoding characteristics of the data set in a variety of ways. Some pieces of information are encoded as a numeric value or have a set domain (e.g., Horizontal Positional Accuracy Value, expressed as a distance in meters); these are relatively simple to extract and interpret. Others, however, have a relatively open domain and structure (e.g., Horizontal Positional Accuracy Report), and in practice, a wide variety of content. These are considerably more difficult to parse, because they require extensive analysis of textual information. For obvious reasons, the early focus of the prototype MI has been on the former type of information.

The first technical concern is being able to read the metadata record from the disk. Because the CSDGM standard does not dictate a particular format, data producers have developed a large number of formal and informal methods of structuring the fields in a metadata record. In the prototype MI, the metadata records are expected to be text files formatted according to the FGDC Formal Metadata format developed by FGDC (Schweitzer 1996). This format, which is a Document Type Definition (DTD) conforming to the Standard Generalized Markup Language (SGML) standard, is very well-defined, and produces consistent results. Several of the less formal formats (e.g., tab-indented) are susceptible to unforeseen variations. Thus, the SGML format is much more simple to parse automatically. The *mp parser* software developed at USGS (Schweitzer 1997) can convert other formats to SGML, so it is assumed that users would use this program to convert if necessary, or that the code of mp would eventually become part of the MI.

### *17.3.1. Assisted add theme*

The first tool for which the metadata are used is to assist users in determining whether to use a data set or not (i.e., the third type of task above). This is done using a dialog box which appears when the user wants to add a theme (rather than the standard add theme dialog), as shown in Figure 17.1. From this dialog, the user selects a theme, at which point the associated metadata file is found automatically and parsed. Not all information in the metadata is extracted at this time, just a few of the elements which will be pertinent to the user when deciding whether to use the data set or not, and generally how it should be used. These include the name, source, time period, general descriptions, and projection and coordinate system. These are then displayed in the dialog box for the user's review; he or she also has the option to view the complete record if necessary.

Once the user decides to add the theme, the complete metadata record is parsed. The collection of fields is stored as a table to enable direct access to specific fields by other MI tasks. This is kept in the project, and is also saved as a dBase DBF table on the disk (which is loaded in subsequent sessions rather than the SGML document). For each field in the record, the field name and contents are stored, as well as the context of the field (i.e., the nested set of elements in which it is found). This latter information is vital, because some field names can appear more than once in a record in different contexts and thus the field name alone is not a unique key. For example, the Contact Person field may appear within four contexts: Identification/Point of Contact, Data Quality/Process Contact, Distribution/Distributor, and Metadata Reference/Metadata Contact.

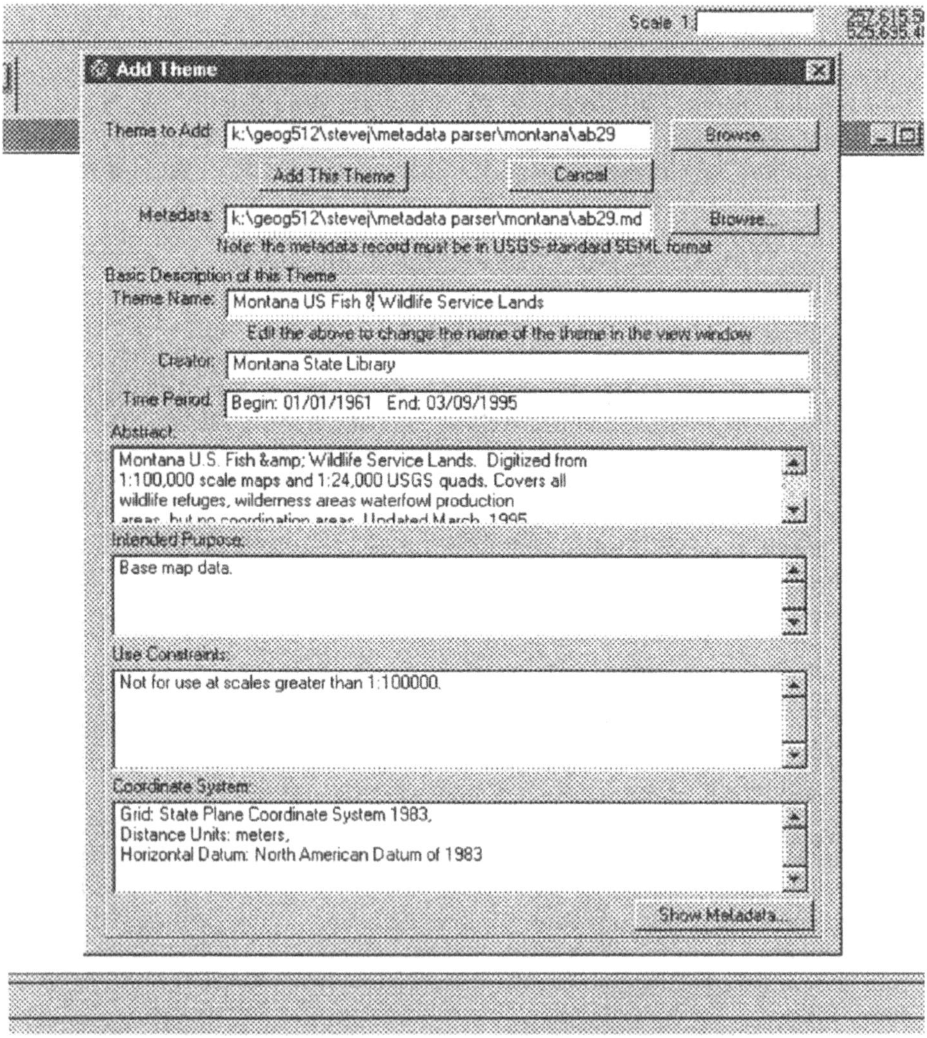


Figure 17.1. Augmented Add Theme function used by the Metadata Interpreter

As the theme is added to a view window, several automated pre-processing tasks can be performed. First, several of the standard properties of the theme can be extracted from the metadata. The name of the theme is extracted from the Title field of the metadata, although the user is given the opportunity to alter it. Also, the *descriptive comments* property of the theme is filled with a formatted representation of the entire metadata record, for the benefit of users who are willing to read it to get the complete picture. More advanced settings would require more semantic knowledge by the interpreter, but would include features such as:

- Setting symbology using the Theme Keyword field and a library of appropriate symbologies (e.g., blue for water, green for vegetation).
- Determining the primary thematic field in the attribute table (especially if there is a nominal feature category available), and classifying the theme automatically.
- Setting the maximum scale limit, which disables the display of the theme if the window scale is too large; this maximum scale would have to be interpreted from a variety of fields, including Source Material Scale, Horizontal Positional Accuracy, and Purpose.
- Converting the data format, if it is not directly readable by Arcview.

If the theme is the first in the view window, some of the characteristics of the view can be set from the metadata for the theme. For example, the early prototype sets the *units of measurement* property for the view from the appropriate metadata field. This property makes scale, area, and distance measurements possible.

When a theme is added to a view that already contains one or more themes, several checks would be performed to test the compatibility of the new theme. The values in various metadata fields from the new theme would be compared to the values for themes already in the view. These checks may include:

- Projection/coordinate system/datum
- Scale/horizontal positional accuracy
- Time period of content

During each comparison, if a discrepancy is encountered (i.e., Problem 2 from Section 17.2), the MI should be able to help the user handle it properly. Generally, the computer can take one of five actions: 1) ignore the problem (i.e., add the theme as is); 2) proceed with the operation, but warn the user of potential problems; 3) end the operation (i.e., do not add the theme at all); 4) process the theme to make it match (e.g., change the projection of the new theme); or 5) warn the user of the concern, then display the choices 1, 3, and 4 and let him or her decide on an action. In all of these actions, any message or dialogs which appear would include buttons for help regarding the discrepancy and its potential results, as well as for directly viewing the relevant metadata fields.

The last choice is likely the best for experienced users, giving them more control over what the system is doing with their data. However, novice users who are not familiar with these concepts may not wish to be bothered with the choice, and want the computer to make a judgment. The computer's choice would be based on factors such as the seriousness of the discrepancy (e.g., a scale difference of 1:2 would be allowed, but 1:100 would be blocked). However, much of the decision-making process here depends on the user's application, about which the computer likely has no information.

### *17.3.2. User preferences*

Each of the actions above will be preferable for different users in different situations. Thus it is not only important to take as much of the effort as possible out of the user's hands, but it is also necessary to return control over these things to the user when necessary. In the MI, this would be accomplished using a simple user preferences dialog. The user here has control over how each type of concern is to be handled, from the five choices above. Users could also control which fields will be displayed in the add theme dialog, as well as more specific settings for each type of conflict (e.g., the minimum discrepancy of scales between a new theme and the rest of the view which constitutes a conflict).

### *17.3.3. Other conflict resolution tools*

Inconsistencies between two themes (Problem 2 from Section 17.2) can cause a problem in situations other than when a theme is added to a view. In fact, conflicts which seem minor when simply displaying two themes together become much more serious in other operations. Probably the most important example of this would be an overlay operation (such as the Select By Theme operation in Arcview). Here, minor discrepancies can seriously corrupt resultant information, causing problems such as sliver polygons.

Thus, the MI design includes a conflict-resolution tool which augments any operations which involve two or more themes. This tool checks for discrepancies, in the same way as in the add theme tool. The only difference is that it will generally be less forgiving in its judgments, and users would be able to specify a different set of action preferences for each type of operation.

### *17.3.4. Tools for identifying inappropriate uses*

Solving problems of the first type discussed in Section 17.2, in which someone tries to use a data set for a purpose for which it will not suffice, is more difficult to handle automatically than the second type. While some applications of the question "is this a good operation to perform with this data set?" can be covered with a set of general rules, many are grounded in the context of the application to be judged by the MI.

One of the most common examples where this can be valuable it is when changing scales. Based on several factors including accuracy, source data, and user purpose, the theme would likely have a maximum scale at which it should be viewed or analyzed. This maximum scale would be used in several operations such as when designing layouts and when zooming in or out in a view. Again, the options for action would be similar to those used before, such as warning the user or preventing the user from doing the operation.

## 17.4. Difficulties in using standard metadata

Even though the MI is only in the early stages of implementation, several difficult obstacles have already arisen. Some require a greater degree of knowledge to be built into the decision-making process of the tools, but others are serious shortcomings of the CSDGM standard in the feasibility of using a computer to interpret its content (not just its structure).

### *17.4.1. Free text fields*

In the CSDGM standard, several fields are allowed to be filled with free text, which means that the data producer can insert any text which may be appropriate, as opposed to other fields which have a specified domain of possible values. This can be valuable for the metadata writer, allowing for flexibility in the content to accommodate the wide variety of data sources which may exist. However, it can cause serious problems when the text needs to be interpreted by the computer. The contents of these fields are not predictable, so the strategies used in interpreting other elements, by comparing the contents of the field based on what the MI expects to find there, cannot be used. It may be possible to use natural language techniques to parse the text in these fields; however, their content usually appears as fragmented notes or structured outlines and thus cannot be parsed as would complete sentences.

In many of the metadata records which were reviewed for the prototype implementation, some of the fields, such as Supplemental Information, have become repositories for entire essays about the data set, including information belonging in the rest of the record. Then in other fields, the metadata producer inserts comments such as "see supplemental information." These cross-references are only meaningful when the data user reads the metadata manually.

### *17.4.2. Scale/resolution/accuracy*

It is apparent from the design above that several of the decisions which the MI needs to make concerning appropriate use involve the related issues of scale, accuracy, and resolution. This involves making judgments such as "should this theme be displayed on a map of this scale?" or "are these two themes similar enough in their level of accuracy to allow them to be overlaid?" A system which would be able to make these judgments automatically would be invaluable, since most GIS users have only a vague notion of how to deal with these concepts in a practical sense.

The problem is that there are very few rules at all about how to use these concepts in a practical setting. A great deal of research has been done in the GIScience field, trying to understand these concepts and the relationships between them (e.g., Goodchild and Gopal 1989), but to date there are only the beginnings of standardized methods for interpreting accuracy information (Montgomery and Schuch 1993: 138). For example, if a theme has a minimum positional accuracy of 100 meters, it is apparent there is an inherent maximum map scale at which it should be displayed. However, there is not a generally accepted formula for converting horizontal positional accuracy into maximum map scale.

The other obstacle related to scale and resolution is that most of the existing metadata records do not include adequate information concerning this area, even though the metadata standard dictates a great deal of content which would support the MI tools described above. For raster data, this information can be obtained from the resolution of the theme. For vector data sets, the CSDGM standard includes an extensive Horizontal Positional Accuracy element to include this type of information, but no records have been found to date which actually have a value in this field. Some include a description of the data quality controls which were used in the development process; others may have a statement of assurance that "the coordinates are as accurate as possible," while many simply have "unknown" in this field.

This is not necessarily the result of laziness on the part of metadata producers, although the general reluctance to date of data producers to produce metadata with sufficient detail has resulted in metadata records which are inadequate for automatic interpretation. Generally, the reason is that the level of accuracy has not been determined. There are relatively well-known sampling methods for assessing positional accuracy (Montgomery and Schuch 1993: 136–43), but in many cases they are not used, usually due to the expense involved. Most organizations can use rather informal methods to satisfy themselves of the accuracy of their data, without having to derive an actual value. However, these informal judgments cannot be communicated adequately to new data users.

### *17.4.3. Application context*

The last major obstacle to a useful metadata interpreter is that many of the judgments of the appropriateness of data depend heavily on the applications for which the data are being use. For example, using socioeconomic data from 1950 to describe the current state of affairs in a city is wholly inappropriate, although using geological data from 1950 is usually acceptable for describing the current geology of an area.

The main difficulty in this high-level interpretation is that information about the use of the data is not contained in the metadata, and currently there is no mechanism for formally describing the characteristics of the user's application as there is for describing the data used (Bishr et al., this volume). Thus it is impossible for the computer to make these kinds of judgments automatically. One would expect that informed, experienced GIS users would be the ones with the knowledge necessary to judge the appropriateness of the data set in a particular application area, but this cannot be guaranteed. A partial solution may be to give the computer some general rules, such as "any data set over 20 years old is suspect," so that if a data set from 1950 were requested, a warning message would appear, but the user would have to make the final decision on what to do with the theme.

The other obstacle for high-level interpretation is that current metadata standards are primarily focused on the *syntactics* of data sets, including things like accuracy, projection, time period, and data structure. Interpreting appropriate applications for using the data would require more information about the *semantics* of the data set, for example, what it is about the nature of socioeconomic data that makes it go out of date more quickly than geological data. The only elements in the CSDGM standard which could be considered truly semantic would include the *abstract*, which is a free text field with no real structure, the *Entity/Attribute Information* section, which contains at most a list of the types of entities and attributes contained with very little about what they mean, and the *Theme Keywords*, which could actually be very useful if the interpreter had some kind of encyclopedia to tell it how to interpret these keywords (e.g., what are the ramifications of a data set covering "geology"?).

## 17.5. Conclusions

Although the system is still under development, early results have been promising. Some information (primarily syntactic information, rather than semantic) can be interpreted and used in the GIS. With a functional Metadata Interpreter, the advanced GIS user will be able to use unfamiliar data sources more intelligently, without having to read and understand the entire metadata records. Novice users, who may not understand many of the technical concepts discussed in the metadata, may be able to use data sources more correctly, often without even realizing it.

While the former goal can be realized by simply presenting the textual metadata content to the user in a more readable and timely form, the latter may be asking too much of metadata. In some cases the concepts are difficult to operationalize, at least enough to make the interpretation transparent to the user. In other cases it is a shortcoming of current metadata standards which focus more on encoding the syntactics of the data set than the semantics. The last is a problem of metadata production, since many of the metadata which has been produced to date fulfill only the barest requirements for completeness and do not include much available information that could be useful to interpretation by the GIS software.

It is therefore expected that an automatic metadata interpreter can be a useful tool within the GIS, to work in concert with the GIS user, giving additional information when necessary, and making suggestions or friendly warnings in places. However, it is unlikely that it will ever be able to take most of the decision-making process of dealing with unfamiliar data sets out of the hands of users. Thus intelligent GIS results will for the foreseeable future require intelligent GIS users.

## Note

The Avenue scripts which make up the currently available portions of the Metadata Interpreter can be obtained free of charge from the second author.

## References

Clinton W 1994 Coordinating geographic data acquisition and access: the National Spatial Data Infrastructure, Executive Order 12906. *Federal Register* 59(71): 17671–4. *http://www.fgdc.gov/Communications/GenInfo/execord.html*

DiBiase D 1996 Rethinking laboratory education for an introductory course on geographic information. *Cartographica* 33(4): 61–71

Frank A U 1992 Acquiring a digital base map: a theoretical investigation into a form of sharing data. *Journal of the Urban and Regional Information Systems Association* 4(1): 10–23

Gersmehl P J 1985 The data, the reader, and the innocent bystander—a parable for map users. *Professional Geographer* 37(3): 329–34

Goodchild M F 1995 Sharing imperfect data. In Onrud H J, Rushton G (eds) *Sharing Geographic Information.* Center for Urban Policy Research, New Brunswick, NJ: 413–425

Goodchild M F, Gopal S (eds) 1989 *Accuracy of Spatial Databases.* Taylor & Francis, London

Meredith P 1995 Distributed GIS: If its time is now, why is it resisted? In Onrud H J, Rushton G (eds) *Sharing Geographic Information.* Center for Urban Policy Research, New Brunswick, NJ: 7–21

Montgomery G E, Schuch H C 1993 *GIS Data Conversion Handbook.*, GISWorld, Ft. Collins

Schweitzer P 1996 A document type declaration for formal metadata. In *Tools for Creation of Formal Metadata.* U.S. Geological Survey, Reston. *http://geochange.er.usgs.gov/pub/tools/metadata/tools/doc/dtd.html*

Schweitzer P 1997 A compiler for formal metadata. In *Tools for Creation of Formal Metadata.* U.S. Geological Survey, Reston. *http://geochange.er.usgs.gov/pub/tools/metadata/tools/doc/mp.html*

# 18 INTRODUCTION: SYSTEMS AND EXPERIENCES

Clifford A. Kottman

The following dozen chapters expose the full complexity of the geospatial interoperability challenge: there are both *problems* and *meta-problems*. The problems stem from the special functional needs of any particular GIS application. Each particular geospatial application, such as those found in each of the twelve chapters of this section, needs a number of unique services that must be specially designed. Examples include custom portrayals, special data access and retrieval capabilities, information parsing and restructuring, persistent and transient storage requirements, communication needs, and many others. Here, the tools so carefully crafted for the general case are often revealed to be too slow, too weak, or too narrow for immediate use in particular real world applications. That is, each project exposes the need for new software implementations for what should be routine applications.

The meta-problems are at the heart of this section. They center on the architecture of software component technology, or *middleware*, and are even more daunting than the problems. While there is general agreement that software components are the key to the future, today's implementers quickly find that there are few guidelines for component architecture. Sometimes, our systems vocabulary for component concepts seems to be too weak to express the shortcomings of today's technology and too weak to express the requirements of the next generation of systems. We find ourselves in a maze of real and conceptual middleware without an overall architecture to make sense of it.

The simple fact is that real working systems need to provide access to a plethora of geospatial information services, and open commerce in geospatial services cannot happen without a rational approach for partitioning these services into clumps of functionality with industry-wide accepted interfaces. This clumping process is at the heart of the componentware revolution, and at the heart of this section.

Table 18.1 gives a quick thumbnail view of the following twelve chapters, which are presented in alphabetical order of first author.

| First Author | Acronym | Focus | Application Area |
|---|---|---|---|
| Gennady L. Andrienko | IRIA | Portrayal and analysis | Land information |
| Oleg Balovnev | GeoToolKit | Software component design | Geology |
| Clifford Behrens | GeoLens | Software component design | Earth observation |
| Sang Kyun Cha | SDBC | Software component design | Korean national GIS |
| Gilles Clement | OGCI | Data enabling | Canadian GDI and DND |
| Liping Di | DIAL | WWW component design | Earth observation |
| Frederico T. Fonseca | (none) | Software component design | General |
| Donatus Kvedarauskas | GeoLib | Software component design | Toulouse tourism |
| Zakaria Maamar | SIGAL | Agent-oriented frameworks | Georeferenced digital libraries |
| Robert G. Raskin | ESRAT | Data enabling | Earth observation |
| Carel van den Berg | (none) | WWW component design | Dutch cadastre, cable TV |
| Chuanchu Wang | VGIS | Data enabling | Land information |

Table 18.1. A thumbnail view of Systems and Experiences

Most telling is the column *Focus*. It shows that seven of the twelve chapters are directed at the challenge of component design. Four others are aimed at specific components in a larger component architecture: portrayal and data enabling. The fourth column of Table 18.1 is also revealing. It shows where the *early implementers* are: in government, land information, Earth observation, and a few selected commercial initiatives. As interoperability becomes a more powerful reality, the fourth column will become dominated by commercial interests.

In summary, the Systems and Experiences reported in the following twelve chapters provide a record of the first footpaths into the wilderness of software component technology for the geospatial sciences. It will be a fascinating study in market development to watch these paths become highways that will carry the growing commerce in geospatial information and processing.

We close this introduction with a brief synopsis of the twelve papers in this section.

Gennady L. Andrienko and Natalia V. Andrienko, in *IRIS: A tool to support data analysis with maps*, focus primarily on the presentation of geospatial information. They gather a number of rules that have been developed for the purpose of guiding the selection of one type of interactive diagram construction (examples: juxtaposition, divergence, segmentation) over another. The contribution of IRIS in automated graphics and cartographic visualization is lies in its ability to calculate and rely on meta-information about relationships among data components, and thus suggest superior interactive portrayals.

Oleg Balovnev, Martin Bruenig, and Armin B. Cremers, in *GEOTOOLKIT: Opening the access to object-oriented geo-data*, argue that there is so much geospatial middleware to be developed that a *meta-tool* to simplify component-based design is in order. They address this need with *GeoToolKit*, a component class library for the efficient storage and retrieval of spatial objects. They apply their creation in the context of simplicial complexes for the modeling of three- and four-dimensional geologic structures including strata and faults.

Clifford Behrens, Leon Shklar, Chumki Basu, Nancy Yeager, and Edith Au, in *The geospatial interoperability problem: Lessons learned from building the GeoLens prototype*, address the special needs of the global Earth scientist who confronts huge, diverse, heterogeneous, and distributed information and metadata holdings such as those of NASA and NIMA. The situation calls for a careful partitioning of the overall problem into manageable software components such as a Client, a Catalog Server, a Schema Server, an Information Access Server, and others that form the GeoLens prototype. Their lessons learned make a powerful case for catalog federation, schema integration, and other weapons for the fight against semantic ambiguity, and to cope with the needs of EOSDIS scientists who wish to discover, access, and extract remotely-sensed information into an environment where it can be mingled with standard raster and vector data formats.

Sang Kyun Cha, Ki Hong Kim, Chang Bin Song, Joo Kwan Kim, and Yong Sik Kwon, in *A middleware architecture for transparent access to multiple spatial object databases*, tackle the central meta-problem of this section straight on. They argue that the central challenge in spatial information is found in data access and management. Using a general object-oriented model, they address transparent access to multiple spatial databases with careful attention to rich semantics, thus providing a powerful data enabler. The chapter provides a roadmap for the addition of application-specific components using CORBA and client–server concepts.

Gilles Clement, Christian Larouche, Paul Morin, and Denis Gouin, in *Interoperating geographic information systems using the Open Geospatial Datastore Interface (OGDI)*, describe a commercially-mature application that is a data enabler. OGDI provides application programming interfaces between GIS client environments and an impressively large variety of data stores. The interface, when invoked, creates an integrated, but transient, collection of features (vector or raster) in a common internal structure. This chapter demonstrates the importance of a powerful data enabler in modern geospatial information architectures.

Liping Di, R. Suresh, K. Doan, Doug Ilg, and Ken McDonald, in *DIAL: A web-based interoperable scientific data distribution system*, consider the needs of today's geoscientists who would like to access global geospatial information using tools and services available on the World Wide Web. The overall capability of formatting, documenting, and distributing data from the desktop is broken into five major components: catalog interoperability, a scientific data server, an open database, a user interface, and a suite of data analysis tools. Realization of the scientist's wishes is shown to be heavily dependent on interoperability not only in the sense of data enabling (that is, data object interoperability), but also at the level of federation of scientific catalogs where a single catalog interoperability protocol must be in place.

Frederico T. Fonseca and Clodoveu A. Davis, Jr., in *Using the Internet to access geographic information: An Open GIS prototype*, consider an architecture with two major components—the Extractor and the Manipulator—and both are prototyped using the Open GIS model as a basic framework. The current Extractor depends on the presence of a compatible GIS client, selects features that satisfy stated conditions, and provides them to the Manipulator. The features are held in a transient store in a form modeled after OpenGIS structures. The Manipulator provides human interface functionality, as well as interfaces for information analysis and portrayal services.

Donatas Kvedarauskas, Patrice Boursier, Xavier Culos, Thierry Deltheil, and Sylvie Iris, in *GEOLIB: A software component for making GIS tools interoperable*, demonstrate another componentization in answer to the meta-problem of this section. Software components selected and implemented include the Gateway (a data enabler), the Dispatcher (query pre-processor), Visualization, Core (basic geospatial functions), and Extensions (advanced geospatial functions).

Zakaria Maamar, Bernard Moulin, and Yvan Bedard, in *Software agent-oriented frameworks for the interoperability of georeferenced digital libraries on the World Wide Web: The SIGAL project*, focus on the problem of georeferenced digital libraries (DGLs). They select a novel approach to the automation of advanced geospatial library services that uses software agent-oriented frameworks. Their architecture identifies three major components: data providers, service providers, and service consumers. The result is a one-stop query engine where the data or metadata of multiple DGLs (heterogeneous and distributed) are queried from a single user interface.

Robert G. Raskin and Elaine R. Dobinson, in *Interoperability with the Earth science remote access tool (ESRAT)*, cope with data-enabling issues. ESRAT potentially enables a host of NASA data formats to be accessible across the Internet, with subsetting carried out on the server side. Interoperability of data is achieved with an intermediate data format that includes four constructor types: arrays, grids, sequences and functions. ESRAT fills a niche in a previously determined EOSDIS component architecture.

Carel van den Berg, Frank Tuijnman, Tom Vijlbrief, Co Meijer, Peter van Oosterom, and Harry Uitermark, in *Multi-server Internet GIS: Standardization and practical experiences*, also exploit the Internet for accessing geospatial information. Their key elements are well-known structures for metadata, a common query language, transfer formats, and authentication procedures. The authors take a minimum-resistance path, selecting a thin metadata model, and spatial SQL for the query language. The transfer format selected is an object-relational model with geometric types, optimized for speed and efficiency, and off-the-shelf methods are selected for authentication.

Finally, Chuanchu Wang and Jiankang Wu, in *A visual geospatial information server (VGIS) providing transparent access to heterogeneous sources*, introduce a data enabler concept with special attention to the preservation of critical metadata. Their prototype, WebGIS, implements these concepts for a number of commonly-encountered geodata repositories, and at least one uncommon one (Singapore land digital data.) A principal contribution of the VGIS catalog server is its exploitation of metadata structures (provided by data store drivers) that enable automated schema mapping for improved semantic mapping.

# 19 IRIS: A TOOL TO SUPPORT DATA ANALYSIS WITH MAPS

Gennady L. Andrienko and Natalia V. Andrienko

## 19.1. Introduction

The theme of our work is intelligent support to users in exploration of spatially referenced data. Under spatially referenced data we mean various thematic data (data on demography, economy, education, culture, history, etc.) referring to geographical objects or locations (countries, regions, cities and so on) rather than purely spatial data such as geometry of geographical objects and their relative positions with respect to other objects. Spatially referenced data cannot be effectively analyzed without representing them on maps. However, a number of different presentation techniques exist, and selection of proper techniques in accord with data characteristics is crucial for successful analysis. An improperly designed map can be unhelpful or even mislead the user into wrong conclusions.

Currently data mapping is done with the use of GIS. However, despite their power in operation with geometric information, the visualization facilities of GIS need to be improved. We see the major shortcomings in the following:

- Users bear full responsibility for a correct selection of presentation techniques. Consequently, users should (1) know the principles of cartographic presentation; (2) spend time and effort for map production; (3) focus on analysis of data and solution of problems.
- GIS are typically complex and difficult to operate; significant training is required.
- GIS do not sufficiently support the further exploitation of produced maps, i.e. their use in data analysis.

With IRIS we strive to pave the way for overcoming these weaknesses. To better support data visualization, IRIS incorporates the knowledge on map design in the form of generic, domain-independent rules. This allows automatic generation of thematic maps correctly presenting user's data. This solution has several important consequences:

- The user does not need to be an expert in map design: the system guarantees conformity of produced maps to the established principles of graphic presentation of data.
- Automatic map generation takes less time and no effort from the user's side.
- Instead of concern about map building, the user can concentrate directly on data analysis and problem solving.
- The system becomes easier to operate: to receive a map the user only needs to select a data subset for visualization.

However, we assume that obtaining a map is not the final goal of the user. The map is intended to be somehow exploited, and we strive to support also the process of its exploitation. Maps being produced by IRIS on the computer screen are not mere reproductions of paper maps: they are dynamic and reactive to user's actions. We develop an arsenal of tools for dynamic manipulation of maps.

Dynamic manipulation means that certain interactive operations of the user result in the display changing in real time. This can promote data exploration: it is highly probable that the features of presented data initially not clearly visible in a map can become more salient when the display changes. So, effectiveness of data visualization can be significantly increased.

It is also important for exploration that IRIS always offers the user several different presentations of the same data. Different presentation techniques offer different opportunities for analysis. For example, bars allow easy estimation of absolute values and differences, pies are good for seeing proportions and totals, painting areas objects in colors or shades according to values of an attribute gives an integral view on variation of this attribute through the territory. All these opportunities can be useful during exploratory data analysis when the user does not know beforehand the inherent features of the data.

## 19.2. Related work

Recently there has been extensive research on the use of basic knowledge on data visualization for automated graphics generation (see the survey of Murray 1994). The system APT developed by Mackinlay (1986) is recognized as classical: Mackinlay's approach to data presentation design was followed in all later research work. In this approach the system partitions the data set to be presented into subsets with fewer domains (fields) so that it becomes possible to selects some *visualization primitive* for each subset. The visualization primitives used in APT include the seven *visual variables* introduced by Bertin (1983): *position*, *size*, *value* (the degree of darkness), *color*, *texture*, *orientation*, and *shape*. More complex graphical encoding techniques, such as bar charts, pie charts, plots, graphs, are also included into the basic set of primitive graphical languages. The visualization primitives selected for the partitions are combined by merging parts of partial designs that encode the same information (axes) or by implementing them as different visual properties (size, shape, color, etc.) of the same graphical elements.

Primitives are selected depending on the characteristics of data sets: the number of domains, cardinality (number of different values) of a domain, type of values in a domain (nominal, ordinal, or quantitative), and the type of dependency between domains (functional or one-to-many). The selection is governed by the basic principle of graphical presentation stated by Bertin: features of data components should be matched by properties of visual variables selected to represent them. For example, a quantitative data component can be represented only by a visual variable that allows quantitative perception, that is, by *position* or *size*.

Mackinlay's work showed the feasibility of automated graphics design on the basis of generic principles and rules of visualization. The suggested approach was developed in further research on data visualization. In the VISTA system (Senay and Ignatius 1994) more visualization techniques and composition operators are available that allow, in particular, construction of 3D graphics. In the SAGE project (Roth and Mattis 1990) the inventory of data characteristics to be accounted for in graphics design is significantly extended. The design is sensitive to the user's information seeking goals expressed by primitive operations such as accurate lookup of separate data values or comparison of values of two attributes.

Casner (1991) considers more complex information-processing tasks performed by the user. He assumes that a user's task can be specified as a sequence of predefined primitive *logical operators* such as a query for a property of an object or arithmetic operations over values. Casner's BOZ system receives such a sequence and replaces logical operators with primitive *perceptual operators* such as *search* for an object with a given graphical property or *lookup* for a specific property of a given graphical object. Then the system selects visualization primitives that enable these perceptual operators.

An attempt to approach the problem of cartographic visualization is described by Zhan and Buttenfield (1995). The authors built an expert system that consults a GIS user about the selection of visualization techniques for a given data field. The system does not plan presentations for several fields and does not perform visualization. On the basis of this work and the approach of Mackinlay the system VIZARD was created that automates the presentation of spatially referenced data on maps (Jung 1995). This system designs maps on the basis of the techniques conventionally applied in cartography.

Though this research on automated design of graphical presentations has yielded rather impressive results, we should admit that these results refer mainly to design algorithms and the factors that govern design commitments. In all information presentation systems obtaining a graphic is seen as the final goal of the work, whereas this is only the first, preparatory step in data analysis and problem solving.

For effective analysis the user should have an opportunity not only to view graphics. According to Bertin's (1983) and Tufte's (1983) recommendations, graphics designed for exploration should give the analyst the maximum possible freedom of manipulation. Bertin offers techniques for building dynamic graphics on paper, but as a medium for presenting graphics the computer screen gives many more opportunities for dynamic graphics and interactive manipulation. The potential of interactive manipulation is recognized in more recent literature on visual data analysis (Cleveland 1993) and cartography (Monmonier 1996).

A number of widgets for interactive manipulation have been designed recently. An overview of interactive manipulation techniques and systems implementing them is given by Tweedie (1997). Though manipulations are always connected with graphical data display, each system proposes one preset type of visualization to manipulate with. This reduces the potential power of such a system in data investigation that typically requires solution of subtasks of various types: as was persuasively shown by Casner (1991), for different tasks different presentations of the same data are useful. Tweedie (1997) claims that "this limitation applies only to static presentations, it is not relevant when interactivity allows different features of the data to be made salient as and when required", but this claim is not sufficiently supported. At least, one should admit that different presentations require different amounts of effort for acquiring the same knowledge from the data.

The SAGE project (Goldstein and Roth 1994) strives to combine the latest developments in both automated visualization design and interactive manipulation. However, in SAGE these two instruments seem merely to co-exist: supported manipulations do not depend on the presentation techniques selected for the data, on the one hand, and have effect only on the number of objects displayed, on the other hand. In IRIS we suggest a tighter integration of automated visualization design and interactive manipulation for the benefit of data exploration.

## 19.3. Map design in IRIS

Implementing knowledge-based data mapping in IRIS, we were aware that the general principles of graphics design are certainly valid for maps but the latter have their own peculiarities. Geographical objects must be depicted on a map so as to reflect actual geographical positions, relative sizes, and geometry of these objects. This requirement prescribes the way other kinds of information are organized in a map: graphical elements selected to represent spatially referenced data should be placed at the locations of the geographical objects or inside their outlines. The visual variable *position* cannot be used to encode data. Besides, the graphical elements should be clearly visible against the background. The latter includes outlines of the geographical objects referenced by the data and, possibly, other geographical layers such as rivers, forests, or mountains. This geographical information gives the territory a familiar appearance and provides easy identification and location of the objects under analysis.

An arsenal of techniques appropriate for data visualization on maps has been developed in cartography. Most of these techniques were included in IRIS. Among the visualization techniques we distinguish those based on painting and those based on the use of signs. Signs may be either simple or structured. Painting can be applied when outlines of geographical objects are given to the system (the outlines are specified as polygons with coordinates of each vertex indicated). In this case we have *area* objects. Another variant is when geographical objects are specified by pairs of coordinates (*point* objects). Signs may be applied both to area and to point objects.

Each visualization technique exploits some visual variable. Painting techniques may use *color hue* or *color value* (further referred to as hue and value). Simple signs may vary in ***size***, ***shape***, ***hue***, and ***value***; combinations of these variables are limited to ***size + hue***, ***size + value***, ***shape + hue***, ***shape + value***. Size and shape are not combined because of the difficulties in comparing sizes when shapes differ. The variable ***size*** is used in two forms: ***length*** (bars) and ***area*** (circles). Area is better in case of large differences in data values. Structured signs (further called *diagrams*) are composed of graphical elements varying in *size* (length, area, angle), *hue*, or *value*; combinations of variables are not used. To ensure that graphical elements are distinguished within a diagram, they are given different colors or textures or fixed positions.

Consistency of perceptual properties of visual variables (summarized in Table 19.1) with data characteristics is the main principle governing selection of visualization techniques (Bertin 1983). According to this principle, quantitative data can be shown by ***size*** or ***value***; ordered qualitative data by ***value***; and unordered qualitative data by ***shape*** or ***hue***.

| Level of perception | Size | Value | Hue<br>red, green, blue | Shape |
|---|---|---|---|---|
| quantitative | ⊕ | | | |
| ordered | ⊕ | ⊕ | | |
| selective | ⊕ | ⊕ | ⊕ | |
| associative | | | ⊕ | ⊕ |

Table 19.1. Perceptual properties of the visual variables (according to Bertin 1983).

When several data components are to be visualized, presentation of each of them should be consistent with its type, and these presentations should be in some way combined. The opportunities for technique combination in cartography are severely restricted: one can overlay signs upon painting but is not encouraged to place signs of different types on the same map. Simple signs may combine two visual variables, each assigned to a distinct data component. Structured signs themselves are means for combining several data components in one presentation. They can be classified according the methods of their construction from basic graphical elements: *juxtaposition*, *divergence*, *segmentation*, and *inclusion* (Figure 19.1).

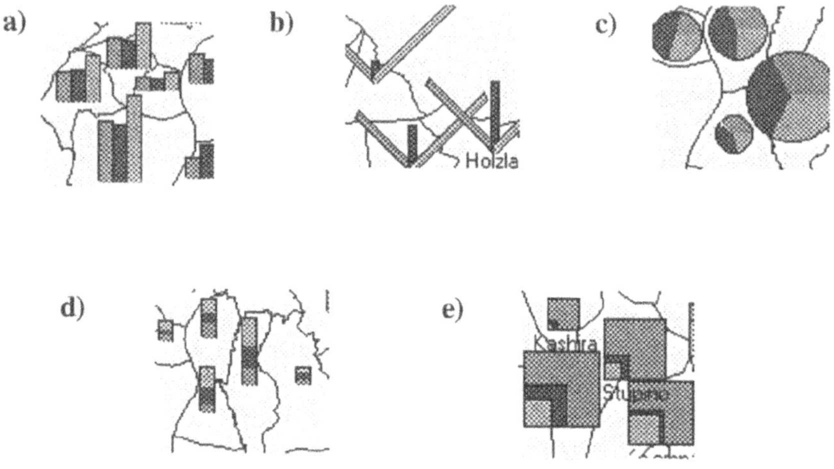


Figure 19.1. Different techniques of diagram construction: *juxtaposition* (a), *divergence* (b), *segmentation* (c, d), *inclusion* (e).

The basic rule of diagram type selection is the following: the construction method must be consistent with relationships among the components of data to be presented. Thus, juxtaposition encourages comparisons between elements of a diagram and therefore may be chosen for *comparable* data components, such as numbers of births and deaths per 10000 population in countries of Europe. Parallel bars, as in Figure 19.1a, give a good opportunity to estimate differences between birth and death rates. But this feature is undesirable when we wish to analyze together birth rates and sizes of national product per capita in dollars: differences of values of these two attributes have no sense. For *incomparable* data components diagrams based on divergence are used (Figure 19.1b). In this case direct comparisons are obstructed but variations of the shapes of the resulted signs (whether there are typical patterns and, if so, whether they are nearly even or substantially asymmetric) allow judging about relatedness of the attributes.

Segmentation is applicable when data components added together make some meaningful whole, for example, numbers of young, middle-aged and old people in in a country's population. Well-known segmentation-based diagrams are pies and segmented bars. The latter may, in their turn, be juxtaposed or diverge. Diagrams based on inclusion are applied when there are inclusion relationships among data components; for example, number of population in total, number of students, and number of university students.

So, in visualization design IRIS significantly relies upon metainformation about relationships among data components. To select a proper presentation, the system groups the data fields selected for visualization by relationships. The inclusion relationships have a priority over the comparability relationship. For unification, we assume that a field unrelated with all others forms a group consisting of one element. For each group the system finds appropriate visualization techniques, according to Table 19.2.

| N | Type(s) of field(s) | Relationship | Visual variables | Presentation methods |
|---|---|---|---|---|
| =1 | nominal (including logical) | | color or shape | area coloring, colored signs, shape signs |
| =1 | ordinal | | value | area shading, shaded signs |
| =1 | numeric | | value or size | area shading, standalone bars |
| >1 | logical | always comparable | color | structured signs varying by presence or absence of colored sectors |
| >1 | numeric | comparable | size | parallel bars |
| >1 | numeric | included in a common total | size | pies, standalone segmented bars, parallel bars |
| >1 | numeric | ordered inclusion (a>b>...) | size | nested squares, parallel bars |

Table 19.2. Selection of visualization methods for groups of data components (N stands for the number of selected fields).

When the number of groups is more than one, the system combines the selected presentation techniques. The basic combination techniques are superimposition of signs over area painting, variation of different visual properties of simple signs, and divergence of parts of diagrams. Table 19.3 summarizes the variants of these combination techniques applied in IRIS.

| Description of groups | Combination technique |
|---|---|
| One nominal or ordinal field + one numeric field | Colored or shaded standalone bars |
| One nominal field + a group of numeric fields | Area coloring with superimposed diagrams |
| Two groups of comparable numeric fields | Two-sided bar diagrams (2 diverging groups of parallel bars) |
| Several groups of numeric fields with inclusion in a common total in each group | Parallel segmented bars if the groups are comparable, radial segmented bars otherwise |
| Several incomparable numeric fields | Radial bars |

Table 19.3. Selection of combination techniques according to contents of field groups and relationships among them

According to Bertin (1983) and Tufte (1983), a powerful instrument to examine relationships between two or more data components is comparison of several maps each presenting one of the components. Our experiments with the earlier versions of IRIS showed that this way of data analysis was much easier for the users and gave a larger number of interesting findings than with presentation of the same information in one map by complicated diagrams. The area painting technique was especially good for these purposes. This observation inspired us to add presentation by multiple maps to the arsenal of combination techniques used in IRIS. Each of the multiple maps presents one of the selected data components by area painting (for contour objects) or by simple signs (for point objects). These maps are united within a single panel, and the system supports their simultaneous scaling so that they always have the same size and show the same territory; this simplifies their joint analysis.

In producing data presentations IRIS keeps a somewhat different strategy from those adopted in other systems for automated visualization design: it does not strive to produce the best presentation. Instead, it aims to supply the user with *multiple different presentations* of the same data whenever possible. We assume that in the course of data exploration the user needs to perform tasks of different types, such as general view of the spatial distribution of values, locating objects with specific properties, comparing values or distributions, and so on. It is impossible to determine beforehand which tasks will emerge and in what sequence. So, to our opinion, it makes no sense to ask the user for the type of task to be performed and then generate a presentation that best suits exactly this task. As MacEachren (1994) has claimed, "the concept of selecting an optimal map, although possibly relevant for presentation, becomes less relevant (and perhaps even counter-productive) as we approach the exploration end of the continuum" (here the author means the continuum of possible map applications from communication to data exploration). For this reason it is very important to give the analyst the opportunity to work with several complementary presentations of the same data that could support the variety of subtasks emerging during the analysis. Accordingly, IRIS always produces all visualizations allowed by the characteristics of given data.

## 19.4. Data exploration support by interactive manipulation tools

Presentation of spatially referenced data on a map is a necessary prerequisite for the analysis of such data, but this is just a prerequisite. The real work of an analyst begins only at this moment. The conventional GIS do not support this work. Our goal has been to develop an environment for visual data exploration, and therefore we should include in IRIS some tools that support map exploitation. For this purpose we designed dynamic manipulation facilities. We intentionally avoided the more widespread term *direct manipulation* because some authors understand it as "a literal replication of physical behavior in the real world" and distinguish it from indirect manipulation (Tweedie 1997). For our purposes this distinction has no importance.

Our approach to supporting data exploration by interactive manipulation with graphics consists in the creation of *specialized* manipulation tools. It differs, on the one hand, from the approach adopted in other systems exploiting interactive manipulation (Tweedie 1997): we combine interactivity with design of different types of visualization depending on data characteristics rather than exploit one pre-selected presentation. On the other hand, our interactive tools differ from those offered by Goldstein and Roth (1994). The latter ones are designed to control the scope of data presented and do not account for peculiarities of different visualization techniques.

Our approach can be explained as follows. As is commonly recognized, each presentation method offers its own means for analysis, that is, due to certain properties of a given presentation type, some analytical tasks can be done more easily and more effectively than with others. Our idea is to *reinforce* these potential capabilities by adding dynamics to data displays employing this presentation method. The dynamic changes to be enabled *depend on the properties* of this type of presentation. Hence, the tools allowing such dynamics need to be specialized, that is, designed individually for each method. Given below are some examples of interactive manipulation facilities offered in IRIS.

### *19.4.1. Interactive visual comparison with a number*

This tool is designed for choropleth maps representing values of a numeric attribute by the visual variable value, that is, by color shades in which objects are painted: the greater is the value of the attribute, the darker is the shade. A choropleth map is good for the study of the spatial distribution of attribute values: colors are promptly perceived by a human; similarly colored adjacent spots tend to be perceived together as larger figures (images), and this favors finding interesting spatial patterns.

The visual comparison tool (Figure 19.2) allows selection of some number *N* between the minimum and maximum values of the attribute shown. In response the map is repainted so that values grater than *N* are depicted by shades of green and those less than *N* are shown by shades of cyan. The greater the difference between some value and *N*, the darker the shade used to represent it. The values exactly equal to *N* are shown in light yellow. The map is immediately repainted after any change of the basis for comparison. There are several ways to control *N* (see Fig. 2): entering an exact number, moving the slider, selecting an object in the map or in the list (the attribute value associated with this object becomes the reference value for the comparison), automatically locating the object with the previous or with the next value of the attribute.

So, visual comparison adds *color hue* to the expressive means used in the map. This encourages visual grouping of objects: neighboring objects painted in the same color tone tend, despite differences in shades, to be perceived together as a single figure. This evidently favors revealing spatial patterns. Thus, in the example map in Figure 19.3 it is clearly seen that the smallest percentage of children (less than the lower quartile) in Bonn is in the center of the city.

Another argument for the use of dynamic comparison is that perception of the degree of darkness strongly depends on the context. A shade seems lighter in a dark surrounding and darker in light surrounding, as may be seen from Figure 19.3. Therefore, when two non-adjacent area objects are painted seemingly in the same shade, one should not rely on the capabilities of the human eye in deciding whether one of the corresponding values equals, is greater than, or is less than another. However, with the dynamic choropleth map the user needs only to point with the cursor to one of the objects and see whether another object has become yellow, green, or cyan.

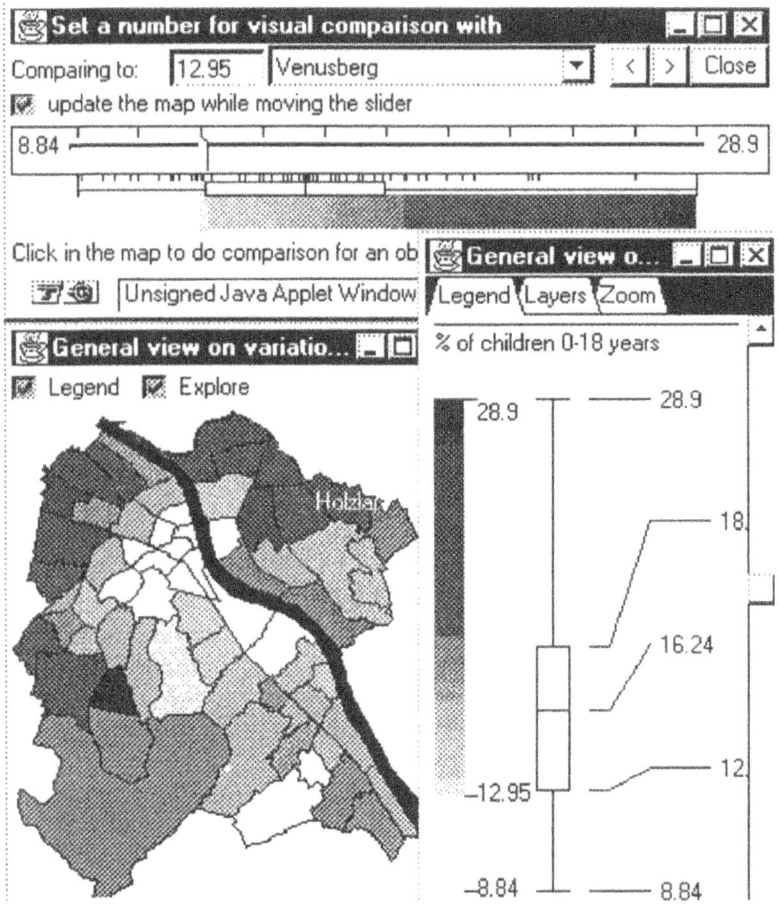


Figure 19.2. Visual comparison with an interactively selected number on a choropleth map

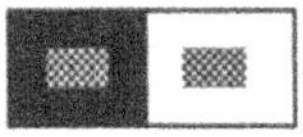

Figure 19.3. The two gray rectangles have the same degree of darkness (value), but the right one seems darker due to the light surrounding

A similar tool can be applied as well to maps encoding values of a numeric attribute by heights of bars. In this case bar heights are changed so as to be proportional to the differences between values of the attribute and the basis for comparison *N*. So, this tool operates with the visual variable *size* on which the presentation method is based. For better legibility, positive differences are shown by green bars and negative by cyan ones.

To illustrate better the idea of *specialization*, we would like to give here an example of a completely different dynamic manipulation tool.

### *19.4.2. Dynamic separation of qualitative values*

This tool works for a map presenting values of a qualitative attribute by colors or shapes. It allows the user to switch off depiction of some values temporarily. Corresponding colors or shapes are removed from the map. This helps to concentrate the attention on the remaining values and to see their distribution more easily. With shifting of the focus from one group of values to another, the user changes the selection of values to be depicted, and the map is immediately redrawn. The work of the dynamic separation tool is illustrated by Figure 19.4.

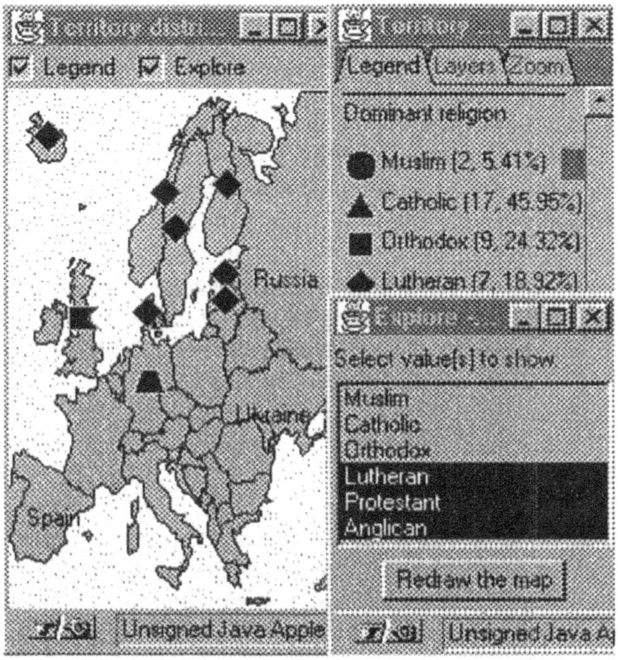


**Figure 19.4. Dynamic separation on a map representing qualitative values by shapes**

Dynamic separation facilitates study of groups and their spatial distributions with presentations based on the use of associative visual variables (*hue* or *shape*). This is especially important for ***shape*** that is, according to Bertin, only associative but not selective. Dynamic separation helps to isolate particular qualitative categories or groups of categories visually. After this has been done, the associative potential of shape or color promotes seeing the geographical distribution of these selected categories as a single image. By contrast, when all categories are shown in a map, it is difficult to focus on some of them and disregard the others.

In addition to visualization and related interactive operations, IRIS contains such facilities for data analysis as querying, calculations in spreadsheet manner, and generation of derived attributes.

### 19.5. Implementation

The latest version of IRIS is designed for running in the WWW; the URL is *http://allanon.gmd.de/and/java/iris*. It has a client-server architecture. The server part is written in C++ and operates on a Unix-based (Solaris or Linux) WWW server. It communicates with the client part written in Java and running on any WWW client computer under Netscape, Internet Explorer, or other WWW browser enabling Java. The server performs all operations with data as well as presentation design. The results are sent to the interface part for displaying on the screen. The result of map design, in the form of a map specification, is realized by the interface part.

This approach differentiates IRIS from other tools providing cartographic presentations on the WWW (Grossley 1996): they are based on CGI interface, that is, building on a server and transferring a static raster picture via the Internet. The latter approach cannot provide such a wide range of interactive actions that we considered essential to implement in IRIS. A little more interactivity can be provided by the use of the Tcl/Tk plug-in software working with WWW browsers, as in the Argus project (Dykes 1997). Still, this is not sufficient for extensive manipulation of data and graphical displays.

The implementation of the user interface in Java provided the opportunity of access to the system with the use of standard Internet browsers. At the same time platform independence was obtained: at the present moment Java programs can work on all widely used hardware and software platforms. Realization in Java allowed wide access to the system via the WWW.

### 19.6. Conclusions

IRIS is designed to be an intelligent environment for visual exploration of spatially referenced data. Two complementary instruments are employed to achieve this goal: *automatic data visualization* and *dynamic manipulation* of produced displays. Data visualization is done on the basis of a knowledge base on map design. Due to this the maps being generated conform to the principles of graphical presentation of information that prevents misinterpretation of the displays by the user.

Automation of map generation gives many benefits to the user. The user is required only to select data subsets for analysis, and after this can concentrate directly on problem solving. The user does not need to care about what presentation methods to select and which GIS operations to employ for producing the map, saving time and effort. Work with the system is easy. The user is not required to be an expert in cartography.

It is important that IRIS, whenever possible, provides the user with several maps and auxiliary displays showing the same data. Each of these presentations gives a somewhat different view on the data and allows seeing different features. In this way conditions are created for comprehensive study of the data.

Dynamic manipulation tools are supposed to support map exploitation for data analysis. Our approach is to create for each presentation method a specific tool that takes advantage of the principal features of the method and promotes the kind of analysis this method is suitable for. We expect that in this way the user can utilize the potential of each presentation more completely and effectively.

The system was tested in various domains such as accounting of forest resources of Russia, analysis of the economic situation in regions of Russia, population and economic indices for European countries, and demographic information for Bonn City. In all cases the maps built by the system were consistent with characteristics of data under analysis. The maps made apparent the peculiarities of the analyzed phenomena previously known to domain experts. This illustrates the correctness of reflection of the data. In some cases the experts uncovered interesting facts and relationships that were earlier unknown to them. This makes us believe that instruments offered by IRIS do support the main goal of data exploration: to gain new knowledge about data.

The work on reinforcing visualization by dynamic manipulation tools is to be continued. We also plan to undertake more strict and objective evaluation of the tools from the perspective of their productiveness for data exploration.

## References

Bertin J 1983 *Semiology of Graphics. Diagrams, Networks, Maps.* The University of Wisconsin Press, Madison

Casner S M 1991 A task-analytic approach to the automated design of graphic presentations. *ACM Transactions on Graphics* 10(2): 111–151

Cleveland W S 1993 *Visualizing Data.* Hobart Press, Summit NJ

Dykes J A 1997 Exploring spatial data representation with dynamic graphics. *Computers & Geosciences* 23(4): 345–70

Goldstein J, Roth S F 1994 Using aggregation and dynamic queries for exploring large data sets. *Proceedings of CHI'94, Boston.* ACM Press: 23–29

Grossley D, Boston T 1996 A generic map interface to query geographical information using the World Wide Web. *World Wide Web Journal* 1: 723–35

Jung V 1995 Knowledge-based visualization design for geographic information systems. *Proceedings of the 3rd ACM International Workshop on Advances in Geographic Information Systems, Baltimore.* ACM Press: 101–108

MacEachren A M 1994 *Some Truth with Maps: A Primer on Symbolization and Design.* Association of American Geographers, Washington DC

Mackinlay J 1986 Automating the design of graphical presentation of relational information. *ACM Transactions on Graphics* 5(2): 110–41

Monmonier M 1996 *How to Lie with Maps.* The University of Chicago Press, Chicago

Murray B S 1994 Intelligent information presentation systems. *The Knowledge Engineering Review* 9(3): 269–286

Roth S M, Mattis J 1990 Data characterization for intelligent graphics presentation. *Proceedings of SIGCHI'90: Human Factors in Computing Systems Conference, Seattle.* ACM Press: 193–200

Senay H, Ignatius E 1994 A knowledge-based system for visualization design. *IEEE Computer Graphics and Applications* 14(6): 36–47

Tufte E R 1983 *The Visual Display of Quantitative Information.* Graphics Press, Cheshire

Tukey J W 1977 *Exploratory Data Analysis.* Addison-Wesley, Reading Mass.

Tweedie L 1997 Characterizing interactive externalizations. *Proceedings of CHI'97, Atlanta*: 375–82

Zhan F B, Buttenfield B P 1995 Object-oriented knowledge-based symbol selection for visualizing statistical information. *International Journal of Geographical Information Systems* 9(3): 293–315

# 20 GEOTOOLKIT: OPENING ACCESS TO OBJECT-ORIENTED GEODATA STORES

Oleg Balovnev, Martin Breunig, and Armin B. Cremers

## 20.1. Introduction

Today a typical geographic information system (GIS) is a complex software package which inevitably inherits the software-engineering practices of the past. Application programming within such systems is extremely complicated. Data structures and functions are often completely hidden from the user. As a result, they are hardly extensible to meet the requirements imposed, for instance, by 3D and 4D modeling. The next generation GIS should benefit from modern software engineering technologies, among which one of the most promising is component-based design. Software building organized in libraries with consistent programming interfaces will enable the fast assembly of special-purpose applications for particular domains. An application-specific component can be customized and re-used (with necessary extensions and modifications, if required) for the development of related applications. Following this approach, today's general-purpose GIS will be replaced by a family of specialized sub-systems which, due to their common design basis, are well suited for inter-communication and mutual data exchange.

An object-oriented programming environment can provide the necessary support for the integration of diverse software components. In general, we are strongly convinced that object-orientation (and object-oriented data management in particular) will play a key role in the development of future generations of GIS because it has the ability to dissolve the boundaries between programs and databases, allowing data structures used in numerical models, for example, to be persistent objects, maintained directly by object-oriented database management systems. Complex data structures optimized for efficient computations in main memory need no longer be pre-assembled from numerous relational tables. An integrated representation for algorithms and data within object-oriented databases enables access not only to the data but also to the data processing methods associated with the data. Consequently, object-oriented geodata stores could play the central role in data and service exchange between heterogeneous geoscientific applications: uniformly stored database objects can serve as mediators between diverse application-specific representations. However, direct database-level interoperation is not always possible because of the extreme heterogeneity of existing applications, software environments, and hardware platforms. A necessary infrastructure to deal with the heterogeneity and remote access in the object-oriented context can be provided by a Common Object Request Broker Architecture (CORBA).

In this chapter we present our experience with *GeoToolKit*—a component software for the development of 3D and 4D geoscientific applications. Its backbone is a class library for the efficient storage and retrieval of spatial objects within an object-oriented database. Common data types inherited by diverse applications from the GeoToolKit spatial class hierarchy establish favorable conditions for the database-level integration of heterogeneous geoscientific data. The key for the data integration is a common object model built on top of general spatial object class hierarchy provided by GeoToolKit. GeoToolKit has been successfully used in the development of several geoscientific applications. We briefly introduce two of them, *GeoStore* and *GeoDeform*, which interoperate via a shared geodata store. In conclusion we present a CORBA-based technique for opening geodata stores to external applications.

## 20.2. GeoToolKit

Within the collaborative research center SFB350 at the University of Bonn we have developed a component software called *GeoToolKit* (Balovnev et al. 1997) which is intended to facilitate the design and implementation of 3D and 4D geo-applications. The idea is to provide the application developer with a set of geo-oriented software building blocks involving DBMS-based spatial data maintenance, special support for efficient spatial retrieval, visualization, graphical interfaces, and communication, which the user may assemble in ready-to-use applications. GeoToolKit is not a closed GIS-in-a-box package; it is rather a library of C++ classes that allows the incorporation of spatial functionality within an application under development. It is primarily oriented to software engineers with C++ experience involved in the development of geo-applications for tasks which can be hardly performed within general-purpose GISs. At the same time GeoToolKit is not restricted to the class library. It also assumes a certain design methodology evolved on the basis of the experience we gained during the development of diverse applications on top of GeoToolKit. Some interactive tools under development involving a universal spatial data browser are expected to make GeoToolKit-based data stores available for the geoscientists without the mediation of software engineers.

Special attention was paid to the generality and extensibility of the class library so that it could be easily adopted for diverse requirements. At the same time it was obvious to us that a toolkit should be more than just an empty interface specification, that is, it should do something. GeoToolKit addresses primarily the efficient storage and access of 3D spatial objects within a database. To achieve this GeoToolKit is tightly coupled with the object-oriented DBMS *ObjectStore*. We did this in spite of potential disadvantages in binding to the concrete DBMS: integration with still non-standardized object-oriented DBMS may demand a cardinal re-design of both interfaces and function bodies. We preferred from the very beginning to benefit maximally from ObjectStore's specific features; for example, we exploited its low-level clustering control facilities intensively while developing DBMS-maintained spatial indices. Though GeoToolKit is built on top of ObjectStore it is organized in such a way that its users do not need to know anything special about ObjectStore: all necessary data management facilities are incorporated in the GeoToolKit classes.

Currently GeoToolKit offers classes for the representation and manipulation of simple (point, segment, triangle, tetrahedron) and complex (curve, surface, solid) 3D spatial objects. We consciously avoided 2D issues as far as possible. Following the object-oriented modeling technique, not only abstract geometric primitives, such as points, curves, and surfaces but real-world entities such as drilling wells, geological sections, or strata, can be modeled and maintained. Applications developed with GeoToolKit simply inherit geometric functionality from GeoToolKit, extending it with application-specific semantics.

### *20.2.1. Class hierarchy*

GeoToolKit is primarily oriented to managing real-world entities modeled as bodies with arbitrary complex shapes. However, such objects as curves, surfaces, lines, and planes (though they do not correspond to any real-world object) also need to be maintained in GeoToolKit since these very useful geometric abstractions are utilized in various modeling and analysis tools. Figure 20.1 presents their relationships within the data model of GeoToolKit shown in OMT-like notation (Rumbaugh et al. 1991).

The abstract *SpatialObject* class specifies the interface which is to be inherited by all concrete spatial objects. Any concrete spatial object must be defined as a specialization of the abstract SpatialObject class. A concrete class provides an appropriate representation for the object as well as the representation-dependent implementation for the geometric functions. For example, since a minimal bounding box is permanently used for efficient computations, the SpatialObject class declares a function which returns a bounding box of the object. This function must be redefined and implemented in every direct specialization of the SpatialObject class. Thus GeoToolKit guarantees that every spatial object has at least the functionality of the most general class. Every particular spatial object class can additionally have geometric functions characteristic to only that class.

A real-world object usually containing additional non-spatial data is easily modeled as a specialization of a concrete spatial object, complemented with data members and methods provided by the application developer. Following this approach, a user-defined class automatically inherits the whole geometric functionality of its geometric parent. Sometimes it may be beneficial to introduce a completely new type. Then, instead of re-using GeoToolKit's types, an application developer creates his or her own class as a direct specialization of the abstract SpatialObject class. In this case he or she is responsible for providing the implementation of the geometric functionality specified in the root class. Newly defined data types can be included in the GeoToolKit class library for further utilization.

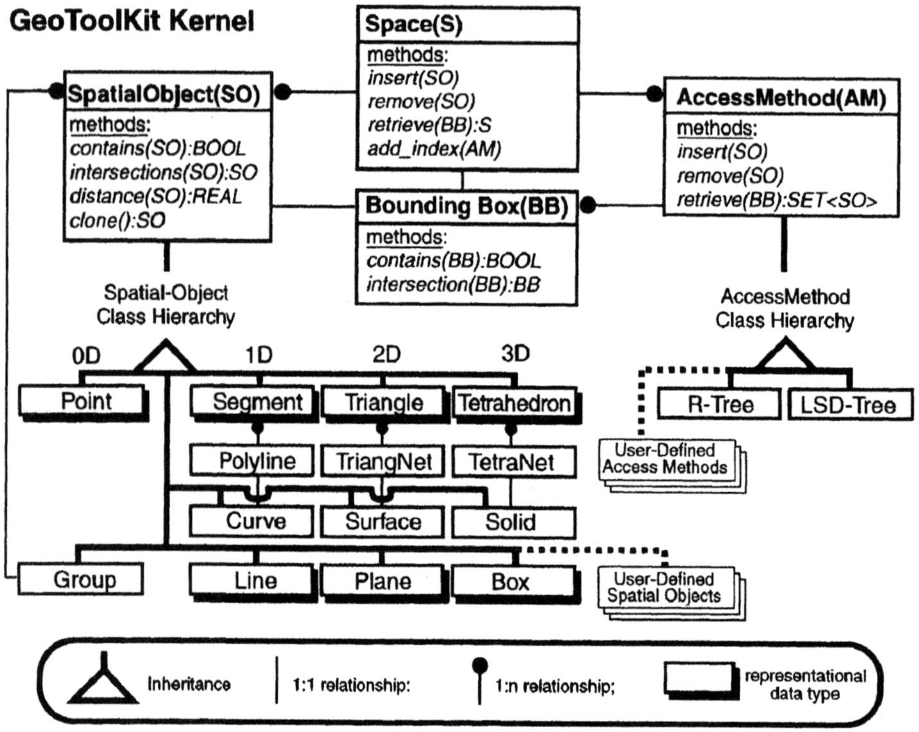


Figure 20.1. GeoToolKit object model

The GeoToolKit class hierarchy is complete: any spatial object can be modeled either directly by one of the built-in spatial classes or as a composition of these classes within a group. A *Group* is a heterogeneous collection of spatial objects which are further treated as a single object. It may contain simple and complex objects, as well as objects of different dimensions including other groups. Operations applied to a group are simply forwarded to separate objects contained in the group. Object grouping enables the organization of data in high-level abstractions without any loss of spatial functionality. Groups are typically used to represent the results of geometric operations.

The representation of spatial primitives—topological simplexes—is straightforward. A point is specified through three coordinates (floating point numbers). Other simplexes are specified via their definition points. The constructor of a simplex automatically checks the consistency of its data; for example, in the case of a triangle the definition points should not lie on the same line. A simplex incorporates a corresponding representational data type. Instances of a representational data type are not independent standalone objects. They exist in a database only as parts of spatial objects. They disappear together with the spatial objects that incorporate them. Representational data types are used primarily to permit a more compact storage of data.

### 20.2.2. Spatial functionality

The number and types of diverse geometric predicates and operations which can be applied to 3D objects depends on the application domain. We started with the operations primarily required in 3D interactive geological modeling (Balovnev et al. 1997), trying to keep the interface as general as possible to permit reuse in other domains. Currently GeoToolKit supports the following operations: intersection of spatial objects, partitioning of a space or spatial object by a plane, clipping a part of a space or spatial object by means of bounding box, and equality and containment checks.

GeoToolKit distinguishes between geometric predicates returning *true* or *false* (*equal, intersect, contains*) and geometric operations (*intersection, division*) returning a new spatial object. Geometric operations are algebraically closed: the result of a geometric operation is a new spatial object, usually a group which can be stored in the database or used in other geometric operations. Geometric operations can produce an object which is not contained in the original class hierarchy (e.g., an intersection of two triangles can result in a quadrangle). In this case the resulting object is automatically decomposed (depending on the dimension), either into a triangle or a tetrahedron network. An operation applied to a complex object often results in a set of separate disconnected objects which cannot be transformed into a single complex: in this case the result is represented as a group of objects.

### 20.2.3. Spatial retrieval

Spatial objects are maintained within special containers (*spaces*) capable of the efficient retrieval of elements according to their location, which can be specified either by a point or by a spatial interval (bounding box). A family of *retrieve* methods provides a convenient program interface for the spatial query manager. In the simplest case a retrieve member function takes a bounding box as a parameter and returns a set of spatial objects contained in or intersected by this bounding box. A user can formulate his or her query in the ObjectStore style as well.

Since all operations in the Cartesian coordinate system are considerably faster for cuboids than for other objects, the approximation of spatial objects by their bounding boxes is intensively used as an effective pre-check in geometric operations and spatial retrieval. To provide an efficient retrieval, a space utilizes spatial indices. In addition to the embedded spatial indices, the user can implement and integrate his or her own indexing method within GeoToolKit. In order to make an arbitrary user-defined index known to GeoToolKit, it must be defined as a specialization of the abstract *AccessMethod* class.

A spatial retrieval involving *indirect* spatial predicates (e.g., intersects) is usually decomposed into two sequential steps. On the first step (pre-selection) the query manager retrieves all objects which intersect the bounding box of the given object. On the second step it checks whether the pre-selected objects really intersect the given object.

### *20.2.4. Representing complex objects*

Simplexes *per se* are rarely used for the representation of real-world objects. They are more interesting as primitive building blocks for the representation of complex 3D shapes. According to the simplicial complex approach (Egenhofer et al. 1989) an arbitrary shape can be approximated and then decomposed into a set of adjacent elementary components, or simplices of the same dimension. The main advantage of such representation is that geometric operations on arbitrary complex shapes are reducible to the operations on the restricted set of primitives. Such representation is especially beneficial for the maintenance of arbitrary non-regular shapes which are typical for the majority of geo-applications (Breunig et al. 1994). As a default representation for a curve we use a polyline. A surface is approximated as a connected set of adjacent triangles, organized in a special data structure referred to as a triangle network. A solid can have two alternative representations: either a bounding surface approximated by a triangle network or a tetrahedron network. The first one, widely used in computer graphics, is more compact. However, for geo-applications a tetrahedron network has certain advantages because each internal point incorporates features characteristic of the whole solid which are always preserved after a series of diverse geometric manipulations and transformations. As default representations for the classes *Curve*, *Surface*, and *Solid*, GeoToolKit uses classes *Polyline*, *TriaNet*, and *TetraNet*, respectively.

As mentioned above, an arbitrary geometric operation on two complexes is reducible to the sequence of operations on pairs of simplices. Obviously, in the case of two large-scale complexes the straightforward application of the operation to every combination of simplices will fail. The optimization is concerned primarily with the restriction of simplices which participate in the operation. In other words, a special organization of simplices is required which could provide a fast pre-selection only of such pairs of simplices which are really involved in the operation. For this purpose, some operations use the neighborhood information stored explicitly with every network element. Additionally complexes are supplied with internal spatial indices which allow an efficient pre-selection of simplices located in a given subspace. Using inheritance and function overloading mechanisms, the developer can easily substitute the default optimization strategy with his or her own.

Naturally, the application developer can still prefer to use a closed boundary surface to model a solid; then the built-in surface type should be employed. However, the semantics of geometric operations and predicates will be different from those expected for a filled solid.

In general GeoToolKit does not restrict users to its representations. Complex spatial objects contain a reference to a dependent data structure referred to as the object representation. Abstract classes (*Surface, Solids*) simply delegate the functionality to the current representations (*TriangleNet, TetraNet*). Such a two-layer architecture allows multiple representations for the same object (e.g., one representation for compact storage, another more redundant one for efficient computations). A complex object can change its representation without changing the object identity. This feature is of extreme importance in the ObjectStore context since an object can be referred to from multiple sources. User-defined representations may be used to increase the efficiency of particular geometric operations.

## 20.3. Developing applications on top of GeoToolKit

GeoToolKit has been successfully used for the development of several applications. The first and the most advanced one is *GeoStore*, an information system for the management of geologically-defined geometries (Bode et al. 1994). Initially it was developed without GeoToolKit. Our GeoStore experience influenced to a large degree the architecture and functionality of GeoToolKit. Before launching the development of new applications, we decided to re-design GeoStore on top of GeoToolKit in order to appreciate the benefits this approach would bring (Balovnev et al. 1997).

The intention of GeoStore was to supply geoscientists with a tool which would provide a consistent storage and efficient access for the data involved in all stages of interactive 3D modeling using a modern database technology. The starting point for the interactive geological 3D and 4D modeling is the digitization of geological *sections* gained from open-cast workings. A section is a geological abstraction obtained as the result of an intersection of a vertical plane with geological strata and faults. Geological strata are sheet-like structures in the Earth with different mineral composition, texture, or grain size. A fracture in the Earth's materials along which the opposite sides have been displaced is known as a fault. A stratum is modeled as a list of layered bodies which are usually specified by their bounding surfaces. A fault is modeled as a simple surface. Within a section, strata and faults are represented as point sets grouped in stratigraphic and fault lines. Every point in a section contains 3D coordinates complemented with geologically-specific data such as stratigraphy. The second step in the interactive 3D modeling is the generation of *triangulated surfaces* spread between sections. The final step which is a part of our current work is the transition from stratigraphical bounding surfaces to *volumes*.

The main design principle we followed was to inherit from the geometric kernel as much functionality as possible. For every geological entity we found its geometric entity which served as its ancestor in the class hierarchy. For example, a geological stratum is represented now as a specialization of the geometric entity *Solid* which was extended with geology-specific features (e.g., stratigraphy) and relationships (e.g., the *Hanging* stratum). By using inheritance, we obtained a representation in which a geological entity could be treated as a geometric object. GeoToolKit's geometric kernel was able to operate directly on the collections of geological objects without intermediate transformations. Topological relationships (e.g., neighborhood) provided very useful hints not available earlier in the pure geometric world. In order to benefit from this additional information we simply re-implemented virtual functions. An overridden function either replaced the geometric function completely or (more often) performed a kind of pre-selection for the geometric function, calling it with the restricted subset of spatial objects.

Another application built on top of GeoToolKit which is worth mentioning in the current context is *GeoDeform*, which performs advanced geological modeling and animation (Alms et al. 1997). The GeoToolKit-based database maintains time-varying geological strata and faults. They have a static geometry as defined in GeoStore with special dynamic extensions to deal with time. Direct access to the database from GeoDeform was not possible because of both hardware and software incompatibilities. We had to organize their interoperation using standard UNIX-sockets and a remote procedure call mechanism.

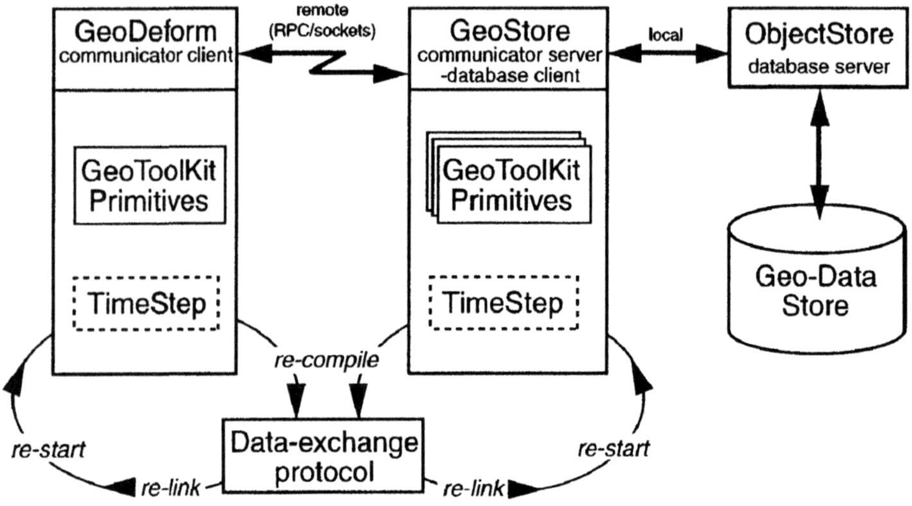


Figure 20.2. Interoperation between GeoStore and GeoDeform

## 20.4. Remote access to GeoToolKit-based data stores

Common underlying data structures provide a necessary basis for the successful integration and interoperation of geoscientific systems built on top of GeoToolKit. Shared object-oriented geodata stores can serve as mediators for data and services exchange. However, as mentioned above, a direct database-level client–server communication with GeoToolKit-based data is not always possible because of the extreme heterogeneity of the already-existing applications and hardware platforms we have to deal with. Many applications simply are not available for the modifications required for direct integration with ObjectStore. Database-level communication facilities supported within DBMSs are not usually flexible enough for the advanced navigation in the network. Apart from this in ObjectStore all data processing (including queries) is carried out on the client site. In some cases this can lead to an unjustifiable overload of the network. For example, when we need to pick a single object from a large collection, the whole collection has to be transmitted through the network from the server to the client in order to test the selection predicate. The alternative solution is to perform the selection on the server site and transfer to the client only the selected object. That means that instead of transferring raw data through the network it is preferable to ship an operation which would be applied to the corresponding data on the server site.

Standard Unix sockets and RPC mechanisms turned out to be very efficient for the direct communication between two particular applications. However, these services are optimized for performance, rather than ease of programming, reliability, portability, flexibility, and extensibility. The most serious limitation is the lack of generality. Communication conventions are hidden from the other world. Another application cannot be made aware of these internal conventions other than after a painful re-design. To avoid this we need a kind of standard distributed-object computing platform. Taking into account the object-oriented nature of data the most suitable solution is *Object Management Architecture* (Object Management Group 1997) which promises to become a worldwide standard. Using this approach any other CORBA-compliant application can get open access to GeoToolKit-based data stores.

The advantage of CORBA is that it delegates much of the tedious and error-prone complexity associated with low-level socket-layer programming to its reusable infrastructure. An application needs only hold a reference to a target object. The ORB is responsible for automating other common activities which usually include locating a suitable target object, activating it if necessary, delivering a request to it, and returning any response back to the caller. Developers can focus on more essential things than analyzing diverse networking problems. Parameters passed as part of the request are automatically and transparently marshaled by the ORB, ensuring a correct interoperation between objects residing on different platforms.

CORBA object interfaces are described using an Interface Definition Language (IDL). Since the IDL specifications are automatically translated into the programming languages, potential inconsistencies between the client stubs and the server skeletons are significantly reduced, providing a higher degree of type safety. For large distributed systems the loss of micro-level efficiency that is inevitable for such universal systems is compensated by the increased extensibility, robustness, and maintainability; that is, macro-level efficiency.

CORBA defines a flexible distribution model for transient objects. However, we have to deal primarily with persistent objects, maintained within object-oriented DBMSs. These facilities are still not standardized. To capture persistence we used a method based on the replacement of the *Basic Object Adapter* (BOA) by a specialized *Object Database Adapter* (ODA), which extends BOA by extra functions for persistent object management and provides a tight integration between the ORB and persistent object implementations. We chose *Orbix* as an implementation platform because it already provided a special ObjectStore ODA which is completely responsible for the management of persistent objects within ObjectStore.

The fact that we have to deal with large data stores which are permanently in use by multiple geoscientific applications imposes additional requirements on the interoperation architecture. First of all, making these data stores CORBA-compliant should not disturb already-existing applications. Since a schema evolution may be a very exhausting process for large data volumes, in the ideal case an arbitrary data store should be made CORBA-compliant without changing where and how the data have been already stored. To achieve this, we used the database wrapping technique. A wrapper encapsulates the underlying data and mediates between data stores and diverse geoscientific applications. In our case the wrapper encapsulates access to all GeoToolKit classes. While reproducing the GeoToolKit class hierarchy in CORBA's

IDL we tried to keep the interfaces as close as possible to the original C++ ones (Figure 20.3).

Within the TIE binding technique a skeleton generated by the IDL compiler corresponds to a TIE object. According to the proposed approach, every TIE object class gets a reference to a *mediator* class which in turn contains a reference to the real database object. The goal of the mediator is to provide a correct mapping between CORBA-compliant objects and database objects. When a mediator class is activated by a TIE object it converts the parameters and forwards the function call to the corresponding database object. When the operation is completed, the mediator converts the result back to the CORBA-compliant representation and returns control to the TIE object. This enables the reuse of the entire functionality of GeoToolKit with minimal expense. A request for the creation of a new object at the client site results in the creation at the server site of the triple of objects: TIE object, mediator object, and database object.

```
exception GTK_ObjectExists { } ;
exception GTK_ObjectNotFound { } ;
  ...
interface GTK_Space : GTK_Object {
// operations on spatial objects
void insert (in GTK_SpatialObject
  raises (reject);
void remove (in GTK_SpatialObject
  raises (GTK_ObjectNot Found) ;
GTK_SpatialObject
  retrieve (in GTK_BoundBox bb) ;
GTK_Space
  intersect (in GTK_SpatialObject
  ...
};

interface GTK_SpatialObject : GTK_Object
// spatial predicates
boolean intersect (in GTK_SpatialObject
  ...
// spatial operations
GTK_BoundBox getBoundBox ( ) ;
GTK_SpatialObject
  intersection (in GTK_SpatialObject
};
interface GTK_Line : GTK_SpatialObject
// functions specific for the lines
  ...
};
```

Figure 20.3. Fragments of IDL specifications for GeoToolKit classes

Spatial objects are usually not maintained as separate database entities with external names. They are typically encapsulated in large collections, in spaces which usually serve as entry points in spatial databases. A particular spatial object is accessed indirectly via diverse retrieval functions associated with the space class. Our technique is based on the complete delegation of the spatial functionality to the GeoToolKit class library. That means that on completing retrieval we have at the server site a set of database objects which satisfy the selection predicate. However, for the communication with the client we need not database objects but TIE objects. The straightforward method is to extend every database object with the additional reference to its TIE counterpart. However, for the exiting databases this will require the modification of the database schema which is not desirable.

To avoid an exhaustive schema evolution we introduced a persistent dictionary with a single key, the database object identifier (OID). When a TIE object is created it is automatically inserted in this dictionary (Figure 20.4). Having obtained a database object as a result of a spatial query, the wrapper retrieves the required TIE object in the dictionary and forwards it to the client. Since OIDs are unique, the retrieval is very efficient. If a TIE object with the specified key is not found in the dictionary (this may happen when a database object is created by a native application which communicates with the database, bypassing the wrapper), the wrapper creates a TIE object together with the related mediator object and inserts it in the dictionary. A lazy strategy of dictionary filling encourages a concurrent operation on data for both native and CORBA-compliant applications. Since the native database stays unchanged, all local applications can continue to work with the data store as before. Moreover, any already-existing data store can be made available for CORBA-compliant applications at any time. Due to the well-defined separation of database and communication objects, the CORBA compliance can be easily eliminated when not required.

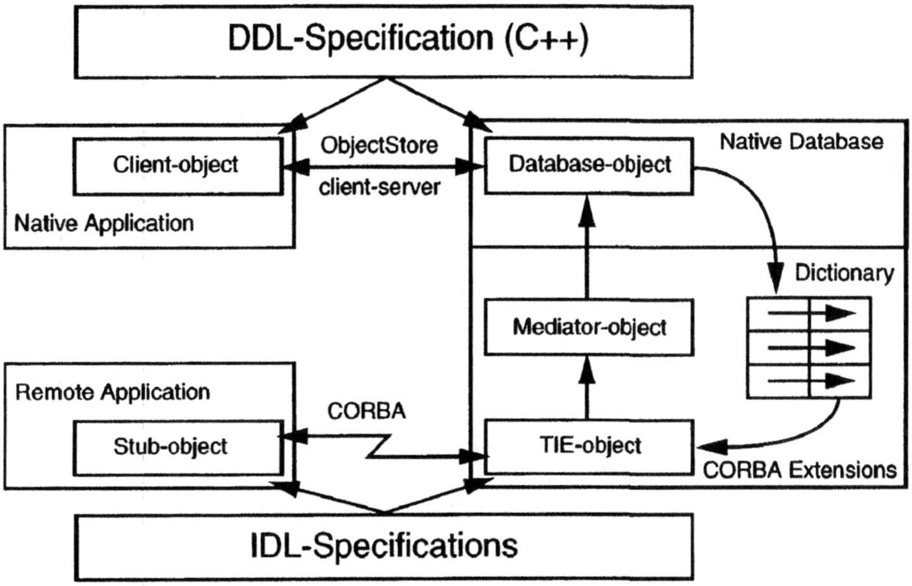


Figure 20.4. Wrapper architecture

## 20.5. Conclusions and outlook

Our experience with the development of various applications on top of GeoToolKit demonstrated that an application developer could focus on the application semantics instead of such uncreative tasks as optimal assembly of the spatial objects from multiple relational tables, or the implementation of routine geometric algorithms. This results in a considerable reduction in the code written and makes the sources more understandable. Data and functions which were previously hidden in a particular application become available for all other applications built on top of GeoToolKit. Common underlying spatial data types encourage a consistent and non-redundant maintenance of data within the framework of GeoToolKit-based applications. Using IDL specifications of the GeoToolKit classes, any CORBA-compliant remote application can obtain access to services associated with a database object. However, the introduction of a conventional integrated object model is still a task of the highest priority. Without a standard model, opening of application-specific data stores cannot be considered fully accomplished: every new external applications will still need to be tuned to the GeoToolKit data model.

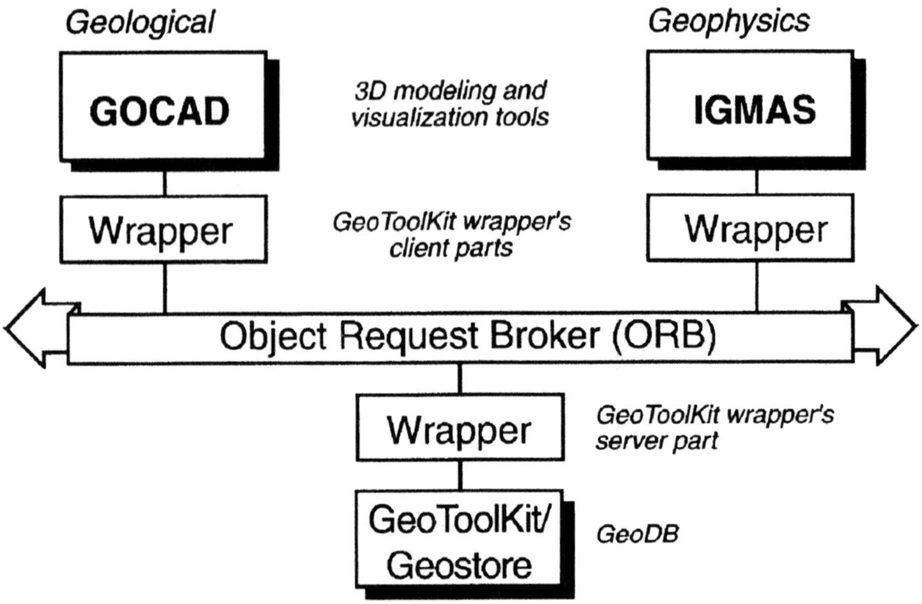


Figure 20.5. Integrated environment for 3D applications

In a recently-started project, GeoToolKit is intended to be used as a platform for a distributed environment that will provide an open access to the GeoStore database for various remote geo-services involving the GOCAD-based 3D modeling application located in Bonn and the geophysical modeling tool IGMAS residing in Berlin. A free data exchange via a common database gives geoscientists an opportunity to perform a cooperative adjustment of geophysical density and geological stratigraphic models.

## Acknowledgments

This work was done in close cooperation with the Geological Institute at the University of Bonn. We thank Prof A Siehl, R Alms and T Jentzsch who provided necessary geological backgrounds for the work. Special thanks to Norbert Klein, Dirk Moebius and Wolfgang Mueller who made a significant contribution in the implementation of software.

## References

Alms R, Balovnev O, Breunig M, Cremers A B, Jentzsch T, Siehl A 1997 Space–time modeling of the Lower Rhine Basin supported by an object-oriented database. *Physics and Chemistry of the Earth.*

Balovnev O, Breunig M, Cremers A B 1997 From GeoStore to GeoToolKit: The second step. *Proceedings of the 5th International Symposium on Spatial Databases.* Lecture Notes in Computer Science 1262. Berlin, Springer: 223–37

Bode T, Breunig M, Cremers A B 1994 First experiences with GeoStore, an information system for geologically defined geometries. *Proceedings IGIS'94.* Lecture Notes in Computer Science 884. Springer, Berlin: 35–44

Breunig M, Bode T, Cremers A B 1994 Implementation of elementary geometric database operations for a 3D GIS. *Proceedings of the 6th International Symposium on Spatial Data Handling, Edinburgh*: 604–17

Egenhofer M J, Frank A, Jackson J 1991 A topological data model for spatial satabases. *Proceedings SSD'89.* Lecture Notes in Computer Science 409. Springer, Berlin: 271–85

Object Management Group 1997 *CORBA 2.0/IIOP Specification.* OMG technical document formal/97-02-25. http://www.omg.org/corba/corbiiop.htm

Rumbaugh J, Blaha J, Premerlani W, Eddy F, Lorengen W 1991 *Object-Oriented Modeling and Design.* Prentice Hall, Englewood Cliffs, NJ

# 21 THE GEOSPATIAL INTEROPERABILITY PROBLEM: LESSONS LEARNED FROM BUILDING THE GEOLENS PROTOTYPYE

Clifford Behrens, Leon Shklar, Chumki Basu, Nancy Yeager, and Edith Au

## 21.1. Introduction

In 1994 NASA issued a Cooperative Agreement Notice to support new research on digital library technology that would enable broader public use of its Earth science data over the Internet. As a response to this CAN, the Universal Spatial Data Access Consortium (USDAC) was formed and it proposed to prototype the *GeoLens* system that would not only give broader public access to NASA's Earth observation data, but also made these data interoperate with other geospatial data served by the Federal government. Part of the challenge of the GeoLens Project has been to decompose the larger ***geospatial interoperability problem*** into constituent issues. This chapter will address these issues and describe solutions implemented in our GeoLens prototype. The purpose of this exercise is to support an end-to-end scenario, beginning with geospatial data discovery and ending with conflation of geospatial data extracts from extremely heterogeneous sources. The larger goal is to investigate the opportunity for new information processing standards and innovative digital library technology to play a key role in the realization of this scenario.

At the moment there exists growing demand for robust geospatial information services, ones capable of federating massive repositories of distributed heterogeneous data and metadata, and of integrating data extracted from these repositories. For example, BAAs issued by the National Imagery and Mapping Agency (NIMA) have sought commercial catalog services for federating geospatial data in its archives (e.g., DMA 1996; NIMA 1997). Also, the Open GIS Consortium (OGC) recently approved a set of implementation specifications to support access of simple geospatial features, and issued an RFI, concerning geospatial catalog services, as part of its process to develop a consensus-based interoperability specification that can be implemented by commercial GIS vendors and others (OGC 1996). Evidence of this demand can also be found in the increasing activity within Federal (e.g., FGDC and NASA EOSDIS) and International (e.g., ISO TC211) standards groups to define geospatial metadata content standards as part of mission-critical information infrastructure needed to implement National and Global Geospatial Information Infrastructure frameworks.

While it would seem desirable to have everyone adopt a single metadata standard to catalog their image and map data holdings, this is not realistic. In fact, already the FGDC metadata content standard (currently in Version 2) has undergone several revisions in response to comments from user communities (FGDC 1994, 1997), NIMA has proposed its own metadata standard which extends the FGDC and CIO-SPIA (CIO 1995) schema to better support its users (NIMA 1996), and Hughes has published a revision (Release B) of its EOSDIS Core System metadata standard which also extends the NASA Global Change Master Directory's (GCMD) Directory Interchange Format (DIF; NASA 1997) and FGDC schema for NASA Earth science data users (Hughes 1996). There exist other similar schema and extensions (e.g., ISO 1997; National Biological Service 1997). Hence, we assert that (1) *data schema evolution is inevitable and even healthy* and, therefore, (2) *federation-enabling geospatial information infrastructure must be extremely flexible and adaptive*.

The core capability of the GeoLens system is to provide a viable solution to federating distributed, heterogeneous geospatial data and metadata by supporting multiple dynamic standards. This approach is entirely consistent with the notion of *information communities*, as introduced in the *OpenGIS User's Guide—Abstract Specification*, and the desirability of supporting each community's *standardized* view on data while also providing translations between these views (Buehler and McKee 1996). As an illustration, this chapter will examine the manner in which the current GeoLens prototype implements the FGDC and EOSDIS metadata content standards, the FGDC Spatial Data Transfer Standard Vector Profile (SDTS-VP), and Hierarchical Data Format-EOSDIS Core System (HDF-EOS) data archive types, and other de facto and incipient data and information processing standards (e.g., the Open Geospatial Interoperability Specification) to achieve greater interoperability. We will further discuss how our implementation of these standards serves to achieve greater interoperability.

## 21.2. The geospatial interoperability problem

What makes geospatial data difficult to use? Part of the challenge of the GeoLens Project has been to decompose the larger geospatial interoperability problem into constituent, more tractable issues. These issues and their solutions in the GeoLens system are represented in the pyramid shown in Figure 21.1.

### *21.2.1. Heterogeneous computing environment*

Geospatial data are created and exist in extremely heterogeneous computing environments. For example, a variety of hardware and software platforms are employed to serve geospatial data by federal and commercial providers; and users have their own hardware and software to access and process these data. To neutralize the effect of this heterogeneity, we have implemented our GeoLens system on Web-Internet infrastructure, principally exploiting HTTP (Hypertext Transfer Protocol) for client–server communications.

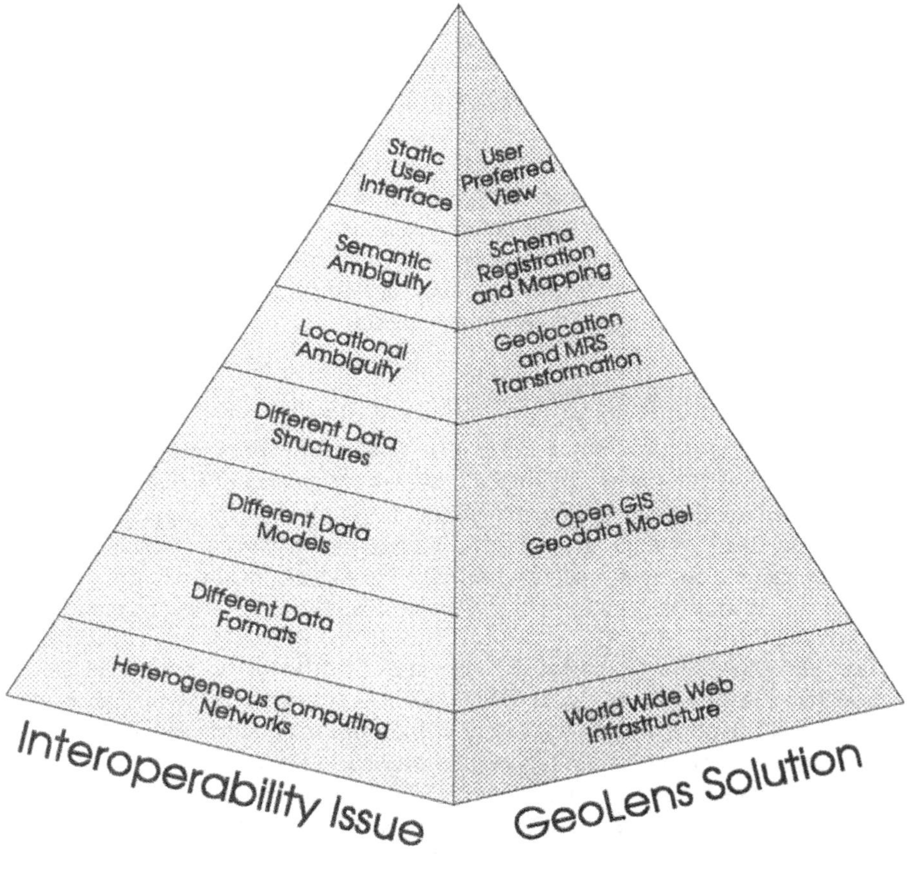


Figure 21.1. Geospatial interoperability issues and their solutions as implemented in the GeoLens prototype

### *21.2.2. Heterogeneous data (and metadata)*

Data are often acquired in *different formats*, which may make them inaccessible to an application. For example, absence of CR/LF as end-of-record delimiters in DEM data is a common problem for those GIS applications that require these delimiters. Other incompatibilities may be attributed to the use of *different models* for representing a geospatial theme (Burrough 1986; Clarke 1990).

For instance, elevation data may be modeled as grids (e.g., a DEM), or vectors (e.g., a Digital Terrain Model). To complicate matters further, the same model may store data in *different structures*. As an example, if the data are of type *vector*, then are they stored as TIGER, SDTS-VP, ARC or some other structure? These problems are exemplified in framework datasets used for our GeoLens testbed. Landsat MSS imagery is of type *raster* stored in HDF-EOS containers (Fingerman 1997), the DEM data are type grid stored in DEM format (as flat ASCII files), and the DLG data are of type vector stored in SDTS-VP (Spatial Data Transfer Standard Vector Profile) distributions (DOI 1995). We address these issues in *GeoLens Data Access Servers* by implementing OGIS-like interfaces which wrap the native file formats and access libraries as OGIS-like objects and translate data to OGIS-like formats.

Support for multiple schema standards introduces other heterogeneity in metadata. Existing metadata content standards have different topologies for their attribute representations, and these may even be extended with the introduction of new standards. To account for metadata evolution in a uniform manner, the GeoLens encapsulates metadata-describing content into hierarchies of attributes and their values. These attributes are further grouped according to whether they describe individual data objects or collections of data objects.

### *21.2.3. Locational ambiguity*

Many applications require only a *subset* of a geospatial data object, for example, a subscene of a Landsat image, to provide coverage for an area of interest. So there is also the need to discover efficiently all of the current image, topographic, and cartographic data available for an area, browse the metadata for these data to determine their usefulness, and retrieve only the highest quality data required to cover the study area but in a form, and with sufficiently accurate georegistration, so that they can be used together.

Thus, another serious problem that can make it difficult for geospatial data to interoperate is *locational ambiguity*, due to poor (or no) georegistration. This is often the case with imagery whose areal extent or *footprint* may be crudely approximated to support discovery, but whose registration may be so inaccurate as to make its use with other georectified data inappropriate. A less severe problem exists when geospatial data are registered to different Map Reference Systems, for example, the 1 degree DEMs mentioned above are registered in geographic coordinates while the Landsat MSS imagery is registered to the Universal Transverse Mercator grid. *GeoLens Data Access Servers* solve this problem by requiring geolocated data, but also provide translations between map reference systems, when necessary (DMA 1983; Snyder 1987).

### *21.2.4. Semantic ambiguity*

Semantic ambiguity exists when different meanings are associated with the same term (*polysemy*), or when different terms mean the same thing (*synonymy*). For example, *floodplain* may have different meanings to a civil engineer who views this feature as an area that may need protection from inundation, and an insurance claims adjuster whose notion may only include the area of financial liability. Terms used to attribute meaning to geospatial objects, and the semantic relationships between terms, may be expressed formally as schema (MacEachren 1995; Medyckyj-Scott and Hearnshaw 1993). The FGDC and ECS metadata content standards, along with the DLG-3 feature schema, provide useful examples, each with somewhat unique vocabularies and grammars.

Translations between these schemas are required to support data discovery and access by the GeoLens. The Landsat MSS collection is compliant with the EOSDIS Core System Metadata Standard (Release A); while both the USGS DEM and DLG metadata are catalogued compliant with the FGDC Content Standard for Digital Geospatial Metadata (Version 1). In addition, the DLG data are attributed with the DLG-3 feature schema.

The design of the GeoLens system is driven by the need for semantic interoperability through better metadata support. We recognize that (1) metadata schemas represent particular views and capture the unique semantics of different groups, or information communities of geospatial data users, (2) multiple schemas exist for describing the content of geospatial data, and (3) these schemas will evolve over time as geospatial data are increasingly distributed and as users become more specialized and sophisticated in their applications of these data. Much of the support for multiple schema and schema translation resides in the *GeoLens Catalog Server.*

### *21.2.5. Static, non-tailored information presentation*

Finally, the manner in which information about the content of geospatial data is presented to a user can significantly affect their understanding of how it might be used (Hearnshaw and Unwin 1994; Medyckyj-Scott and Hearnshaw 1993). Many map and image browsers currently exist on the Web that allow a user to browse geospatial metadata. However, most of these present a static view of metadata; the labeling and formatting of information about a geospatial data object does not adapt well to data content or the preferred view of the user. By supporting multiple schemas, we can exploit a variety of *views* to drive an interactive presentation by the *GeoLens Client*, that is, metadata may be organized and tagged consistent with a different schema than was employed by the data provider who catalogued these metadata.

## 21.3. GeoLens solutions

Figure 21.2 illustrates the distributed architecture of the GeoLens system. It includes a *GeoLens Client*, a *GeoLens Catalog Server*, a *Schema Mapping Server*, *Data Access Servers* and, potentially, other servers to process geospatial data or their metadata (Shklar et al. 1997).

### *21.3.1. Client and graphical user interface*

The GeoLens Client is implemented as a framework of powerful Java applets that exploit the unique full-feature capabilities of GeoLens catalog servers. Using the server-side support of multiple schemas to drive the browser-side client makes it possible to achieve greater customization of the presentation. The overall effect is a graphical presentation of metadata that preserves both the structure and semantics of a user's preferred schema or metadata content standard. Queries of GeoLens

Catalogs may be spatial or formed by combining any of a query schema's attributes with logical operators.

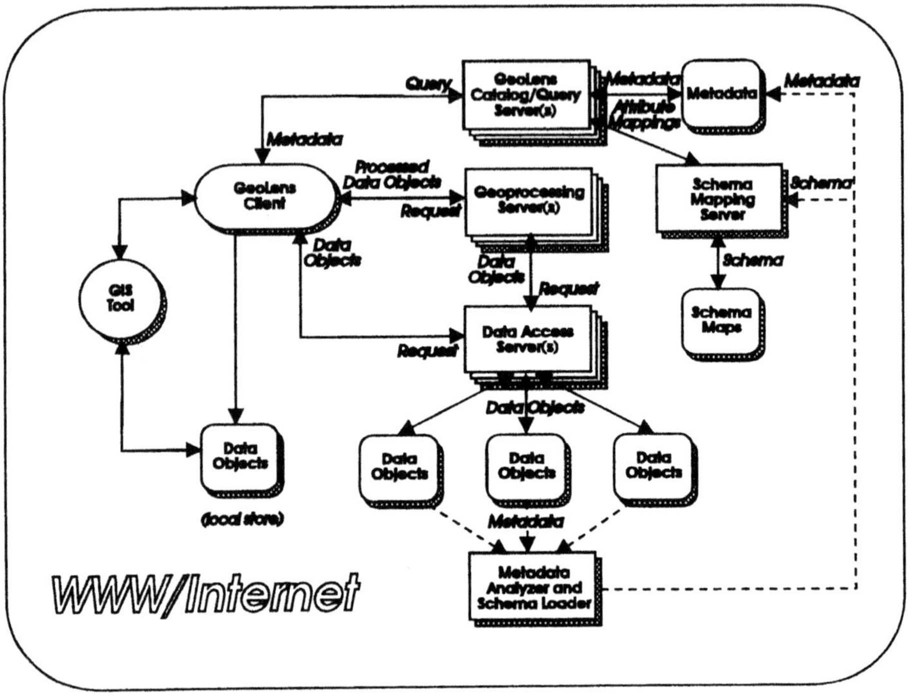


Figure 21.2. GeoLens system architecture

Soon after a GeoLens Catalog is accessed, a user is presented with a geospatial data *Collection Tree* by the GeoLens Client. As shown in Figure 21.3, the tree, represented in indented-outline form, displays classes of data objects, for example, Digital Elevation Data Collection or Satellite Imagery and Aerial Photography, and subclasses of these, for example, AVHRR and Landsat for the latter class. A user can follow down branches of this tree to collections of data objects, for example, Landsat Thematic Mapper or Multispectral Scanner, by clicking on nodes in the tree to open their subtrees, eventually reaching leaves containing these collections. Having selected either a collection or leaf-level object, the user can browse the collection-level or inventory-level metadata, which may include encapsulated data ranging from plain text to a browse image, if one exists for the target data object, illustrated in Figure 21.4. The GeoLens Catalog Server processes requests based on encapsulation attributes, and so the same text or images may be presented differently depending on their encapsulation type. Thus, the GeoLens Catalog Server does not perform any format conversions of the original information. Instead, metadata are passed directly to the GeoLens Client, which either presents them directly, or uses them to retrieve the information.

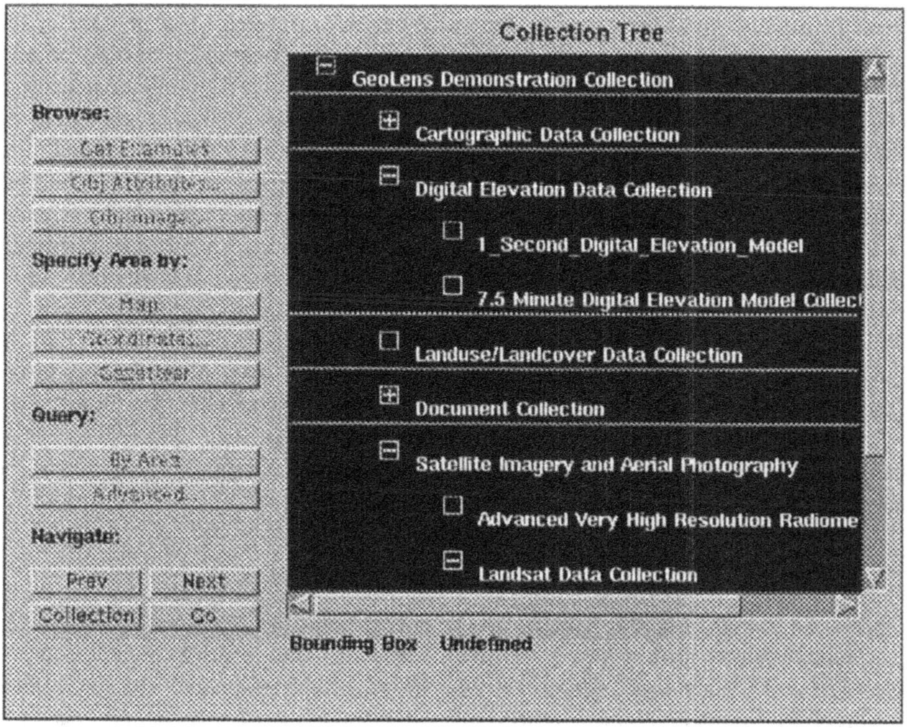


Figure 21.3. Collection-tree representation of GeoLens federation

Special facilities exist to query GeoLens Catalogs spatially with a user-supplied bounding rectangle. The coordinates for this rectangle may be defined in several ways: interacting graphically with a map, a dialog box, or search on a placename with the USGS Geographical Name Information System gazetteer service. The last method also nicely demonstrates the manner in which the GeoLens Client's framework easily provides plug-in support for third-party, external services. Attribute schemas may also be queried to aid users in defining queries. A user can profile a preferred schema and all or just some of its attributes for building queries of GeoLens Catalogs. Currently a simple dialog box displays a list of a query schema's attributes from which a user may select member attributes and appropriate conditions to form complex queries that may include conditions for searching full-text indices.

Other facilities are provided in the GeoLens Client to help a user navigate through the Collection Tree. A presentation history is cached so that a user may easily browse collection-level and inventory-level metadata by traversing the Collection Tree.

### *21.3.2. Catalog services*

The GeoLens Client locates geospatial data on the Internet through a particular GeoLens Catalog Server, which also serves as a transparent proxy for other GeoLens

Catalog Servers. Metadata loading and analysis routines extract metadata descriptions and build the catalog. These metadata are analyzed for their schema and a Schema Mapping Service verifies their properties for compliance to that schema. Finally, the metadata are added to a catalog, which is currently implemented as an O2 database (O2 Technology 1995; however, our design makes it very easy to substitute other commercial products). A user launches a search for geospatial data by querying the contents of metadata catalogs with the help of one or more GeoLens Catalog Servers, even though the use of multiple catalogs is completely transparent to the user. Catalogs may be queried for the schema used to structure and document metadata for a collection of data objects, or for any of the metadata attributes used to catalog data objects. Queries may be spatial, temporal, or composed of conditions on arbitrary metadata attributes used to catalog data objects. Since a query may be issued to one or more catalogs, each with metadata that might be organized consistent with different metadata schemas, the query processor may contact the Schema Mapping Service to translate between schemas with different attribute names and structures. This feature enables a recursive search of all (or some) of the catalogs known to GeoLens Catalog Servers in the federation.

### 21.3.3. Data access servers

Once a list of candidate data objects has been returned to the GeoLens Client, a user can select objects from this list for data retrieval. The GeoLens Client sends a message to a Data Access Server with the ID of a data object and a user-provided Minimal Bounding Rectangle applied by the Data Access Server to clip the data object. The data extract is remodeled as an OpenGIS-like *Well-known Structure* (WKS) and returned to the GeoLens Client where it may be stored locally, or immediately exploited by an application, for example, a commercial GIS, that implements OpenGIS interfaces. In addition, the GeoLens architecture allows for integration of other services, for example, geoprocessing or map production, that may be requested by the GeoLens Client or Data Access Servers.

## 21.4. Lessons learned

As might be expected in any prototyping effort, numerous technical obstacles arose, particularly ones involving the integration of geospatial information processing standards. Table 21.1 lists the standards leveraged in the GeoLens system, their type and application. This section will discuss the manner in which some of the most severe difficulties in implementing these standards were resolved in the design and implementation of GeoLens.

| **Standard** | **Type** | **GeoLens Implementation** |
|---|---|---|
| FGDC CSDGM | metadata content | DEM & DLG Catalog Servers |
| ECS Metadata Standard | metadata content | MSS Catalog Server |
| DLG-3 | Metadata content | DLG Catalog and Data Access Servers |
| HDF-EOS | data archive format | MSS Data Access Server |
| DEM | data archive format | DEM Data Access Server |
| SDTS-VP | data archive format | DLG Data Access Server |
| HTTP | Messaging | client-server communications |
| Common Gateway Interface | Messaging | client-server communications |
| Java Development Kit 1.0.2 | Development language | distributed software applications |
| gif | image format | browse image metadata |
| jpeg | image format | browse image metadata |
| OGIS-like *Grid Coverage* | well-known structure | DEM & MSS Data Access Servers |
| OGIS-like *Simple Features* | well-known structure | DLG Data Access Server |
| RDF/XML (see text) | schema representation | Schema Mapping Service (planned) |

Table 21.1. Standards integration in the GeoLens prototype

### *21.4.1. Catalog federation and schema integration*

The approach to federating heterogeneous catalogs within GeoLens Catalog Servers assumes that metadata are just like actual physical data and need not be stored together in the same physical repository. Using the same mechanism that enables the GeoLens Catalog Server to encapsulate physical location and processing information for different data objects, information about the network location of catalogs may be stored as access-descriptive attributes. In this way, catalogs in part (or in their entirety) may be referenced by other catalogs, each residing on separate servers. This not only demonstrates the flexibility of catalog access and presentation, but also the generalizability of the GeoLens encapsulation mechanism.

In our opinion, not only should the physical existence of individual catalogs be hidden from a user, but also true federation of heterogeneous catalogs implies that users are able to navigate through different catalogs without being burdened with the complexities of querying separately each of these catalogs for pertinent information. Such *transparency* touches on the key issue of *semantic ambiguity*. It is not reasonable to assume that any single metadata schema can capture the full extent of meaning held by the name of any given metadata attribute. Therefore, some mechanism is needed to resolve ambiguous situations. Because there are many different types of these situations, it has been useful to elucidate cases when schema translations are required, from most simple to most difficult.

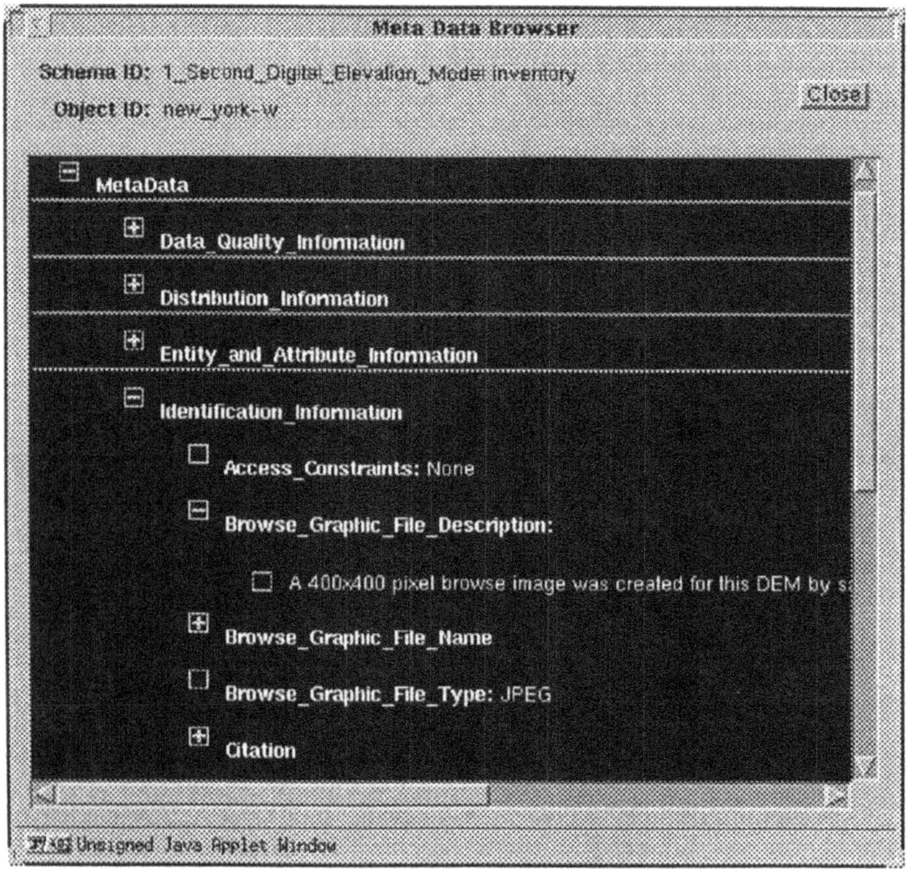


Figure 21.4. GeoLens metadata browser

Perhaps the simplest situation requiring cross-schema mapping of metadata attributes is one-to-one attribute naming translations. For example, translation of

FGDC bounding box coordinates to their counterparts in the ECS schema follows below:

Identification_Information:

Spatial_Domain:

Bounding_Coordinates:

East_Bounding_Coordinate=

maps onto

SingleTypeCollection:

Spatial:

SpatialDomainContainer:

HorizontalSpatialDomainContainer:

BoundingRectangle:

EastBoundingCoordinate=

In situations like these, it is sufficient to know which attribute names in two schemas carry the same meaning.

The next level of complexity might involve one-to-many attribute translations. A slightly contrived example of this type might be:

Spatial_Domain:

Bounding_Coordinates:

Southeast_Pair:

maps onto

SingleTypeCollection:

Spatial:

SpatialDomainContainer:

HorizontalSpatialDomainContainer:

BoundingRectangle:

SouthBoundingCoordinate=

AND

SingleTypeCollection:

Spatial:

SpatialDomainContainer:

HorizontalSpatialDomainContainer:

BoundingRectangle:

EastBoundingCoordinate=

In these situations not only are attribute names different between two schemas, but the values assigned to several attributes in one schema may correspond to only a single attribute in another schema.

Still a third level of complexity would involve translations of meaning for different attribute values, rather than just the attribute names themselves. Some of these may only require simple kinds of conversions, for example, between coordinates expressed as *DD.MM.SS* (Degrees, Minutes and Seconds) and *DD.DD* (Decimal Degrees), or geographic coordinate to UTM grid conversions. Other kinds of translation can be extremely difficult, such as negotiation between a property owner's notion of the term *parcel* and the one held by a county tax assessor.

Current approaches to dealing with semantic interoperability of heterogeneous schema usually rely on some form of human input to resolve conflicts. The GeoLens approach exploits key attribute mapping information from schema providers. We further define concepts that are primed for specific meaning (e.g., consider the concept of a bounding box), and the constituent elements necessary to define a concept (e.g., what minimal amount of information describes a bounding box). This core information is made available to the Catalog Server and is used at query time both to retrieve permissible mappings between schema standards, and to fill the necessary *slots* associated with a given concept.

In the course of our work, we have come to appreciate the complexity of this problem, especially with regard to achieving a general solution. The derivation of new concepts requires an understanding of the domain in question and each concept entails additional knowledge, which would need to be prebuilt. Even simple one-to-one attribute name mappings raise interesting questions. For instance, are the mappings transitive? In other words, suppose that two schema standards share no registered or standardized mappings, but there is a third schema which maps to both of them. At the present, this kind of transitivity is not implemented in the GeoLens system. However, as the analysis reveals, there are different levels of complexity to semantic translation, and much can be accomplished by at least solving the simplest situations as we have begun to do in the GeoLens.

### *21.4.2. Representation of schema standards*

An important objective of the GeoLens Project was to provide a proof of concept for supporting multiple schema standards instead of trying to enforce a single one. An additional complication is a *lack of uniformity* in representing standards. Until recently, standard-setting activities concentrated exclusively on defining syntax and semantics of individual standards but not on defining a common representation for these standards. As part of the project, we have defined the most important characteristics of such a representation, where a standard is composed of attribute specifications that include *data types*, *applicability*, *generality*, *topology* and *extensibility* (Shklar et al. 1997). Attribute data types determine processing of query conditions and serve to support schema verification (type mismatch is likely to represent an inconsistency). Examples of data types include *strings*, *integers*, and *geospatial coordinates*.

Attribute applicability determines whether an attribute is *mandatory*, *mandatory-if-applicable*, or *optional* (cf. FGDC 1994), while attribute generality determines whether it belongs to the common part of a standard or to a named extension. If an attribute is defined as optional for the common part of a standard, it may still be defined as mandatory or mandatory-if-applicable for one or more named extensions. If the attribute is not defined for the common part of a standard, its specifications for different named extensions don't have to match. A mandatory-if-applicable attribute may be further characterized by a list of other attributes, the presence of which would change its status to mandatory.

Attribute topology is defined by specifying the component-of relationships with parent attributes. Extensibility is only defined for composite attributes and determines whether all their child attributes (or components) are already known. This is important because we are considering two sources of information for constructing a schema standard: *schema standard specifications* and *incoming metadata entities*. If a composite attribute is marked as non-extensible, an encounter of its unknown components is considered an error.

In the absence of a common approach, we have invented a proprietary standard representation syntax, but we consider it only an interim solution. Our hopes are with the World Wide Web Consortium, which has initiated several standard-setting activities around the so-called *Resource Description Framework* (RDF). We are strongly encouraged by the working documents that emerged from this body and stand ready to comply to RDF specifications when they stabilize (Lassila and Swick 1997). Already the richness of the RDF model seems sufficient for expressing geospatial schema. RDF specifications utilize the XML syntax, which is of course a promising common idiom for such specifications.

### 21.4.3. Geoprocessing services

The initial design of the cooperation between the GeoLens Metadata Browser and an external server, such as the Data Access Server, is rather simple—the GeoLens Metadata Browser invokes an external server based on the encapsulated information of a single data object. However, there are situations where a more complex higher-level layer is required to process multiple data objects. For example, a user may want to exploit network services to create a visualization by overlaying a DLG object on a DEM object, each extracted by different data servers. Alternatively, this user may want to extract data and store them locally for use by his or her image classifier and GIS. These operations require the GeoLens Metadata Browser to send multiple data objects to an external service (e.g., a visualization service). While not currently implemented, our new design will utilize a *shopping cart* to collect and send multiple data objects to external geoprocessing services capable of handling multiple requests.

### *21.4.4. Data model implementation*

A key objective in the GeoLens Project has been to develop proof-of-concepts for ideas expressed in OpenGIS documents and Working Groups. While we have taken primarily a breadth-first approach to support the end-to-end scenario above, nonetheless we have learned some important lessons related to current limitations of the OGIS data model when processing and retrieving very large geospatial data objects. For example, while the OGIS data model provides APIs (Application Programming Interfaces) for requesting data, the model stops short of suggesting how data should be returned to an application. This leaves open the possibility for many different implementations of the model. Moreover, in a distributed (and potentially low-bandwidth) environment such as that provided by the Internet, it is not feasible for a client application to invoke fine-grained operations (such as those specified for OGIS features) on an object that resides on a remote data server. Instead, a transfer syntax is required for transmitting a representation of the object (or part of it) to the client, allowing the client applications to operate directly on a copy of the object. At the moment, GeoLens Data Access Servers return only small data extracts to the client in the form of gifs and flat ASCII files for demonstration purposes. But industrial-strength data access servers will need to do a better job of packaging data extracts for Internet transfer.

Other implementation issues exist regarding the location and support of massive collection-level metadata such as feature schemas. While one might conclude on logical grounds that the distinction between collection and inventory metadata is arbitrary, such a distinction may be critical to make for performance or other reasons, for example for information management. For example, the DLG-3 feature schema is needed to support both data discovery and access; yet it is too large to transfer between catalog servers, clients, and data access servers to support these functions. Moreover, we discovered that only a small subset of features were actually supported by data in any particular DLG data object, and the size of this subset actually varied from object to object. Our manner of addressing this issue in the *DLG Data Access Server* was to store the DLG-3 feature schema as an *Entity_and_ Attribute_Information.Detailed_Description* attribute in FGDC collection-level metadata (though one could alternatively store a reference to a DLG-3 feature schema service). The DLG-3 features actually supported by each DLG data object were precomputed and stored as other inventory-level metadata on the Data Access Server. This information is used to build an interface for users so that they select for extraction only those relevant features in a data object discovered by the GeoLens Catalog Server. This approach provides the user with copies of the feature schema for reference *and* access purposes, but minimizes the need for transporting feature schema information. It also enables data providers to update their DLG repository in a timely manner without having also to update feature schema entries in the DLG catalog.

This last point touches on another significant accomplishment of the GeoLens Project, that is, developing a relatively inexpensive approach to federating data repositories, but one that also permits maximum autonomy among data providers and encourages their participation. Our system allows independent data providers to participate in the federation with very few hooks. There is no requirement that their metadata need map to any single preexisting metadata content standard. This allows data providers to make available as many (or as few) mappings as desired. While the advantage of having a common schema *interlingua* permits multiple standards to communicate with one another, deriving and enforcing widespread adherence to such an interlingua is currently neither practical nor desirable. Instead, the understanding supported here is that the data provider has the flexibility to choose which mappings are more readily accessible and meaningful, and the GeoLens would support these.

## 21.5. Conclusions

As this chapter has shown, the geospatial interoperability problem actually consists of many smaller, but in themselves extremely complex problems. We have used the GeoLens Project to expose these problems better, analyze them in detail, and experiment with different software approaches to solving them. This experimentation has yielding software with several distinguishing features. A user is provided with a consistent, unified view over a federation of distributed, heterogeneous geospatial catalogs. This user need not be aware of the location (or even the existence) of individual catalogs in the federation; they merely appear as a single local source of metadata. Catalog servers make available both content-descriptive and access-descriptive metadata; the latter references data access servers, which are needed either to acquire additional metadata or to extract information from data objects referenced by the catalogs. User queries are recursively applied over individual catalogs that are linked within a collection tree. Furthermore, a single query is distributed over all (or some subset) of catalogs in the federation without the user having to be cognizant of the metadata standard used to organize the catalog, and consequently of which queries are appropriate. To support this type of seamless query, an infrastructure is provided that maintains multiple, dynamic metadata content standards within a single catalog hierarchy, along with translation or mapping services between catalogs at query and data extraction time. In addition to attribute-based SQL queries, full-text indices on specific attributes within a catalog may also be queried and then the combined results of these two query methods presented to the user.

Among the most significant lessons that we have learned over the last three years is that data modeling is an approach to federation that offers both the power and flexibility required to ensure the autonomy desired by data providers, the seamless accessibility to data required by users, and the potential for technology to evolve with changes in the requirements of these communities. Finally, we believe that our work has demonstrated the importance of geospatial information processing standards to solving geospatial interoperability issues, and the manner in which one might design and implement sophisticated catalog and data access services on the Internet to support multiple, dynamic geospatial data standards more effectively.

## Acknowledgments

The Universal Spatial Data Access Consortium consists of Bell Communications Research, Camber Corp, OpenGIS Consortium, Rutgers Center for Remote Sensing and Spatial Analysis, NASA's Space Science Data Operations Office, and the California Resources Agency's CERES Program. This work was funded by NASA's Information Infrastructure Technology and Applications (IITA) initiative of the High Performance Computing and Communications (HPCC) program, Digital Library Area of Interest (Dr. Nand Lal, Program Manager), Cooperative Agreement NCC5-102. The authors would like to thank Michael Lesk, Ed Runnion, Ronald Underwood, Kent Wittenburg, Eric Sigman, Gomer Thomas, Charles St. Charles, and Tim Muradian for their contributions to the GeoLens Project.

## References

Buehler K, McKee L (eds) 1996 *The OpenGIS Guide: Introduction to Interoperable Geoprocessing. Open Geodata Interoperability Specification (OGIS), Part I.* Open GIS Consortium Inc, Wayland, MA

Burrough P A 1986 *Principles of Geographical Information Systems for Land Resources Assessment.* Clarendon Press, Oxford

Central Imagery Office 1995 *Standards Profile for Imagery Access, Version 2.* CIO-2020. Central Imagery Office, Vienna, VA

Clarke K C 1990 *Analytical and Computer Cartography.* Prentice Hall, Englewood Cliffs, NJ

Defense Mapping Agency 1983 *Geodesy for the Layman (DMA TR 80-003).* Defense Mapping Agency, Washington, DC

Defense Mapping Agency 1996 *Global Geospatial Information & Services (GGI&S) and Data Architecture and Gateway Services (DAGS).* Solicitation released on June 27, 1996

Department of the Interior 1987 *Digital Line Graphs from 1:100,000-Scale Maps—Data Users Guide 2.* U.S. Geological Survey, Reston, VA

Department of the Interior 1990 *Standards for Digital Line Graphs, Part 3: Attribute Codes.* U.S. Geological Survey, Reston, VA

Department of the Interior 1997 *DLG-3 SDTS Transfer Description, Draft. ftp://sdts.er.usgs.gov/pub/sdts/datasets/tvp/dlg3/*

Federal Geographic Data Committee 1994 *Content Standards for Digital Geospatial Metadata.* Federal Geographic Data Committee, Washington, DC

Federal Geographic Data Committee 1997 *Content Standards for Digital Geospatial Metadata, Version 2.0).* Federal Geographic Data Committee, Washington, DC

Fingerman P W 1997 *HDF-EOS 2.00 Version Description Document (VDD) for the ECS Project, Version 1.00* (814-RD-009-001). Hughes Information Technology Systems, Upper Marlboro, MD

Hearnshaw H M, Unwin D J 1994 *Visualization in Geographical Information Systems.* John Wiley and Sons, New York

Hughes Information Technology Corp 1996 *Release-B Science Data Processing Segment (SDPS) Database Design and Database Schema Specifications for the ECS Project.* Document No. 311-CD-008-001

International Standards Organization 1997 *Geographic Information—Metadata, Version 2.0.* ISO/TC211 Working Group 3, Working Document No. 1997-01-20

Lassila O, Swick R R (eds) 1997 *Resource Description Framework (RDF) Model and Syntax*. WWWC Draft Specification (WD-rdf-syntax-971103). *http://www.w3.org/Metadata/RDF/Group/WD-rdf-syntax/*

MacEachren A M 1995 *How Maps Work: Representation, Visualization, and Design*. Guilford Press, New York

Medyckyj-Scott D, Hearnshaw H M (eds) 1993 *Human Factors in Geographical Information Systems*. Belhaven Press, London

National Aeronautics and Space Administration 1997 *Directory Interchange Format (DIF) Writer's Guide, Version 5.0a*. NASA Goddard Space Flight Center, Global Change Data Center, Code 902

National Biological Service 1997 *National Biological Information Infrastructure Metadata Standard*. National Biological Service, Washington, DC

National Imagery and Mapping Agency 1996 *Geospatial Metadata: DoD Geospatial Data Standardization Project Report, Vol. 3*. National Imagery and Mapping Agency

National Imagery and Mapping Agency 1997 Geospatial Information Integrated Product Team (GI IPT) Geospatial Information Infrastructure GII 97 Requirement—SOL BAA. *Commerce Business Daily* (January 17–23): A6–A8

O2 Technology Inc 1995 *Technical Overview of the O2 System*. Technology Technical Report No. 9. O2 Technology Inc, Palo Alto, CA

Open GIS Consortium 1996 *Request for Information: OGIS Catalog Service Interfaces*. OGC Request 3, Open GIS Services Working Group (August 30, 1996). Open GIS Consortium, Wayland, MA

Shklar L, Behrens C, Basu C, Yeager N, Au E 1997 *New Approaches to Cataloguing, Querying and Browsing Geospatial Metadata*. Paper presented at the 2nd IEEE Metadata Conference. NOAA, Silver Springs, Maryland, Sept. 16–17, 1997

Snyder P 1987 *Map Projections, A Working Manual*. Geological Survey Professional Paper 1395. U.S. Government Printing Office, Washington, DC

# 22 A MIDDLEWARE ARCHITECTURE FOR TRANSPARENT ACCESS TO MULTIPLE SPATIAL OBJECT DATABASES

Sang Kyun Cha, Ki Hong Kim, Chang Bin Song, Joo Kwan Kim, and Yong Sik Kwon

## 22.1. Introduction

The need to access multiple databases arises frequently in geographic information processing because a single spatial object database may not contain all the information at the desired level of abstraction, completeness, and accuracy. For example, in planning the extension of underground urban utility networks such as gas and telecommunication lines, it is necessary to access databases of existing and planned utility networks. Such databases are usually maintained independently by individual operating companies. It is also common that these independent databases maintain much geographic information redundantly with different levels of abstraction, completeness, and accuracy. Many high-level decision making processes can take advantage of such redundancy to unify the content of one database with those of others. This unification of spatial objects in multiple databases is expected to extend the solution space that would otherwise be very limited or nonexistent.

Two keys to the unification of spatial objects in heterogeneous databases are the ability to connect to multiple databases and manipulate them transparently and consistently, and the ability to resolve various types of semantic heterogeneity among multiple databases. Much work is going on to provide interoperability among geographic information systems (Buehler and McKee 1996; Nittel et al. 1996; Nittel et al. 1997). Especially, the ongoing OpenGIS activity defines a spatial data model for representation of geographic information, a service model for establishing a set of interoperable component services, and an information communities model for practically dealing with the semantic heterogeneity issues (Buehler and McKee 1996).

This work presents an ongoing effort of developing a middleware for providing transparent access to multiple spatial object databases. Called SDBC (Spatial

DataBase Connectivity), it gives programmers a single, object-oriented view of database and programming while relieving them from the burden of knowing the implementation-specific details of individual databases such as the physical location and the ODBMS process architecture. The key to such transparency is the extension of the ODMG object model (Cattell and Barry 1997) with a set of spatial data types and operators based on OpenGIS (Open GIS 1997). In addition, SDBC incorporates the global transaction management function to ensure the consistency of SDBC transactions accessing multiple databases. Finally, a framework for spatial integrity constraint definition and checking is being incorporated into a part of SDBC to ensure the quality of spatial databases.

The immediate goal of developing SDBC is to provide transparent connectivity to multiple spatial object databases. A longer-term objective is to provide an experimental framework for research on semantic heterogeneity resolution and spatial object unification for high-level decision making processes. This chapter is organized as follows. Section 22.2 describes the design goals in developing SDBC. Section 22.3 presents the functional architecture of SDBC. Section 22.4 describes the data and programming model of SDBC. Section 22.5 presents the implementation layers of SDBC in the client–server environment. Section 22.6 concludes the chapter.

## 22.2. Design goals

### *22.2.1. Object-oriented model*

In addition to the ordinary DBMS features (Elmasri and Navathe 1994), a spatial DBMS should support at least the following features (Egenhofer 1997; Medeiros and Pires 1994): (1) spatial data types for representation of commonly occurring geometric information such as point, line, and polygon; (2) generic spatial operators (such as overlap and intersect) among these types; (3) spatial indexes like R-tree (Guttman 1984), R*-tree (Beckmann et al. 1990), and R+-tree (Sellis et al. 1987). These spatial features are best implemented in object-oriented database systems because many possible domain-specific spatial types can be derived from a set of basic ones through the type inheritance mechanism. Implementation of spatial operators may be encapsulated as methods within the object classes defining their operand types. Many of these operators such as the distance between two spatial objects are overloaded and the run-time polymorphism is useful for resolving the specific operator implementation to invoke. Finally, application-specific complex data types involving spatial data as their components can be easily defined by referring to these spatial data types through relationship links.

Much research has been done on object-oriented database systems and many variants of object model have been proposed (Cattell and Barry 1997; Kim and Lochovsky 1989). Among these, the ODMG model is taken as the reference model because it supports seamless integration with object-oriented programming languages and provides source-code interoperability among ODBMS implementations. SDBC extends this ODMG model with spatial data types, operators, and spatial indexes because such features are not included in the ODMG model. The actual implementation of these spatial data management functions can be on top of a commercial ODBMS (Balovnev et al. 1997; Günther and Riekert 1993; Scholl and Voisard 1992). While it is rare that commercial ODBMS implementations support spatial data management functions, we expect dedicated spatial ODBMSs conforming to the ODMG standard to appear in the near future.

### *22.2.2. Transparent access to multiple spatial databases*

Although the ODMG model helps to resolve the interoperability issues among multiple ODBMSs, these ODBMSs offer variations in implementation details such as process architecture, clustering, and initialization procedure. As a result, programs interacting with multiple ODBMSs may look very awkward. The SDBC solves this problem with middleware providing an identical programming interface for interacting with different ODBMSs. To facilitate hiding the ODBMS-specific implementation details, the current SDBC implementation uses CORBA (Common Object Request Broker Architecture; OMG ORBTF 1992). However, the SDBC API hides even this CORBA layer from the programmer, thus avoiding the burden of CORBA programming.

Accessing multiple databases in the distributed environment may leave them in inconsistent states if global transaction management is not supported. To avoid this inconsistency problem, SDBC supports global transactions, each of which may wrap local transactions to multiple databases (Bernstein et al. 1987; Breitbart et al. 1992; Zhang and Elmagarmid 1993).

### *22.2.3. Room for rich semantics*

Since one goal of developing SDBC is to provide an experimental framework for research on semantic heterogeneity resolution and spatial object unification, SDBC intends to leave room for capturing various types of knowledge for resolving semantic heterogeneity and sharing them among multiple geographic information systems. One example of such knowledge is the distinction between inherently continuous objects such as road segments and discrete ones such as electric poles. Spatial object databases tend to record continuous physical objects as piecewise linear approximations. Although two database systems use the same data type for representing these approximations, they are likely to record different sampled approximations for an identical physical object. However, inherently discrete objects such as the power distribution lines hanging over electric poles do not have this problem unless errors are introduced in the measurement of pole positions.

Another type of important knowledge in the spatial database domain is integrity constraints. Spatial data are prone to many types of errors. Some of these errors such as GPS measurement error are difficult to avoid. One way to deal with this problem is to filter inconsistent data with spatial integrity constraints. For example, in urban areas many types of underground utility backbone networks tend to be located together in parallel with major road segments. Any utility backbone network located off the road area is suspected to have significant errors in its positional information, and the database system may reject such suspected information based on the stored integrity constraint.

It is desirable that the above types of knowledge should be captured somewhere in the system, and shared among the applications accessing databases. SDBC leaves room for capturing these knowledge types in its architecture.

## 22.3. Functional architecture

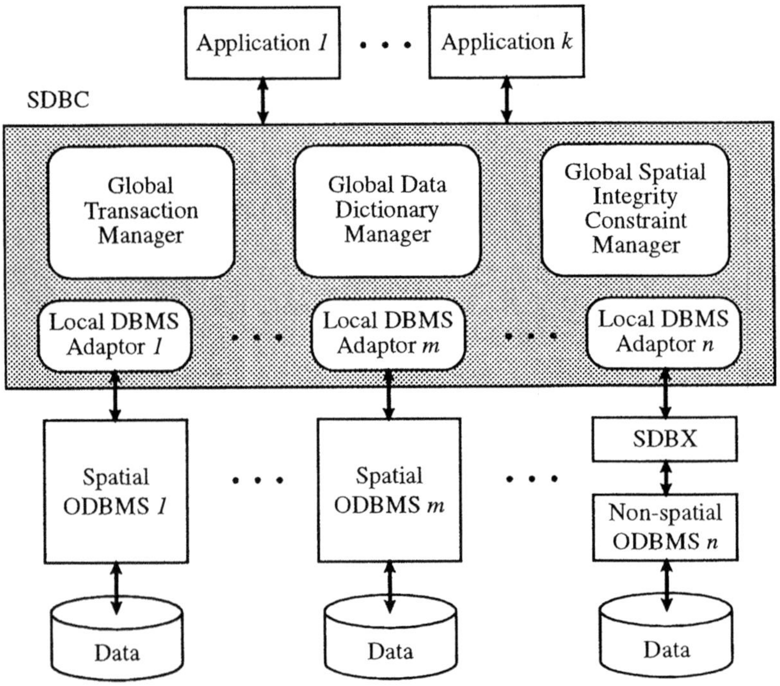


Figure 22.1. SDBC functional architecture

Figure 22.1 shows the functional architecture of SDBC. SDBC assumes that the underlying ODBMS conforms to the ODMG standard and supports a set of spatial data types and operators defined by SDBC. In case the ODBMS does not provide these spatial data management functions as in most commercial implementations, the *Spatial DataBase eXtension* (SDBX) layer is provided. SDBX implements SDBC-defined spatial data types, operators, and spatial indexes on top of commercial ODBMSs.

SDBC consists of four types of functional modules.

1. *Local DBMS Adapter* (LDA): This module translates SDBC application requests into corresponding requests to a target ODBMS, thus hiding ODBMS-specific implementation details from SDBC applications.

2. *Global Data Dictionary Manager* (GDDM): This module maintains metadata. It understands the SDBC-defined spatial data types, operators, and indexes.

3. *Global Transaction Manager* (GTM): This module coordinates the consistent execution of local transactions in a global transaction.

4. *Global Spatial Integrity Constraints Manager* (GSICM): This module maintains the quality of spatial databases by invoking spatial integrity constraints.

## 22.4. Spatial data and programming model

The current SDBC object model extends the ODMG object model to deal with multiple spatial databases. This section describes the spatial data model, the transaction model, and the programming model of SDBC.

### 22.4.1. Spatial data model

In general, the SDBC spatial data model follows the OpenGIS geodata model (Open GIS Consortium 1997). The current SDBC prototype implements a subset of this model for verifying the utility of SDBC concepts. Figure 22.2 shows the inheritance hierarchy for this subset. It provides three subtypes of *Geometry*: *Point*, *Curve*, and *Surface* for zero-, one-, and two-dimensional objects, respectively. With the exception of *Point*, these types are abstract and therefore cannot be instantiated. Derived from *Curve* is *LineString*, which models a piecewise linear abstraction of curve with a sequence of *Point* objects. *LineString* has two subtypes: *Line* and *Ring*. *Line* is a straight segment represented by two boundary points and *Ring* is a piecewise linear abstraction of closed curve. *Surface* also has two subtypes: *Polygon* and *Rectangle*. *Polygon* represents two-dimensional objects with area and has *Ring* objects to represent its outer and inner boundaries. *Rectangle*, which is absent in the OpenGIS model, is supported in SDBC because it is frequently used to represent approximate boundaries of spatial objects in coarse-grained maps.

In terms of operators, SDBC supports ten topological ones, thirteen directional ones, and four metric ones. The list of supported topological operators is *equal*, *disjoint*, *meets*, *overlaps*, *contains*, *inside*, *covers*, *covered_by*, *intersects*, and *crosses*. The first eight operators are defined by Egenhofer (1991). The *intersects* operator is the negation of *disjoint*. The *crosses* operator is applicable to two *Curve* objects. Except for *equal*, *disjoint*, and *intersection*, these operators are applicable to particular combinations of spatial data types. Table 22.1 shows the applicability matrix of these operators. Note that the matrix is not symmetric: rows represent left operands and columns right operands. For example, a *Surface* object can contain a *Curve* object but the inverse cannot occur.

SDBC also supports directional operators such as *north_of*, *south_of*, *right_north_of*, *right_south_of*, *north_east_of*, and so on (Frank 1992). The metric operators supported are *distance*, *length*, *perimeter*, and *area*. *Length* is applicable to one-dimensional geometry types like *LineString* and returns their length. *Perimeter* and *area* are applicable to two-dimensional geometry types like *Polygon* and return perimeter and area respectively.

### 22.4.2. Programming model

Developing SDBC applications involves the processes of schema definition, method implementation, and application code development. SDBC provides extensions of ODMG ODL and C++ OML for these processes.

**22.4.2.1. SDBC transaction model.** SDBC extends the ODMG transaction model to support accessing multiple databases within a single SDBC transaction. Two SDBC-specific interface classes *GlobalTransaction* and *LocalTransaction* replace the ODMG interface class *Transaction.* A pair of *begin()* and *commit()* (or *abort()*) operations indicates the boundary of each global or local transaction. An arbitrary number of local transactions can exist within the boundary of a global transaction. The argument of the *LocalTransaction::begin()* operation binds the local transaction variable to a specific target database.

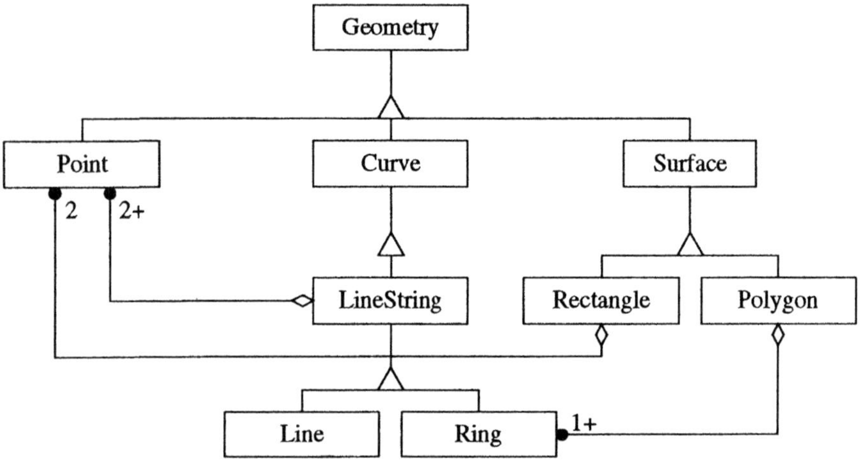


Figure 22.2. Current SDBC-defined spatial data types

A global transaction commits if all nested local transactions do. However, a global transaction may be programmed to commit by its own logic even if some of its local transactions abort or fail to precommit within a certain time limit. Of course, no local transactions can commit if the global transaction aborts.

**22.4.2.2. SDBC ODL.** Figure 22.3 shows a sample schema in SDBC ODL for two public transportation databases: bus line and subway. The schema of each database, called a local schema, is wrapped with *BEGIN LOCAL_SCHEMA* and *END LOCAL_SCHEMA*. These local schemas are named *BuslineDB* and *SubwayDB*, respectively.

The schema of the bus line database shows the definition of two persistent types, *BusStop* and *BusLineLink*, and two transient types, *PathQuery* and *NNQuery*. The *BusStop* type defines the name and location of a bus stop as its attributes anda bidirectional relationship to a set of *BusLineLink* objects. Here the location attribute is a spatial attribute of the *Point* type, and is defined as the spatial key of the *BusStop* extent. The *BusLineLink* type models links between two adjacent bus stops. It defines an attribute named *line_type*, a bidirectional relationship to up to two *BusStop* objects, and a unidirectional relationship to the underlying *RoadSegment* objects. For the simplicity of presentation, the definition of *RoadSegment* is not shown in Figure 22.3. The *BusLineLink* type also defines a method *cost()*, which computes the time for traveling the link based on the traffic condition of the underlying road segments.

| | Point | Curve | Surface |
|---|---|---|---|
| Point | | inside covered_by | inside covered_by |
| Curve | contains covers | inside covered_by contains covers meets overlaps crosses | inside covered_by meets overlaps |
| Surfa ce | contains covers | contains covers meets overlaps | inside covered_by contains covers meets overlaps |

Table 22.1. Applicability matrix for topological operators. Note that *equal, disjoint,* and *intersects* are always applicable

The transient types *PathQuery* and *NNQuery* define the method *execute()*, which implements a shortest-path-finding algorithm and a nearest-neighbor-finding algorithm, respectively. These two methods as well as the *cost()* method of the *BusLineLink* type are by default implemented on the server side and executed by the method-shipping mechanism. In addition, SDBC will support the object-shipping mechanism, which requires the management of an object cache on the client side.

Similarly, the schema of the subway database shows the definition of four types, *SubwayStation*, *SubwayLineLink*, *PathQuery*, and *NNQuery*. To resolve possible naming conflicts among multiple databases, SDBC allows prefixing the local schema name to the type name as in *BuslineDB::PathQuery* and *SubwayDB::PathQuery*.

**22.4.2.3. SDBC C++ OML.** Figure 22.4 shows a simple application program written in SDBC C++ OML that accesses the bus and subway line databases. This program computes the shortest travel path given the position of an origin and a destination. Since the traveling time on the subway is a lot more predictable than on a bus in the metropolitan area with heavy traffic, maximizing the subway portion of the traveling path is a preferred strategy for nontrivial cases. Following this strategy, the program first finds the subway stations nearest to the origin and the destination, and computes the optimal route between these two stations. If the nearest subway stations are located far from the origin and the destination by walking, this subway route is complemented by the bus travel routes between the subway stations and the bus stops nearest to the origin and the destination.

```
BEGIN LOCAL_SCHEMA BuslineDB
   interface BusStop
   (   // type properties
       extent busStops
       key location )
   {   // instance properties
       attribute string name;
       attribute Point location;
       relationship set<BusLineLink> line inverse BusLineLink::stop;
   };

   interface RoadSegment ( ... ) { ... };

   interface BusLineLink ( ... ) {
      attribute unsigned short line_type;
      relationship set<BusStop> stop inverse BusStop::line;
      relationship set<RoadSegment> roads;
      // instance operations
      float cost( );
   };

   interface PathQuery() : transient {
      list<BusLineLink> execute( in BusStop, in BusStop );
   };
   interface NNQuery() : transient {
      BusStop execute( in Point );
   };
END LOCAL_SCHEMA


BEGIN LOCAL_SCHEMA SubwayDB
   interface SubwayStation ( ... ) { ...... };
   interface SubwayLineLink ( ... ) { ...... };
   interface PathQuery ( ) : transient { ...... };
   interface NNQuery( ) : transient { ...... };
END LOCAL_SCHEMA
```

Figure 22.3. A sample schema in SDBC ODL

Figure 22.4 shows four SDBC-provided types prefixed with *s_*: *s_Database*, *s_GlobalTransaction*, *s_LocalTransaction*, and *s_Ref<T>*. To access a local database, a variable *s_Database* needs to be declared. Variables *s_GlobalTransaction* and *s_LocalTransaction* need to be declared to indicate the boundary of global and local transactions, respectively. Note that a bound database variable is needed as the argument of the local transaction's *begin()*. The role of the template type *s_Ref<T>* is the same as that of *d_Ref<T>* in the ODMG model. Namely, it is used to reference a persistent object of type *T*.

```
void main() {
   s_Database busDB, subwayDB;
   s_GlobalTransaction globalTrans;
   s_LocalTransaction busTrans, subwayTrans;
   try {
      // B1: open databases and begin transactions
      busDB.open("BusDB", s_Database::read_only);
      subwayDB.open("SubwayDB", s_Database::read_only);
      globalTrans.begin();
      busTrans.begin(busDB); subwayTrans.begin(subwayDB);
      // B2: create ad-hoc query objects
      s_Ref<BuslineDB::PathQuery> busPath = new
BuslineDB::PathQuery;
      s_Ref<BuslineDB::NNQuery> busNN = new BuslineDB::NNQuery;
      s_Ref<SubwayDB::PathQuery> subwayPath = new
SubwayDB::PathQuery;
      s_Ref<SubwayDB::NNQuery> subwayNN = new SubwayDB::NNQuery;
      // get user's input here
      s_Ref<Point> origin, destination;
      ...
      // B3: Find the nearest stations from origin and destination
      s_Ref<SubwayStation> station1 = subNN->execute(origin);
      s_Ref<SubwayStation> station2 = subNN->execute(destination);
      s_Ref<SubwayLineLink> subwayPath =
            subwayPath->execute(station1, station 2);
      // check if the subway stations are far from the origin and
      // the destination here
      ......
      // B4: find the nearest bus stops from origin and the
      // starting subway station which is far from the origin
      s_Ref<BusStop> stop1 = aBusNN->execute(origin);
      s_Ref<BusStop> stop2 = aBusNN->execute(station1->location);
      s_Ref<BusLineLink> busPath = busPath->execute(stop1, stop2);
      // report the found path here
      ......
      // B5: commit transcations and close databases
      busTrans.commit(); subwayTrans.commit();
      globalTrans.commit();
      busDB.close(); subwayDB.close();
   } catch(...) {...}
}
```

Figure 22.4. A sample application program in SDBC C++ OML

Note that the application program in Figure 22.4 accesses two databases transparently. The application programmer is relieved from knowing the physical database locations and the variations of underlying local DBMS implementations.

## 22.5. Client–server implementation

### *22.5.1. Layered SDBC implementation architecture*

Figure 22.5 shows the layered architecture of the current SDBC implementation, which employs the OMG Object Request Broker (ORB) for transparent network-level access. Right above this ORB layer lies the SDBC access layer, which implements the transparent access provider (TAP) library. An application program is linked to this TAP library and runs as a client process. The shaded ovals in the figure represent process boundaries.

Below the ORB layer lies the SDBC server layer, which implements the LDA (Local DBMS Adapter) and the SDBC Coordinator consisting of GDDM (Global Data Dictionary Manager), GTM (Global Transaction Manger), and GSICM (Global Spatial Integrity Constraints Manager). Either an LDA or an LDA with an ODBMS library runs as a process depending on the process architecture of a specific ODBMS. The SDBC coordinator exists as another process.

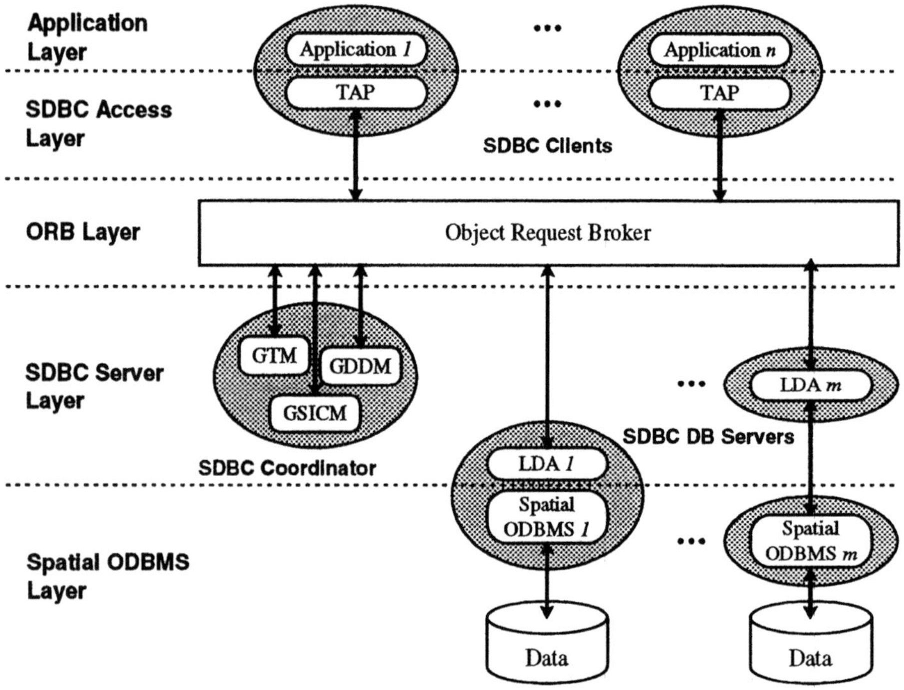


Figure 22.5. Layered SDBC implementation architecture

Four functional modules of the SDBC server layer, LDA, GDDM, GTM, and GSICM, are implemented as CORBA objects and registered in the ORB. LDAs hide ODBMS-specific implementation details. The interfaces of these objects, called the SDBC server layer interface (SLI), are described in the OMG interface definition language (IDL), and the ORB directs a client request to an appropriate object (Siegel 1996). The method shipping is supported by the ORB itself.

Although an application programmer can access SLI directly through the ORB, this approach has the following limitations. First, an application programmer has to know how to deal with CORBA. Second, because IDL does not support function

```
ss_DataDict_var DataDictRef; // Reference to GDDM object

class s_Database {
private:
   ss_Database_var DBref;
   ...
public:
   void open(string dbname, access_status status)
   {
      // Look up LDA location.
      string hostName = DataDictRef->lookup_host(dbname);
      string LDAName = DataDictRef->lookup_LDA(dbname);
      string DBName = dbname + ":" + LDAName;

      // Bind Database object and open it.
      DBref = ss_Database::_bind(DBName, hostName);
      DBref->open(dbname, status);
   }
   ...
};
```

**(a) *s_Database* class defined in the schema-neutral TAP library**

```
class ss_Database_i {
private:
   d_Database localDBref; // Reference to ODBMS's database
   ...
public:
   void open(string dbname, access_status status)
   {
      localDBref->open(dbname, status);
   }
   ...
};
```

**(b) *ss_Database_i* class defined in the schema-neutral LDA library**

Figure 22.6. Mapping of Database Class Definitions between TAP and LDA

overloading, runtime polymorphism, parameterized types, and so on, an application programmer cannot take advantage of the full expressive power of popular ODBMS programming languages like C++. Third, the performance of certain applications is bound to be limited by the method-shipping mechanism. The TAP in the SDBC access layer has been introduced to overcome these limitations.

Hiding the existence of ORB, the TAP shields an application programmer from the burden of knowing ORB programming. It solves the mismatch in the expressive power between SDBC C++ OML and OMG IDL. It provides facilities for managing the object cache in the client, and supports the object-shipping mechanism. Applications that have frequent access to a set of objects can benefit from this mechanism.

Figure 22.6 demonstrates the mapping provided by the TAP with the class definitions of the local database in TAP and LDA layers. The *s_Database* class shown in (a) is implemented in the schema-neutral TAP library. It accesses the GDDM object of the *ss_DataDict* type and the local database object of the *ss_Database* type, both of which are implemented in the SDBC server layer. Here, *ss_DataDict* and *ss_Database* are IDL interface names. The actual implementation uses variables of the *ss_DataDict_var* and *ss_Database_var* types, which are object reference classes automatically generated by the IDL compiler. All client requests are forwarded automatically to their corresponding implementation classes in the SDBC server layer.

Figure 22.6(b) shows the *ss_Database_i* class included in LDA. It implements the *ss_Database* interface used by the *open* method of the *s_Database* type. Since the SDBC application program references a database with its logical identity, the open method of *s_Database* first looks up the GDDM object to find the physical location and name of the target database. It then binds the database object reference variable (named *DBref*) of the *ss_Database_var* type using the *_bind()* method in CORBA (Orbix 1997). On the other hand, the *ss_Database_i* class defines a local variable *localDBref*, which references a local database object of *d_Database* type. Its *open* method implements the details of the database open operation specific to a underlying DBMS.

For the user-defined persistent classes such as *BusStop*, the SDBC ODL preprocessor automatically generates corresponding IDL definitions such as *ss_BusStop,* and the corresponding implementation class in the server such as the *ss_BusStop_i* class shown in Figure 22.7. As in the case of *s_Database*, the IDL compiler generates object reference classes such as *ss_BusStop_var* for these IDL definitions. The implementation of *s_Ref<T>* then uses variables of these object reference types. For example, the *delete* operator of *s_Ref<BusStop>* calls a *ss_BusStop::delete_object()* method. On the SDBC server side, an *ss_BusStop_i* object references a persistent *BusStop* object in the target database through the *d_Ref<BusStop>*.

### *22.5.2. Application development process*

Figure 22.8 shows the process of developing SDBC schemas and applications based on the current SDBC implementation. Four shaded ovals represent the application developer's portion of the work: schema definition in SDBC ODL, method implementation on client or server sides, and application code development in SDBC C++ OML. All the rest of the work is done automatically by the SDBC ODL preprocessor, a CORBA IDL compiler for C++, and the preprocessors of target ODBMSs. Figure 22.8(a) shows the process of schema development. The SDBC ODL preprocessor translates an SDBC ODL schema into three types of declarations: CORBA IDL definitions such as *ss_BusStop,* their implementation classes such as *ss_BusStop_i,* and class definitions of a target database such as *Bus_Stop*. It also extracts metadata and stores them in the GDDM repository. The IDL declarations are again translated into the stub and skeleton code such as *ss_BusStop_var* in C++ by the IDL compiler.

On the server side, the preprocessor of the target ODBMS processes its ODL declarations, generates C++ header and implementation code files for user-defined types, and stores metadata (Object Design 1995; Objectivity 1996; Versant 1993). The next step is to implement schema methods in C++ OML of the target ODBMS. Finally, these implementations, the generated C++ code, and the skeleton code are compiled and linked with the target ODBMS library and the schema-neutral LDA library to build a schema-dependent LDA. Note that the schema-neutral LDA library is dependent on a specific ODBMS.

On the client side, the programmer may choose to implement some of the schema methods in SDBC C++ OML instead of the server side. These method implementations, based on the object shipping mechanism, are compiled together with the IDL stub code and linked with the schema-neutral TAP library to build a schema-dependent TAP library. The schema-neutral TAP library and schema-dependent TAP libraries are independent of participating ODBMSs.

```
class ss_BusStop_i
{
public:
    ss_BusStop_i(d_Ref<BusStop> aRef) : ref(aRef) {}
    ~ss_BusStop_i(){};

    string name() { ... };
    void name(string aName) { ... };
    ss_Point * location() { ... };
    void location(ss_Point * aPoint) { ... };
    ss_BusLineLink * line() { ... };               // relationship
    void line(ss_BusLineLink * aBusLineLink) { ... };

    void mark_modified() {}
    void delete_object() { ref.delete_object(); }

protected:
    d_Ref<BusStop> ref;        // Reference to Persistent Object
};
```

Figure 22.7. *ss_BusStop_i* class definition included in schema-dependent LDA library

Figure 22.8(b) shows the development flow of an executable application. The code written in SDBC C++ OML is compiled and linked with a schema-dependent TAP library to build an application.

## 22.6. Summary and implementation status

The need for unifying spatial objects in heterogeneous databases arises frequently in processing geographic information. One of the keys to unification is the ability to connect to multiple databases and manipulate them transparently and consistently. This chapter has presented an ongoing effort of developing middleware, called SDBC, for providing transparent access to multiple spatial object databases. Although the name *SDBC* is similar to *ODBC* (ODBC 1997) or *JDBC* (Hamilton and Fisher 1997), it differs significantly from the latter two in the level of abstraction provided to application programmers. Programmers develop applications accessing multiple spatial databases with a standard object-oriented model based on ODMG and Open GIS. The SDBC middleware hides DBMS-specific implementation details, and fills the gap between SDBC and local ODBMSs.

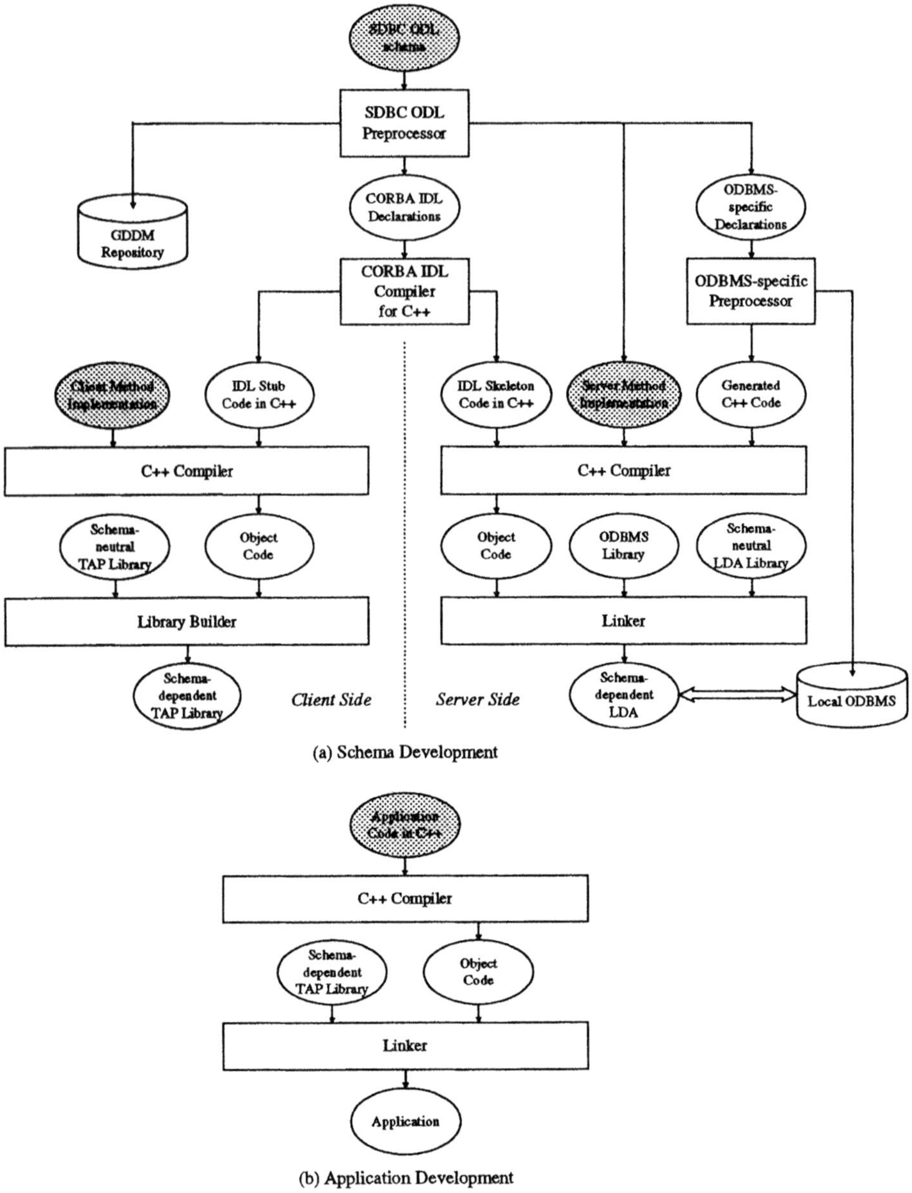


Figure 22.8. SDBC schema and application development processes

Sponsored by the Korean government as a part of the National GIS Database tool development initiative, SDBC has been under development at Seoul National University since December 1996 together with SDBX. An SDBC prototype is currently being implemented in SUN Solaris 2.5 with C++, Orbix (Orbix 1997), and SDBX on top of Objectivity/DB (Objectivity 1996). The current prototype system includes a working prototype of LDA for Objectivity/DB, TAP, and a simple coordinator. Ongoing research focuses on the spatial integrity constraint manager, the recovery manager for global transactions, and the SDBC ODL preprocessor. The object caching in TAP is being implemented to support the object shipping mechanism. A demo application for finding traveling paths from multiple transportation databases has been implemented to prove the utility of SDBC.

## References


Balovnev O, Breunig M, Cremers A B 1997 From GeoStore to GeoToolKit: the second step. *Proceedings of the 5th International Symposium on Spatial Databases, Berlin*: 223–37

Beckmann N, Kriegel H-P, Schneider R, Seeger B 1990 The R*-tree: an efficient and robust access method for points and rectangles. *Proceedings of the 1990 ACM SIGMOD International Conference on Management of Data, Atlantic City*: 322–31

Bernstein P, Hadzilacos V, Goodman N 1987 *Concurrency Control and Recovery in Database Systems*. Addison-Wesley

Breitbart Y, Garcia-Molina H, Silverschatz A 1992 Transaction management issues in a failure-prone multidatabase system environment. *VLDB Journal* 1(1): 1–39

Buehler K, McKee L (eds) 1996 *The OpenGIS Guide: Introduction to Interoperable Geoprocessing*. Open GIS Consortium, Inc

Cattell R G G, Barry D K (eds) 1997 *The Object Database Standard: ODMG-2.0*. Morgan Kaufmann: 251–5

Egenhofer M J 1991 Reasoning about binary topological relations. *Proceedings of the Second International Symposium on Spatial Databases*: 143–60

Egenhofer M J 1997 Spatial relations: models and interface. *Tutorial in 5th International Symposium on Spatial Databases*

Elmasri R, Navathe S B 1994 *Fundamentals of Database Systems*. Benjamin-Cummings

Frank A 1992 Qualitative spatial reasoning about distances and directions in geographic space. *Journal of Visual Languages and Computing*: 343–71

Günther O, Riekert W–F 1993 The design of GODOT: an object-oriented geographic information system. *Bulletin of the Technical Committee on Data Engineering* 16(3): 4–9

Guttman A 1984 R-trees: a dynamic index structure for spatial searching. *Proceedings of ACM SIGMOD Conference, Boston*: 47–57

Hamilton G, Cattell R, Fisher M 1997 *JDBC Database Access with Java: A Tutorial with Annotated Reference*. Addison-Wesley

Kim W, Lochovsky F H 1989 *Object-Oriented Concepts, Databases, and Applications*. Addison-Wesley

Medeiros C B, Pires F 1994 Databases for GIS. *SIGMOD Record* 23(1): 107–15

Nittel S, Muntz R R, Mesrobian E 1997 GeoPOM: a heterogeneous geoscientific persistent object system. *Proceeding of 9th Int'l Conference on Statistical and Scientific Database Management, Washington*

Nittel S, Yang J, Muntz R R 1996 Mapping a common geoscientific object model to heterogeneous spatial data repositories. *Proceedings of the Fourth ACM Workshop on Advances in GIS*: 51–8

Object Design 1995 *C++ Programming with ObjectStore v4.0*. Object Design, Inc., Burlington

Objectivity 1996 *Using Objectivity/C++, Version 4*. Objectivity Inc

ODBC 1997 *Microsoft ODBC 3.0: Software Development Kit and Programmer's Reference*. Second Edition. Microsoft Press

OMG ORBTF 1992 *Common Object Request Broker Architecture*. Object Management Group, Framingham

Orbix 1997 *Orbix Programming Guide*. IONA Technologies Ltd

Open GIS Consortium 1997 *OpenGIS Simple Features Specification for CORBA*. Revision 0. Open GIS Consortium, Inc

Scholl M, Voisard A 1992 *Geographic Applications: An Experience with $O_2$*. In Bancilhon F, Delobel C, Kanellakis P (eds) 1992 *Building an Object-Oriented Database System: The Story of $O_2$*. Morgan Kaufmann: 585–618

Sellis T K, Roussopoulos N, Faloutsos C 1987 The R+-tree: a dynamic index for multi-dimensional objects. *Proceedings of the 13th International Conference on Very Large Data Bases, Brighton*: 507–18

Siegel J 1996 *CORBA: Fundamentals and Programming*. John Wiley & Sons

Versant 1993. *Versant Reference Manual*. Versant Object Technology, Inc., Menlo Park, California.

Zhang A, Elmagarmid A K 1993 A theory of global concurrency control in multidatabase systems. *VLDB Journal* 2(3): 331–60

# 23 INTEROPERATING GEOGRAPHIC INFORMATION SYSTEMS USING THE OPEN GEOSPATIAL DATASTORE INTERFACE (OGDI)

Gilles Clément, Christian Larouche, Paul Morin, and Denis Gouin

## 23.1. Introduction

Interoperability of spatial systems is a goal sought by many and, by most measures, attained by none. If systems were truly interoperable, it would be straightforward to share geographic information across software platforms and between databases collected for different purposes by different organizations. In contrast, a variety of technical, institutional, and physical barriers inhibit the fluid exchange of geographic information.

The geomatics industry has grown continuously over the last two decades, and the number of application domains in which geographic information systems (GIS) and geospatial information are used has increased every year. However, growth in the industry has not lived up to the predictions made 10 years ago, especially compared with that of other information technologies such as the Internet, office automation, or sales automation. The major reasons for slow growth are the lack of digital geospatial data and their high cost, the geospatial data barrier, problems integrating GIS into existing systems, and the complexity of existing GIS software (Larouche et al. 1996).

The geospatial data barrier is probably the most significant problem: geospatial data products are offered in a large variety of different and incompatible formats, many using different coordinate systems or cartographic projections. In response, GIS vendors have developed geospatial data translators, but this solution results in poor interoperability, which leads to 60% to 85% of the implementation cost of a GIS going into data conversion and integration (see Clément et al. 1997; Gouin et al. 1997). This high cost is a prime cause of the lack of geospatial information and consequently the slow growth of the GIS industry. According to experts, the geomatics industry would generate ten to twelve times more revenue per year if the geospatial data barrier could be eliminated.

Although the standardization of the format of geospatial data is a better solution to the problem, it is very unlikely that industry will move to a single standard. At least a half dozen significant standards can be expected, in addition to the proprietary formats used in the commercial data products already gaining momentum in the marketplace. Standardization efforts alone will not solve the geospatial data conversion and integration problem.

This chapter on interoperable system experiences describes a solution to the geospatial data access and integration problem, the Open Geospatial Datastore Interface (OGDI). OGDI is an application programming interface (API) that sits between a GIS software package (the application) and various geospatial data products and, using a standardized access method, provides solutions to some of the most difficult geospatial data integration problems.

This chapter first describes the main characteristics and architecture of OGDI (Section 23.2). Section 23.3 demonstrates how to integrate OGDI into an application. Successful integration of OGDI into a GIS application (GRASSLAND), in a Web viewer (WebMap) that provides geospatial data browsing capability and in a large geospatial data warehouse is then discussed. The same section shows how OGDI has influenced the development of major national infrastructures such as the Canadian Geospatial Data Infrastructure (CGDI) and the Canadian Earth Observation Network (CEONet). Section 23.3 also demonstrates how OGDI becomes a key element for the Department of National Defence (DND) in its mission of providing improved access to geospatial data suitable for worldwide military operations. Conclusions and plans for future development are presented in Section 23.4. OGDI (including its source code) is available free in order to improve the interoperability of GIS software. Section 23.5 provides information about obtaining the source code of this freeware.

## 23.2. Open geospatial datastore interface (OGDI)

As has been described in the previous sections, the geospatial data barrier problem exists because various geospatial data products, in different formats and cartographic projections, coexist and need to be integrated. In a way, the situation is analogous to the multiplicity of spoken languages that coexist. Multilingual people can share information with each other without having to translate into their mother tongue. For example, if English-speaking people were to learn to *read* Chinese and vice versa, nobody would have to learn to *write* Chinese; the Chinese would know how to read English.

Applying this "learn foreign languages instead of being ethnocentric" theory to geospatial data has led to the Open Geospatial Datastore Interface (OGDI). Using OGDI, different geospatial data formats can be read without any translation or conversion.

The following subsections define OGDI, describe its main characteristics, present the data model it uses, describe its client–server architecture based on a new geospatial Web protocol, and list the geospatial data formats that it understands.

### 23.2.1. Definition

OGDI is an application programming interface (API) that sits between a GIS software package (the application) and various geospatial data products (or drivers), providing a standardized access method to the latter. The open style of OGDI is similar to that of Microsoft Open DataBase Connectivity (ODBC), a programming interface that allows applications to access data in a database management system using Structured Query Language (SQL; Microsoft Corporation 1994). Like ODBC, OGDI uses a driver approach, achieving interoperability by accessing each geospatial data format through an appropriate driver.

### 23.2.2. Main characteristics

OGDI is an evolving concept that is simple and easy to maintain. OGDI is already available. It is presently in its third generation implementation, which implies robustness and maturity. OGDI is based on the TCP/IP protocol and consequently can easily use the Internet as a medium to distribute geospatial data products. OGDI facilities can be accessed either through an ANSI C or a Tcl/Tk (Ousterhout 1994) API, which can be called from Win32 or UNIX applications.

OGDI provides tools for some of the most difficult geospatial data integration tasks, including:

- the conversion of various formats into a uniform transient data structure;
- the transparent adjustment of coordinate systems and cartographic projections;
- the transparent transformation of platform-dependent representations; and
- the retrieval of geometric and attribute data.

### 23.2.3. Transient data model

The transient data model supports both geometric and attribute data. Geometric data can be vector (area, line, point, text) or raster (line or tile access). Each area feature is made up of one or more rings. Metadata include geographic coverage, cartographic projection, and sources. Table 23.1 shows the C data structures used to describe vector elements in the transient data model. The DIGEST data model largely inspires this data model. Each feature has a corresponding C data structure, used by the C language API and by all servers.

The *c len* variable indicates the number of coordinates that describe this linear feature. Each coordinate *c val* is defined by the following C substructure:

```
struct ecs_Coordinate {
    double x;
    double y;
};
typedef struct ecs_Coordinate ecs_Coordinate;
```

The current model can support only two-dimensional vector representations. In addition, OGDI uses a number of supporting C data structures to describe the metadata information.

| **Point Feature** | **Line Feature** |
|---|---|
| Point features are composed of a single instance of ecs Coordinate.<br><br>struct ecs_Point {<br>ecs_Coordinate c;<br>}; | Line features are composed of two or more coordinates. Line features must be homogeneous in direction.<br><br>struct ecs_Line {<br>struct {<br>u_int c_len;<br>ecs_Coordinate *c_val;<br>} c;<br>}; |
| **Area Feature** | **Area Feature (cont.)** |
| Area features are composed of one or more rings. Rings are similar to line features except that the last coordinate is always equal to the first.<br><br>struct ecs_FeatureRing {<br>ecs_Coordinate centroid;<br>struct {<br>u_int c_len;<br>ecs_Coordinate *c_val;<br>} c;<br>};<br><br>typedef struct ecs_FeatureRing<br>ecs_FeatureRing; | struct ecs_Area {<br>struct {<br>u_int ring_len;<br>ecs_FeatureRing *ring_val;<br>} ring;<br>}; |

Table 23.1. OGDI transient vector structure

**Text Feature**: Text features are very similar to Point features except that they also hold a text string.

```
struct ecs_Text {
    char *desc;
    ecs_Coordinate c;
};
```

**Matrix Feature**: Matrices (rasters) are accessed on a line-by-line basis. Each raster line is described as follows:

```
struct ecs_Matrix {
        struct {
            u_int x_len;
            u_int *x_val;
        }x;
    };
```

### *23.2.4. A client–server architecture with a new geospatial web protocol*

Figure 23.1 illustrates the components of the OGDI architecture. A dynamically loadable C language library is normally used to access OGDI facilities. First, an application that needs to perform geospatial processing calls functions through the C API or Tcl/Tk API and retrieves the results. Drivers are used to access various geospatial data formats, one for each format. Each driver is a dynamically loadable library that processes C language API requests for a specific datastore. Once a driver is loaded, it receives requests, fetches data from the datastore, translates the data into the transient data structure, and returns the results to the application. All the APIs are available for UNIX (Solaris and Linux) and for Windows NT and Windows 95.

For a programmer using OGDI, there is no difference between a local datastore and a remote one. Drivers can be accessed either directly, for local datastores, or remotely, for external datastores. For remote procedure calls, other components are needed. The *gltpd* (geographic library transfer protocol deamon) and the network driver take care of the communication protocol transparently, and automatically transform data between incompatible processor architectures.

The gltpd module is a small utility program that mimics the behavior of the C API on a remote computer. When the gltpd process receives its first request from an application, it creates a new thread (a fork) of itself, which loads the requested driver type, takes control of the communication process with the network driver, and serves all subsequent OGDI calls coming from the application. The network driver is a dynamically loadable library that relays calls from the C API to a gltpd process running on a remote computer. The gltpd process and the network driver together link the OGDI application to a remote driver through a TCP/IP network, allowing the application programmer to access remote drivers as if they were local, using a client–server paradigm. *Client* is used here to describe a connection made by an application and *server* to describe an instance of one driver resident on a data server connected to one application. In the current implementation, the gltpd process and the network driver are based on the ONC RPC 4.0 protocol. The combination of the gltpd process and a specific driver becomes a server to the client (i.e., the application's connection).

The gltp component is a necessary substitute for the HTTP or FTP transfer protocols. Like HTTP and FTP, gltp gives direct access to a local database or remote access to an external database. Unlike HTTP, gltp is a *state aware* protocol, which preserves knowledge of the last completed transaction. *Stateless* protocols (as is HTTP) process each query and call independently of previous and successive queries and calls. An HTTP server has no memory, which precludes such functions as

« GetNextObject » and « GetObject » (see Table 23.2). The advantage of a stateless protocol is that only one server is sufficient to communicate with many users. However, OGDI needed a protocol with memory to handle successive queries in the context of earlier ones; this led to the strategy of implementing the gltp architecture, which spawns a server process for each client.

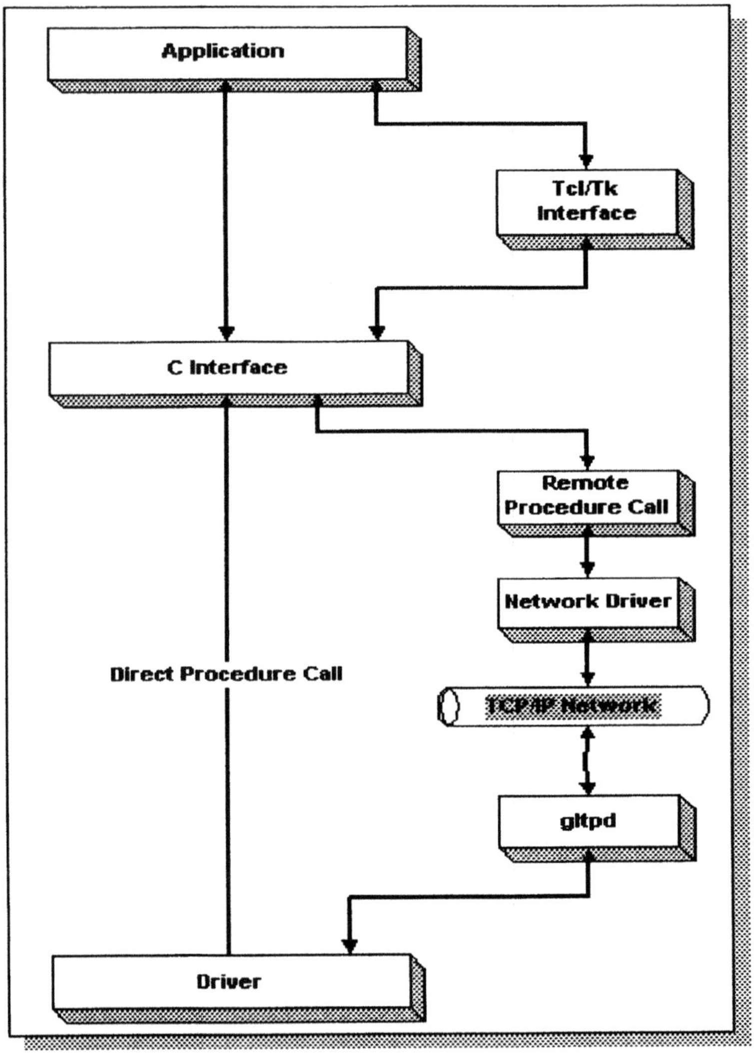


Figure 23.1. OGDI architecture

| Function | Description |
|---|---|
| CreateClient | creates a client (connects to a geographic datastore). |
| DestroyClient | deletes a client and unloads the associated driver from memory. This terminates the communication with the geographic datastore. |
| SelectLayer | specifies the current coverage or layer. |
| ReleaseLayer | releases a layer. |
| SelectRegion | selects the current geographic region. |
| GetDictionary | retrieves an [incr Tcl] applet from the driver. The applet describes the contents of a geographic datastore. |
| GetAttributesFormat | specifies the attribute format of the currently selected layer. |
| GetObject | specifies the attributes of the selected geometric object. |
| GetNextObject | specifies the next object in the currently selected coverage. |
| GetRasterInfo | gathers information on the currently selected raster coverage. |
| GetObjectIdFromCoord | retrieves the object in the current layer that is nearest to the set of specified coordinates. |
| UpdateDictionary | returns an updated list that describes the content of a datastore. |
| GetServerProjection | returns the server's current projection. |
| GetGlobalBound | specifies the driver's global geographic region. |
| SetClientProjection | specifies the client's projection. |
| SetServerProjection | specifies the projection of the driver. |
| SetRegionCaches | specifies the geographic region occupied by caches. |
| LoadCache | loads data for the region set by the ecs SetRegionCaches command. |
| SetServerLanguage | specifies the language in which the server returns information. |
| ReleaseCache | deletes the cache related to a coverage stored by the cln LoadCache command. |

Table 23.2. List of OGDI functions

### *23.2.5. OGDI vs. ODBC*

Like ODBC, OGDI uses a driver approach, achieving interoperability by accessing each geospatial data format through an appropriate driver. This approach presents two main advantages:

- the client learns one API, but can connect to many data sources; and
- new drivers can extend existing clients with upgrading or recompilation.

It has been decided to develop a new driver framework because the current version of ODBC is not well suited for geospatial applications. For instance, ODBC does not provide suitable data types for geospatial data and does not handle cartographic projection and coordinate transformations. Another important argument was that few legacy GIS formats are compatible with the query language used by ODBC (SQL). Due to this impedance mismatch, ODBC drivers to legacy GIS formats would be very difficult to implement. The need to have a solution working on the Web with a widespread portability is another factor that influences the decision of a new approach. The middle layers of the OGDI implementation (projection and coordinate transformation and simple spatial operators) provide a lot of functionality. This keeps the cost of implementing a driver low. OGDI drivers can be developed in about 2–3 weeks effort.

### *23.2.6. Available drivers*

OGDI drivers are already available for several geospatial formats. Drivers have been developed for the various formats of the DIGEST standard, including Vector Relational Format (VRF) products, ARC Digitized Raster Graphics (ADRG) products, and the Raster Product Format (RPF). A driver for Digital Terrain Elevation Data (DTED) matrix products has also been developed.

A driver for the National Topographic Data Base (NTDB) of Canada represented in the CCOGIF format and a driver for the different types of satellite imagery in the CEOS format are also available. Drivers for several proprietary formats such as GRASS, PAMAP, DGN, DWG and DXF, drivers for USGS SDTS and DOQ formats, and drivers for original TIFF and geo-referenced TIFF (GeoTIFF) image formats are also ready. Drivers for the different ESRI ARC/INFO profiles such as the native coverage and INFO files, the shapefiles and the exchange E00 files are also available. Geospatial data drivers for the IDRISI and the MapInfo MID/MIF formats can also be used. OGDI drivers, such as the Common Operational Picture (COP) driver, have been developed to access tactical data. Finally, the grid driver was developed to create grid line or interval points or text, in any projection, given a pre-selected region. Drivers can easily be developed for other geospatial formats.

## 23.3. Using the OGDI API in applications

The underlying philosophy of OGDI is to encapsulate all the tasks related to geospatial database access in a simple and standard API. With OGDI, an application is isolated from the details of integrating a new kind of datastore. The task of navigating through a datastore falls to the drivers themselves, with the C API providing a uniform way to retrieve information, regardless of the datastore format.

The C-language API is the heart of OGDI. It is a dynamically-loadable C library that programmers use to access OGDI facilities. It is composed of 20 functions that:

- manage and load the geospatial datastore driver;
- provide an entry point to OGDI functions for each driver;
- allocate storage for geometric and attribute data;
- perform garbage collection of previously allocated storage;
- provide parameter validation and sequence validation for OGDI calls; and
- make any necessary coordinate and cartographic projection transformations.

A complete list of these functions is given in Table 23.2 and a complete set of instructions for programmers is provided by LAS (1996a).

To interact with a datastore, a simple application goes through the sequence of steps shown in Figure 23.2. First, it establishes a connection or creates a *client*, the term used to describe each instance of a connection. Second, it selects a geographic region. Then it selects a map layer and extracts objects from it, either sequentially or randomly. It processes the results and finally terminates the connection.

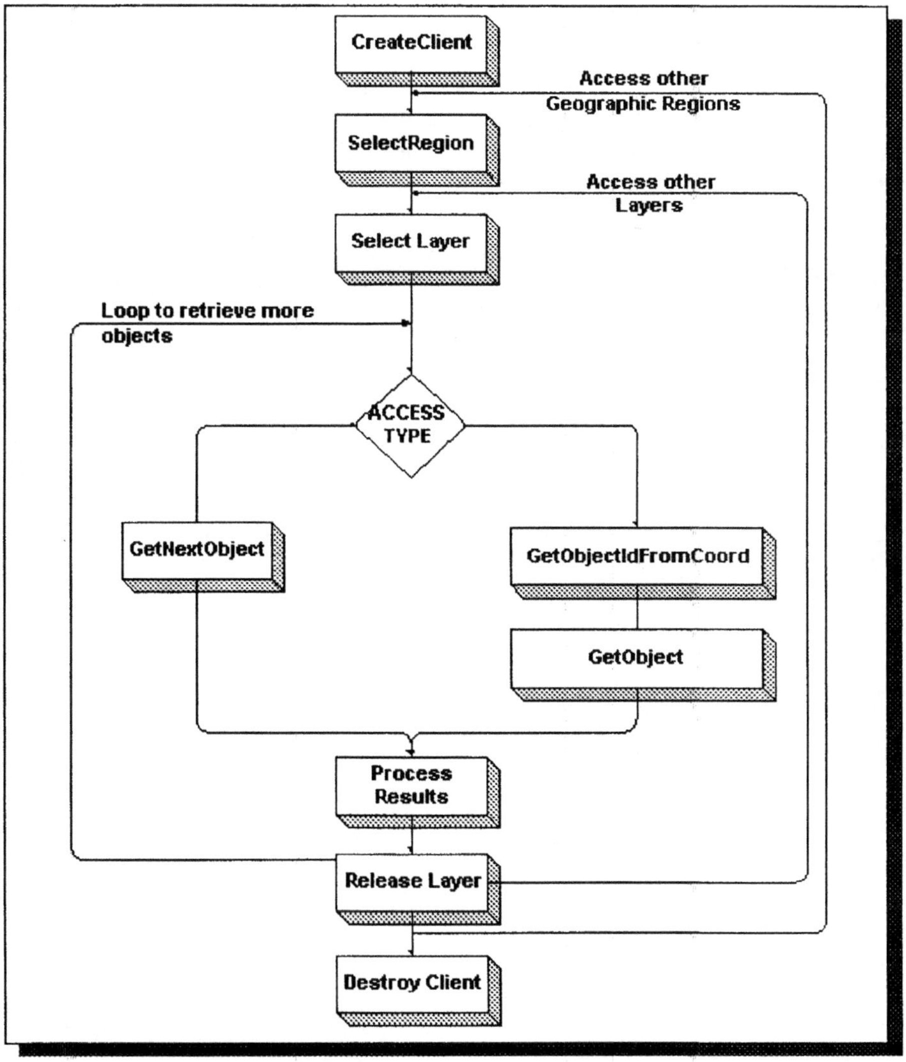


Figure 23.2. Using the OGDI API in an application

### *23.3.1. In a GIS application: GRASSLAND and WebMap*

The OGDI source code with complete documentation (LAS 1996a) is available free of charge. However, some basic knowledge of computer programming is needed to integrate the library of functions into a particular application. For those without that knowledge, all the OGDI facilities have been compiled into GRASSLAND GIS software (Larouche et al. 1996; LAS 1996b) and into a geospatial Web viewer called WebMap.

This section illustrates how OGDI is used by GRASSLAND and WebMap to access different data formats. Figure 23.3 shows the component of GRASSLAND called *Librarian*, which manages datastores that can be local or accessed either through a local network or via the Internet. The Librarian uses OGDI to access data formats such as DTED, ADRG and VRF, as seen in Figure 23.3.

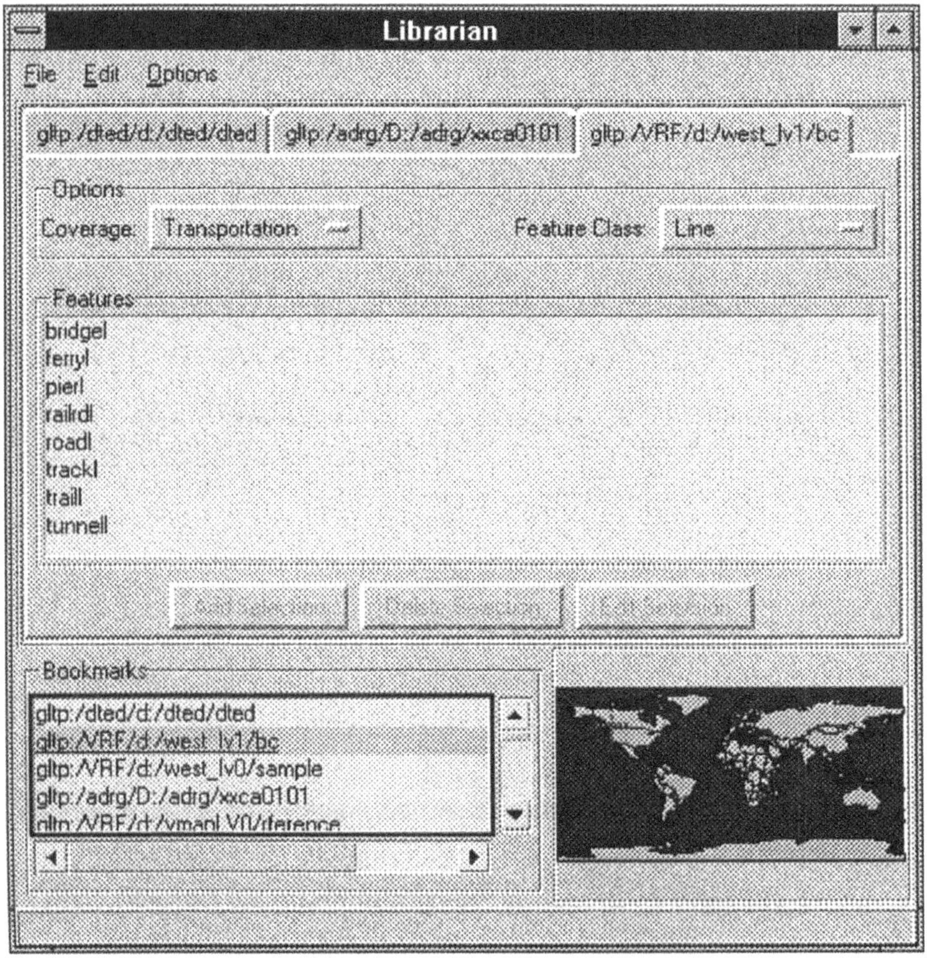


Figure 23.3. Use of OGDI in GRASSLAND Librarian

Each connection between the application (a *client*) and a driver (a *server*) is defined by an ASCII string similar to the Uniform Resource Locator (URL) used by the World Wide Web. Each string is prefixed by *gltp* (analogous to the URL prefixes HTTP or FTP). The prefix is followed by a host name for remote driver access, a driver descriptor, and then a file path-name that indicates the location of the datastore. The general form is:

gltp:[//<hostname>]/<name of driver>/<pathname>

The hostname is not used when accessing a local datastore. The presence of the hostname string indicates that a connection to a remote driver is being made, using gltpd. The following is an example of a connection string:

gltp://copernic.las.com/grass/las3/gis/spearfish/PERMANENT

This string describes a *GRASS* datastore named */las3/gis/spearfish/PERMANENT* located on the host computer *copernic.las.com.* In Figure 23.3, one of the connections is

gltp:/vrf/d:/west_lv1/bc

which describes a *DIGEST-VRF* datastore named `/d:/west_lv1/bc`, on a local host computer.

The Librarian gives the list of map layers available in a datastore. Its interface will offer different sets of buttons, depending on the data format. In the example above, for instance, the vector format VRF is active. With this type of data, the software allows queries to be posed (by pressing the *Add Selection* button) to select particular information. For example, secondary roads only might be selected, rather than the full layer.

The Librarian displays a World map to indicate the geographic location of a database. In Figure 23.3, geospatial data of the Vancouver area is shown. From the Librarian, map layers can be dragged and dropped onto other components of the software, such as the Map Viewer. Figure 23.4 shows GRASSLAND Map Viewer, which displays several map layers of the Vancouver area obtained from OGDI. An ADRG raster map layer is displayed in the background, with different VRF map layers (e.g., thicker vectors representing coast lines and thinner vectors representing roads) overlaid. The data was extracted from the 1995 Canadian Geomatics CD-ROM (DND-Canada and Geomatics Canada 1995).

Even though GRASSLAND uses the GRASS database format as internal database structure, it can read, via OGDI, the spatial information of other formats and display this information on its Map Viewer without any data conversion.

WebMap, a sub-product of GRASSLAND, is a Helper App called from an Internet browser to display geospatial data accessed through OGDI. For data producers, WebMap can be used to give their current or potential users an access to their geospatial data server. Data producers need to install an OGDI-based server package that contains all needed drivers and the gltpd. On the other side, users need to install WebMap to access the data producer server and visualize the available data on this server. Users can even query the server database and select exactly the information (coverages or layers) they need for viewing.

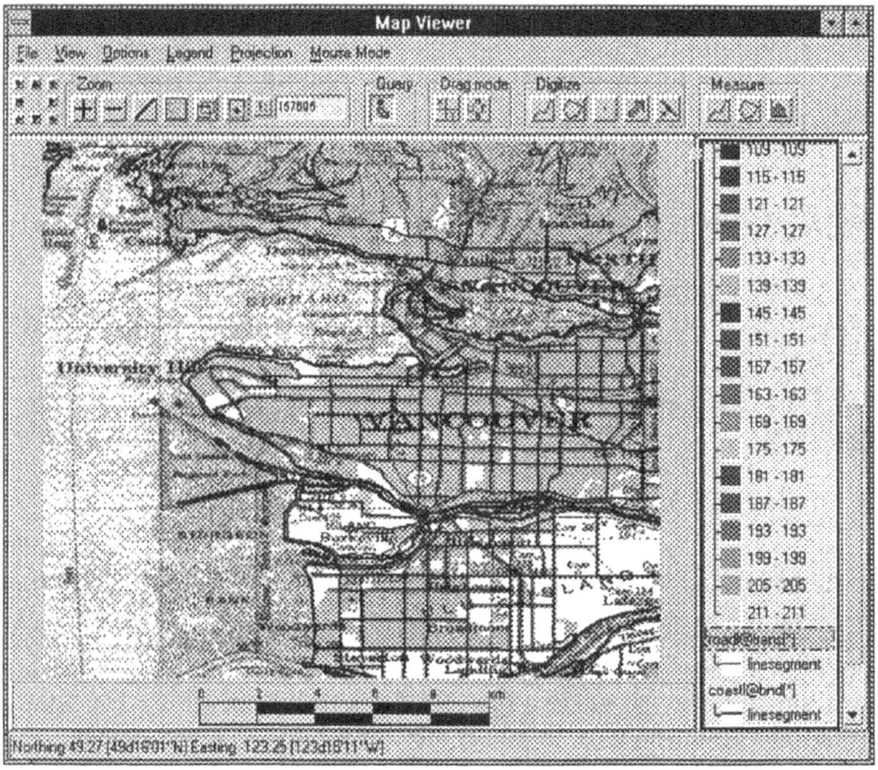


Figure 23.4. Visualizing geospatial information obtained with OGDI using the GRASSLAND Map Viewer

### *23.3.2. In a large geospatial data warehouse*

OGDI has been used in a geospatial data warehouse demonstrator project to provide open access to multiple heterogeneous geospatial data sources (local or remote data products and local or remote GIS native data) and integrate these data within a single seamless data warehouse. A data warehousing approach is used in which integrated multi-dimensional multi-scale databases, legacy data, and stores of standardized data products residing in various computer environments are assembled through a distributed server architecture designed to service an information community (St. Laurent et al. 1997).

A *geospatial data warehouse* is used to support the storage, processing, analysis, and access to foundation data such as imagery data, elevation data, and feature data, and their integration within a seamless and continuous geospatial data information architecture. The data warehouse includes legacy geospatial data already available within most organizations or information communities.

A geospatial data warehouse provides a *human interface* architecture that includes the following components:

- Any geospatial data applications software used for processing, analyzing and presenting geospatial data.
- Geospatial data interface software that provides open access to the geospatial data warehouse such as OGDI.
- Geospatial data services consisting of customer profile, query facility, ordering capability, administrative and data security components.

We tested the performance of the implemented geospatial data warehouse demonstrator using two data sets. Using DTED Level 0 format and Vector Data in VMAP Level 0 format at a 1:45,000,000 scale, it took 1.5 minutes to process the 1468 partitions of the 150 Meg matrix data and 2.4 minutes to process the three vector coverages (OCEANSEA: 18.9 Meg, COASTL: 8.8 Meg and POLBNDL: 780 Kb). Using DTED Level 0 format and Vector Data in VMAP Level 0 format but at a 1:1,000,000 scale it took 2.2 seconds to process the 6 partitions of the 614 Kb matrix data and 3.2 seconds to process the 295 Kb of the six vector coverages.

The OGDI software was proven to be very efficient not only to provide a transparent access to disparate native geospatial data structures maintained in flat file systems, but also an excellent database interface software to an integrated geospatial data warehouse (St. Laurent et al. 1997). As improvement, the OGDI API could be extended to support the SQL Call Level Interface (CLI) in order to provide access to catalogs developed using the spatial information modeling facility. This would allow OGDI compliant applications to access information in flat and object-relational files, providing a greater degree of interoperability than is possible today.

### *23.3.3. In the Canadian Geospatial Data Infrastructure (CGDI)*

In Canada, governments and industry are collaborating to develop the infrastructure to collect, manage, and disseminate geospatial data using the emerging information highway. The Inter Agency Committee on Geomatics (IACG) has taken the lead to guide federal and provincial governments and the commercial sector to create a Canadian Geospatial Data Infrastructure (CGDI). The CGDI is accessible through the Web at URL: *http://cgdi.gc.ca*. To accomplish its mandate, the IACG is using the following five thrusts:

1. *Partnerships* between governments and industry to make geospatial data available, to share the development, and to capitalize on emerging technology;
2. *Supportive Policies* to foster harmonized access to, and use of geospatial data;

3. *Geospatial Standards* participation to simplify access, quality, and data integration, and to encourage development of compliant software technology;
4. *Geospatial Data Access* through cooperative development of technical components of the CGDI;
5. *Framework Data* for the provision of a reliable foundation of geospatial data for adding value, application development, and more specific detailed data collection.

OGDI enters especially in item 4 to solve the geospatial data access problem of the large amount of data available from different organizations in Canada. Among these data are the Earth observation data distributed through another initiative: CEONet.

### 23.3.4. In the Canadian Earth Observation Network: CEONet

The Canadian Earth Observation Network (CEONet) is an initiative by the Canadian government to create a national infrastructure for providing access to Earth observation archives and other complementary spatial databases. This initiative is being driven by the requirements of Canadian users for better access to Earth observation data, and by the opportunities for Canadian industry afforded by the rapid growth of the international market for Earth observation data, services, and network systems.

Earth observation applications have reached a level of sophistication where simultaneous access to widely distributed data sources is often required. The profile of the typical user of Earth observation data is changing to that of an expert in a field who needs to be able to simply use Earth observation data without becoming an expert in the Earth observation field itself. Accessing and distributing spatial data needs to become a process where the transition from data discovery to data use is enabled in such a way that the complexity inherent to the spatial data organizations and structures is the responsibility of the mechanism and not the user.

CEONet takes into consideration a number of important technology trends and related issues. Some areas of importance are: finding information in networks; integration of heterogeneous data; search languages; dictionary, vocabulary, and directory issues; information exploration; service qualities; standards; and general trends in hardware and software. Where possible, CEONet system development takes advantage of existing and emerging technologies and standards as the technologies evolve and mature as was the case for OGDI.

Interoperability is implemented in CEONet within the data discovery process. At the current stage of development, CEONet is investigating mechanisms for direct access to Earth observation data. As a demonstration for data suppliers, a prototype geospatial data warehouse is accessible through the CEONet interface. The Data Supplier Prototype is meant to be an implementation of a directly accessible geospatial data repository. Access to the prototype repository of global data (1:1M) is enabled using OGDI. An OGDI driver for the different types of satellite imagery in the CEOS format (as seen in Section 23.2.5) has been developed for the CEONet infrastructure.

### *23.3.5. In the Department of National Defence (DND)*

As part of their mandate to insure national and international security, the Canadian Forces have always been very active in a full range of multilateral operations, including the preventive deployment of forces, peacekeeping, observer missions, and assistance in natural disasters. These missions require rapid intervention in various parts of the world and very often necessitate work as part of a coalition force.

Military forces rely heavily on geospatial information to conduct their missions successfully. In the era of the Information age, traditional paper maps have been replaced with digital products. In order to facilitate interoperability, the NATO countries have defined a geospatial standard known as DIGEST (Digital Geographic Information Exchange Standard). This standard (Version 1.2) was issued in 1994, and approved as a NATO (STANAG 7074) and National Standard of Canada for use by DND and the Canadian Forces.

As a commitment to the provision of the next generation of geomatics data and support, the Defence Geomatics has layed out, in cooperation with the U.S. National Information and Mapping Agency (NIMA), the Global Geospatial Information & Services (GGI&S) Initiative (McKellar 1995). This initiative draws the road map for Defence Geomatics to satisfy its customers in the near, mid, and long term, by enhancing their production facility to:

- populate more extensive, higher resolution, and consistently accurate databases,
- reduce the production time,
- improve the access to the data, and
- provide data suitable for world-wide military operations.

The privileged geospatial products are those which are DIGEST-compliant or typical military digital products produced by NIMA. However, as rapid access to data is essential for many deployed operations, and as geospatial data production is costly in time and money, provision of terrain data in any geospatial format (legacy datasets) must be supported.

OGDI is a key element of the solution for interoperability and for the GGI&S initiative. It provides the needed access to both local and remote sites and a transparent access to multiple heterogeneous datastores. Moreover, it can easily be embedded in commercial geographic information systems.

NATO defines interoperability of command and control systems in terms of four levels:

- Level 1 refers to unstructured message exchange such as e-mail.
- Level 2 refers to structured message exchange such as AdatP-3.
- Level 3 refers to shared information through common data models and data structures.
- Level 4 refers to shared applications.

Let us review the four levels of interoperability and see how they apply to geospatial data. Interoperability Level 1 would be to exchange unstructured geospatial data, mainly in the form of paper maps or digital images of maps. This is easily achievable but does not provide support to all the GIS functions a command and control information system (C2IS) would require. We will come back to Interoperability Level 2. Interoperability Level 3 requires the use of the same data format, basically, the DIGEST format. It will take time before DIGEST is accepted as the sole standard and millions of dollars have been invested in the production of geospatial data in various formats, and C2ISs should be able to exploit these existing data. For Level Four of interoperability, the United States have specified a Common Operating Environment which defines the set of standard applications, including the GIS package, to be used across services and countries. This seems to be much more achievable at the level of national systems. OGDI corresponds to Interoperability Level 2 (and in part to Level 4) as it provides a structured mechanism (Application Programming Interface) to access data and as it provides a set of tools available to GIS vendors and application integrators.

To even achieve better interoperability in command and control systems, terrain data is often managed differently than tactical data. Terrain data (e.g. roads, rivers) are voluminous but change slowly with time. On the other hand, tactical data are small in volume but change rapidly with time. Conversely, terrain data is often held locally in each system and only tactical data are exchanged. Some OGDI geospatial drivers, such as the Common Operational Picture (COP) driver, have been developed to access tactical data.

## 23.4. Conclusion and future plans

The need for data pervades every aspect of information systems. To minimize duplication of data collection efforts and to capitalize on the digital database legacy, effort must concentrate on mechanisms for exchanging and assimilating data and information. The previous sections have described an interoperable system experience called OGDI that provides direct and transparent access to a large number of heterogeneous geospatial data formats. OGDI was designed to reduce the barriers to spatial data exchange. It solves the immediate problem of file-based data exchange, and aims to address the larger issue of standardized multi-database exchange.

Currently, OGDI provides:

- retrieval functions for geometric and attribute data;
- the transparent adjustment of coordinate systems and cartographic projections;
- the transparent transformation of platform-dependent representations;
- a growing number of geospatial data drivers (currently 20 formats with different profiles);
- the use of the Internet as a medium to distribute geospatial data products with a new Web protocol;
- compression and decompression facilities to improve communication between the server and the client;

OGDI has already been integrated into several geospatial systems and infrastructures:

- in GRASSLAND, a GIS software package;
- in WebMap, a World Wide Web map viewer Helper-App;
- in a large geospatial data warehouse;
- in the Canadian Geospatial Data Infrastructure (CGDI);
- in the Canadian Earth Observation Network CEONet;
- in the Department of National Defence (DND).

In the near future, OGDI will be improved in the following ways:

- it will be enhanced to improve remote communications;
- encryption facilities will be provided;
- authentication services will be developed;
- spatial analysis services will be applied to geospatial data in its native format;
- new drivers will be developed.

Finally, there was strong interest shown in OGDI from important international committees such as ISO/TC 211 (International Standards Organization/Technical Committee 211) and OGC (Open GIS Consortium). Also, the OGDI Research Institute (OGDI-RI) has been founded in fall 1997 by a group of private and public organizations. OGDI-RI is a not-for-profit organization initiated to foster global geospatial data interoperability through the Open Geospatial Datastore Interface (OGDI) by coordinating development, standardization, maintenance, education, and dissemination of OGDI technology. Other organizations interested in OGDI are invited to join OGDI-RI as either participating or associate members.

## 23.5. How to get the source code

Documentation and source code for OGDI can be downloaded from the Defence Geomatics server (*http://www.j2geo.ndhq.dnd.ca*) or the LAS server (*http://www.las.com*).

## Acknowledgments

The authors wish to thank Mr. Dave McKellar of Defence Geomatics (Canada), who was instrumental in the adoption of the DIGEST standard and in the elaboration of the GGI&S initiative and the DIGEST Software Tools project, and whose vision has led to many of the concepts in OGDI. The authors are also thankful to all the developers at Logiciels et Applications Scientifiques Inc., for their contribution in the design and development of OGDI. Finally, the authors thank Ms. Louiselle St-Laurent of GeoAccess Division, Canada Centre for Remote Sensing, who kindly provides valuable information to write different subsections of this document.

## References

Clément G, Larouche C, Gouin D, Morin P, Kucera H 1997 OGDI: Toward interoperability among geospatial databases. *SIGMOD Record* 26(3)

DND-Canada and Geomatics Canada 1995 Canadian Geomatics CD-ROM. *GEOMATICS 1995 Conference, Ottawa, Canada, June 11–15*

Gouin D, Morin P, Clément G, Larouche C 1997 Solving the geospatial data barrier. *Geomatica* 51(3): 278–287

Larouche C, Clément G, Moore T 1996 Adapting GIS technology for decision-makers. *Proceedings of the 1996 Canadian Conference on GIS, Ottawa, Canada, June 4-10*

LAS 1996a *Open Geographic Datastore Interface (OGDI), Version 1.0: Programmer's Reference*. Logiciels et Applications Scientifiques Inc

LAS 1996b *GRASSLAND User's Guide for Windows 95 and Windows NT, Version 1.1.* Logiciels et Applications Scientifiques Inc

McKellar D 1995 The Global Geospatial Information and Services Initiative. *GEOMATICS 1995 Conference, Ottawa, Canada, June 11-15*

Microsoft Corporation 1994 *Microsoft ODBC 2.0 Programmer's Reference and SDK Guide: for Microsoft Windows and Windows NT*. Microsoft Press, Redmond, Washington.

Ousterhout J K 1994 Tcl and the Tk Toolkit. Addison-Wesley

St. Laurent L, Kucera H, Keighan E 1997 Geospatial data warehousing: an architecture. *Proceedings of the International Symposium: Geomatics in the Era of Radarsat (GER'97), May 25-30, Ottawa, Canada*

# 24 DIAL: A WEB-BASED INTEROPERABLE SCIENTIFIC DATA DISTRIBUTION SYSTEM

Liping Di, R. Suresh, K. Doan, Doug Ilg, and Ken McDonald

## 24.1. Introduction

In recent years, one of the most exciting developments in information technologies is the proliferation of Internet connectivity and the popularity of World Wide Web services. Internet and Web technologies are providing the general public with a convenient way to access vast quantities of information. Almost all computer users with an Internet connection can use Web browsers to access information. These technologies and enhanced hardware and software products enable users with moderate levels of computer knowledge to establish information distribution sites and to participate actively in the evolving national information infrastructure.

Although currently most information accessed through the Web is descriptive, the possibility of distributing scientific data through the Internet using the Web will immediately offer scientific datasets to millions of potential users with little effort.

The unrivaled increases in computer network capabilities have raised questions about the cost-effectiveness of the traditional centralized scenario for processing and distributing Earth science data. The scenario involves three parties—data producers, data managers or distributors, and data users. Both data producers and data users have diverse academic backgrounds and a wide range of capabilities to produce and examine data. In most cases, data producers are also data users. Data managers and distributors focus their efforts on organizing the diversity of data and on developing and applying standards to enhance data usability, data management, and user support.

Today's data users, managers, and producers agree that it is data producers who are the most familiar with the data and the processes by which they were created, and who generally have the best insight into the data's use and improvement. Because of enhanced capabilities and the fact that data producers have the best knowledge about the data, we envision that future scientists will not only be satisfied data users who can obtain reliable, ready-to-analyze data in their favorite form easily and promptly, regardless of where the data are physically archived, but will also be effective data producers and distributors, all through their desktop computers. In order to realize this vision, it is important to move the data management activity closer to the source and let data producers distribute their data directly to users. This alternative approach will also facilitate the creation of a federation of data producers.

What has inhibited the formation and proliferation of nodes in a science data federation is the huge amount of resources required to create, manage, and distribute standard datasets and supporting documentation and to provide costly follow-up support to data users. If some compact and easy-to-use data management and distribution software tools were available, however, many data producers would choose to manage and distribute their own data. This new model is more streamlined by removing intermediate data managers and distributors from the data flow and providing direct contact between data users and data producers through a federation of data producers. Therefore, the data will be more widely used, and cost savings in data distribution will be achieved. The key to the success of this two-party data access model is to provide a set of easy-to-use software tools for data producers to format, document, and distribute the data through their sites rather than through data distribution centers. Recent advances in software and Web technologies make the development of such tools possible. Given the right tools, data producers can also be data managers and distributors, and a federation of data producers can be formed. Use of a common set of tools also promotes standardization of interfaces, communications, data formats, and documentation, which further facilitates data exchange and sharing. Also, it brings data users closer to data producers and should result in an enhanced likelihood of data users obtaining the data they need directly from the people who collected or produced them.

This paper describes an interoperable Web-based scientific data distribution system called Data and Information Access Link (DIAL; Data Engineering Group 1998). The server is designed specifically for scientists and small data producers to distribute their data through the Internet. Architecture, components, and interoperability of the system are presented.

## 24.2. Objectives

The major objective of DIAL is to provide scientists and small data producers with an easy-to-set-up and easy-to-use system for formatting, documenting, and distributing data through their desktop computers. The system also allows data users, by using common Web browsers, to search for and manipulate the data interactively in online archives and to download data in their favorite forms in terms of spatial and temporal coverage and resolution, parameters, and data formats.

## 24.3. Technical approach

We view DIAL as a specific type of what could be generally termed a *scientific data repository*, which is designed to provide scientific data over networks. There will soon be many thousands of such repositories, varying in scale from large global archives to science teams or even individual scientists. These repositories must all interoperate as seamlessly as possible, while operating in heterogeneous computing environments. Groups of interoperating servers will form one or more federations of repositories. The key software technologies to create the DIAL system are catalog interoperability and efficient access to complex scientific data over networks.

The fully developed DIAL system, as planned, will consist of five parts:

1. A catalog interoperability layer to support major protocols for interoperation among DIAL servers and between DIAL and other data systems.
2. A scaleable scientific data server (SDS) to *serve complex scientific data in multiple formats* and to allow data users to manipulate the selected data interactively so that they can obtain the data in their favorite form in terms of spatial and temporal coverage and resolution, parameters, and format.
3. An open database layer to interface with Open Database Connectivity (ODBC)-capable metadata catalog databases for powerful data search and finding. In case the data producer does not have an expensive commercial database system, DIAL has a search engine with a file-based catalog database to provide basic metadata database capability.
4. A user interface layer to support user-friendly interaction between client and server.
5. A suite of software tools to help data producers to ingest the data into the system and to help data users to download and analyze the data.

Figure 24.1 shows the architecture of the DIAL system.

### *24.3.1. Catalog interoperability*

Users can discover the contents of a particular repository by querying its catalog (inventory) service. With relevant scientific data located at hundreds or thousands of repositories around the world, it is crucial to be able to locate data wherever it may exist—a one-stop shopping approach. This requires that related catalog services interoperate to form a single, federated information space, which would enable users to search all sites in the federation by querying just one of them. The catalog interoperability layer of DIAL is designed exactly for this purpose. It is planned that DIAL will support four major Earth science catalog interoperability protocols: Z39.50 Geo Profile and the Catalogue Interoperability Protocol (CIP), the Earth Observing System Data and Information System (EOSDIS) Version 0 interoperability protocol, and the EOSDIS Core System (ECS) interoperability protocol.

The ANSI Z39.50 standard for information retrieval is a generic application program interface (API) to interact with any database or information service (ANSI/NISO 1995). A Z39.50 profile is a document that describes in detail which Z39.50 services are required and how to interpret those services in order to support interoperability among clients and servers of a particular system type. The Committee on Earth Observation Satellites (CEOS) Working Group on Information

Systems and Services Protocol Task Team (WGISS/PTT) has defined CIP to make it possible to locate Earth observation data from any data provider (CEOS 1996). In addition, the Federal Geographic Data Committee (FGDC) has also developed a profile of Z39.50 called *Geo Profile* for achieving catalog interoperability among FGDC Clearinghouse nodes (Nebert 1997).

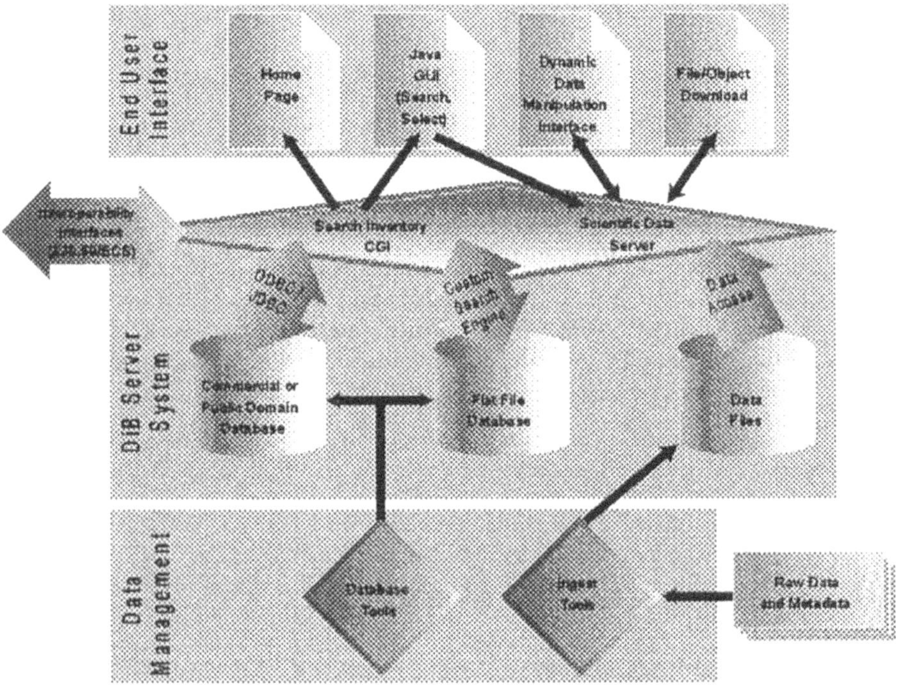


Figure 24.1. Data and Information Access Link (DIAL) system architecture

In addition to CIP and Geo Profile, both EOSDIS Version 0 and ECS interoperability protocols will also be implemented in the DIAL system. The EOSDIS protocols are designed to be used for catalog interoperability among NASA Distributed Active Archive Centers (DAACs) so that a user can search the entire EOSDIS data archive by querying any one of the DAACs. Implementing those protocols will allow a DIAL site to interoperate directly with NASA EOSDIS.

### *24.3.2. Scientific data server*

As stated earlier, one of the objectives of the DIAL system is to make it easy for data users to find data and to retrieve the data exactly as desired, regardless of where the data are physically located and how the data are stored on the server site. The catalog interoperability layer of the DIAL takes care of finding data; SDS handles the rest.

The collection of data that a data producer or federation member may want to share via the Web is often diverse in its formats. Ideally, a universal data format standard is enforced so that DIAL needs to serve data in only one format. In reality, a format used as standard in one site is not necessarily a suitable standard for another site. In order for DIAL to be widely used, we should not adhere only to one format. Thus, SDS has 1) to present each type of data to users in a unified form, hiding all physical storage characteristics of the diverse data in servers; 2) to allow users to browse the data interactively to determine their usefulness before downloading; 3) to permit users to subsample and subset the data in spatial, temporal, and parameter domains, and possibly reproject the data spatially; and 4) to let users download the data in their favorite formats so that the data can be imported into their analysis system easily. The key software technologies to be addressed for SDS are to develop a framework for the server to access data in multiple formats in a unified way and to serve scientific data objects to users dynamically and interactively through the Web.

In earlier work, Raythoen STX and the National Center for Supercomputing Applications (NCSA), with NASA support, developed the concept of the Scientific Data Browser (SDB) to address these problems. SDB implements a dialog, allowing users to discover the contents of datasets, and to select, examine, view, and finally retrieve the desired subset in various data formats. Some facilities for data analysis, such as plotting tabular data, are also provided. SDB is a Common Gateway Interface (CGI) program.

The current SDS is based on SDB with the following extensions:

1. Modularize the data access layer in the server so that the server can access data in multiple formats. Each data format will have its own data access module. The data access interface will be object-based so that a greater portion of the software is reusable and programming efforts required to create new access modules are decreased. To support a new format we need only to develop a new access module and plug it into the server.

2. Develop access modules for Hierarchical Data Format (HDF), HDF-EOS, Common Data Format (CDF), netCDF, and various ASCII and binary data formats. The access modules for simple ASCII and binary formats will be implemented in a generalized way so that variations of the formats can be accessed through the same modules. Such a generalization would greatly expand the scope of data a science team could readily make browsable and distributable via the Web. One implementation approach for this work would be to incorporate data description languages such as FreeForm (Habermann and Miller 1994) into SDS.

3. Modularize presentations (serving) of generic scientific objects (such as raster images and multidimensional arrays, vector, polygon, and point data, in geocoded, georeferenced, or non-georeferenced forms). The basic idea is similar to the modularized data access listed in item 1.

4. Serve dynamic/executable data to users. A set of generic operations, which can be executed with the data object, will be developed for each data object (such as resampling and subsetting for raster and array data, reprojecting for geocoded data, and formatting for all data objects). These operations will be executed at either the server side or the client side, depending on the situation. The dynamic/executable data serving gives data users more flexibility, interactivity,

and control on what they can get from the server and how they can get it than does traditional static data serving. New network programming languages, such as Java, make the dynamic/executable data serving possible. The formatting function will format the data object on the fly to allow users to download the data in their favorite formats. Formats supported by the formatting function will include all formats supported by the data access modules plus several commonly used GIS and image processing formats, such as ERDAS, PCI, and ARC/INFO.

The generic scientific object model and access protocol can be implemented in many technologies: as a conventional Web server, with Web object technology, with OpenGIS objects, or with a CORBA distributed-object framework. At this time, we are not certain exactly which technology will survive. There are several candidate technologies, such as Java, ActiveX, and CORBA. All these technologies are supposed to interoperate in the near future, but this is far from certain. There are also at least two important frameworks for modular, network-based objects: Java (e.g., JavaBeans, Hamilton 1997; RMI, Farley 1998; Sun Microsystems 1997), and Servlets (JavaServer Product Group 1997) and ActiveX with OLE/COM (Chappell 1996). It is our intention to evaluate the prototype server and client software as it evolves and use whatever the best standard technology becomes.

### *24.3.3. Open database interface, search engine, and file-based catalog system*

In order to search data archived in a DIAL site, there is a need for a searchable database to store the metadata (catalog). Because the intention of DIAL is to provide scientists and small data producers with the capability of distributing their data at low cost, it is unreasonable to require an expensive commercial database in order to run DIAL. Therefore, DIAL will provide a simple yet powerful file-based database system with an SQL-capable search engine instead. Tools will also be developed for populating the database with metadata. This homemade catalog system will provide basic database capabilities to support full DIAL functionality.

However, data producers may already have a database populated with metadata, or they may want to use other databases instead. Because there are many database systems, it is unrealistic for DIAL to have individual interfaces for each system. Therefore, DIAL provides an open database interface layer to interface with database systems. The interface is based on the ODBC protocol with JDBC (Reese 1997). With such an interface, any ODBC-compatible database can be directly plugged into DIAL.

Because existing metadata databases can be plugged into DIAL, the metadata keywords might be different between the metadata in the database and the interoperable metadata standards. In order to search the catalog database, we need to have a metadata mapper sitting between the catalog interoperability layer and the open database interface layer to map the metadata automatically. DIAL uses equivalence tables to handle the mapping between metadata systems. The equivalence table technology has been successfully used in GeoVu (NGDC 1997) to handle the automatic mapping of the metadata. DIAL also provides a tool for data distributors to create an equivalence table interactively.

### 24.3.4. The user interface

DIAL is a Web client–server-based data browsing and distribution system. Data producers and distributors can use this system on their desktop computers to distribute data over the Internet. Data users can interact with this system to get data in their favorite form through a Web browser. Strictly speaking, there are two interfaces in the system: one deals with data producers and distributors on the server side for ingesting data into the system, and the other, described below, deals with data users on the client side.

The DIAL system provides two user interfaces, one based on Java and the other on HTML. The interfaces permit data users to enter search criteria and view the search results and to browse, subset, subsample, reformat, and download the selected data interactively with a DIAL server. The Java-based interface runs in Web browsers on the client side as applets, providing fast interaction and prompt response. One drawback of applets is the time needed for loading applet codes and companion data from server to client. This will increase the network traffic, but it reduces the processing burden on the server. If more functionality is executed on the client side as applets, more data must be sent to the client before applets start to run. Although applet cache technologies can remedy part of the problem, for a slow network the initial applet downloading by the browser might take a long time. In the interface design, we have considered both the network traffic and applet functionality to achieve optimal balance of interface interactivities, functionality, workloads between server and client, and the network traffic. In addition, some data users might only have applet-incapable browsers or a very slow network connection. For this type of users, DIAL provides an HTML-only interface. Although it is less user friendly than the Java-based one, an HTML-only interface allows more people to access the data.

Because DIAL is based on Web technology, there are no installable components required on the client side. Unlike the traditional X-windows-based client–server technology, which has to have a client-side application installed before the server can be accessed, the data user can access data directly in the server side through a regular Web browser. Therefore, data users can get started with DIAL more easily. The Web-based user interface also saves significant software development time because the cross-platform issues are handled by Web browser developers, and almost all computer platforms have some kind of Web browsers available. Therefore, DIAL needs to have only one good platform-independent interface, which will automatically work on almost all computer platforms.

### 24.3.5. Tools for data ingestion and documentation

One of the important objectives of DIAL is to make it easy for data producers to distribute their data from their desktop computers. To achieve this objective, DIAL provides two types of tools to help producers ingest data into the DIAL system: metadata ingestion tools and data ingestion tools. The metadata ingestion tools will make it easy for data producers to create the catalog database for search. These tools will provide the capability for creating a catalog database from metadata in ECS,

FGDC, and ISO standards. In addition, an interactive software tool is also planned for documenting scientific data in standard metadata formats. This tool will enable data producers to document their data easily in a standard metadata format without having to know standard details. The metadata generated by this tool will be free of syntax errors. These tools almost automate metadata documenting and ingestion, saving significant time and money for data producers to prepare their data for distribution. Another feature of DIAL is that it allows data users to use their existing software as helper applications of Web browsers, permitting them to incorporate their analysis tools with DIAL's data serving capability.

In addition to DIAL's ability to serve data in multiple formats, DIAL also provides an interface for supporting additional formats by developing an access module for the format. To put scientific data into DIAL for distribution, data producers either produce data in DIAL-supported formats or write an access module to support their own format. However, the data might already exist in formats not supported by DIAL, or the production system cannot produce data in a DIAL-supported format and the producer does not have the resources, experience, or willingness either to modify the production system or to write an access module. In such cases, producers can use DIAL-provided reformatting tools to translate data into a DIAL-supported format. Generalized two-way data translators between ARC/INFO and HDF-EOS, which are currently being developed, are an example of such translators.

## 24.4. Interoperability and the federation of scientific data servers

There are two levels of interoperability in DIAL: catalog interoperability and data object interoperability. Because DIAL is intended to be used by scientists and small data producers to distribute their data, it is most likely that the data needed by a user will reside in several DIAL or non-DIAL sites. It is very tedious for data users to search individual sites to obtain all the data they want.

With catalog interoperability, a federation of data providers can be formed to allow data users to search the whole federation by querying just one of the federated sites. Because of the catalog interoperability layer in DIAL, a DIAL site can join an existing federation easily as long as the federation uses the same catalog interoperability protocol that DIAL uses. The catalog interoperability layer also allows interoperation among DIAL sites to form a DIAL-based federation.

Because of the data object interoperability of the DIAL system, data users can retrieve data from the federation in a uniform way without noticing the personality of individual DIAL sites even if the sites use different metadata protocols, database systems, or data formats. The feature of hiding the difference between DIAL sites in a federated environment will greatly enhance data interoperability and interuse and will significantly reduce the time needed for data users to find, download, and assemble data for analysis. Therefore, DIAL will work not only as a standalone data distribution system for individual data producers to distribute their data, but also as a federated system for a group of data producers to distribute their data as a federation.

Figure 24.2 shows an example of hierarchical-type federation configurations with DIAL. Figure 24.3 shows an example of a DIAL-based Web-like federation.

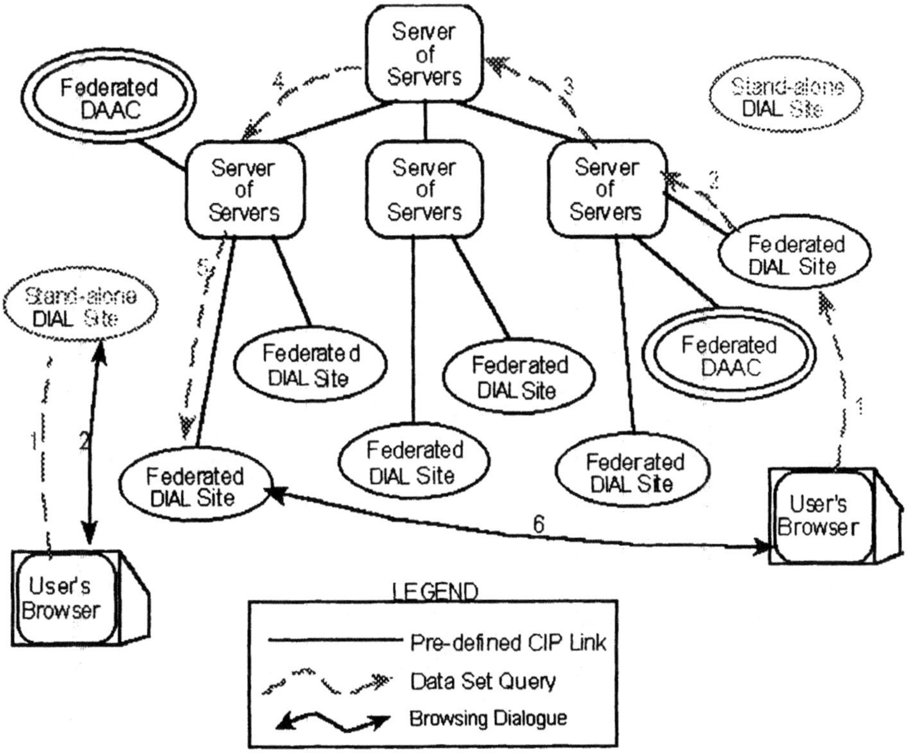


Figure 24.2. An example of hierarchical-type federation configurations with DIAL

## 24.5. Current implementation

The development of the DIAL system is an ongoing process. Currently, a prototype of the DIAL system has been developed (Data Engineering Group 1998). The prototype system has all of the above-mentioned components in some stage of development. The prototype DIAL is a compact yet powerful Web client–server-based data browsing and distribution system, which provides low-cost online data access to users through the Internet and enhances data access and interoperability through a simple Web interface.

The current version of DIAL permits data producers to set up a low-end workstation as a data server for distributing their data and metadata quickly and easily. Users can access the data server through any Web browser to search and query the data archive based on geographic location, time, and other relevant parameters. Once the data in the archive are found through a query, users can view the metadata associated with the data and browse, view, subsample, subset, and download the data to their local disk. Data can be downloaded in one of three formats (HDF, ASCII, or binary), or displayed on the screen as a GIF image, HTML text, or tables. The server currently supports data in HDF, HDF-EOS, and netCDF formats. DIAL supports array coordinate-based and record ID-based subsetting and subsampling for data in all supported formats as well as geographic coordinate-based and time-based subsetting and subsampling if the data in the server are in HDF-EOS format. Both Java- and HTML-based user interfaces are provided in the current version of DIAL. The open database Interface layer, search engine, file-based catalog system, and several data ingesting tools are also provided.

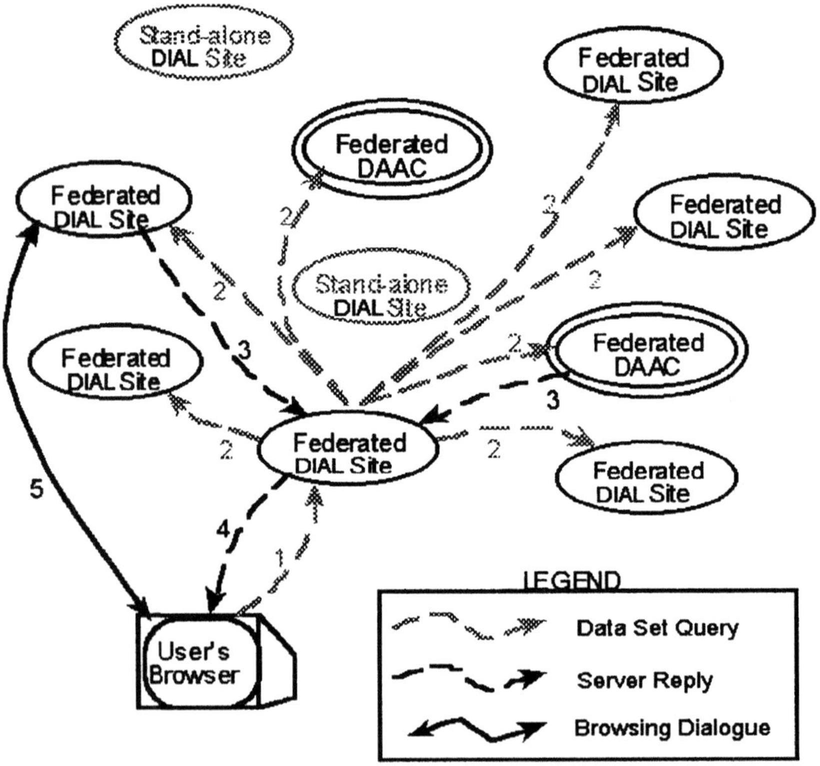


Figure 24.3. An example of DIAL-based Web-like configuration

Binary executables of the prototype software are free to the public, and available on all major UNIX platforms, and will soon be available on the Windows/NT platform. The source code of the software is also available freely for noncommercial uses. The address for DIAL demo and program download is *http://hops.stx.com/dialhome.html.*

## 24.6. Future work

In our current release version of DIAL, one important component, the catalog interoperability layer, is not implemented yet. Therefore, the current version of DIAL can only serve as a standalone scientific data server. Our current focus is on developing the catalog interoperability layer by implementing the four catalog interoperability protocols mentioned previously. Currently, we are implementing the EOSDIS Version 0 catalog interoperability protocol. This protocol was selected as the first to be implemented because EOSDIS Version 0 is already an operational system with large data archives. We can test DIAL's interoperability and federation capability by directly plugging DIAL into the EOSDIS Version 0 system after we implement the EOSDIS Version 0 interoperability protocol.

We will carry out another major project to implement data object serving to facilitate data interoperability. Currently, only part of what we intended to do with the SDS component of DIAL is implemented. By implementing object serving, all scientific data would be presented to the user or an application system as scientific data objects. As we have already pointed out, there are many competitive protocols for data object interoperability. Currently, we are evaluating those competitive technologies and will select a best one to be implemented in DIAL.

## Acknowledgments

This prototype work of DIAL was funded by the NASA Earth Science Data and Information System (ESDIS) project. The following individuals also contributed their time to the prototype DIAL through creative discussion and computer programming: Mike Folk, Robert E. McGrath, and Nancy Yeager of NCSA.

## References

ANSI/NISO 1995 *Information Retrieval: Application Service Definitions and Protocol specification: Z39.50.* American National Standards Institute, New York

CEOS 1996 *Catalogue Interoperability Protocol (CIP) Specification Release B.* CEOS/WGISS/PPT/CIP-B, European Space Agency/ESRIN, Frascati, Italy

Chappell D 1996 *Understanding ActiveX and OLE.* Microsoft Press, Redmond

Data Engineering Group 1998 *DIAL Release Version 1.0.* Raytheon STX Corporation, Greenbelt, Maryland. *http://hops.stx.com/dialhome.html*

Farley J 1998 *Java Distributed Computing.* O'Reilly, Sebastopol

Habermann T, Miller T 1994 *The FREEFORM Tutorial.* NOAA National Geophysical Data Center, Boulder, Colorado. *ftp://ftp.ngdc.noaa.gov/Access_Tools/FREEFORM*

Hamilton G 1997 *JavaBeans.* Sun Microsystems, Inc., Mountain View, California. *http://splash.javasoft.com/beans/docs/beans.101.pdf*

JavaServer Product Group 1997 *The Java Servlet API.* Sun Microsystems Inc., Mountain View, California. *http://jserv.javasoft.com/products/java-server/documentation/webserver 1.0.2 servlets/api.html*

Nebert D D 1997 *Z39.50 Application Profile for Geospatial Metadata or GEO.* Federal Geographic Data Committee, Reston. *http://www.fgdc.gov/Clearinghouse/Tutorials/ GEO21a.html*

NGDC 1997 *GeoVu Software Package*. NOAA National Geophysical Data Center, Boulder, Colorado. *http://www.ngdc.noaa.gov/seg/geovu/*

Reese G 1997 *Database Programming with JDBC and Java*. O'Reilly, Sebastopol

Sun Microsystems Inc. 1997 *Java Remote Method Invocation (RMI)*. Sun Microsystems Inc., Mountain View, California. *http://java.sun.com/products/jdk/rmi/index.html*

# 25 USING THE INTERNET TO ACCESS GEOGRAPHIC INFORMATION: AN OPEN GIS PROTOTYPE

Frederico T. Fonseca and Clodoveu A. Davis Jr.

## 25.1. Objectives

This project started with the objective of creating middleware to allow users to gain access to geographic data from several different sources, using a simple object-based model. The intention was to provide the widest possible range of users with easy and inexpensive access to public GIS data. The initial concept called for the development of an interface between the user and the GIS. Initially, this interface would only provide basic data access, and should evolve in the future to support the definition and implementation of more specialized services. However, the ability to connect simultaneously to different GIS platforms has been considered important to the success of the project.

The Java language has been chosen as a development platform. Java includes features that provide for easy client–server computing, Internet connections, and standard database access, in a multiplatform, object-oriented environment. Furthermore, using Java allows for the development of highly interactive visual interfaces (Morris 1988; Shneiderman 1992), including the possibility of visualizing geographic objects as vectors or images. This can be used to overcome some of the limitations of current GIS interfaces: Egenhofer and Richards (1995) observe that, for a class of systems that has visual characteristics as striking as a GIS, the number of non-visual interfaces is impressive.

Java applets can be made available through the Internet (or any Intranet), cutting down on software maintenance and distribution costs. The Internet also provides an excellent medium for distributed applications, as required by the project's objectives.

## 25.2. GIS interfaces

Oliveira and Medeiros (1996) identify four essential items in an architecture for a GIS interface: the integration of the interface with the GIS; the description of the main architecture modules, specifying their functionality and interoperability; the intermediate representation model and its mapping to the GIS model; and the division of tasks between the GIS and the interface. Each of these items will be discussed next.

### *25.2.1. The integration of the interface with the GIS*

An interface to access geographic data generally is backed by a GIS, and therefore must communicate with it somehow. The classification of integration between the interface and the GIS as strong or weak, as proposed by Voisard (1994), clarifies this point and concludes that weak integration is better, since it ensures that the interface is independent from the GIS. Considering weak integration, Oliveira et al. (1997) observe the independence from the GIS as a feature present in most modern architectures (Abel et al. 1992; Oliveira and Medeiros 1995; Pissinou et al. 1993; Rigaux 1995; Voisard 1991, 1994). That is, the interface is stable even when the GIS changes, or when an additional GIS becomes available to the user. Shneiderman (1992) warns that even small changes in the interface can lead to unpleasant and dangerous mistakes. Greater changes, which normally happen when a GIS is replaced by another, will require heavy training and will disturb the user in many ways (Mark and Frank 1992). Oliveira and Medeiros (1995) emphasize other benefits from GIS independence, including the worldwide tendency towards open systems, the improvement of functionality in each component of the solution (interface and GIS) and also the possibility of progressively adding new services and functions. The necessity for the development of a conversion module between the GIS and the interface can be perceived as an implementation difficulty. This undertaking can be complicated, and depending on the GIS the development of low-level communication routines can be necessary. Oliveira and Medeiros (1995) also highlight the necessity for a rich data model, representing concepts that are present in several GIS.

When strong integration is used, the interface becomes itself a part of the GIS, and therefore there is a very strong dependence between them. As an advantage, there is the fact that the interface is able to manipulate the objects in the same way they are stored, leading to greater efficiency in manipulation and exhibition operations. But the disadvantages are great, such as the enormous difficulty in incorporating new GIS to the system, and the dependence on the (not always available, and often not good enough) user interface tools provided by the GIS developer.

### *25.2.2. Functionality and interoperability of the main architecture modules*

The architectures of the interfaces presented by Voisard (1994), Oliveira and Medeiros (1995), and Peloux and Rigaux (1995) show some similarities. All of them have a user interaction module, a database connection module, and a module dedicated to converting geographic objects from the GIS's format to the interface's format and back. The user interaction module generally depends on the operating system, or on operating system tools, in order to be able to control user keyboard or mouse actions. This module also generates the presentation of the results of user queries. The conversion module translates the data format used by the GIS into the interface's representation format. It is also the responsibility of this module to convert user actions into queries formulated in a language accepted by the GIS. The connection task is also executed by this module (Oliveira and Medeiros 1995; Voisard 1994). Oliveira and Medeiros (1995) present a special module to deal specifically with the user's view of the data.

### *25.2.3. Mapping to the GIS model*

Most of the time the interface has its own representation model, and each GIS that supports the interface, if there is more than one, certainly will have its own model. Therefore, it is necessary to create a mapping between the representations contained in the GIS to the corresponding representations in the interface. Voisard (1994) states that the mapping should be made at a higher level, and must allow for the conversion of one representation (GIS) to the other (interface). This must be accomplished through a model that perceives the vision both sides have of the problem. It is also observed that this mapping is not 1 to 1, but *m* to *n*. Voisard (1995) suggests that this model can be defined using a functional language such as FGD (Breunig and Perkhoff 1992). Oliveira and Medeiros (1995) present a module specifically for this function, the data model module, which is responsible for providing the user with a vision of the underlying GIS, according to the data model in use. This also stresses the idea that the data model should be broad enough to accommodate concepts used in different GIS.

### *25.2.4. The division of tasks between interface and GIS*

It is very important to be able to separate clearly the functions of the interface itself from the tasks that are the exclusive responsibility of the GIS. Consider the following situation: a set of objects is currently represented in the user's display screen, and he or she needs to obtain specific data (spatial or not) about one of those objects. Since it has been displayed, the object could also, depending on the underlying architecture, be present in some intermediate data structure, in memory, and therefore the data needed by the user could be readily accessed.. The temptation to leave this task to the interface is great, but it must be observed that the responsibility for identifying objects on the screen is the GIS's responsibility, not the interface's. The interface must again request the object that has been identified by, say, a mouse click, to the GIS. Even though this operation is transparent to the user, it is fundamentally important to establish the limits of the interface's functionality. There is no consensus around this point in the literature. Voisard (1995) identifies a tendency towards building faster, more efficient, and more intelligent tools, and therefore as many tasks as possible should be designated to the interface, even though the general lack of standards for the representation and interchange of geographic data constitutes a great impediment to the wide adoption of this alternative. On the other hand Oliveira and Medeiros (1996) consider that not having a clear division of tasks generally causes code redundancy, since the interface functions are repeated several times for each type of analysis. Furthermore, this creates difficulty in the maintenance and evolution of the GIS. Ferreira (1992) also observes that GIS developers are starting to rewrite their code in order to create an isolation between the user interface and the algorithms and processing functions.

### *25.2.5. Summary*

It is clear that the choice of architecture will strongly influence the ease of development of the interface, as well as its interoperability. As to the development, it can be emphasized that, in a weakly integrated architecture, several user interface building tools can be used, resulting in an interface that is more flexible and therefore more adaptable to the user's needs. As to interoperability, a desirable feature in the GIS of the next generation (Câmara 1995), it becomes clear that an architecture that adequately defines the roles of each interface module, as well as the role of the GIS, and also incorporates a representation that is adequate to the user's mental model, has a great chance of being successful. Such an architecture will isolate the user from the specific implementation problems of the underlying GIS.

## 25.3. The architecture

A general architecture for the interface will be proposed next. The idea is to create mechanisms to translate user actions into geographic database operations, treating information coming from the database to allow for visualization and other forms of manipulation. These mechanisms assume a clear separation between the part of the system that deals directly with user interaction, and the part that communicates with the GIS (Figure 25.1).

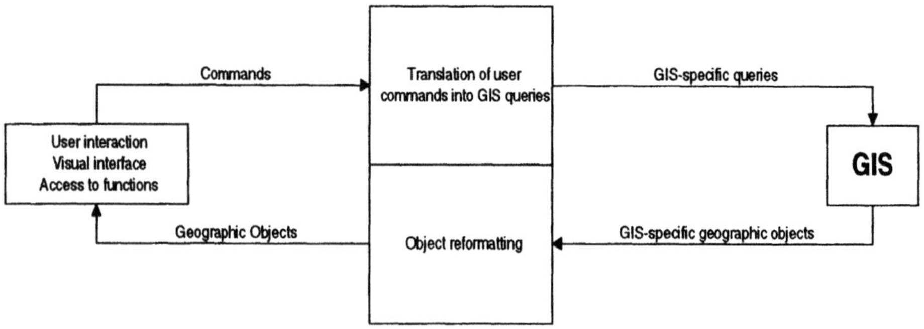


Figure 25.1. Conceptual model for the interface

In a conventional database, information can be displayed basically in the format used for storage, with no need for changes in structure or abstraction level. GIS, on the other hand, stores information in more complex structures, that need to be shaped into a more convenient display format (Peloux and Rigaux 1995). Furthermore, different GIS use different geometric structures to represent objects, thus requiring geographic information retrieval code that must be specifically developed for each GIS. Likewise, there is no single standard for geographic query languages, and a certain degree of translation is required in order to achieve generalized access to GIS data.

The conceptual schema presented in Figure 25.1 evolved into the physical model presented in Figure 25.2, with the incorporation of other guidelines for this work: development using Java, in order to accomplish platform independence, and access to the geographic database through the Internet. Therefore, in the physical model it was decided that the interface would be implemented with two modules. The first one, called *Manipulator,* is responsible for direct user interaction. This module will receive commands from the user, and will send these commands to the other module, called *Extractor*. The latter will execute all necessary communication with the GIS, and will return any retrieved objects to the Manipulator, which in turn will present them to the user. The objects that are returned from the Extractor are standardized, in a way that is similar to the object model that is defined by the OpenGIS Consortium (McKee and Buehler 1996). This standard is independent from the GIS that is used to manage the geographic data, and it is the Extractor's task to implement methods that will interpret and decode information retrieved from the GIS, and to rearrange everything into the standard object format. With this, the Manipulator becomes completely independent from the GIS from which objects are being extracted, while the Extractor needs to be specifically developed to interact with a specific GIS. In this conception, the Extractor functions as a filtering and standardizing layer, acting on data coming from or going to the GIS.

Extending this concept a bit further, it becomes feasible to access several geographic databases, either separately or simultaneously, without the need to inform either the Manipulator or the user. This effectively incorporates in the architecture the possibility of interoperating different GIS, since the user gets indistinguishable, simultaneous access to information maintained by each of them.

According to the classification proposed by Voisard (1994), notice that the Manipulator is loosely coupled to the GIS, since it is isolated from the GIS by the Extractor. The Extractor, on the other hand, is strongly coupled to the GIS, since it needs to know details about the way the GIS works, in order to fulfill the translation of commands and the reformatting of objects. With this, it becomes possible to combine the advantages of each integration model, while preserving the desired isolation between user interface and GIS retrieval code.

It is also important to notice that in the proposed architecture, the interface does not take over geoprocessing tasks, since those are the responsibility of the GIS, and belong to the GIS's functionality. These functions can be accessed through the mechanisms already proposed, keeping the interface simple and compact. Considering that the intention is to use the Internet as a medium to access the GIS, even with all its congestion and response time problems, it is fundamentally important to avoid the execution of spatial processing and information retrieval tasks by the interface (which is a really great temptation, considering the considerable processing power and memory capacity of today's average client computer). This option may lead to some performance loss, but it can be counterbalanced by the advantages that emerge from the stability and independence of the interface.

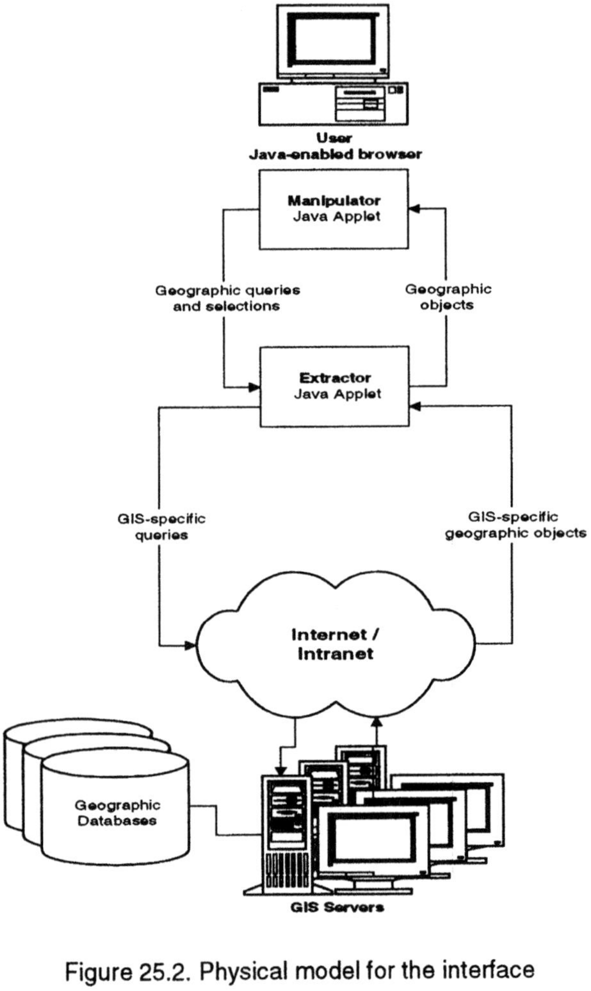


Figure 25.2. Physical model for the interface

The high interoperability potential of this solution could only be achieved using the object-oriented software development principles recommended by Meyer (1988). That article indicates that routines containing an oversized share of knowledge about the overall system have a very high potential for producing problems. If this happens, small modifications can produce an impact over a big part of the system. Meyer (1988) quotes "*Ne sutor ultra crepidam*, the shoemaker should not look into anything beyond the shoe. This is one of the software project principles." As an example, observe that a routine that performs a rotation on objects of the type *Shape* does not need to know anything about the classes that derive from *Shape*. This principle becomes clear in the geographic object drawing routine that will be described in detail. That routine can be understood as the capability that each object has to draw itself on the screen. The implementation is specific for each type of geographic object, following the method established by the superclass that defines geographic objects.

### *25.3.1. The Extractor*

The Extractor module has been developed as a Java class. This class is responsible for extracting geographic objects from the GIS, and sending them over to the Manipulator. Since the object formats used by the GIS and by the Manipulator differ, this class is also responsible for converting objects from the GIS into the Manipulator's format.

As implemented, the Extractor is totally dependent on the source of data, and it must be rewritten, modified, or specialized for each update that occurs on the underlying GIS. Its implementation can be extended to provide access to more than one GIS, using a set of parameters that indicate the source GIS from which each layer of information should be retrieved. So, when called by the Manipulator, the Extractor will return a instance of the desired class according to the parameters. Let's say that the Manipulator needs a layer that is stored on *GIS_03*. The Extractor is called, with the name of the layer and the number of the GIS as parameters. As a result, the Extractor will return to the Manipulator a class that is derived from itself, *Extractor_03*, that knows how to deal with *GIS_03* and will provide all the service that the Manipulator needs. Since *Extractor_03* is a specialization of the Extractor superclass, it provides all the services that its ancestor provides, and therefore the Manipulator doesn't have to know if it is dealing with the Extractor itself or with one of its subclasses.

**25.3.1.1. General operation.** The Extractor needs to obtain some information before initiating its operation:

- What is the current geographic space, defined as a rectangle using two geographic points, and corresponding to the current zoom window limits?
- What are the currently active layers, that is, what layers will be visible?

The first type of information can be retrieved from the class *geographic_space*, that contains all information on the coordinate domain, as well as the definition of the current geographic space. The second type is kept by the Manipulator, and will be sent over to the Extractor whenever geographic object retrieval services are requested.

In order to execute fundamental operations, such as zooming, two different kinds of selection are combined: a spatial selection of objects within or over the border of a given rectangle, and a selection of objects that belong to one of the active layers.

The Extractor implements the following selection services:

- retrieving and sending all geographic objects from all requested layers within a given rectangle;
- given a coordinate pair, retrieve all objects from the currently active layers that are located on that point or that contain that point;

- given an alphanumeric identifier, retrieve the associated geographic object.

**25.3.1.2. Caching**. When the Internet is involved, it is always necessary to consider developing mechanisms to avoid excessive network traffic. The proposed architecture facilitates the implementation of local buffers or caches that will increase the performance of the application. The necessity for these buffers at the client side of an Internet application is mentioned by Câmara (1995). Oliveira et al. (1997) mention that this task is usually performed by the DBMS, but in the case of a GIS application it should be performed by the interface.

Since extracting geographic objects from the database is the Extractor's task, it should also be responsible for caching. The factors that can influence the architecture of such a cache are many, among them the storage capacity of the client machine and the probability of repeated requests for objects, based on proximity or relationship to already retrieved objects. A very simple cache, based only on proximity, and taking advantage of the spatial indexing logic, was implemented in order to demonstrate the necessity for this kind of service, and to show how it can be implemented without further impact on the Manipulator, while preserving the interoperability potential of the proposed architecture. As a result, we achieve the benefits of weak integration, keeping the Manipulator unaware of the data retrieval mechanisms, and the benefits of strong integration, since the Extractor can benefit from its specific knowledge of the GIS to implement the caching mechanism.

### *25.3.2. The Manipulator*

The Manipulator is responsible for the interaction with the user. Through the Manipulator the user will be able to execute operations like object selection, zooming, searching, thematic mapping, and so on. The user can double-click a mouse button and select a geographic object, and can also use the mouse to select a menu command or perform a zoom (Figure 25.3).

The user requests are sent by the Manipulator to the Extractor. The Extractor translates the requests to a format that the GIS can understand. When the Extractor receives the answer from the GIS, it prepares a response to the Manipulator, in the form of geographic objects, instances of a class called *geographic_object*. The exhibition of the results will be produced by the Manipulator, using methods available within the *geographic_object* class. As a result, the Manipulator code needs to deal only with user interaction and object display.

The clear distinction between the tasks of the Manipulator (user interaction and exhibition of results) and the Extractor (communication between the GIS and the Manipulator) follows one of the basic principles mentioned by Oliveira and Medeiros (1995): a greater degree of specialization of the software modules will facilitate the integration of new modules in the future, and the progressive incorporation of additional services and functions.

### *25.3.3. Open GIS and geographic objects*

Standards facilitate the sharing, integration, and transference of data. Although the most common standards are the ones directed to data transfer, there are others that deal with specification languages and metadata (Câmara et al. 1996). One of these standards is under development by the Open GIS Consortium. The OpenGIS data schema (McKee and Buehler 1996) defines two basic types of data: *feature* and *coverage*. These types correspond to the ones described by Couclelis (1992) and Câmara et al. (1994): geo-fields and geo-objects. We will focus here on the *feature* (geo-object) type, that corresponds to the geographic objects that have been implemented. It contains three components (McKee and Buehler 1996):

- *geometry*: a description of the geometry of the phenomenon with an associated spatial or temporal reference model;
- *semantics*: a description of how the phenomenon is defined in the lexicon of a particular group;
- *metadata*: contains information needed to position the phenomenon in the context of the application environment user community.

The geographic object actually implemented follows partially the OpenGIS guidelines. Some simplifications were made, such as making use of a single reference system and disregarding a temporal reference system.

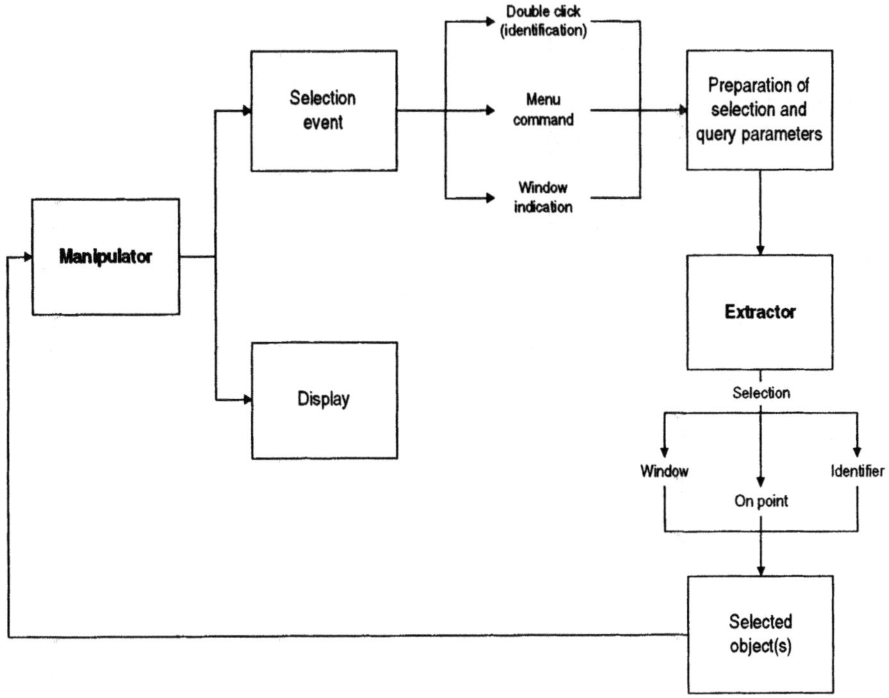


Figure 25.3. General Manipulator working schema

The geographic objects are implemented as an abstract class called *geographic_object*. To reflect OpenGIS *features*, the class *geographic_object* was specialized as *vectorial_geographic_object*. The latter was further specialized in three new classes: *Polygon*, *Line*, and *Point*. Basic services, provided by the *geographic_object* class, include methods that enable a geographic object to draw itself on the screen, to draw its identifier on the screen, and to return a list of its attributes. These methods are only declared in the abstract class, and are actually implemented by the subclasses.

When a new GIS is made available to the interface, *geographic_object* needs to be specialized, and the defined methods must be implemented. The *extractor* class must also be specialized to communicate with the new GIS. Of course, if any of the available classes can fulfill the needs of the new GIS, they can be used.

So, we can see that the integration between the Manipulator and the Extractor is achieved through the use of geographic objects. The Manipulator just has to know how to work with them, or one of the subclasses. Thus, interoperability is achieved and the stability of the interface (the Manipulator) is assured.

## 25.4. Conclusion

Results achieved with the initial implementation of the interface indicate that the proposed architecture, even though it carries the burden of translating data between the GIS and the user interface, has benefits that overcome its limitations. Among the main benefits, we can mention the adherence to open systems concepts, the concentration on the basic functional roles of each component (Manipulator, Extractor, and GIS), and the possibility to progressively incorporate new services and functions. Furthermore, the strong integration between the Extractor and the GIS brings the opportunity for optimization and performance gains, due to specific knowledge of the GIS internal routines, while respecting the basic interoperability rules through the weak integration between the Manipulator and the Extractor.

The prototype of the interface presented here can be tried out at *http://www.pbh.gov.br/prodabel/geo*.

## Acknowledgments

The authors would like to thank to Prodabel for support of Frederico Fonseca's Masters Degree program work, resulting in the development of the interface described here, and also the Computer Science Department of the Federal University of Minas Gerais for funds for the presentation of this chapter.

## References

Abel D, Yap S, Ackland R, Cameron M, Smith D, Walker G 1992 Environmental decision support system project: an exploration of alternative architectures for geographical information systems. *International Journal of Geographical Information Systems* 6(3): 193–204

Breunig M, Perkhoff A 1992 Data and system integration for geoscientific data. *Proceedings of the Fifth International Symposium on Spatial Data Handling, Charleston, South Carolina*: 272–81

Câmara G 1995 *Models, Languages and Architectures for Geographic Databases*. PhD Thesis, National Institute of Spatial Research (INPE), Brazil (in Portuguese)

Câmara G, Casanova M A, Hemerly A S, Magalhães G C, Medeiros C M B 1996 *Anatomy of Geographic Information Systems*. Computing Institute, University of Campinas, Brazil (in Portuguese)

Câmara G, Freitas U, Souza R, Casanova M A, Hemerly A, Medeiros C M B 1994 A model to cultivate objects and manipulate fields. *Proceedings of the Second ACM Workshop on Advances in GIS, Gaithersburg, Maryland*: 30–7

Couclelis H 1992 People manipulate objects (but cultivate fields): beyond the raster-vector debate in GIS. *Proceedings of International Conference on GIS – From Space to Territory: Theories and Methods of Spatial Reasoning*. Springer-Verlag Notes in Computer Science 639: 65–77

Egenhofer M J, Richards J 1995 Visual map algebra: a direct-manipulation user interface for GIS. In Spaccapietra S, Jain R (eds) *Visual Database Systems 3: Visual Information Management, Proceedings of Third IFIP 2.6 Working Conference on Visual Database Systems*. Chapman & Hall, London: 235–53

Ferreira J 1992 User interfaces for geographic information systems. In Mark D M, Frank A U (eds) *User Interfaces for Geographic Information Systems: Report of the Initiative 13 Specialist Meeting*. Technical Report 92-3. National Center for Geographic Information and Analysis, Santa Barbara: 63–66

Mark D M, Frank A U 1992 *User Interfaces for Geographic Information Systems: Report on the Initiative 13 Specialist Meeting*. Technical Report 92-3. National Center for Geographic Information and Analysis, Santa Barbara

McKee L, Buehler K (eds) 1996 *The Open GIS Guide*. Open GIS Consortium, Inc, Wayland, Massachusetts. *http:www.opengis.org/guide*

Meyer B 1988 *Object Oriented Software Construction*. Prentice-Hall, New York

Morris B 1988 CARTO-NET: graphic retrieval and management in an automated map library. *Special Libraries Association Geography and Map Division Bulletin* 152: 19–35

Oliveira J L, Medeiros C M B 1995 *A Direct Manipulation User Interface for Querying Geographic Databases*. Technical Report DCC-95-08. Computer Science Department, University of Campinas, Brazil

Oliveira J L, Medeiros C M B 1996 *User Interface Issues in Geographic Information Systems*. Technical Report IC-96-06. Institute of Computing, University of Campinas, Brazil

Oliveira J L, Cereja N, Medeiros C M B 1997 Intermediate interface model for geographic information systems. *Proceedings of GIS Brasil 97* (in Portuguese)

Peloux J, Rigaux P 1995 A loosely coupled interface to an object-oriented geographic database. In Hartmanis J, Goos G, van Leeuwen J (eds) *Spatial Information Theory (International Conferences COSIT, Semmering, Austria)*: 123–37

Pissinou N, Makki K, Park E 1993 Towards the design and development of a new architecture for geographic information systems. *Proceedings of the Second International Conference on Information and Knowledge Management*: 565–73

Rigaux P 1995 *Interfaces Graphiques pour Bases de Données Spatiales: Application à la Représentation Multiple*. PhD thesis, Conservatoire National des Arts et Metiers, France

Shneiderman B 1992 *Designing the User Interface: Strategies for Effective Human–Computer Interaction*. Addison-Wesley, New York

Voisard A 1991 Towards a toolbox for geographic user interfaces. *Proceedings of the Second Symposium on Spatial Database Systems.* Springer Verlag Lecture Notes in Computer Science 525: 75–97

Voisard A 1994 Designing and integrating user interfaces of geographic database applications. *Proceedings of the 1994 ACM Workshop on Advanced Visual Interfaces, Bari, Italy*: 133–42

Voisard A 1995 Mapgets: a tool for visualizing and querying geographic information. *Journal of Visual Languages and Computing*

# 26 GEOLIB : A SOFTWARE COMPONENT FOR MAKING GIS TOOLS INTEROPERABLE

Donatas Kvedarauskas, Patrice Boursier, Xavier Culos, Thierry Deltheil, and Sylvie Iris

## 26.1. Introduction

Geographical information systems (GIS) are now in common use in many bodies interested in the use of spatially-referenced data (Laurini and Thompson 1992). The new challenge concerns the extension of the use of GIS data in client–server environments, particularly through the Internet or within Intranet networks (Duda 1998; Fonseca and Davis, this volume; van den Berg et al., this volume).

One of the main objectives addresses the possibility for a large public to access geographical information, for example for finding tourist or practical information. This will give rise to new GIS applications through the Internet that will democratize the use of GIS data (Grossley and Boston 1996).

GIS tools are still too complex, and consequently there is a need for new very simple tools for accessing and visualizing geographical information (Aufaure-Portier and Trépied 1996; Boursier and Mainguenaud 1992; Voisard 1994). It is already possible to access raster-based geographical data as GIF or JPEG images, but their use is limited to map visualization, and interaction is limited since buttons and links must be explicitly established by the author of hypermedia applications or Internet pages (Boursier et al. 1996; Kraak 1996).

It would be interesting to allow the user to click on a vector-based map, for having access to any kind of geographical entity, for example for knowing the availability of places in a theater or a parking. Such tools are currently being developed, with two main approaches, depending on both software and computer environments used, the pure Microsoft approach and the generalized Java approach. This results in two development strategies: ActiveX components on the one hand, and Java applets on the other hand. Both approaches have advantages and drawbacks.

ActiveX components may be developed with various languages, and built from existing components. Performance is good since executable code is transferred for execution on the client. On the other hand, ActiveX can only be run under Microsoft Windows environments, and preferably using Microsoft's proprietary browser, Internet Explorer.

Java applets may be run on various platforms, for example Unix or Windows, using various browsers. Java components (*Java Beans*) are being developed by a large community of users, but their performance is not very good because of Java code interpretation on the client.

The aim of the GEOLIB project was to develop a library of geographic functions, portable and interoperable with application development environments, database management systems (DBMS), multimedia and hypermedia authoring systems as well as external GIS, and also exploitable in an Internet or Intranet environment. Consequently, it can be used either in standalone, or in client–server and Internet or Intranet environments. Designed as a software component, GEOLIB is extensible (new data formats, new GIS functionalities). It makes it possible to integrate more easily the spatial dimension within existing interactive multimedia applications, as well as office or management applications. It allows manipulation of maps, including computation of itineraries, but also to address spatial queries to existing spatial databases, without having to know how to use GIS tools.

GEOLIB has been developed in the framework of a contract funded by the French agency for technology transfer between research and industry (ANVAR: Agence Nationale de Valorisation de la Recherche), in collaboration with the computer science research laboratory at the University of La Rochelle (L3i: Laboratoire d'Informatique et d'Imagerie Industrielle). It has benefited from the work that has been done in the framework of the European MAGIC TOUR project, which aimed at developing an authoring system for tourism with geographic functionalities (Boursier et al. 1995).

In Section 26.2, we review the design options that have been selected for the development of GEOLIB, and we list the functions which have been implemented in the first release. The architecture of GEOLIB is described in Section 26.3, showing the subdivision into specific software modules and the linking of these modules. Section 26.4 explains how queries are processed, depending on the architecture of the system and the query processing schema.

## 26.2. Design choices and functionalities

GEOLIB has been designed as an extensible software component, exploitable either as an OCX component to be linked to such software tools as database management systems and multimedia or hypermedia authoring systems, or as an ActiveX component to be run in an Internet or Intranet environment. In the latter case, it is compatible with Microsoft Internet Explorer and Netscape Communicator browsers.

GEOLIB is built as a set of C++ modules and OLE-2 controls. Because of performance requirements, a low level socket mechanism has been designed and implemented for client–server communication when using GEOLIB in Intranet- or Internet-based environments.

Adaptive query processing is also used for improving performance. It depends on main memory and processing power available on the client, information which is stored in a configuration file. This configuration file also lists all geographic data sources exploited by the application that uses GEOLIB, along with the way these data are going to be used. It is then used for optimizing access to spatial data, and reducing data transfer.

With the same objective of performance, a software cache mechanism is included in the client–server version of GEOLIB. It can be considered as some kind of differential data loading mechanism, since it aims at transferring to the client the data which are required for processing a query only when they are not already present on the client. For that purpose, specific data structures are kept in the main memory of both the client and the server for keeping track of previous queries and data uploading on the client.

A limited number of GIS formats are supported in the first release of GEOLIB, namely MapInfo MIF/MID files and ESRI shapefiles. The first release of GEOLIB, which has been launched in September 1997, contains functions for:

- data acquisition and handling;
- exploitation and querying;
- display and visualization;
- geographic data processing.

The following set of basic functions has been implemented for *data acquisition and handling*:

- import of a vector or raster map layer and associated descriptive non-spatial data;
- creation of a vector map layer from a drawing layer;
- creation of a map from the superposition of raster or vector layers;
- map updating (layers insertion, deletion, and reorganization);
- map projection and coordinate systems.

Functions that deal with *exploitation and querying* concern spatial indexing on the one hand, and spatial and non-spatial queries on the other hand :

- vector layer indexing;
- non-spatial queries;
- simple spatial queries, based on distance, adjacency, intersection, and inclusion.

Basic functions have also been implemented for map data *display and visualization*:

- 2D map display;
- physical zoom on raster or vector layer;
- logical zoom on raster or vector layer (access to more detailed maps);

- panning in four directions;
- management of links between spatial and multimedia data (creation, modification, deletion).

Simple functions are proposed for *geographic data processing*:

- map overlay;
- length and area measurement (lines, polygons);
- distance between objects;
- buffer zone management.

## 26.3. Architecture

GEOLIB is made of several modules (Figure 26.1), in order to be easily portable and linkable with other software tools, and usable in a client–server (Internet or Intranet) environment.

The *Gateway* module:

- imports existing geographic data files;
- loads data within object structures in main memory, in order to be processed by the other modules of GEOLIB.

The *Dispatcher* module:

- analyzes and pre-processes queries;
- determines the optimal strategy for answering the query, depending on the processing and result of previous queries: access to data stored on the client, access to data stored on the server, extraction of data from files stored on the server, data retrieval, data processing, and preparation of the answer.

The *Visualization* module:

- translates user interaction into queries;
- addresses the query to the *Dispatcher* module;
- displays results on the client.

The *Core* module :

- receives queries addressed by the *Dispatcher* module;
- executes basic geographic functionalities;
- passes on results to the *Dispatcher* module.

The *Extensions* module :

- receives queries addressed by the *Dispatcher* module;
- executes specific geographic functions;

- passes on results to the *Dispatcher* module.

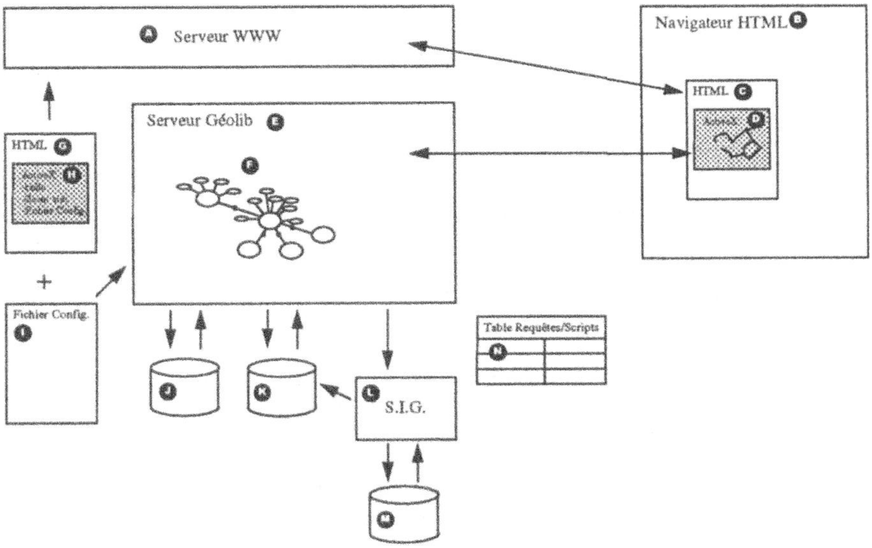


Figure 26.1. GEOLIB architecture. (A) Web server (sends HTML pages to the client); (B) Web browser (Microsoft Internet Explorer, Netscape Navigator): (C) Current HTML page on the client; (D) ActiveX component (or Java applet) handles interaction, map display, and communication with the GEOLIB server; (E) GEOLIB server, answers queries using data present in memory or querying the GIS; (F) Image of geographic information present on clients; (G) Initial HTML page stored on the server, to be sent to a client at the first connection; (H) ActiveX component, part of the application HTML page; (I) Configuration file; (J) Geographic database accessed by the GEOLIB server (permanent data); (K) Geographic data generated as an answer by the GIS (temporary data); (L) External GIS (e.g., MapInfo or ARC/INFO); (M) GIS proprietary database; (N) Queries/scripts table

## 26.4. Query processing

Queries are expressed on the client side, and they can be processed in two different ways: either locally when they require data which are already available on the client (In that case, some functions of the GEOLIB component are activated for processing geographical data locally, using only the processor on the client), or on the server when all the data needed for answering the user query are not present on the client, and new data need to be accessed on the server (in that case, a new query will be transmitted via URL).

Three different scenarios have been identified and considered for query processing, depending on the architecture of the system and the query processing schema:

- query processed on the client;
- query processed on the server;
- query addressed to an external GIS.

These are detailed and illustrated in the following sections.

### *26.4.1. Query processed on the client*

The simplest way of processing a query is the local mode:

1. Following a user interaction, the *Visualization* module on the client creates a query object that is sent to the *Dispatcher* module.
2. The *Dispatcher* module analyzes the query and concludes that it can be processed locally. Consequently, the query is sent to the *Core* module on the client.
3. The *Core* module takes charge of the query processing.
4. Once the query has been processed locally by the *Core* module, the answer is sent to the *Visualization* module to be displayed.

### *26.4.2. Query processed on the server*

The processing of a query may require access to the GEOLIB server. The execution scenario is then the following:

1. The *Visualization* module translates a user interaction into a query object, and sends it to the *Dispatcher* module.
2. The *Dispatcher* module on the client analyzes the query and concludes it must be sent to the server for further processing, either because all required data are not present on the client, or because functions required to process the query are not available on the client. Consequently, it initiates a communication between the client and the server, and the query is sent to the server to be processed.
3. The GEOLIB server receives the query and restructures it, before sending it to the *Dispatcher* module on the server.
4. The *Dispatcher* module analyzes the query and decides which processing module must be called, either the *Core* module if the query is simple enough, or the *Extension* module if a more specific function is required.
5. Once the answer is elaborated, the GEOLIB server sends it back to the client.
6. The *Dispatcher* module on the client receives the answer and reformats it before it is displayed.
7. The *Visualization* module can display the result to the user.

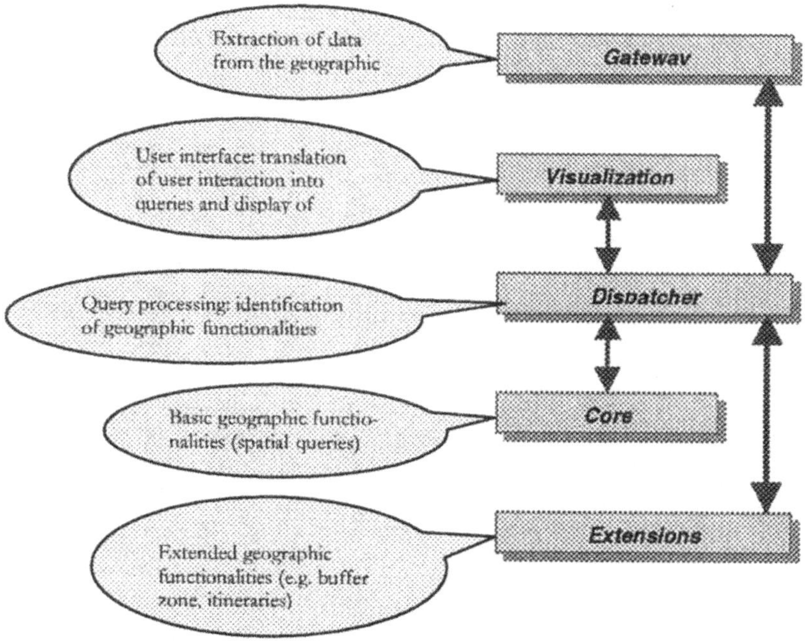


Figure 26.2. Query processed on the client

### *26.4.3. Query addressed to an external GIS*

When the query concerns external data (not already available, neither on the client, nor on the server), query processing is a little more complex:

1. The *Visualization* module translates a user interaction into a query object, and sends it to the *Dispatcher* module on the client.
2. The *Dispatcher* module on the client analyzes the query and concludes it must be sent to the server for further processing, either because all required data are not present on the client, or because functions required to process the query are not available on the client. Consequently, it initiates a communication between the client and the server, and the query is sent to the server to be processed.
3. The GEOLIB server receives the query and restructures it, before sending it to the *Dispatcher* module on the server.
4. The *Dispatcher* module notices that the query corresponds to a script that concerns an external GIS. The execution of this script is then activated.
5. A communication process is established between the GEOLIB server and the external GIS. Data are formatted and exchanged.

6. Once this operation is terminated, the GEOLIB server sends the answer back to the client.
7. The *Dispatcher* module on the client builds the result and displays it to the user.

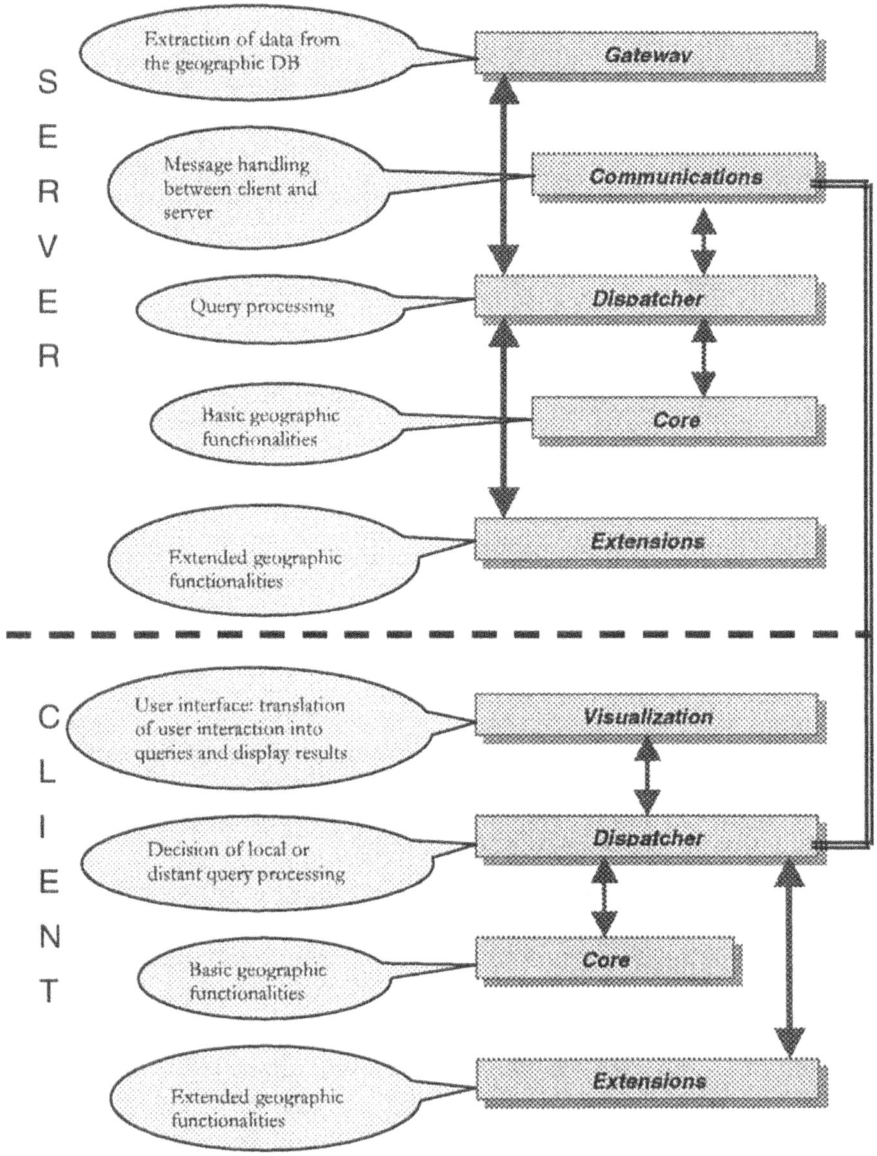


Figure 26.3. Query processed on the server

## 26.5. Conclusion

The GEOLIB software component has already been used for different kinds of application, either in a standalone mode, or in a client–server environment. It already supports ARC/INFO shapefiles and MapInfo MIF/MID formats. It is currently being extended in order to support other GIS formats. New functionalities are also being added, in order to be able to deal with more complex applications. Evaluation and validation of GEOLIB have already been conducted in various environments, and for different kinds of applications.

A tourism-oriented standalone prototype application has been developed for the city of Toulouse. For that purpose, GEOLIB has been used as a DLL linked with Macromedia Toolbook. A Web-based real-size application has also been developed for urbanism and tourism with real-time access to an existing MapInfo database on the city of Toulouse (see *http://athena.silogic.fr/DemoGeolib/default.html*). Both standalone and client–server real-size applications have also been developed for regional water management, using GEOLIB as an OCX within a Borland Delphi environment. Ongoing work concerns the availability of new GIS formats (e.g. Arc/Info or Star). We are also extending GEOLIB with new functionalities, and attempting to optimize performance by generalizing the use of spatial indexes. A Java version of GEOLIB is also under development, to be used with Microsoft-based platforms, but also with UNIX- and CORBA-based ones.

## References

Aufaure-Portier M A, Trépied C 1996 A survey of query languages for geographic information systems. *Proc. 3rd Int. Workshop on Interfaces to Database Systems, Edinburgh*

Boursier P, Mainguenaud M 1992 Spatial query languages: extended SQL vs. visual languages vs. hypermaps. *Proc. 5th Int. Symposium on Spatial Data Handling (SDH '92), Charleston*

Boursier P, Kvedarauskas D, Iris S, Guilloteau A 1995 MAGIC TOUR: the integration of multimedia and geographic information technologies in an authoring system. *Proc. 1st Int. Conference on Spatial Multimedia and Virtual Reality (SMVR '95), Lisbon*

Boursier P, Dbouk M, Kvedarauskas D 1996 Querying and visualizing geographical data sets with dynamic maps. *Proc. 2nd ACM Workshop on New Paradigms for Information Visualization and Manipulation, Rockville*

Duda K A 1998 *Database Query and Map Generation via the WWW : Online Resources and Examples. http://www.ssc.msu.edu/~geo/wwwgis.html*

Grossley D, Boston T 1996 A generic map interface to query geographical information using the World Wide Web. *World Wide Web Journal* 1

Kraak M J 1996 Integrating multimedia in geographical information systems. *IEEE Multimedia* 3(2): 59–65

Laurini R, Thompson T 1992 *Fundamentals of Spatial Information Systems*. Academic Press

Voisard A 1994 Designing and integrating user interfaces of geographic database applcations. *Proc. ACM Workshop on Advanced Visual Interfaces (AVI'94)*

# 27 SOFTWARE AGENT-ORIENTED FRAMEWORKS FOR THE INTEROPERABILITY OF GEOREFERENCED DIGITAL LIBRARIES ON THE WORLD WIDE WEB: THE SIGAL PROJECT

Zakaria Maamar, Bernard Moulin, and Yvan Bédard

## 27.1. Introduction

Within an organization managing spatially referenced data, several types of documents are used to describe the land and its resources (for instance, topographic maps, land use maps, aerial photographs, satellite images, and so on). Managing such documents is a complex task, each document being characterized by its own content, spatial reference system, quality, sources, mode of distribution, and format. Hence, Georeferenced Digital Libraries (GDLs) can be very useful and helpful to describe the available geodocument resources in an organization. For instance, GDLs provide information about the physical location of the documents and identify who does what, when, and how. GDLs are nowadays implemented on the World Wide Web (WWW), mostly to facilitate the distribution of spatial data; these are the GDLs of interest for the present chapter although the research is not limited to the WWW.

There exists a number of such GDL projects, for example Alexandria Digital Library (University of California, Santa Barbara, *http://alexandria.sdc.ucsb.edu*), the digital library project at the University of California, Berkeley (*http://elib.cs.berkeley.edu*), and GEOREP (Laval University, *http://sirs.scg.ulaval.ca*). They aim at helping users to identify data which may be useful to them, at helping producers to increase accessibility to their spatial data for potential users, and at facilitating the sharing of spatial data between organizations. GDLs can be defined as a "computerized index map used to locate georeferenced documents, to describe their content and nature with metadata, and to visualize their content" (Létourneau 1998).

Users are accustomed to use such GDLs independently from each other. Yet, a more complex task is to use different GDLs, generally distributed and heterogeneous, which may or may not overlap on a given geographic area. This new reality requires new alternatives for GDL interoperability and for a *one-stop* query engine (i.e., a single place where someone can go and query multiple GDLs from a unique user interface in spite of the heterogeneity of the source GDLs queried in the background, similarly to meta-search engines on WWW).

A survey conducted in October 1995 and updated in July 1996 (Proulx et al. 1997) sheds light on the characteristics of current GDLs. If a person visits these GDLs, he or she will easily notice that the content of GDLs differs from one to another; the data standards are quite different, or even absent in some cases; different words are used to represent the same concept or the same word can be used to represent different concepts; the user interfaces always differ; and so on (see Létourneau 1998 for a detailed description of these discrepancies). All these differentiating elements force users to adapt to each GDL's requirements and understand their different information and structural characteristics, which is a time-consuming and almost impossible task.

At Laval University Center for Research in Geomatics, we are working on several GDL projects (Figure 27.1). These are the Georep project (Proulx et al. 1997), the SIGAL project (French acronym of Geographic Information System and Software Agent; Maamar 1997), and the data warehouse-based and fuzzy logic-oriented SOS-DS project (French acronym of System for the Optimized Selection of Spatial Data; Létourneau et al. 1997). In the SIGAL project, which is the topic of the present chapter, we work on an interoperable environment for GDLs. Interoperability is a process that allows cooperative interactions between several systems in general and GDLs in particular. Such an interoperable environment will provide users with services which will not require them to know individual characteristics of the interconnected GDLs (a concept similar to the concept of meta-engine used to search on the WWW).

Interoperability requires process has to keep the local systems, in general, and GDLs, in SIGAL, autonomous and independent from this process (Maamar et al. 1997; Wiederhold 1989). In order to reach this goal, we suggest the introduction of specialized components, called software agents, based on development techniques borrowed from Distributed Artificial Intelligence (Moulin and Chaib-draa 1996). These agents are the front end of the systems being integrated, have the ability to act on their own behalf, and can help users fulfill their needs in a single and homogeneous stop. However, in order to reduce the complexity of managing operations in distributed and heterogeneous contexts, agents can be gathered into teams, which in turn evolve within what we call software agent-oriented frameworks (Maamar and Moulin 1997b). A software agent-oriented framework is an environment composed of a supervisor program and one or several teams of sofware agents that fulfill services offered by this framework to users and to other frameworks.

In this chapter, we describe the use of these software agent-oriented frameworks in the SIGAL project. The chapter is organized as follows. Section 27.1 presents an overview of our theoretical and practical research study. Section 27.2 presents the basic notions of software agent-oriented frameworks. Section 27.3 describes the architectural and functional characteristics of the SIGAL environment. A prototype of the SIGAL environment is presented in Section 27.4. Finally, Section 27.5 summarizes our work.

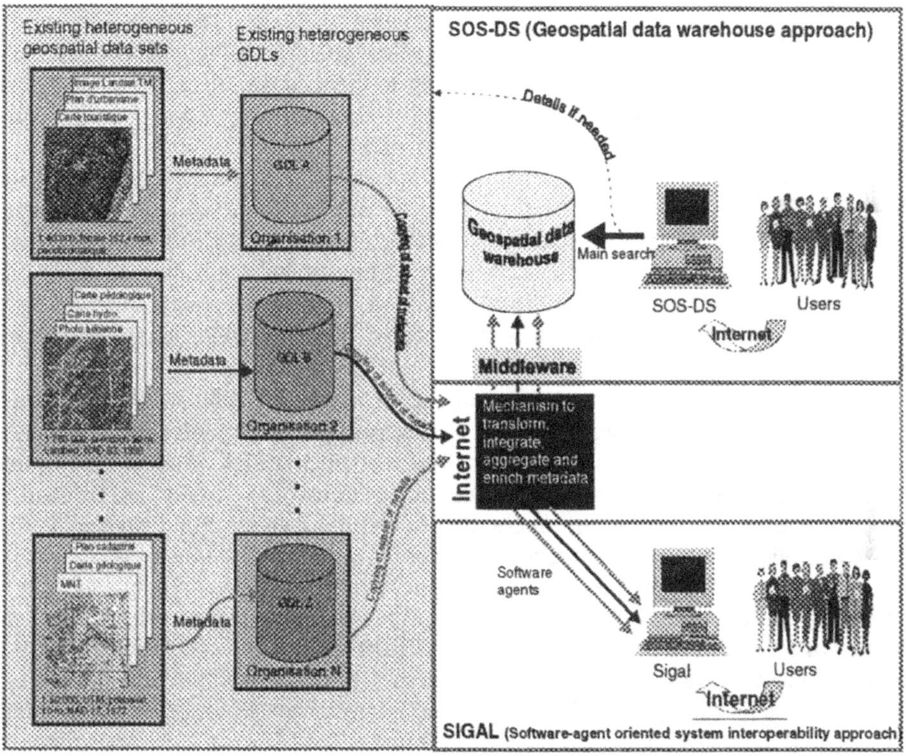


Figure 27.1. Presentation of the SIGAL and SOS-DS research projects (adapted from Bédard et al. (1997)

## 27.2. Software agent-oriented frameworks

Designing software agent-oriented frameworks is an iterative activity that is applied in turn to frameworks, teams of agents, and software agents (Maamar and Moulin 1997a,b).

A software agent-oriented framework offers a set of services that can be requested either by users or by other frameworks. A framework is an environment composed of a supervisor and one or several teams of software agents. These teams are composed of agents selected from a bank of software agents. This bank contains several agents having different functionalities which are typical of the application to be developed. In the SIGAL project, we aim at developing an interoperable environment for Web GDLs which involves three types of frameworks (*server, client, local-source*). Those frameworks are composed of different kinds of agents which play specific roles. For instance, the local-source framework integrates the following agents: *Coordinator, Interaction* and *Knowledge Agents*.

Teams of agents are set up by the framework and are structured differently according to their responsibilities. A team is also characterized by a supervisor and a set of roles that agents must fulfill according to their capabilities. Teams of agents can be set up in a unique framework or in several frameworks leading, in the latter case, to the creation of a multiframework environment.

Services provided by a framework satisfy specific user needs such as information retrieval. When a service is invoked by a user or another framework, the framework supervisor agent activates a realization scenario, which specifies the operations required in order to carry out the service. According to that realization scenario, the framework supervisor creates pertinent teams. At their own level, team supervisor agents activate realization scenarios in order to coordinate the activities of their software agents.

A realization scenario is composed of the following elements: software agents' types and roles, scheduling sequences indicating in which order agents should perform their plans, the list of information resources which provide knowledge required by agents when achieving their operations, and finally an evaluation procedure used to monitor the operations of various agents. For instance, in the SIGAL environment, a data request directed to a GDL is defined by an access scenario (Figure 27.2) which involves two Interaction Agents, each one respectively belonging to a client framework and to a local-source framework. The client framework's Interaction Agent has the knowledge structures needed to transform a user's request into a data request expressed in a form compatible with the GDL's data manipulation language. The local source's Interaction Agent processes the data request and activates the access plan which queries the GDL. It, then, transmits the answer to the client framework's Interaction Agent.

## 27.3. The SIGAL project, an application of software agent-oriented frameworks

In the SIGAL project, we are developing an interoperable environment for GDLs on the Web. In the remainder of this chapter, we describe the SIGAL environment in terms of architecture (Section 27.3.1), agents and frameworks (Section 27.3.2), services (Section 27.3.3), ontology (Section 27.3.4) and prototype (Section 27.4).

### *27.3.1. Architecture of the SIGAL environment*

The design of the SIGAL architecture applies principles elicited in the field of information systems interoperability (Ambite and Knoblock 1995; Genesereth and Ketchpel 1994; Hsu 1996; Kerschberg 1997; Sycara et al. 1996), namely that an interoperable environment must: maintain the autonomy and independence of the systems that will cooperate; reduce the informational disparities of the various interconnected systems; and help users satisfy their needs.

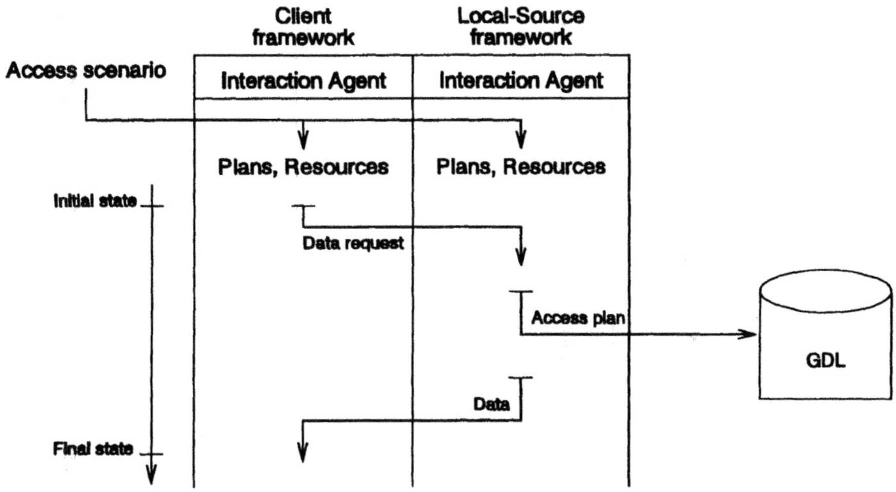


Figure 27.2. Representation of a realization scenario example

Based on these considerations, we have proposed an interoperable architecture for the SIGAL environment (Figure 27.3). It is characterized by the use of a client–server approach, by the introduction of mirror sites, and by the creation of three types of frameworks (server, client, local-source; Maamar and Moulin 1997a).

The local-source frameworks maintain the autonomy and the independence of the GDLs in the interoperable environment. Therefore, local-source frameworks interface the GDLs with communication networks and process data requests sent by the client frameworks.

The server framework monitors all the operations needed to support the services offered to users and to other frameworks. In order to avoid overloading the server framework, we suggest duplicating it on mirror sites. Setting up mirror servers is used to orient users towards less-loaded servers. However, in such a case, it is important to maintain the coherence of the information duplicated on the server frameworks and consequently to define reliable update protocols.

When users need information from several GDLs, they invoke relevant services from the server framework. The invocation of such services initiates the creation of a client framework on the user's machine. Hence, the server framework delegates the performance of required services to the client framework and limits its involvement to the monitoring of the client framework's operations. After completion of these operations, the client framework can be either deleted or recorded for future use.

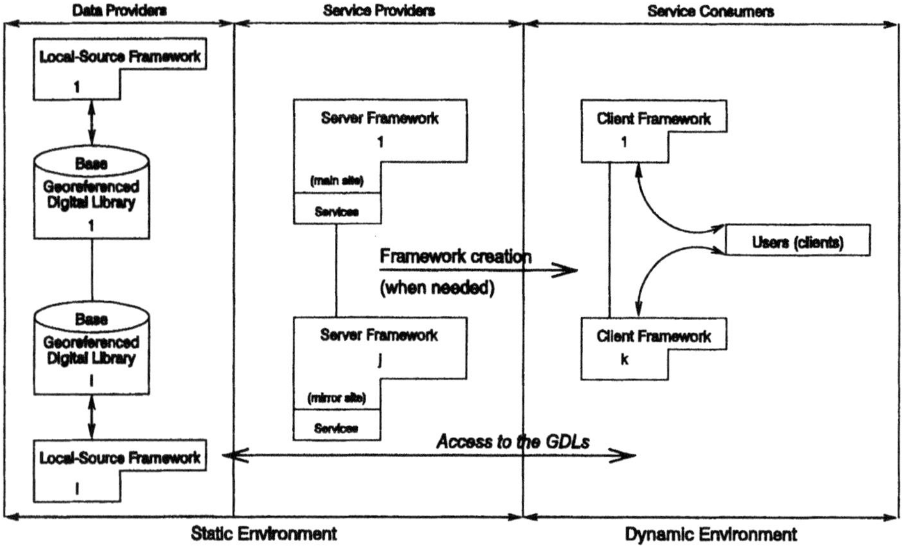


Figure 27.3. Multiframework architecture of the SIGAL environment

### *27.3.2. Agents and frameworks of the SIGAL environment*

In order to meet SIGAL's architectural requirements (Section 27.3.1) and to perform the various services offered by SIGAL (Section 27.3.3), we have identified different types of agents:

1. Coordinator Agent: monitors the performance of operations needed to fulfill the various services within a framework and between frameworks.
2. Domain Agent: resolves knowledge disparities by using the common knowledge which is recorded in the Ontology Base
3. Help Agent: presents the services which are offered by server frameworks to users.
4. Interaction Agent: handles interactions between frameworks.
5. Interface Agent: assists users when formulating their needs.
6. Knowledge Agent: knows the protocols through which a GDL accepts requests. This agent also monitors changes occurring in the GDL during local knowledge updates.
7. Learning Agent: provides new knowledge to the agents of the framework.
8. Resolution Agent: processes user queries that may require a decomposition into a set of sub-queries.

9. Scenario Agent: manages the realization scenarios that carry out the services offered by the SIGAL environment.

Agents described above are used to create the various framework teams of the SIGAL architecture. In order to represent a software agent-oriented framework graphically (Figures 27.2, 27.3, and 27.4), we have used the following formalism: a rounded corner rectangle represents the studied framework; a rectangle represents a team or an agent; an oval indicates other frameworks that interact with the current studied framework; a rectangle with an oval on the top represents an informational resource; and finally, a line with an arrow identifies an interaction within a framework or between frameworks. Three types of frameworks are proposed: server, client, and local-source.

The *server* framework (Figure 27.4) is responsible for carrying out services required by users, or needed to maintain SIGAL's architecture. The server framework contains a single team composed of Coordinator, Help, Scenario, Interaction, Domain, and Learning Agents. Moreover, this team has access to three informational resources: the Service, Scenario, and Ontology Bases. The Service Base is accessible through the Help Agent and contains a list of user and SIGAL services offered by the server framework. The Scenario Base is accessible through the Scenario Agent and contains the realization scenarios available to the team.

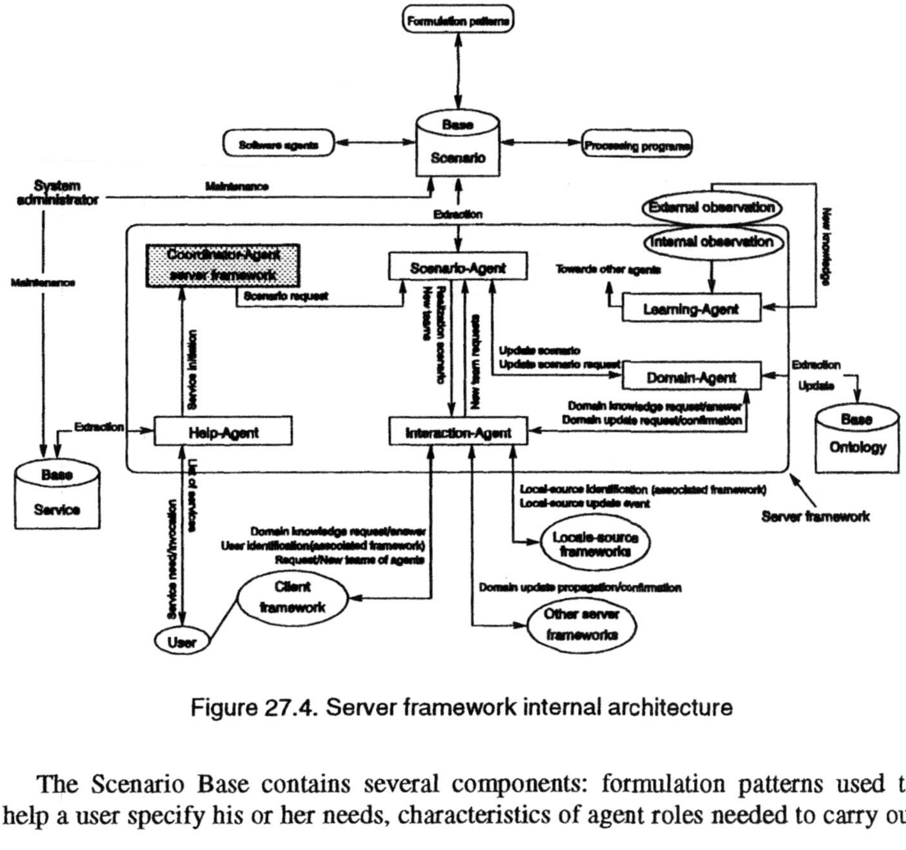


Figure 27.4. Server framework internal architecture

The Scenario Base contains several components: formulation patterns used to help a user specify his or her needs, characteristics of agent roles needed to carry out the scenarios, and procedures used to build the plans that are executed by the agents during the scenario. Finally, the Ontology Base provides a detailed description of the knowledge contained in each GDL.

The *client* framework is created by a server framework when a user requires a service (Figure 27.5). The client framework is composed of a Coordinator Agent and of one or several teams according to the number of GDLs needed to process user queries (Section 27.3.3.3). The Coordinator Agent creates the main team of software agents and assigns it the activities prescribed by the realization scenario which identifies the invoked service.

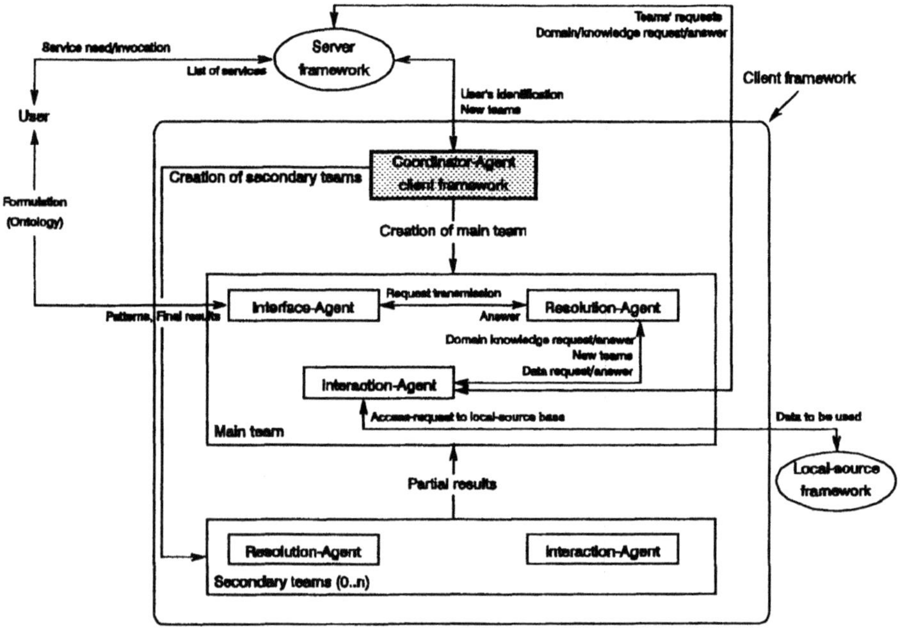


Figure 27.5. Client framework internal architecture

The main team is composed of Interface, Resolution and Interaction Agents and is responsible for processing the user's query as well as for identifying which GDLs must be accessed. Based on the number of GDLs needed, the main team asks the Scenario Agent of the server framework for additional agents. One or several secondary teams are created, each containing an Interaction Agent and a Resolution Agent. Each team (main or secondary) interacts with a specific local-source framework in order to solve the sub-problem at hand.

Each local-source framework interfaces with a GDL of the SIGAL environment (Figure 27.6). It is assigned to a specific server framework in order to keep it informed about the informational changes occurring in the GDL. When such a change occurs, the Domain Agent of the server framework updates its Ontology Base; these updates are then propagated to the other mirror servers. This mechanism preserves the coherence of the various copies of the Ontology Base through the network. All update operations are specified by update scenarios, that belong to SIGAL services. A local-source framework also processes data requests directed to its associated GDL.

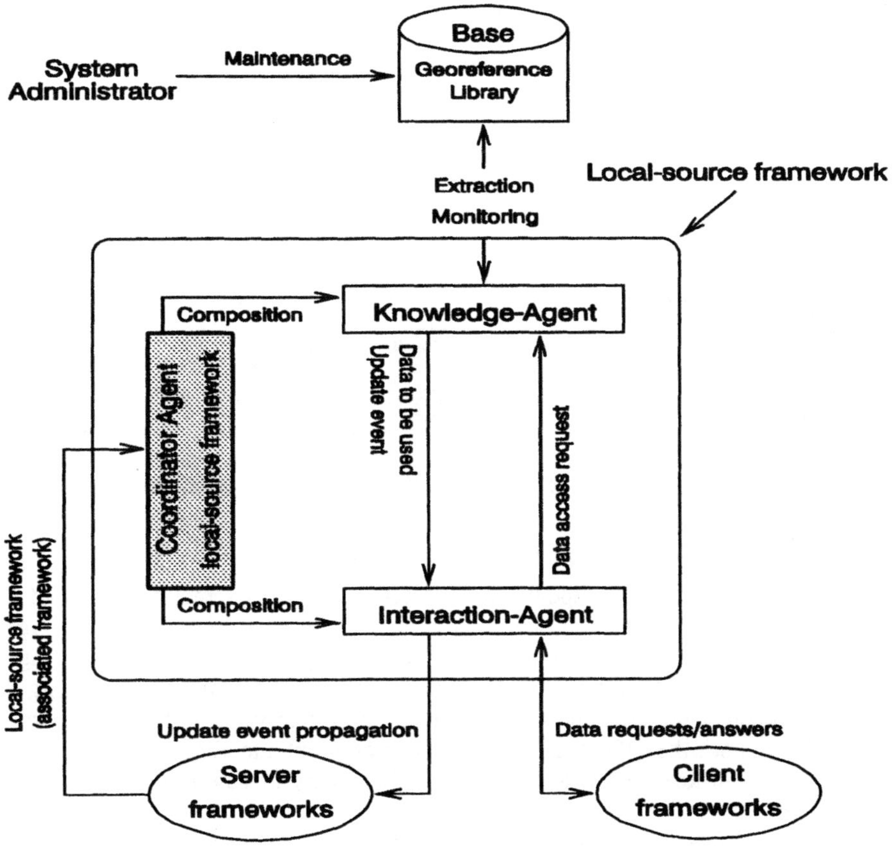


Figure 27.6. Local-source framework internal architecture

### *27.3.3. Services of the SIGAL environment*

Two categories of services are offered by server frameworks: SIGAL services and user services. SIGAL services support operations that allow insertion or deletion of a GDL in the interoperable environment, as well as the modification of GDL informational content. A user service fulfills a need that involves several GDLs. Users formulate their information queries independently from the GDLs distribution and heterogeneity.

Each service is described by a realization scenario and is graphically represented by an interaction diagram. The representation formalism for interaction diagrams is as follows (Maamar and Moulin 1997b): a rounded corner rectangle represents a framework; a filled rectangle represents an agent role; an oval indicates an external resource whose interactions (dashed lines in Figures 27.7, 27.8, and 27.9) with the framework are performed by agents; a solid line identifies a service offered by an agent (the arrow indicates the agent which supports the service); and finally, numbers indicate the chronology of operations.

**27.3.3.1. SIGAL services: modification of GDL-informational content**. Any change occurring in a GDL must be pointed out by its local-source framework to the assigned server framework (Figure 27.7). The objective is to keep a total coherence between the informational content of the GDLs and that of the Ontology Base of the server framework.

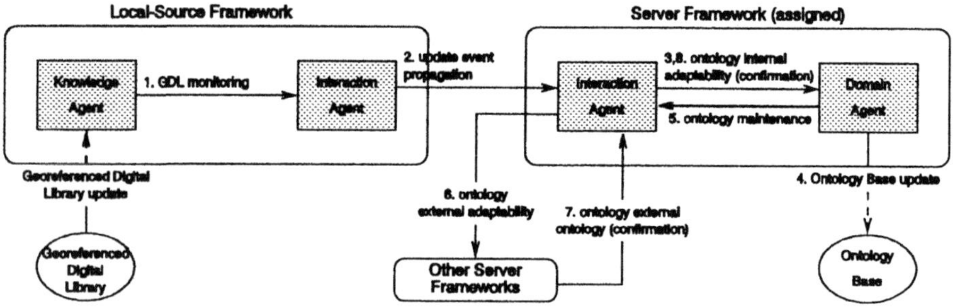


Figure 27.7. Interaction diagram of SIGAL services, modification case

The Knowledge Agent of the local-source framework monitors the behavior of a specific GDL. During each update, it contacts (1, 2) the Interaction Agent of the associated server framework. Once it gets this information, the Interaction Agent sends it (3) to the Domain Agent of the server framework. Subsequently, this agent performs the maintenance operations (4) and transmits a request for spreading the update to the other server frameworks (principal or mirror sites) using the Interaction Agent (5, 6). Since the SIGAL environment brings about a total replication situation, it is important to keep the Ontology Base coherent by managing confirmations of update operations (7, 8).

**27.3.3.2. SIGAL services: new GDL insertion**. Any new GDL that has to be integrated into the SIGAL environment is identified by a local-source framework assigned to a server framework. The insertion of a new GDL is triggered by the invocation of the appropriate service of the server framework, by the person (system administrator) responsible for managing this framework (Figure 27.8). According to this service, the Help Agent (1) has to contact the Coordinator Agent to inform it about the need for setting a scenario modifying SIGAL's architecture (2). This scenario is managed by the Scenario Agent and is available in the Scenario Base (3). In this context, the realization scenario is characterized by two elements: processing programs and formulation patterns for the Domain Agent of the server framework, and processing programs and software agents for the new local-source framework.

After sending the components of the new local-source framework and after its assignment to a server framework (4, 5), the Scenario Agent sends the update scenario to the Domain Agent in order to update the Ontology Base (6,7). When these update operations have been executed, the Domain Agent sends a request for spreading the update to the other server frameworks (8, 9) using the Interaction Agent. Since the SIGAL environment brings about a total replication situation, it is important to keep the Ontology Base coherent by managing confirmations of update operations (10, 11).

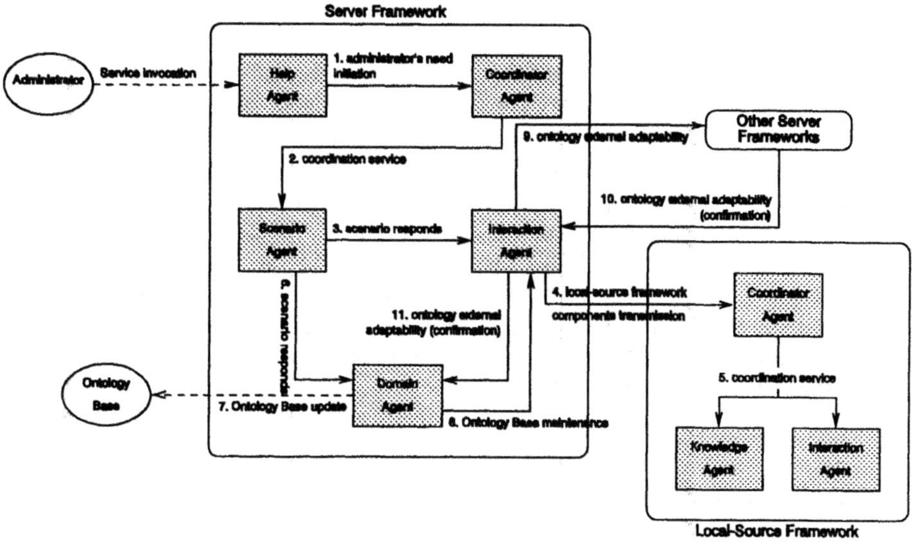


Figure 27.8. Interaction diagram of SIGAL services, insertion case

**27.3.3.3. User services.** In an interoperable environment, such as SIGAL, a user's service (also called a global query) is usually decomposed into a set of sub-queries. Each sub-query is sent to the system which contains data required to satisfy it. The sub-queries and their syntactic and semantic forms depend on the functional type of systems to interoperate. Therefore, we propose autonomy and complementarity criteria to define how the sub-queries are generated and related. These two criteria are defined as follows:

1. Autonomy criterion: local systems are independent and do not need to cooperate. A sub-query, identical to the user query, is generated for every system which is able to satisfy the query. Hence, the sub-queries obtained from the query decomposition are semantically identical, but use different concepts in their formulation.

2. Complementarity criterion: local systems are not independent and need to exchange data. A sub-query is generated for each system which contains part of the information needed to satisfy the query. Hence, the sub-queries are semantically different and use different concepts in their formulation.

The SIGAL environment has been built to satisfy the autonomy criterion. In the SIGAL environment, a user query processing is decomposed into four steps (Cheung and Hsu 1996). The first step specifies user needs by using the concepts of the ontology. The next step identifies the needed GDLs according to the autonomy criterion and provides adequate sub-queries. The third step processes these sub-queries by requesting data from their associated local-source frameworks. The last step provides the user with answers to his or her needs. All these steps particularly rely on the knowledge of the Ontology Base, which considers semantic disparities and implementation structures of the concepts used by each GDL. In the SIGAL environment, a user service fulfills a need that involves several GDLs (Figure 27.9). The user consults the list of services proposed by the Help Agent (1). When a service is identified, this agent initiates its realization (2) by informing the Coordinator Agent of the server framework. The Coordinator Agent performs the operations characterizing the service and asks the Scenario Agent to select the relevant scenario (3). After the scenario has been identified the client-framework components are selected (4) and sent to the client's platform by the Interaction Agent (5). The client framework is composed of a Coordinator Agent and of a main team of agents (6). The client framework must fulfill the user's request. Once this request has been formulated (7), the Interface Agent sends it to the Resolution Agent (8), in order to determine which GDLs are needed to satisfy the request. To perform this operation, the Resolution Agent uses the information provided by the Domain Agent of the server framework (9, 10, 11, 12, 13, 14).

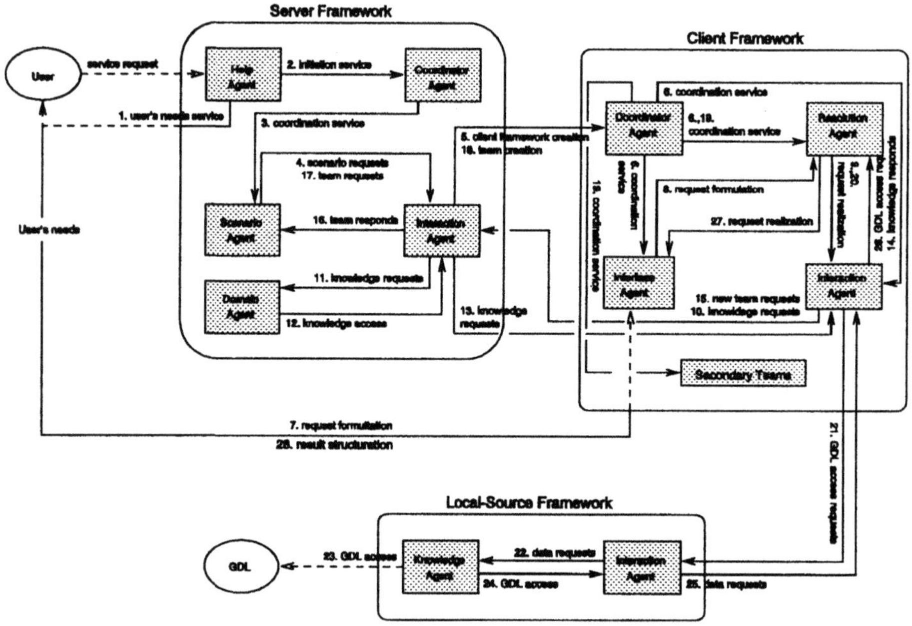


Figure 27.9. Interaction diagram of SIGAL services, user services case

Depending on how the selected GDLs are distributed, the Resolution Agent asks the Scenario Agent for new teams for the benefit of the client framework (15, 16, 17, 18, 19). The number of agent teams set up in the client framework depends directly on the number of GDLs to be queried; in fact, each team (including the main team) is assigned to only one local-source framework. The goal is to request relevant data from these selected GDLs (20, 21, 22, 23, 24, 25, 26). After processing these data, all the teams send their results to the Resolution Agent of the main team (27). The final answer is sent to the user by the Interface Agent services (28).

### *27.3.4. Ontology of the SIGAL environment*

By establishing an ontology, we offer a common terminological basis for the various interconnected systems, hence reducing the risks that users get inconsistent information.

In SIGAL's case, ontological disparities exist at different levels. First, being generally developed independently, GDLs may use different terms to describe their data: these different ontologies make it difficult for users to consult several GDLs simultaneously. Second, current GDLs do not help users when formulating their requests. Moreover, a user has to express needs according to his or her own vocabulary and understanding of the application domain. Unfortunately, a user is not always able to get the information he or she is looking for, because GDLs cannot adapt to these terminological differences (disparities between user ontology and GDL ontology). In SIGAL, the teams of the client, server and local-source frameworks cooperate to solve the problems resulting from ontological disparities.

When defining the SIGAL ontology, we used the concept of metadata (Hsu 1996), defined as data which describe data of the analyzed domain. Metadata serve at least three purposes (Morgenstern 1996): (1) cataloging and then locating the relevant data when needed, (2) interpreting the form and format of the data, and (3) understanding the data: their meaning and semantics, and the context in which they are relevant and valid. We developed a metadata model (Figure 27.10) which provides a concise description of the information manipulated by each GDL. This model is called the Ontology Base and is managed by the Domain Agent of the server framework. The metadata model is illustrated in Figure 27.10 based on OMT (Rumbaugh et al. 1991) formalism. The model is composed of meta-entities and meta-associations. For example, the meta-entity Entity describes all the existing entities in the GDL data models. Also, the meta-entity Attribute refers to the attributes that belong to the entities and the associations of the GDL level. (since the data processed by SIGAL are GDL data, which already are metadata from the usual GIS perspective, then SIGAL metadata correspond to meta-metadata from the usual GIS perspective; in other words, SIGAL metadata are at one level of abstraction higher than what is presented in ISO and FGDC spatial metadata standards).

## 27.4. Prototype of the SIGAL environment

Before we started the implementation of the SIGAL environment, we defined its logical architecture, using agent-interaction graphs (Maamar 1997). We decomposed the SIGAL architecture into a set of sub-systems, each corresponding to a type of framework in this architecture. Furthermore, each agent of a framework has been considered as a distributed object. Finally, all the offered services between frameworks or between users and frameworks became the operations (described with agent-interaction graphs). For instance, we present the realization operations of user services (Figure 27.11). The needed operations are: *needs satisfaction*, *setting up framework*, *needs analysis*, *resources requests*, *GDLs access*, and *results return*. Only the operation *needs satisfaction* is described in this chapter (Figure 27.14). In fact, this operation corresponds to the service *service request* of the Help Agent as illustrated in Figure 27.9.

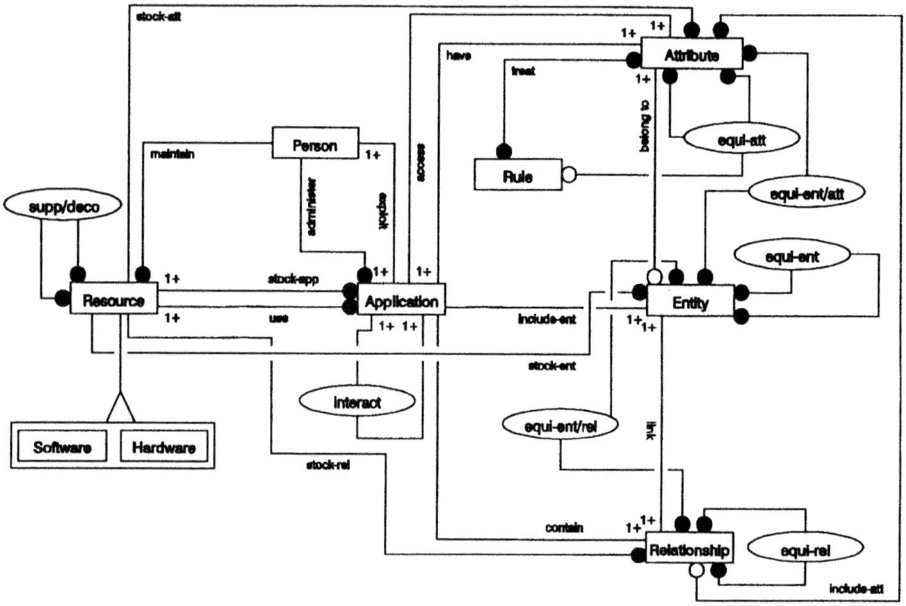


Figure 27.10. The metadata model of the SIGAL environment

In Figure 27.12, the operation *needs satisfaction* is triggered by the user via a terminal. The agents involved in this operation are represented by rounded corner rectangles. Agents can access several informational resources which are represented by dashed rounded corner rectangles. Numbers represent the realization sequence of the current operation.

Operation specification: Needs satisfaction

Controller: Terminal;

Collaborators: Help Agent, Coordinator Agent, Interaction Agent, Scenario Agent;

Description:

Operation Terminal.Server: needs satisfaction initiate the operations needed to satisfy user's requests:

(1) ask for the list of services;
(2) select a service;
(3) coordinate the service realization;
(4) ask for the operation components;
(5) get ready to send the client environment's components.

In the development of the SIGAL prototype, we used the Object Request Broker (ORB—An ORB is a software that respects the conventions defined in the CORBA standard by OMG for the development of an object-oriented distributed environment) VisiBroker for Java developed by Visigenic Company (and now owned by Borland), Java as a programming language (particularly, its applet mechanisms), and a JDBC driver in order to connect the available informational resources, for instance the Ontology Base. We implemented the three types of frameworks which compose a distributed system. Figure 27.13 illustrates the SIGAL prototype. Currently, it is dedicated to the realization of user services. Our objective is to use the SIGAL ontology to guide the user when formulating a query. For each field to fill in the user interface, some attributes are proposed, along with their description. The user selects the relevant attributes in order to define the query.

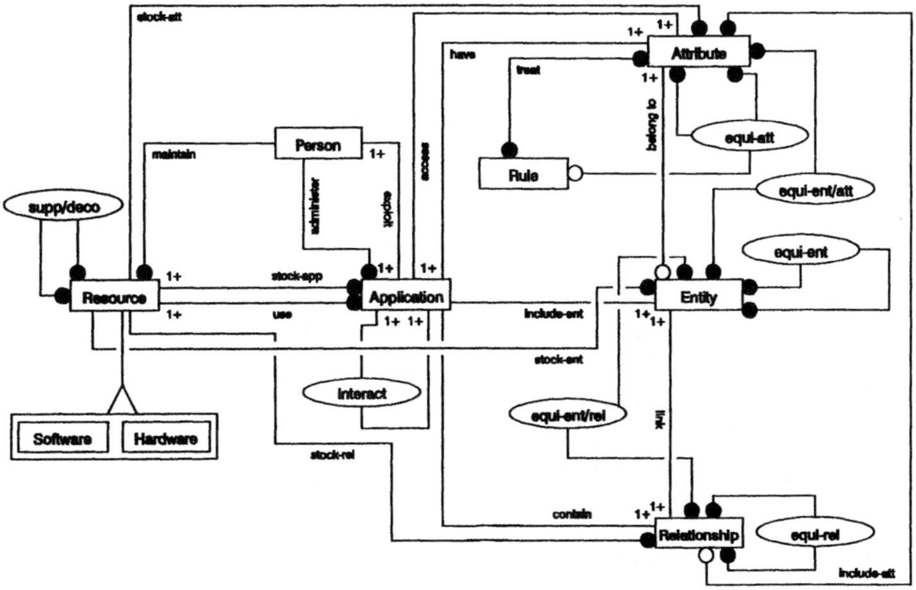


Figure 27.11. Realization operation of user services in the SIGAL environment

Here are relevant characteristics of this prototype:

1. The client applet represents the client framework and runs on the user's computer. This applet, which integrates a JDBC driver, is able to invoke the server framework services and to send user requests to the local-source frameworks after their specifications are obtained from the server framework.
2. The server framework is developed on a Web site. This framework uses a JDBC driver to connect the Ontology Base.
3. VisiBroker for Java is used to establish communications between the client applets, the server frameworks, and the local-source frameworks.
4. Microsoft Access database system is used to manage the Ontology Base.

Figure 27.13 also illustrates the implementation of the frameworks. They constitute a distributed system. Hence, these frameworks need to communicate in order to exchange information about user needs, GDL data, etc. Each exchange operation—we have two types of exchanges: (client–server) and (client–local-source)—is associated to a specific object called Interface (each interface corresponds to a Java class) in the CORBA terminology (see OMG 1997). The file called Sigal.idl shows the instructions needed for the communication between the client and the server frameworks. The file is written in OMG's Interface Definition Language (IDL).

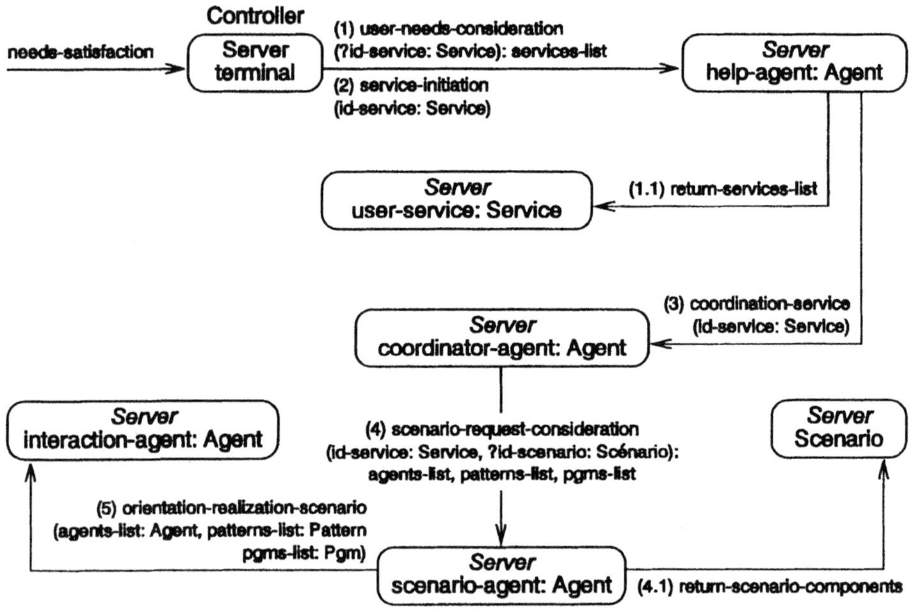


Figure 27.12. Agent interaction graph of the operation *needs-satisfaction*

```
File Sigal.idl
  module Sigal
{
    interface Server
    {
       string locate();
    }
    interface ServerFramework
    {
       Server find(in string user_need);
    }
}
```

When the file Sigal.idl is parsed by the idl2java precompiler, we obtain what is called the client-stub and the skeleton-server. The role of these two programs is to allow programmers to specify all the communication aspects in an abstract way. Such aspects (for instance, How to connect to a server? How to send and get back information? How to detect connection failures? etc.) are handled by VisiBroker for Java. Once the client-stub and the skeleton-server are obtained, we respectively link them to the client and server programs that we have developed. In what follows, we give some examples that illustrate how we have specified software agent-oriented framework concepts in the Java language.

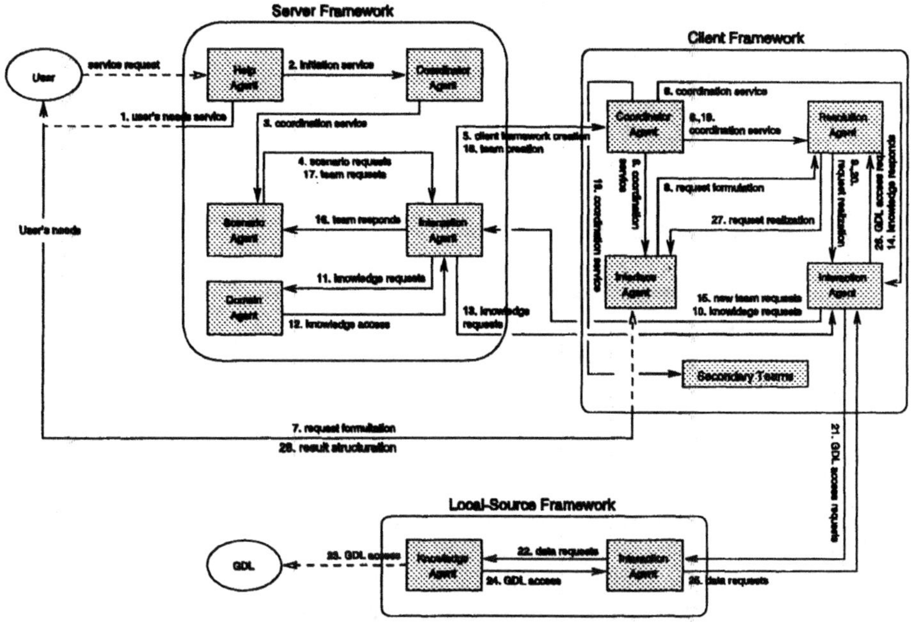


Figure 27.13. Prototype of the SIGAL environment

First, all of the needed agents of a framework are declared as Java classes. For example, the Domain Agent of the server framework public class *Domain-Agent extends Agent*

```
{
    // variables declaration
        String information_return;
        // methods declaration
        // an access method to a database
        String access(String_url, String_request)\\
    {
     Class.forName("sun.jdbc.odbc.JdbcOdbcDriver");
        _connect = DriverManager.getConnection(_url);
        _state = _connect.createStatement();
        _result = _state.executeQuery(_request);
        _result.next();
        return information_return;
```

```
        }
    }
```

Second, all of the communications between frameworks are handled by the VisiBroker for Java. For example, when the client framework requests the local-source framework, the client framework has to bind it before this client framework sends the relevant information.

```
try
{
    // ORB initialization
    CORBA.ORB orb = CORBA.ORB.init(this);
    // locate an object of the local-source framework
    _source_manager= Source.SourceFramework_var.bind("NAME");
    // processing
        .....
}
catch(CORBA.SystemException ex)
{
    ex.printStackTrace();
    System.out.println(ex);
}
```

## 27.5. Conclusion

In this chapter, we presented the major characteristics of the SIGAL project that deals with GDL interoperability on the WWW. A framework is made up of teams of software agents that are able to fulfill services offered to users. These teams can be set up in a unique framework or in several frameworks leading in the latter case to the creation of an interoperable multiframework environment. Interoperability is a process that allows cooperative interactions between distributed and heterogeneous systems. We showed how an interoperable environment can be built with teams of software agents integrated in a framework architecture. We also showed how these components (frameworks, teams of agents, software agents) can play several roles and how scenarios can be used to specify the behavior of these components.

The results are at the intersection of two larger research programs: GDL research as presented at the beginning of this paper, and the C-FOAL research whose objective is to propose a design method for interoperable environments based on software agent-oriented frameworks. Future work involves integrating in a complementary way the best elements of the two solutions experimented in so far by our group to develop a uniform one-stop solution to query heterogeneous GDLs on the web, that is the interoperability-oriented SIGAL project and the data warehouse-oriented SOS-DS project.

## Acknowledgments

The authors wish to thank the collaborators who have provided support and insights for this research, especially Dr. Gilbert Babin from the Computer Science Department and François Létourneau, a graduate student at the Center for Research in Geomatics. The authors also acknowledge the Natural Sciences and Engineering Research Council of Canada and Laval University's Center for Research in Geomatics for their financial support.

## References

Ambite J L, Knoblock C A 1995 Reconciling distributed information sources. *Working Notes of the AAAI Spring Symposium on Information Gathering in Distributed Heterogeneous Environment, Palo Alto*

Bédard Y, Proulx M J, Létourneau F, Moulin B, Maamar Z 1997 Géorépertoires sur Internet: Outils puissants de démocratisation des données à référence spatiale. *Proceedings of Géomatique VI: Un monde accessible.* Canadian Institute of Geomatics, Montreal: 23–37

Cheung W, Hsu C 1996 The Model-Assisted Global Query System for multiple databases in distributed enterprises. *ACM Transactions on Information Systems* 14(4): 421–470

Genesereth M R, Ketchpel A P 1994 Software agents. *Communication of the ACM* 37(7): 48–53

Hsu C 1996 *Enterprise Integration and Modeling: the Metadatabase Approach.* Kluwer Academic Publishers, Boston

Kerschberg L 1997 Knowledge rovers: cooperative intelligent agent support for enterprise information architectures. *Proceedings of International Workshop on Cooperative Information Agents '97 (CIA'97).* Lecture Notes in Artificial Intelligence. Springer Verlag: 77–100

Létourneau F, Bédard Y, Proulx M J 1997 Data warehouse-based system for the optimized selection of spatial data (SOS-DS: Système optimal de sélection des données spatiales couplé à un data warehouse). *D-Lib Magazine: The magazine of Digital Library Research. http://www.dlib.org/dlib/march97/laval/03letourneau.html*

Létourneau F 1998 Analyse du potentiel de l'approche entrepôt de données pour l'intégration des métadonnées d'un ensemble de géorépertoires disponibles sur internet. M.Sc. thesis, Dept. of Geomatics Sciences, Laval University, Québec, Canada

Maamar Z 1997 Contribution à la résolution des problèmes d'interopérabilité des systèmes, une méthode de conception par frameworks orientés-agents logiciels. PhD thesis, Computer Science Department, Laval University, Québec, Canada

Maamar Z, Moulin B 1997a An agent-based approach for intelligent and cooperative systems. *Proceedings of IEEE Knowledge and Data Engineering Exchange Workshop '97 (KDEX'97), Newport Beach*: 19–26

Maamar Z, Moulin B 1997b C-foal: une méthode de conception par frameworks orientés-agents logiciels de systèmes multiagents interopérables. *Proceedings of Third International Symposium on Programming and Systems '97 (ISPS'97, Algiers*: 374–396

Maamar Z, Moulin B, Bédard Y, Babin G 1997 Software agent-oriented frameworks meet georeferenced digital library interoperability. *D-Lib Magazine: The magazine of Digital Library Research. http://www.dlib.org/dlib/september97/laval/09laval.html*

Morgenstern M 1996 Integrating diverse information resources: capturing meaning in a sea of data. In Terashima N, Altman E (eds) *Advanced IT Tools.* Chapman & Hall, London: 471–478

Moulin B, Chaib-draa B 1996 An overview of distributed artificial intelligence. In O'Hare G M P, Jennings N R (eds) *Foundations of Distributed Artificial Intelligence*. Wiley, New York: 3–55

Object Management Group (OMG) 1997 *http://www.omg.org*

Proulx M J, Bédard Y, Létourneau F, Martel C 1997 Catalogage des données spatiales sur le world wide web: Concepts, analyse des sites et présentation d'un géorépertoire personnalisable georep. *Revue Internationale de Géomatique* 7(1): 7–32

Rumbaugh G, Blaha J, Premelani W, Eddy F, Lorensen W 1991 *Object-Oriented Modeling and Design*. Prentice Hall

Sycara K, Decker K, Pannu A, Williamson M, Zeng D 1996 Distributed intelligent agents. *IEEE Expert-Intelligent Systems and Their Applications* 11(6): 36–45 *http://www.cs.cmu.edu/*

Wiederhold G 1989 *The Architecture of Future Information Systems*. Computer Science Department, Stanford University

# 28 INTEROPERABILITY WITH THE EARTH SCIENCE REMOTE ACCESS TOOL (ESRAT)

Robert G. Raskin and Elaine R. Dobinson

## 28.1. Introduction

The Earth Science Remote Access Tool (ESRAT), an enhancement to the Distributed Oceanographic Data System (DODS), is an HTTP-based client–server system that facilitates Internet access to distributed Earth science data. ESRAT addresses the great diversity of data formats used in the Earth sciences by providing a common data model with translators for the models inherent in many standard raster formats. It enables applications normally using local data in a particular format to access remote datasets in that or other supported formats.

ESRAT was developed as a prototype extension of DODS for the Earth Observing System Data and Information System (EOSDIS). It potentially enables a broad range of data sets to be accessible, although at present, the holdings are limited to oceanographic raster data. ESRAT was designed for use across the Internet, with spatiotemporal subsetting carried out on the server side prior to data transmission, reducing bandwidth requirements. This feature is important for EOS applications, where individual data files may exceed 1 gigabyte in size.

ESRAT reflects a distinctly different philosophy from Sequoia 2000 (Dozier 1992), Paradise (DeWitt et al. 1994), and related prototypes that strive to put all data in a common database. It also differs from the standard EOSDIS strategy of storing all data in a common format (HDF-EOS). Experience has shown that Earth Science data providers are resistant to these prescribed approaches. Instead, ESRAT enables applications to be transparent to the data model, format, and location of the data. It permits an application program written for a particular API (such as MATLAB) to access remote data in that or other supported formats. The general model supported has multiple client and server types at distributed locations.

## 28.2. Architecture

The client–server design is displayed in Figure 28.1. The client side includes a Java GUI applet for selection of spatio-temporal regions and display of data coverages. Java was used to increase portability and to enable catalog access from any web browser. The client side also includes a visualization and analysis package linked with software to translate incoming data. The server side includes the data catalogs (currently residing at JPL) and distributed data sites that host DODS servers. A query from a Web browser to the catalog server returns a summary of datasets satisfying the search criteria. The spatial and temporal bounds of available datasets are displayed as layers that can be interactively selected.

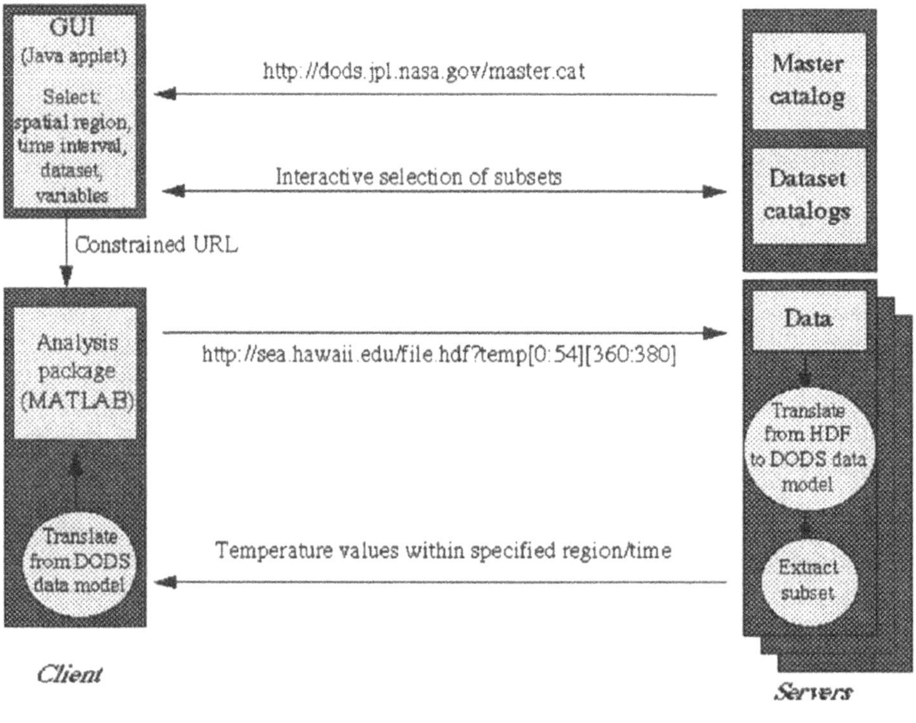


Figure 28.1. ESRAT client-server design. The client includes a Java GUI and a visualization package, linked to translation software. Any single client can access the catalog servers residing at JPL and distributed data archives that host DODS servers.

Figure 28.2 presents the software architecture in greater detail. ESRAT enables users to discover data of interest based upon specified spatial, temporal, and variable name constraints. ESRAT's data catalogs maintain spatial and temporal indexes that allow users to extract only the desired portions of a file, without the need to know about file names, data structures, or array information. The server side includes master catalog, dataset catalog, and *URL creation* servers, implemented as C++ CGI programs. The master catalog contains a list of dataset holdings at local and remote sites and their associated URLs. The dataset catalogs contain the spatiotemporal bounds of each data subset and include examples of swath, grid, and network data. For swath data, each cross-track is individually cataloged in space and time. When a user requests swath data, the server concatenates any contiguous cross-tracks satisfying the selection criteria, and returns a swath polygon to the client. The URL creation CGI assembles a constrained URL that is ultimately passed by the client to the visualization and analysis package via a Netscape helper program.

The client data request is sent in the form of a constrained URL, corresponding to the user's spatiotemporal and variable name constraints. The data server software parses the constrained URL, translates the formatted data, extracts only the desired subset of interest (as specified in the constraint expression), and returns the data as a MIME document (of type binary/x). The data server is described in detail in the following section.

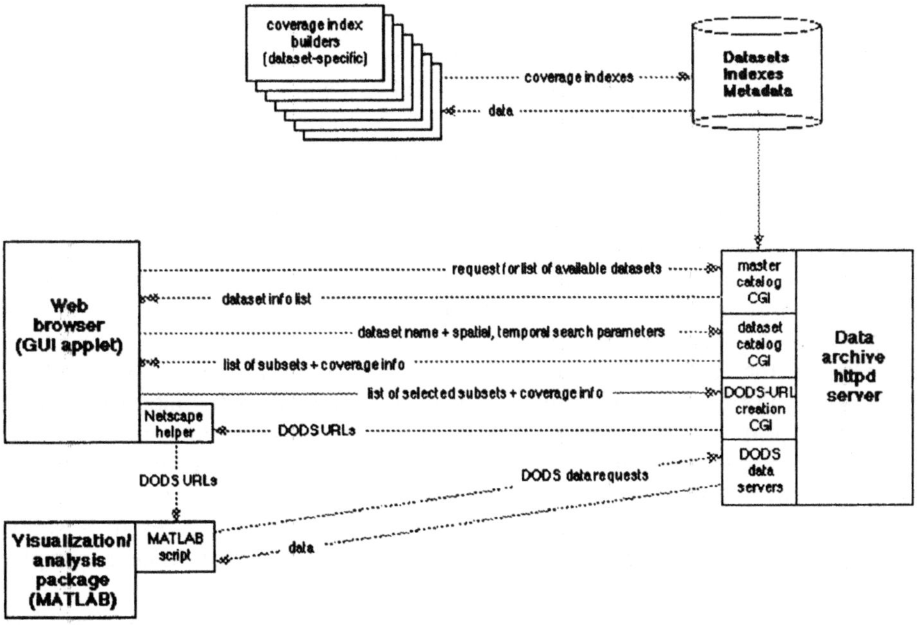


Figure 28.2. Overall software design. Web browsers perform interactive queries with catalog servers implemented as C++ CGIs. DODS data servers send data subsets to DODS client software integrated with MATLAB.

## 28.3. DODS

ESRAT uses DODS as the data server. DODS was developed at the University of Rhode Island and the Massachusetts Institute of Technology (Gallagher and Milkowski 1995; Sgouros 1996) and provides the capabilities to:

- ingest data residing in one of several supported formats;
- translate from the data model of the supported format to the DODS data model consisting of arrays, grids, tables, and structures;
- carry out an HTTP-based data transfer; and
- translate back to a data model and format expected by the client.

Interoperability is achieved by utilizing an intermediate data format. The data model for this format includes the constructor types *Arrays*, *Grids*, *Sequences*, and *Functions*. A Grid is a georeferenced Array and need not refer to equally spaced points, as its name implies. A Sequence holds records in relational tables. A Function is a subclass of a Sequence where independent and dependent variables are explicit. The data model also supports *Structures* to create arbitrary combinations or hierarchies of the basic constructor types, analogous to a *struct* in the C programming language.

Data structure information is stored in the Dataset Descriptor Structure (DDS), a structural declaration similar to the class declaration syntax in C. The server uses the DDS to translate into the DODS data model. Some clients can use this information directly; for example, netCDF can use the DDS to identify variable names. All other metadata are packaged in a Dataset Attribute Structure (DAS). Both the DDS and DAS are ASCII files that are returned with any data request in the form of separate mime documents.

To add a format not presently supported, translation between the format's data model and DODS's data model must be provided. To add a new instance of a currently supported dataset, spatiotemporal mappings must be generated; this is a relatively straightforward task. To add a new client type, the client's API must be expandable to enable linking with DODS remote access and translation software.

One of the lessons learned in the development of ESRAT was that 100% interoperability between all formats is difficult. However, partial interoperability is sufficient for most Earth science applications, as certain translation combinations are not likely, in practice. HDF data structures are richer than those in netCDF, so some mappings from HDF into netCDF lose information. The DODS intermediate data model itself is deficient in its inability to handle common data types such as long integers, short floats, or unsigned variables, and this further hinders complete interoperability.

ESRAT is intended to work with existing application programs. Data visualization and analysis is achieved by linking the client's application package (e.g., MATLAB) with DODS software. Figure 28.3 shows that the user's application program need not require modification. In Figure 28.3a, the standard API is used. With the modified API in Figure 28.3b the user may access either local or remote files, but the application program remains unchanged. The client software enables remote data access and provides translation from the intermediate DODS data structures to those supported by the client. The scheme permits data from multiple sites and in multiple formats to be loaded into a single client application (e.g., a MATLAB plot). Figure 28.4 shows a sample MATLAB output plot, with wind and temperature data overlaid. The wind and temperature data were loaded from different sites and in different formats.

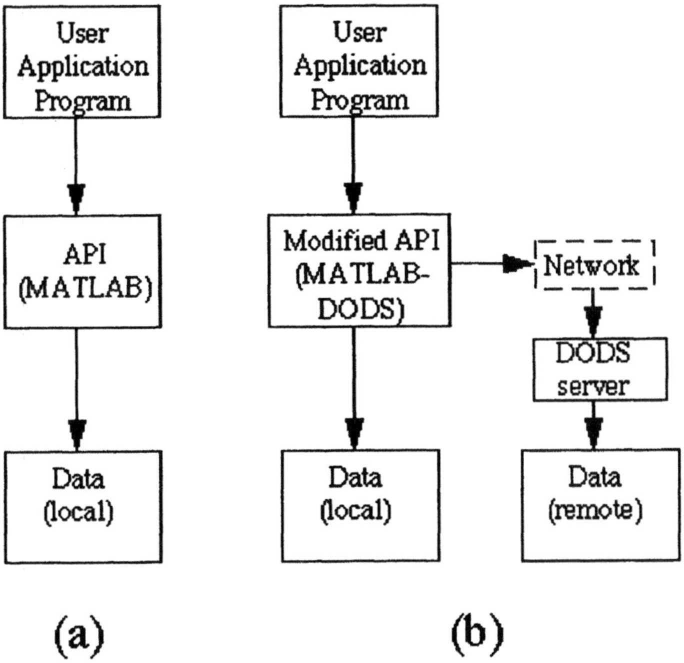


Figure 28.3. Relinking with DODS software. In (a), the user's application program is linked to the standard API to read local data. In (b) the modified API is used that enables either local or remote data access. No change in the application program is required other than relinking it with the modified API.

To read in a local file using MATLAB, a user invokes the MATLAB command:

LOAD filename

When the DODS software is linked, a remote file may be read from MATLAB using the auxiliary command:

LOADDODS URL/filename?variablename[rowrange][columnrange]

The optional text following the question mark constrains the load to the specified variable name, rows, and columns. Boolean operators are available to select multiple variables and constraints in a single statement. ESRAT goes a step further and facilitates selection from a spatiotemporal tool. This frees the user from having to know the file name and the row and column ranges associated with the region and time interval of interest. The ESRAT data catalogs contain this translation information.

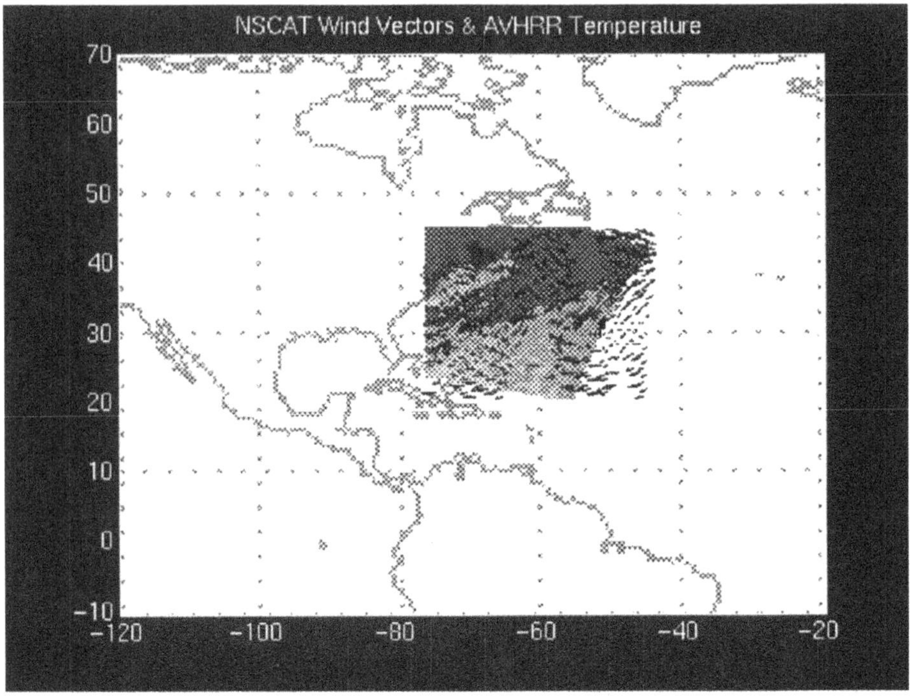


Figure 28.4. Sample MATLAB plot overlaying ocean winds and temperatures. The wind data were loaded from a JPL archive and the temperature data were loaded from a University of Rhode Island archive.

## 28.4. Supported data models

Figure 28.5 shows the currently supported client and server formats and data models. DODS can ingest data in netCDF, DSP, JGOFS, and MATLAB formats. Support for reading HDF data was added as part of this prototype work and has been incorporated as an official DODS capability. The addition of HDF involved mapping the HDF data structures and data types to their corresponding representations in the DODS data model. Server support for HDF Vgroups is now under development.

## 28.5. Spatiotemporal subsetting using constrained URLs

The interactive selection procedure between the GUI and the catalog servers culminates with the construction of a constrained URL. The constraints are transferred as characters following the ? in the URL. In the example of Figure 28.1, the constraint restricts the downloaded data to be the variable *temperature* with rows

0–54 and columns 360–380. These constraints are created by the URL creation CGI using the spatiotemporal selections and the mappings contained in the data catalogs.

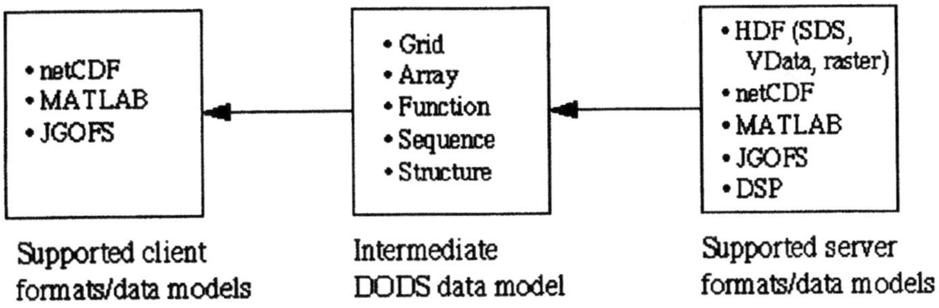


Figure 28.5. Supported client and server data formats and the intermediate data model. DODS provides translation software for converting between the relevant representations.

The type of subsetting operators available depends on the type of the data involved. For Arrays and Grids, spatial subsetting produces a subarray (of the same number of dimensions) based upon a row–column constraint. For Sequences, records can be selected by value using an SQL-like syntax.

## 28.6. Relationship to EOSDIS and HDF-EOS

This work was carried out by the Physical Oceanography Distributed Active Archive Center (PO.DAAC), one of eight DAACs that will archive EOSDIS data. ESRAT was developed as a potential alternative to the standard EOSDIS client interface, the Java Earth Science Tool (JEST). JEST does not provide immediate access to the EOS data, as selected datasets must be ordered and subsequently downloaded via FTP. In contrast, ESRAT enables direct data access, via HTTP transfer into the user's visualization and analysis package.

EOSDIS has adopted HDF as its standard format and created an enhanced version, HDF-EOS, that most EOS data providers are obligated to use for new data products, HDF-EOS adds three spatial data types—the swath, grid, and point (or network), which are implemented as VGroups in HDF. The translation between HDF-EOS high-level and HDF VGroup structures (known as *structural metadata*) is stored in an HDF global attribute *structmetadata.0*. Standard HDF tools can access HDF-EOS data files, but will be unable to carry out spatial operations without an understanding of the structural metadata. At present, only test data are available in the HDF-EOS format; consequently, our ESRAT data catalogs include examples of close surrogates of the HDF-EOS swath, point, and grid data types.

EOSDIS was designed to be a distributed archive that appears to the user as a single data system. Therefore, HDF-EOS uses standard representions for any metadata that will be used for system-wide searches. The searchable metadata reside in the HDF global attribute *coremetadata.0* using a standard naming convention. For example, spatial boundaries can be specified in terms of either a polygon (*G-Ring*), a bounding rectangle, a circle of specified center and radius, or a single point. If a bounding rectangle is used, the bounds must have names such as *LeftBoundingRectangle*. Other non-searchable metadata will be stored in the HDF global attribute *archivemetadata.0*. ESRAT is being designed to read the coremetadata.0 and archivemetadata.0 files and transfer this information into the DAS and DDS files, as appropriate.

## 28.7. Conclusions and future work

In conclusion, interoperability for EOS-type datasets was successfully achieved without using a prescribed format or database schema. The PO.DAAC is now looking into incorporating ESRAT as a standard method of accessing DAAC data.

In the near future, we expect to convert the entire tool to Java. Java provides an unprecedented level of interoperability because both its bytecode and its internal representation of numbers are platform-independent. This feature permits Java's data model (in terms of arrays, classes, and streams) to be used as the abstract intermediate data model. Subclasses for spatial data types will be developed. This approach has the advantage of allowing users to package executable code as metadata that can accompany the dataset (Rew 1997). The code might be used for geolocation, interpolation, development of spatially-dependent error estimates, creation of derived variables, or other purposes. The use of Java also will increase the number of supported clients and servers by leveraging on the work of software vendors developing Java translators and interfaces.

Currently, HDF, netCDF, PV-WAVE, and other vendors have Java class libraries for their data types. Each uses similar representations but a slightly different collection of access methods. Use of these interfaces greatly reduces the amount of translation work required in the DODS software.

We also plan to convert our catalog databases to a commercial object–relational database system (ORDBMS) with spatial data type extensions. This will provide greater GIS functionality, including improved support for spatial operations on swaths. A Java interface using JDBC will be used to access the catalog entries. A global quadtree representation of the catalog entries is being designed to provide efficient global searches when the system is expanded to include large volumes of data.

ESRAT currently lacks any ability to ingest vector data. This deficiency is hoped to be remedied in a subsequent phase of the project, if ESRAT becomes a standard interface for the Physical Oceanography DAAC.

## Acknowledgment

This work was funded initially by the EOSDIS Prototype Office and is continuing under a NASA ESIP award. This work was carried out at the Jet Propulsion Laboratory of California Institute of Technology under contract to the U.S.

Government. Software support was provided by Todd Karakashian, Isaac Henry, and Robert Morris.

## References

DeWitt D J, Kabra N, Luo J, Patel J M, Yu J-B 1994 Client–server Paradise. *Proceedings of the 20th VLDB Conference, Santiago, Chile*

Dozier J 1992 How Sequoia 2000 addresses issues in data information systems for global change. Technical Report 94-14. Sequioa 2000, University of California, Berkeley

Gallagher J, Milkowski G 1995 Data transport within the distributed oceanographic data system. *4th International World Wide Web Conference Proceedings*: 691–700.

Rew R 1997 Georeferencing with Java: An example of executable metadata. *http://www.unidata.ucar.edu/cgi-bin/mfs/01/staff/russ/MapGeoGrid/ams98.html*

Sgouros T 1996 DODS User Guide Version 1.0. *http://dods.gso.uri.edu/DODS/user/guide-html/guide.html*

# 29 MULTI-SERVER INTERNET GIS: STANDARDIZATION AND PRACTICAL EXPERIENCES

Carel van den Berg, Frank Tuijnman, Tom Vijlbrief, Co Meijer, Peter van Oosterom, and Harry Uitermark

## 29.1. Introduction

An approach to an open infrastructure for geographic information on the Internet is presented in this chapter. This infrastructure enables data providers to publish their data independently, while enabling end users to access data from several providers simultaneously, and integrate the data locally in a geographic browser. Our goal is that an end user finds accessing geographic information in this environment as easy as if he or she were working with a state-of-the-art GIS package with all data of interest on the local computer. The key elements that are required are a common format for publishing metadata on each server, a common SQL derived query protocol, standard transfer file formats, and standard certificate-based authentication procedures, for access control and (optionally) billing. An experiment with this approach has been carried out, with three data providers in the Netherlands: the Dutch Cadastre, the municipality of Almere, and the cable-TV company Casema. This chapter presents the major design decisions, the choices for the prototype environment, and the relationship to ongoing specification and standardization processes for geographic data. In particular the relationships with the proposed European CEN standards, and the recently accepted specifications from the OpenGIS consortium, are described.

The current wave of GIS software for the Internet is based either on the file downloading paradigm, or on the picture paradigm (presenting a map as a JPEG picture), or on the client–server paradigm (creating a closed interaction between a client and a single server). None of these approaches can capitalize on the main potential of the Internet for integrated and easy access to a vast amount of geographic information on various servers. In addition, the interaction protocol between client and server is typically proprietary, which means that someone who browses geographic information needs software from the same vendor as is used by the publisher.

To make GIS popular on the Internet one needs to create for geographic information the same level of uniformity as the World Wide Web has done for text. The brilliance of the World Wide Web lies in the combination of the hypertext model with the Internet, together with a formatting standard for text (HTML). The hypertext model, however, does not work for geographic data: it is not particularly useful to jump from one map to the next. So another basic metaphor has to be used.

The standard model for geographic data on the computer is the layer model. The layer model of geographic information systems relates to a paper map like the hypertext model relates to text on paper. To make geographic data on the Internet attractive one has to set a standard for the layer model, so that one can obtain a topographic layer from one source, a pollution layer from another, and a property layer from a third source, and dynamically merge them in a geo-web browser. To achieve this the following standard protocols have to be defined (in addition to support for authentication and billing): (1) a method to inquire which layers are available, and what data they contain; (2) a method to request a particular layer from a particular source; and (3) a standard format for returning this information. Many aspects of these protocols are subject to ongoing standardization efforts. The format for returning geographic information is essentially a description of a transfer file format.

The method for querying the metadata has a clear relationship with metadata standards (CEN-META 1996), and clearinghouse-related activities (Absil van de Kieft and Kok 1997). The protocol for querying geographic information is new. CEN has acknowledged the need for something like this; see (CEN-QUERY 1997), which specifies names and semantics of required spatial operators. The closest relevant specification for the query protocol is the OGC specification for SQL with simple geometric features. Despite the fact that it is relatively easy to identify protocols that can be used to address part of the problem, no comprehensive proposal exists so far that can be used to achieve open access to geographic data publishers on the Internet.

## 29.2. Architecture

Our approach to the design of the metadata structure, the query formalism, and the format for the returned data is based on the object-relational formalism, where we include geometric data types as attributes within a relational table. The motivation for this approach is based on the following considerations:

1. Object-relational database management systems with support for geographic data are now available from most major vendors (Informix, CA-OpenIngres, Oracle). Even if one wants to provide access to a file based collection of geographic data it is not difficult to implement a limited selective capability on top of it, though of course performing such selections will cost more time. So it is technically feasible for any organization to implement this functionality.

2. The object-relational model is currently the only widely available formalism that can deal with geographic data, and in which all three required elements (a

metadata structure, a query formalism, and a format for returned data) are defined in an integrated manner. This is essential from a technical point of view: the metadata do not only describe the data, but also have to provide the *words* that can be used in queries, and it has to be clear which words can replace which syntactic elements in a query. The returned data has to be understood as a response to the query, so there has to be a well defined relationship between the semantics of queries, and the actual data that are returned.

3. It can be mapped easily to the stateless HTTP protocol, because SQL is also stateless (meaning that any request can be handled independently from previous or subsequent requests).
4. It ensures that the browser requires only knowledge about what data are available, rather than detailed knowledge about file naming conventions, and tiles.

In mapping extended SQL to HTTP, in such a way that it can be used effectively for geographic data publication over the Internet, a number of issues have to be tackled:

1. The geometric types and their representation.
2. The metadata structure.
3. The query and the way it is encoded in a URL.
4. The format of the returned data and data compression (efficient data transfer over the Internet).
5. Authentication, and billing mechanisms to support commercial geographic data publication.

### *29.2.1. Geometric types*

A common understanding of geometric types is essential for exchange of geographic information. All file transfer standards that currently exist define their own geometric constructs, and so do the proprietary data formats. Examples are the new CEN standards (CEN-SCHEMA 1996; CEN-TRANSF 1997), the OpenGIS Simple Features specification (Buehler and McKee 1996), and national standards, such as NEN1878 (1993). The basic types that we choose to support are point, polyline, and polygon. A point is represented by its coordinates, a polyline and a polygon as a list of points. In the current system polylines and polygons consists of straight line segments only. The NEN1878 standard, commonly used in the Netherlands for the exchange of geographic data, allows polylines and polygons that are defined as sequences of segments, where each segment can have its own interpolation method. Due to the object-relational formalism in which these types are embedded, it is relatively straightforward to include topological information.

A practical problem that anyone implementing a system like this faces is whether the client software should be sufficiently intelligent to be able to handle also the most complex structures, or whether the data publisher simplifies the dataset, in order to enable more simple client software to use this data. We believe that the only practical approach here is that the data publisher (typically a large organization) takes the trouble to offer information in a number of ways, targeted to both advanced software systems, and simple ones. For example, the data publisher has to make a decision whether or not the client software has to understand topology. Parcels can be represented as a table of lines with references to left and right polygons, and a chain topology (in the Netherlands NPR3611 1997 is a commonly used standard). Parcels can also be presented by the data publisher as a table with polygons with holes. In the first case the client software needs to understand the meaning of the attributes that encode the topological structure, or simply view a line image. In the second case the client software can pass the polygons almost directly to a graphics library for drawing.

### *29.2.2. The metadata*

In our view the primary role of a data publisher is to enable easy access to information, and not to define a particular application. In particular, visualization characteristics, and the structure of the legend, should be customizable (in our implementation: in the HTML page from which the applet is started), without having to change the server. In this way a single data server can support a wide range of applications, and applications may use several servers. On the other hand, someone without a customized browser should be able to view the data in an acceptable manner. For the metadata that we use here we distinguish three levels:

1. The definition of the structure of the geographic data;
2. The definition of the visualization characteristics of a layer; and
3. The aggregation of a number of layers to a single layer, or map, from the user perspective.

The first level describes which tables are available, which attributes are defined and what their types are, the scale for which the information is intended, which projection has been used, and which attributes can be used in the *where* clause of a query. It gives the minimal information necessary for someone to query the data, and to display it. The second level determines the use of colors, line styles, fill patterns, and symbols. It is useful to have default values at the server side, for those situations where someone just wants to take a look. In most practical cases, however, choice of symbols and colors will have to be defined at the client side, by the client application. This makes it possible to define different visualization characteristics for different applications; and applications that use data from several servers will almost always have to define their own visualization characteristics, to avoid conflicting or confusing use of colors, symbols, and line styles. The third level, defining maps as composite layers, is also something that is required both at the server, and at the client side. At the server side a good hierarchical structure of the layers is needed to support browsing. At the client side one can define maps that are composed of layers from several sources, and that are specific to a particular application. It is worth noting that the metadata that we require here are far less than is typically included in a geographic metadata catalog (CEN-META 1996). The information included here is the minimal information necessary to be able to retrieve and visualize information.

### 29.2.3. The query model

The query model is essentially based on spatial SQL, without joins. The reason to exclude joins is that this makes it relatively easy for the server administrator to control the load on the database. If (spatial) joins are required, one can circumvent this limitation by including a view in the server database that constructs the desired join. So the client designer has the freedom to select a subset from a relation, and the designer of the server has to decide (by defining suitable views), which joins on tables are allowed. We believe that this constitutes a fair compromise between flexibility at the client side, and the need to guarantee stable operation at the server side. To use the HTTP protocol the queries have to be encoded in a URL. Below the syntax for that encoding is given (capitals indicate non-terminal symbols for which we have not included the remaining production rules):

<Query>:: <prefix> ? <query>
<prefix>:: http://<sitename>/<directory>/<program>
<query> :: coordsys=<COORDSYS> & database = <DATABASENAME> & format=<FILEFORMAT> & relation= <RELATIONNAME> & attributes=<attributelist> & where=<whereclause>
<whereclause> :: <spatialclause> | <spatialclause> & <ANDOR> & <conditionlist>
<spatialclause> :: WRectangle.intersects( <x1>,<y1>,<x2>,y2> )
<conditionlist> :: <condition> | <condition> & <ANDOR> & <conditionlist>
<condition> :: <ATTRIBUTENAME> <OPERATOR> <VALUE>

The example below shows a query which requests four named attributes (magma_oid, geo_bbox, geo_pgn, owners), for all parcels in *percelen* relation in the *kad4* database, within the selected region, where the owner's names prefix is between *oost* and *oostf*. The coordinates are given in RDM (plane state coordinates in the Netherlands). As a result a list of tuples will be returned, satisfying the *where* clause.

//ooa.kadaster.nl/cgi-bin/magma?
coordsys=rdm&format=javabin&database=kad4&relation=percelen&
attributes={magma_oid,geo_bbox,geo_pgn,owners}&
where=WRectangle.intersects(189000,485000,192000,488000)&and&
owners>='oost'&and&owners<='oostf'

### 29.2.4. The returned data format and data compression

The returned data format consists of a list of tuples, in a single file. Compared to other popular file formats, such as the shape file format by ESRI and the MapInfo exchange format, the main difference is that all information is in just one file: geometric data as well as the non-geometric attributes. An important aspect of the returned data format, not required by most standards, is that each object should have an object identifier, or some other unique key. This is important to enable the browser to determine whether an object has already been received. Receiving objects twice can easily occur after panning, when objects that overlap the previous viewing region may be transferred again. The other reason is that we believe that each object on the map should be traceable to the supplier, and that it should be possible to make inquiries about the object. Geographic data area often based on 8-byte double-precision floating-point numbers. Due to the spatial cohesion a substantial reduction in size can be achieved. In large-scale mapping, differences between points within a polyline can almost always be encoded with 16 bits, and usually with less. A differential encoding scheme has been used, which achieves good compression, with relatively little computing overhead.

### *29.2.5. Commercial services*

Commercial services for geographic data require a mechanism for either authentication, or for direct payment. Authentication is feasible with current Internet technology. The most popular method is still the use of passwords and login names. For a system where one of the main attractions is the ability to access several providers simultaneously, it is cumbersome to have to use (and remember, and administrate) login names and passwords. The use of certificates is essential for three reasons:

1. The data publisher can be confident that data are transferred to the proper users;
2. The user doesn't have to type passwords each time he or she obtains information from another server;
3. The user can trust that the information originates indeed from the source, and not from some impostor.

Though technically feasible, a major commercial problem is that the pricing models that are currently used for geographic data are not targeted to the infrequent and non-professional user. And it is not easy to see how the price structure should be changed. In the Netherlands, for example, a utility company will pay an annual fee to obtain updates of the large-scale maps of the area in which it has cables. If the same map would be offered to infrequent users on the Internet for a low fee per query, the utility companies would be tempted to stop their current annual subscription, and request the information on the Internet instead.

## 29.3. Relationship to standardization activities

The architecture described here can be used to support various standardization approaches. We discuss here two of them: the CEN standards, and the Open GIS Consortium standards. The Open GIS Consortium has defined semantics for geometric types, and (among others) bindings to Extended SQL and to CORBA. To place the Lava/Magma architecture in the Extended/SQL framework all that has to be done is to define the format of the URL that encodes the query string, and to include a description of a compression mechanism (though this step could be omitted). The advantage of extending the current standards in this manner is that an easy-to-use Internet interface would be available to all geographic databases that conform to the OpenGIS standard. A second advantage is that it is possible to write simple interface software mapping the standard OpenGIS model to the model that is actually implemented in the database. A third advantage is (provided that only an HTTP mapping for the extended SQL is defined) that a uniform Internet interface can be created to geographic databases that implement the OpenGIS SQL interface, and those that implement the OpenGIS Extended SQL interface.

The main weaknesses of revision 0 of the OpenGIS specification for CORBA are that the procedure to obtain the metadata about a database is inadequately specified, and the absence of a compression mechanism. In addition, we believe that the CORBA interface is unnecessarily complex for simple retrieval applications.

The new CEN-SCHEMA (1996) standard proposals describe schemas in Express, an object-oriented specification language, for various types of geographic data sets, with a complete topology on one extreme, and *spaghetti* on the other. Mapping these to the architecture described here requires a mapping from the object-oriented Express language to the object-relational paradigm. For a non-topologic data set this is straightforward: the Express types describing the geometry become types in the object-relational paradigm. For topologic datasets the relationships that are expressed directly in the Express formalism would have to be mapped to references through foreign keys. This can be done in a number of ways, and therefore would have to be explicitly standardized. For the returned data format the CEN-TRANSF (1997) standard can be used. The query syntax and semantics would have to be described from scratch.

## 29.4. System architecture

The Lava/Magma system, developed by PGS, implements the architecture with two main components: Lava, a Java-based map viewer and presentation software, and Magma, a middleware program that interfaces with an NTTP server on one side, and a geospatial database on the other side. Figure 29.1 depicts this architecture. The system supports both raster and vector data. Raster images, such as aerial photographs, are stored as tiles in the database. Upon a request from Lava a single JPEG image is constructed without tiles, and with the right number of pixels for the display. Note that this requires that Lava requests the image for the actual screen size of the image. Currently, the images are not cached, so that each zoom or pan operation results in a request to the server.

Vector data are cached by the Lava map viewer. The cache administration maintains a representation of the areas that have already been viewed. So once a particular area has been viewed, a second viewing (after panning or zooming), requires no request to the server.

## 29.5. Trial experiences

A data integration trial with the PGS products Lava and Magma has been carried out in the province of Flevoland, where three organizations, the municipality of Almere, the cable-TV company Casema, and the Cadastre, have implemented a system that allows them to access the data of the other parties directly. Three servers are used, one operated by the Casema, one by the Cadastre, and one by PGS. The data published by the Cadastre consists of the parcel boundaries. The data from the municipality consists of a large-scale topographic map and large-scale topographic plans for new neighborhoods. Almere builds about 3000 new houses annually. The Casema has published the cable locations, both planned and existing. For browsing, the Java-based Lava GIS browser is used, and for the server the Magma GeoData publisher is used to interface between the HTTP requests and various geographic datastores: CA-OpenIngres 2.0 (1995) for the Cadastre, Illustra for the topographic data, and flat DXF files for the Casema. The three partners have installed their own servers, connected to the Internet, providing spatial data on request.

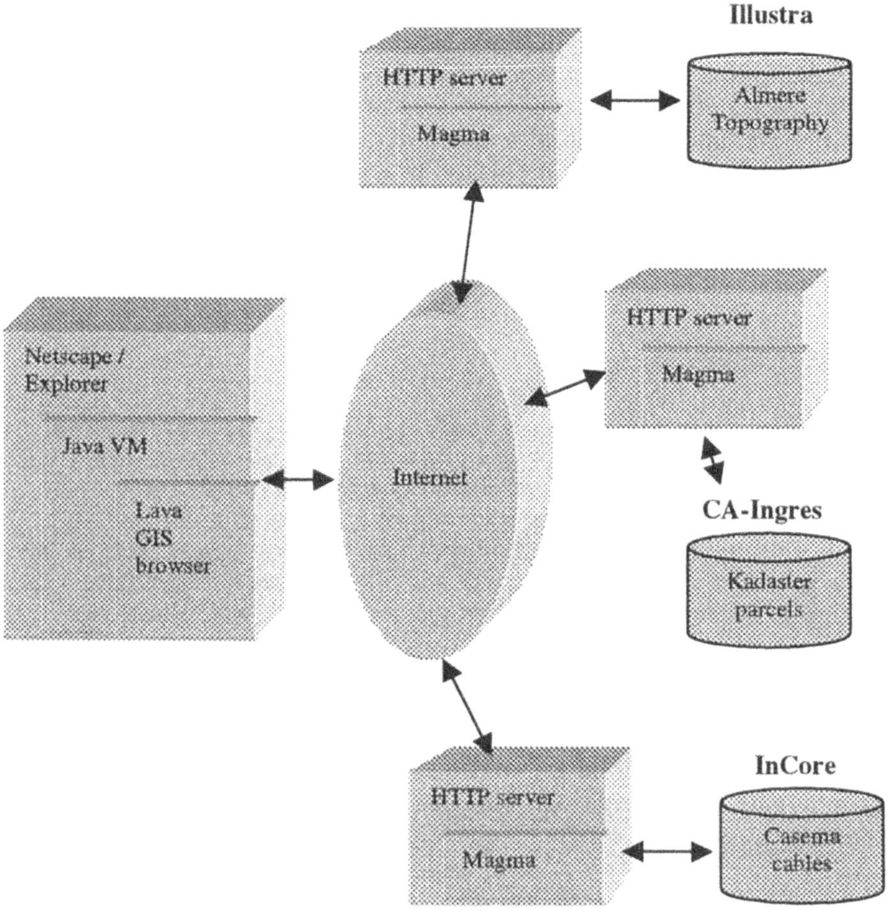


Figure 29.1. Overview of the GeoShop Magma/Lava architecture

In this trial the feasibility of dynamic data integration over the Internet has been demonstrated, supporting both raster and vector data at the client side, and using a

Java-based GIS browser to give everyone direct access. The user interface offered by this system is comparable to a somewhat simplified version of GEO++ (Vijlbrief and van Oosterom 1992). Figure 29.2 shows the Lava user interface as it appears when invoked from *http://ooa.kadaster.nl/*. At the bottom of the screen a message line (left) and the *x,y* position of the mouse (right) are displayed. The overview map provides orientation for the user, and allows selecting an area for viewing. From the icon bar above the main map zoom, pan, and object selection operations can be activated. Object selection results in a window that shows all the attributes which are requested from the server that has that object; see Figure 29.3.

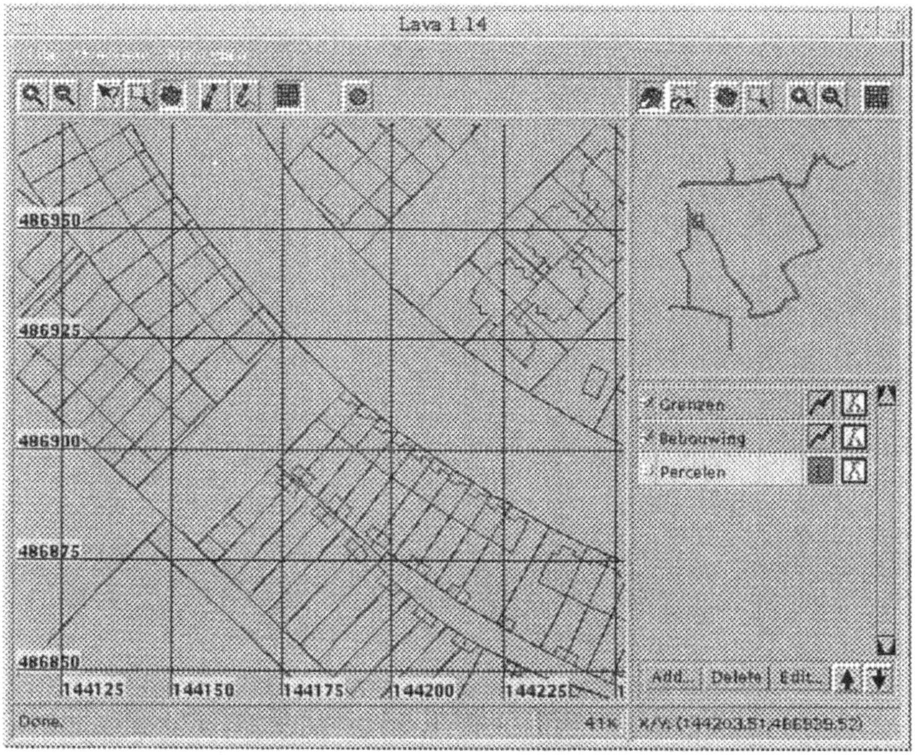


Figure 29.2. The standard GeoShop interface as started from the Kadaster

An important part of the user interface is the legend, to the right of the main map panel. The legend indicates the layers that constitute the map, shows their visualization, and shows an icon to indicate the server from which the layer was obtained. The user can add layers, delete layers, or temporarily switch off visualization of a layer. The ordering of layers can be changed with up and down arrows. After choosing the *Add* button a new screen pops up. In this example a user has selected the database *Almere* from the server *magma.pgs.nl*, and the layer *Gras*. When combining information from several servers, the user has to be able to configure locally colors, line styles, and symbols, to avoid conflicting usage. This is possible by pushing the *Edit* button, which brings up screens for text labels or for line objects. These screens also enables a user to enter additional selection criteria, such as "source='GEOGR' and quality= 'T1'", which selects a subset from the parcel boundaries, shown as black lines in Figure 29.4.

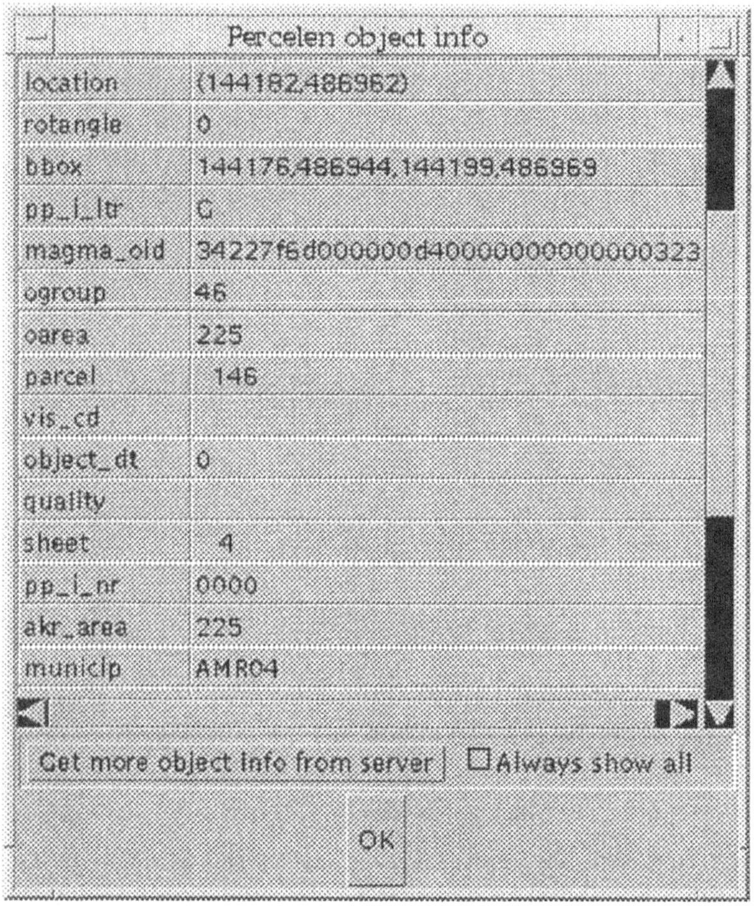


Figure 29.3. Getting more information for a certain object

The trial has demonstrated that it is possible to work interactively with these maps. A typical picture, like the one in Figure 29.4, requires about 100Kb of vector data. With an ordinary telephone modem this takes only 40 seconds, and with an ISDN line about 15 seconds. This is already acceptable. In the near future this will be improved upon substantially because the Dutch utility companies are committed to offer cable modem Internet access to 70% of the country by the year 2000. By then the performance of the server will become more crucial than the transmission delay.

Two non-information-technology aspects contributed to the success of this trial. The first was that all geographic data had a high positional accuracy, and were surveyed with respect to the same reference points (corners of buildings, etc.). As a result, the fences from the topographic map nicely overlap with the cadastral boundaries, and the relative distance between cables (on the Casema layer), and houses (on the topographic layer) can be trusted. So even though these organizations make their own maps, they already collaborate to ensure that these maps can be combined. If this is not the case, a dynamically-combined map becomes far more difficult to interpret, and may easily mislead a casual user.

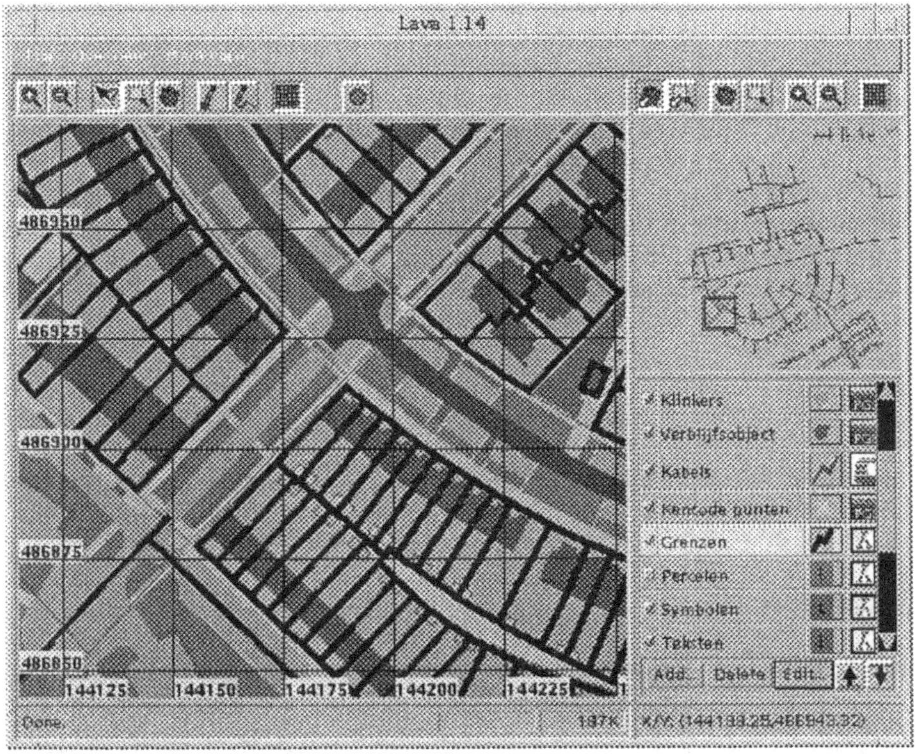


Figure 29.4. Selected parcel boundaries (source='GEOGR' and quality='T1') in bold black lines

A second aspect that contributed to the success of the pilot was that there was no overlap in *themes*. This was avoided by using from the Cadastral map only the property boundaries, and not the buildings, that are present on the Cadastral map on paper. If the houses from the Cadastral maps would have been included as well, they would have overlapped mostly with the buildings on the topographic map, but not completely, due to different classifications, and different criteria for deciding whether a house or a building should be on the map. This is something that does not represent a problem for the professional user, but would be confusing for the casual users, who would be confronted with subtly different meanings attributed to names or concepts that would beconsidered synonyms. If casual users are to make effective use of this type of map, their meaning should be well-standardized at a national level. In the Netherlands a new national standard has been proposed (NEN3610 1995) for modeling terrain information that will be used to address this problem.

## 29.6. Conclusion

Three related protocols need to be defined to support open access to geographic data over the Internet: (1) a method to inquire which geographic data are available from a particular publisher; (2) a method to request a particular layer from a particular publisher; and (3) a standard format for returning this information. The solution proposed in this chapter is based on the object-relational model with geometric types. Simple SQL requests are encoded as URLs for HTTP, and the result is returned as a list of tuples, with geometric attributes.

The trial, with the Lava/Magma software from PGS, has demonstrated that it is effectively possible to implement this approach, and to achieve in this manner open, easy, and integrated access to data from different organizations. As such it can be the basis for a public infrastructure that allows each organization to publish its data independently, while at the same time it enables clients, both professionals and citizens, to have integrated access through the Internet to all available geographic information.

An effective usage of such an infrastructure by infrequent users requires that the maps are made with a well-defined (and sufficiently high) accuracy, and that the meaning of the maps, relative to the actual situation in the terrain, is standardized as much as possible.

## References

Absil van de Kieft I, Kok B 1997 The development of a geo metadata service for the Netherlands. *Third Joint European Conference & Exhibition on Geographical Information (JEC-GI'97)*: 1165–1176

Buehler K, McKee L 1996 *The OpenGIS Guide: Introduction to Interoperable Geoprocessing*. Technical Report 96-001. Open GIS Consortium, Wayland. *http://www.opengis.org/*

CEN-META 1996 CEN/TC 287. Geographic information—data description—metadata. Technical Report draft prEN 12657. CEN, Brussels

CEN-QUERY 1997 CEN/TC 287. Geographic information—processing—query and update. Technical Report draft prEN 12660. CEN, Brussels

CEN-SCHEMA 1996 CEN/TC 287. Geographic information—data description—spatial schema. Technical Report final draft prENV 12160. CEN, Brussels

CEN-TRANSF 1997 CEN/TC 287. Geographic information—data description—transfer. Technical Report draft prEN 12658. CEN, Brussels

INGRES 1995 *CA-OpenIngres. Object Management Extension User's Guide, release 1.1*

NEN1878 1993 Exchange format for data about spatial objects that are related to the earth surface (in Dutch). Nederlands Normalisatie-instituut

NEN3610 1995 Terreinmodel vastgoed. Terms, definitions and general rules for the classification and coding of spatial objects related to the earth surface (in Dutch). Nederlands Normalisatie-instituut

NPR3611 1997 Practical guidance for NEN 1878 and NEN 3610 (in Dutch). Bijlage B - Ketting-topologie. Nederlands Normalisatie-instituut

Vijlbrief T, van Oosterom P 1992 The GEO++ system: An extensible GIS. *Proceedings of the 5th International Symposium on Spatial Data Handling, Charleston, South Carolina*: 40–50

# 30 A VISUAL GEOSPATIAL INFORMATION SERVER (VGIS) PROVIDING TRANSPARENT ACCESS TO HETEROGENEOUS SOURCES

Chuanchu Wang and Jiankang Wu

## 30.1. Introduction

Geospatial information has been realized as a valuable resource for more and more organizations and departments. Recently, developments of new techniques in worldwide high-speed network communications and applications make it increasingly possible for computer-based information systems to access a diversity of autonomous and remote information sources. The availability of various Internet information servers and browsing tools further promote information sharing across departmental, organizational, and national boundaries. However, since the development of the earliest recording and storage technologies in the GIS field, considerable efforts have had to be placed on data format translations and standardized data development so that applications in one organization can make use of geospatial data maintained by other organizations. This is because of the proprietary data models and representation methods of various geospatial information systems used by different communities. With the increasing removal of proprietary restrictions on geospatial data sources by providers and with the information consumer's increasing need to import geodata, the interoperability of heterogeneous geospatial data sources must be achieved.

### *30.1.1. Background*

Recently, many studies and practices have been made aiming at the interoperability of heterogeneous information data sources and systems. Standardization of geospatial data format is the first attempt. The US Spatial Data Transfer Standard (SDTS), the Spatial Data Transfer Specifications (SDTS) by ISO, and the Digital Geographic Information Exchange Standard (DIGEST) are all examples of such efforts. However it is very unlikely that the industry will move to a single standard because proprietary commercial data products have already gained momentum in the marketplace, and there are various needs of those legacy applications. Secondly, some projects have been carried out to develop an open geographic information system to bridge the gaps. An open geographic information system represents an evolution from traditional GIS solutions, in which proprietary data models and monolithic software functions are made interoperable and extensible. One example of OGIS systems is the Open Geographic Datastore Interface (OGDI) by LAS Inc (1996). OGDI provides an application programming interface (API) that uses a standardized access method in conjunction with a GIS software package (the application) and various geospatial data products. OGDI uses client/server architecture to facilitate the dissemination of geospatial data over any TCP/IP network, and a driver-oriented approach to facilitate access to various geospatial data products and formats. GeoLens (Shklar et al. 1997) is another project to make NASA's remote sensing imagery and other geospatial data available to the public on the Internet. It provides a framework for resolving network and data interoperability issues. The Catalog Server is the most important part of this system, which is designed to provide seamless interoperability between different metadata standards. A Metadata Browser then can work with the Catalog Server to aid the user to discover and retrieve geospatial data. However, the current system needs many improvements in order to provide GIS data consumers with transparent access to heterogeneous geospatial data stores over the Internet. The Visual Geospatial Information Server (VGIS) system that will be discussed in this paper is our new effort based on OpenGIS concepts. The Open GIS Consortium Inc (OGC) is an association dedicated to promoting new technical and commercial approaches to interoperable geoprocessing (Buehler and McKee 1996). VGIS follows the specification concerning the definitions of geographic features, coverage, data structure, processing models, etc. defined by OGC (OGIS 1997).

### *30.1.2. Characteristics of VGIS*

The Visual Geospatial Information Server (VGIS) is designed to provide one-step service for public users who want to access heterogeneous data sources transparently over the network. The design objectives of VGIS are as follows:

- to integrate geographic information contained in heterogeneous data stores whose incompatible formats and data structures have been preventing interoperability; and
- to preserve the value of legacy geoprocessing systems and legacy geodata while incorporating new geoprocessing capabilities and geodata sources.

Liu and Pu (1997) have summarized the general requirements for a large-scale interoperable database system as USECA properties:

- *Uniform* access to heterogeneous information sources;
- *Scalability* to the growing number of information sources;
- *Evolution* and *Composability* of software and information sources; and

- *Autonomy* of participants, both information consumers and information producers.

Here for an interoperable GIS integration system, similar requirements can also be applied. VGIS has been built to meet the USECA requirements as well:

- Various repository wrappers provide uniform APIs to the Internet applications. These wrappers act as a mediator between the application program and various geodata stores, which have their proprietary models, data formats, and syntax.
- Transparent access to the various data store sites is supported. Metadata managed by the catalog server can aid the user in browsing all the geospatial data stores that can be accessed by VGIS in the network environment. As it does in GeoLens (Shklar et al. 1997), a catalog server helps the user to discover interesting data and construct a correct SSQL (spatial query language) statement to fetch these data.
- A uniform transient data structure defined and used in VGIS facilitates geoprocessing and data conversion.
- An icon-based visualized spatial query language (VSQL) makes the process of the geospatial data query as easy as clicking icons and buttons on the easy-to-understand GUI of the client Java applet.

In the following sections, we explain the VGIS system architecture (Section 30.2), describe the main components of the system (Section 30.3), discuss query processing in VGIS (Section 30.4), and finally conclude the discussion.

## 30.2. System architecture

The VGIS project is intended to provide an infrastructure to enable public users (government and private agencies, etc.) on-line access to heterogeneous land databases and value-added geographic information services. In the VGIS environment, geospatial data sources may differ in their organization, such as georelational databases and flat data files; by their content, such as attributes, graphical data, and image data; and by query languages used for information access. The data sources to be processed and integrated include (but are not limited to):

- Geospatial data created and managed by ESRI ARC/INFO on Unix;
- Geospatial data managed by Intergraph/MicroStation on Windows NT;
- Statistical data saved in a relational database such as ORACLE, SYBASE, etc.;
- Simple DXF data files exported by AutoCAD and some attribute data in compliant text data files;
- DGN data files with attribute data in a relational DBMS.

We use a client-server network model to realize the VGIS architecture. Components of the network model include data repository wrappers, a catalog server, and a query processing server. Figure 30.1 shows the network model with the mediator services collaborating within the VGIS network architecture.

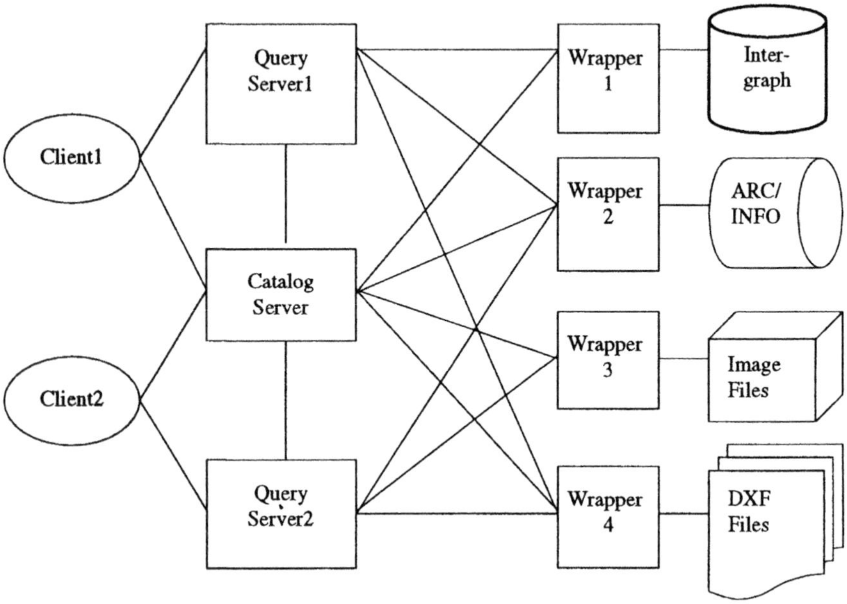


Figure 30.1. The network model of VGIS

As we can see in Figure 30.1, each geodata repository has a wrapper located in the resource site, which acts as a mediator between the Internet services and the local GIS system or data files. This wrapper provides the Internet servers with standard APIs to carry out the manipulations to the local data. For example, an Intergraph wrapper provides information about the Intergraph database, translates global query commands (SSQL) into local query language format, and calls local DBMS functions to query for requested feature data. A DXF/DGN wrapper manages the geospatial data files located in the local computer, executes query commands for the SSQL processor, etc. These wrapper-based mediators are in fact playing the roles of data repository servers for the VGIS catalog server and the Internet query server.

While acting as clients to the various data repository wrappers, the catalog server and the Internet query servers are acting as Internet servers to serve the client user applications. The Internet user uses a Java applet application via an Internet browser (such as Netscape browser or Microsoft Internet Explorer) to utilize the services provided by these servers to carry out geospatial data query and geoprocessing tasks.

By adopting the VGIS model, we have developed a prototype system, WebGIS, which is actually an integrated application with accesses to various kinds of heterogeneous geodata repositories. It can be a spatial database system, a DGN file archive, and a DXF file archive with corresponding attribute text data files. The architecture of this system is shown in Figure 30.2.

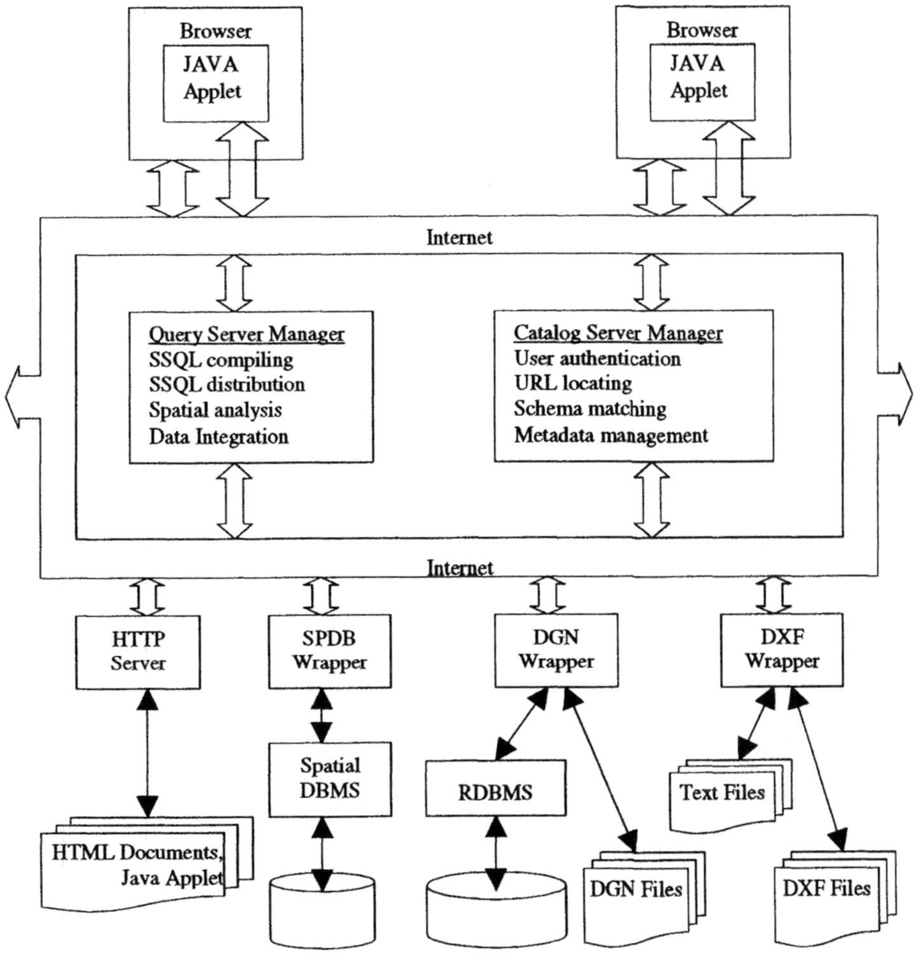


Figure 30.2. Architecture of WebGIS

With WebGIS, the user can browse through the whole contents of these kinds of heterogeneous geospatial data repositories to discover interesting geospatial feature data, issue query commands with VSQL (Visualized Spatial Query Language) GUI, and perform geo-processing functions on the geospatial data.

## 30.3. WebGIS components

We will give a short description for each component shown in Figure 30.2 in this section.

### *30.3.1. SPDB wrapper*

This wrapper provides access to a spatial database system which manages Singapore land digital data. Stored information includes physical and attribute data of lines, such as streets, roads, expressways, highways, and MRT lines; areas, such as apartment buildings, university campuses, schools, shopping centers, temples and churches, theaters, sport complexes; and points, such as bus stops, MRT stations, and other landmarks. The DBMS of the spatial database system provides an SQL interface to access all of these objects. We have built an SPDB wrapper, which has been implemented with improved spatial processing functions and Internet access communication functions to process SSQL commands. In the following aspects, the SPDB wrapper extends the spatial database function:

- *Spatial relationship predicates and spatial algebra operations.* Implemented functions include basic logical operations (intersection and union), metric operations (distance); directional operations; and topological operations (within, contain, overlay, meet, buffer, disjoint, etc.).
- *Spatial SQL.* Spatial relational predicates and operators have been added in the SSQL syntax.
- *Algorithm for path searching.* This is a simple geospatial application to find the shortest path between two nodes in a traffic network. Traffic roads are weighted with cost values indicating selectability according to traffic status. The path found is the one with the minimum sum of costs.
- *Multimedia information.* Spatial objects may have photo images, video clips, or audio records. For example, a shopping center may have images to show outlook and structure, movie data to show the surrounding landscape, sound data to introduce its history, business scope, etc, and text of some kind. The SPDB wrapper can manage all these multimedia data for the spatial objects.

### *30.3.2. DGN wrapper*

Land information includes lot number, ownership, address, and other attribute data besides physical graphic data. The graphic data are created by using MicroStaton software and saved in DGN format files, while attribute data are stored and managed by a relational data base system. The DGN wrapper is implemented with functions to query spatial objects by given physical position or attribute value such as lot number, postal address, or owner name. The implemented functionality of the DGN wrapper is as follows:

- Computing and maintaining the correspondences between object attribute data saved in the relational database system and graphic data saved in DGN files;

- Spatial indexing of the graphic objects so that the spatial query commands such as "query for land lots near a spatial location or within a spatial range" can be effectively implemented;
- Data format translation from DGN format to system internal format of uniform transient data structure; and
- Integration of the result data after local query command processing is finished.

### 30.3.3. DXF wrapper

We have some cadastral data saved in DXF files, created by using AutoCAD software. Similar to the DGN wrapper, geographical features have attribution data and graphical data. However, the attribute data were saved in text files. Functions similar to that of DGN wrappers are provided here:

- Computing and maintaining the correspondences of objects' attribute data in TEXT files and Graphic data in DXF files;
- Key value indexing on attribute values;
- Spatial indexing on geographical positions;
- Data format translation from DXF format to system internal format, that is uniform transient data structure;
- Integration of the result data of local query processing.

### 30.3.4. Client interface

The client application program is implemented in Java so that VGIS provides a device-independent interface to the end users. Users can start a client application program by navigating the VGIS web page through any kind of Internet browser, such as Netscape or Microsoft Internet Explorer. When the VGIS web site is connected, the document will be displayed in the Internet browser window, and the Java applet code will be downloaded and executed in the user's local machine. An Internet TCP/IP link between the client and the Catalog Server is built, and the user is provided with outline information of all the geospatial date stores in this system. The user can navigate through the data stores, fetch interesting data, and draw on the Java window. Once a specified map is displayed on the screen, the user can do basic graphic manipulations and attribute queries:

- Zooming, scrolling, and panning the map in the display window;
- Clicking any object using the mouse to view its attribute value;
- Showing multimedia information for objects which have associated audio records, video clips, photo images, and text documents;
- Simple GIS applications such as searching for the shortest path between two nodes in the traffic network, etc.;

- Using the VSQL GUI to construct and issue a spatial query.

Spatial SQL (SSQL) is used for user to query for specified spatial feature data. In this extended query language, spatial relational predicates (such as *near, within, joint, etc.*) and spatial operational functions (such as *Intersection, Union, Buffering, etc.*) are embedded in the SSQL syntax. A typical spatial SSQL will be like this:

```
FIND_house where unit_price < 2500 and floor_area > 120
WITHIN (FIND ad_area where name = "Clementi")
NEAR (FIND school where catelog = "primary school")
```

This statement will ask the query server to query for all flats with unit price less than $2500 per square meter and construction area less than 120 square meters in administration area "Clementi", and the building should be near a primary school. Here spatial relational predicates *within* and *near* are used to indicate the specified spatial requirements. This is a typical query command for a user who wants to buy a flat. He or she may have small children, so wants the building to be near a primary school. If the user is a sports amateur, he or she may also add more specific conditions to constrain the system to query for available flats with sports complexes nearby.

An easy-to-use icon-based visual query language (VQL) interface has been developed for the user to construct query command by simply clicking on icons. The easy-to-understand GUI in VQL exempts the user of the tedious work of understanding the complex syntax of the SSQL, which otherwise may bring unendurable syntax errors for ordinary system users. In VQL, icons are either object icons representing spatial features, or operator icons standing for spatial relational predicates and spatial graphic operations. Only correct icon sequences can be accepted, and translated to SSQL command automatically by the system.

The metadata information stored in the Catalog Server can be used to construct query conditions. When the user clicks on a feature icon, corresponding attribute names, value types, domains, and unit measurements will be displayed for the user to refer to when he or she is provided with an interface to key in suitable attribute values as the query conditions.

### *30.3.5. Internet communication model*

Unlike some kinds of Internet servers which have been implemented as CGIs (Common Gateway Interface) to HTTP servers, our Catalog Server and various repository wrappers are all standalone background programs. However, the World Wide Web HTTP Server can also play an important role in VGIS.

Firstly, the HTTP server helps to transfer Java applet codes. As showed in Figure 30.2, when the end user starts a Internet navigator and goes to the application web page, the source code will be transferred to the client computer through an HTTP server running on the server computer where the application web page resides. Then the Java applet embedded in the HTML file will be downloaded and run on the Java virtual machine of the user's local computer. A direct Internet connection through TCP/IP protocol between the client applet and the catalog server is created by the client applet. Such connections will also be built up by the Catalog Server to connect to various geospatial data repository wrappers if necessary. All service servers in VGIS use RPC (Remote Process Call) utilities to provide Internet GUIs to their clients (Bloomer 1991).

Secondly, HTTP servers can also be used in transmitting query result data files from VGIS servers and geospatial data repository wrappers to the client. These files are integrations of spatial features from various heterogeneous geospatial data stores fetched through corresponding wrappers. There is no direct Internet connection between the client application program and the wrappers of various heterogeneous geodata stores. Normally, the requested feature data is sent to the query server to be assembled before being sent back to the client application program. In cases where all the result data is from only one repository, neither assembling processes nor special processing actions are needed on the data in the query server. But still two phases of Internet transmissions are needed (one from data store wrapper to the catalog server, and another from the query server to the client). However, by the use of the HTTP server, files created by repository wrappers can be directly sent to the client, with one transition time saved. In this case, instead of sending geographic data to the query server, the corresponding data store wrapper saves the result data into temporary files on the local machine, and sends the URLs (Uniform Resource Locators) of these files to the client. The client then fetches these data files via the HTTP server. The wrapper will remove these temporary data files after the client completes the reading process. This is especially meaningful in the multi-hierarchical mediator architecture shown in Figure 30.1. At higher and higher layers in the hierarchy, the system response time will be longer and longer. Unless the Internet transmission speed can be greatly improved, an Internet multi-hierarchical information system with too many layers will be impractical. That is one reason we implement VGIS in 3-layer structure. We separate the catalog driver from the query server as a standalone server, so that any other servers can access it.

## 30.4. Query processing

The query processing is the key issue for a heterogeneous information system. Many studies have been made and many new methods and algorithms have been proposed. For example, Muth (1997) proposed a transaction model of 3-level transactions for multi-database systems with autonomous component systems, etc. Here in VGIS, our main contributions are in the following three aspects:

- Schema mapping in the catalog server helps to achieve interoperability of various heterogeneous data stores. Different GIS communities have different concept definitions, different data models, and different data schemas. As a mediator service, the catalog server uses the metadata provided by various data store drivers in the semantic mapping process. For example, features and their attribute names, types, values, and domain definitions, etc., are converted between the client application domain and the various data source domains, according to the respective schemas, projection methods, unit definitions, spatial domains, and all other necessary descriptive information defined in the metadata of related domains.

- The icon-based visual query language facilitates the ordinary user in specifying query requirements in the query SSQL construction. With icons, the user need not memorize the object type names, operation names, attribute names, and complex SSQL syntax.
- Spatial reasoning capabilities utilize strong spatial operation functionality in the processing of the spatial query language.

### *30.4.1. Catalog server*

The catalog server provides catalog services to all VGIS applications. In Figure 30.2, we show the catalog server functionality. The catalog server acts as a manager of metadata, both for information consumers and for various heterogeneous data sources, about their spatial data models, schemas, detail definitions, and descriptions. The structure of the VGIS catalog server is shown in Figure 30.3. Concepts such as Catalog, Catalog Entry, Feature, Catalog Schema, Catalog Entry Schema, Feature Schema, Catalog Registry, Feature Registry etc., and their relationships, are specified by the OGIS Project Technical Committee (1996) and Buehler and McKee (1996).

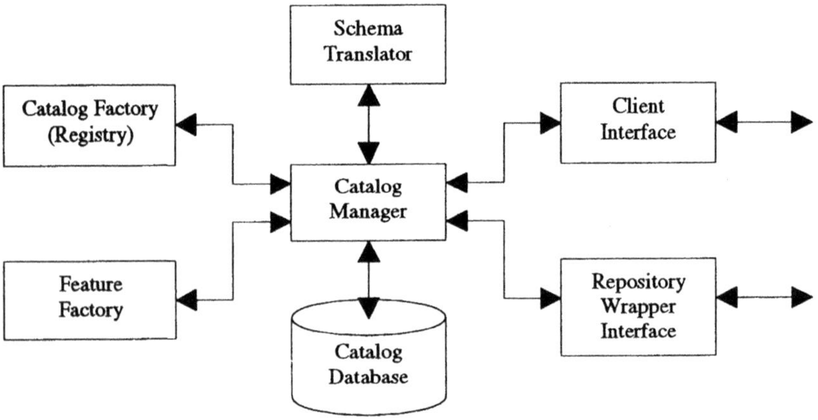


Figure 30.3. Catalog server structure

**30.4.1.1. Catalog and catalog tree**. Catalogs can be classified into three types according to their purposes and existing life duration. The first type is the client catalog. This catalog provides an indexing system for applications. Application developers can configure a catalog entry schema and feature schema to refer to whatever information is considered necessary to the application. Since it is the application developer that defines the client catalog with preferred feature and catalog schemas, the user can follow the conventions of his or her community or GIS environment, about feature names, data types, property definitions, projection method, etc. Since the catalog server dynamically creates client catalogs through the catalog registry with spatial feature and catalog schemas whenever a new application starts in VGIS, the client catalogs are actually dynamic catalogs. Of course, the user of the application can also choose to use the default definitions provided by the system, that is, the system catalogs, as discussed below.

The second type is the system catalog. System catalogs define the schemas, structure and definitions of system internal uniform transit data structure. All contents concerning system catalogs are statically saved in the catalog database and cannot be changed. So we also refer to them as *static catalogs*, which neither the client application user nor the resource geospatial data provider can change. This is also the default to describe the data formats of client applications when the client user does not explicitly specify a choice of catalogs, as mentioned before.

The last type is the resource catalog. Resource catalogs can be regarded as indexes to various heterogeneous data sources. Before a new data repository can be added into VGIS as a component, and expose all or part of its data and services to the VGIS network, it must register itself with the catalog server. The process of registration uses the following information provided by the newly-added data repository to build the resource catalogs for this data store:

- The Internet address of the wrapper that has been built up to intermediate between the catalog server and the data repository;
- Catalog schemas and feature schemas that are needed by the catalog server to build the resource catalogs.

The first one is for the catalog server to locate a wrapper and create an Internet link with it whenever it is necessary. The second one lets the server know what data this repository can provide to the VGIS network and the metadata about these data. So the data provider can choose to provide only a part, or the whole range of its data and services to the VGIS client users.

Catalogs are organized to construct various catalog trees. Three kinds of catalogs are organized into three kinds of catalog trees. Figure 30.4 illustrates the structure of these catalog trees and the relationships between them. The system catalogs are organized to construct a system catalog tree, with the root node named *System Catalog 0*, the descent catalog nodes named *Catalog-a*, *Catalog-b*, etc., descent feature nodes named *Feature-a*, *Feature-b*, etc. Similarly, client catalogs are organized to construct client catalog trees, and resource catalogs construct resource catalog trees. In Figure 30.4, there are a client catalog tree, with the root catalog named *Client Catalog 0*, and three resources catalog trees, with the root catalog named *Resource Catalog 0*, *Resource Catalog 1*, and *Resource Catalog 2*.

VGIS creates and saves the system catalog tree permanently, creates a resource catalog tree whenever a new geospatial data store is added to the system, and creates a client catalog tree whenever a new application is build upon the catalog server. All catalog trees are stored in a metadata database system on the catalog server, managed by a *catalog manager*, as shown in Figure 30.3.

Client catalog trees are defined by the application developers and created before the applications are built up. Whenever an application starts up, the system will create an instance catalog tree for it from one of the predefined client catalog trees or the system catalog tree that is the default when the user does not select a catalog tree explicitly. The client user can make changes to the instance of its client catalog tree.

The preferred type of catalog tree and all the changes the client user made are saved in a *user profile.* So whenever a user starts the client application program (the Java applet), the system first checks if there is a profile for him or her. If yes, the system will use this profile to create an instance catalog tree, which is created from the user-preferred client tree and installed with user definitions saved in the profile. Otherwise, the default catalog tree will be used to create an instance.

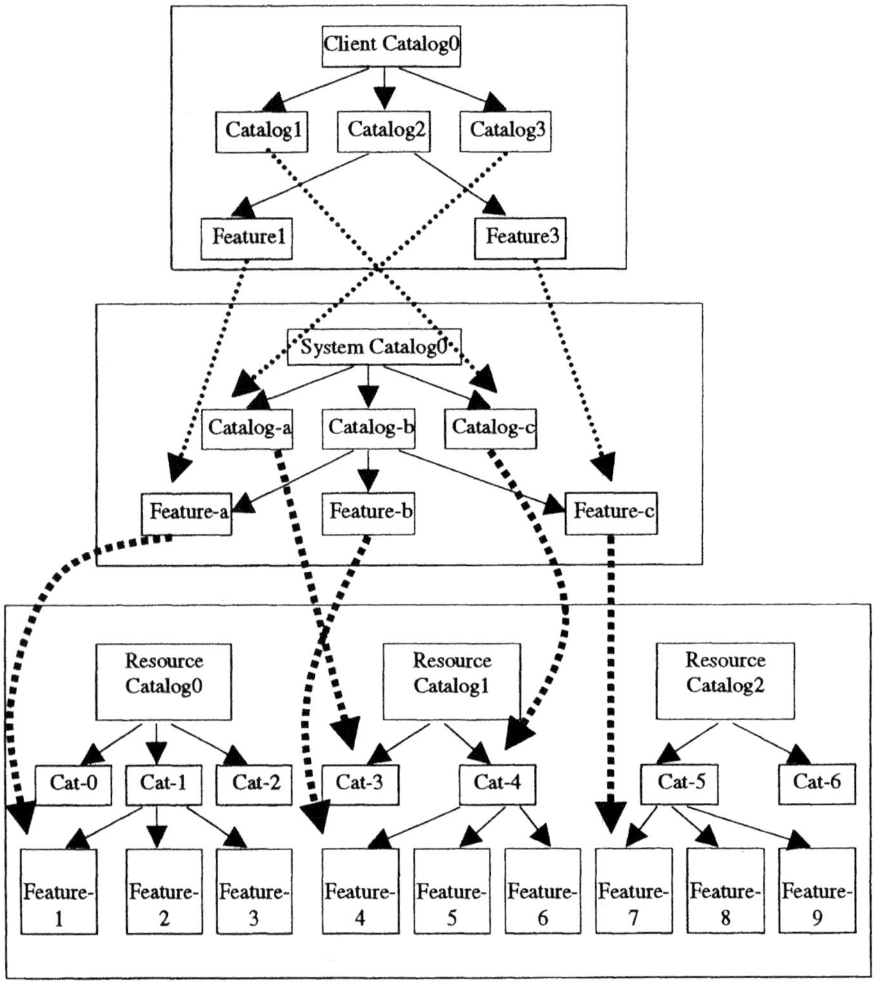


Figure 30.4. Catalog trees and catalog association

**30.4.1.2. Catalog mapping.** With the three-level catalog tree structure illustrated in Figure 30.4, the client user can perform his or her queries and data manipulations with the schemas and conventions of his or her domain while using VGIS. The mapping between the user domain and the resource domain is built and carried out with a *catalog association* process by the catalog manager. In Figure 30.4, the mapping links represented by dotted lines between the client catalog tree and the system catalog tree, and that between the system catalog tree and the resource trees, are the results of two phases of association processes.

The first association process is for building mapping links between the resource catalog tree and the system catalog tree. Whenever a new geospatial data repository calls the catalog server to add itself to the VGIS network, a resource catalog tree is built up with the information provided by the caller. Then this association process is called to build the catalog mapping links between the system catalog tree and the newly created resource catalog tree. These mapping links are saved in the database system until the associated repository removes itself from the VGIS. These mapping links are from the leaf nodes to corresponding nodes in the resource catalog tree.

The second association process is to build mapping links between the client catalog tree and the system catalog tree. When the application developer designs a new type of application program with new domain schema definitions, he or she needs to add a new client catalog tree to the client catalog database of the catalog server with the new user domain information. This association process is started to build mapping links from the leaf nodes of the client catalog tree to some nodes in the system catalog tree.

Every mapping link therefore represents the relationship between two linked nodes in different catalog trees. Following these mapping links, paths from leaf nodes of the client catalog tree to nodes in resource catalog trees will be found. These paths indicate that the geospatial features represented by the nodes in the client catalog tree will be found in remote data repositories associated with the path-linked resource catalog tree. Mapping links keep all necessary information about all the pairs of linked nodes, such as relationships of feature schemas, property names, and value types, which will be used in the schema translation and data conversion. This is the key technique to achieve interoperability of various heterogeneous geospatial data stores in one application environment.

### *30.4.2. VSQL: visual spatial query language*

VSQL provides the system user with a visual facility to compose SSQL query commands for interesting spatial features. The user interface is shown in Figure 30.5. There are four parts in the VSQL window; they are spatial object icons, spatial relation and operation icons, an editing area, and a command panel including system-provided command buttons. We will give each part a brief description in the following sections.

- *Spatial object icons.* In Figure 30.5, these icons are at the top left part of the VSQL window. Every object icon stands for a specific spatial feature, represented by a leaf node in the client catalog tree. A click on one of these icons tells the system that this object is selected in the query command statement. An *Object Attributes Set* window will pop up to let the user input attribute values in those slots. These attribute names and the added values will constitute part of the query conditions of the query statement. Double clicking on the attribute name will pop up another window in which a detailed metadata description about this attribute, such as type, domain, unit values, etc. will be displayed. These will help the user to select suitable values for object attributes. Because the metadata about all of these objects are maintained in the catalog server, each query of object metadata will cause the applet application to trigger a metadata query to the remote catalog server.

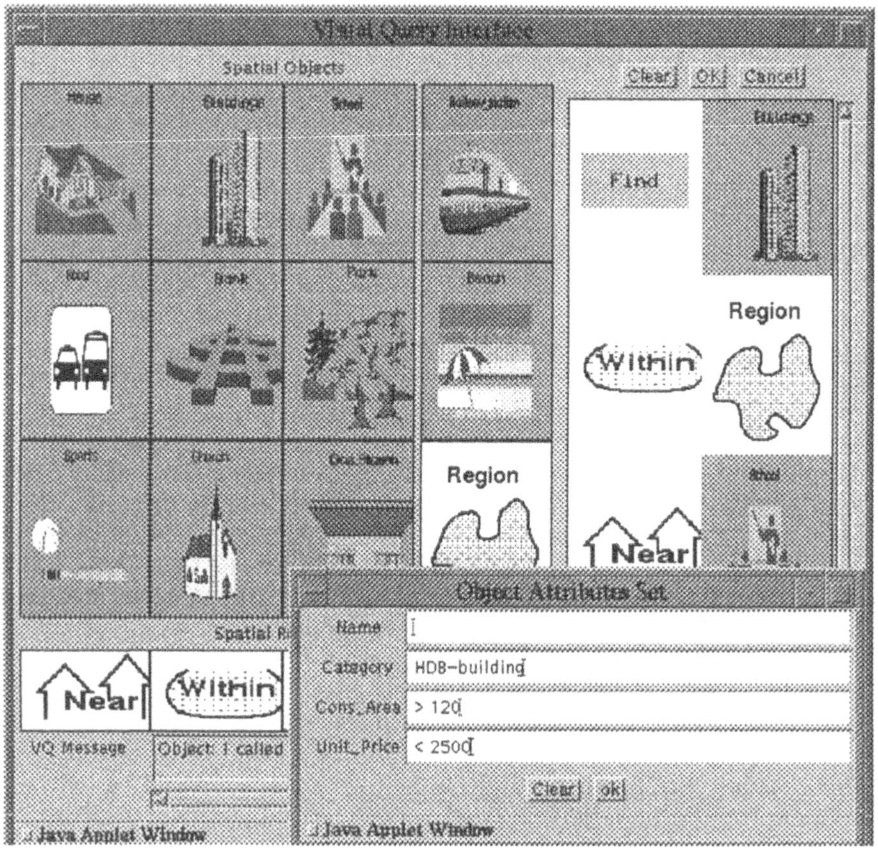


Figure 30.5. VSQL GUI

- *Spatial relation icons and operation icons.* Operational icons are in the bottom left part of the VSQL window shown in Figure 30.5. Each icon stands for a kind of spatial operation that will be performed on spatial objects. Spatial relationship verification functions include *near, within, pass-through,* etc.; spatial algebra operations include *intersection, union, difference, buffer,* etc.

- *Query composition.* The query composition area is on the right of the VSQL window. An iconic spatial query statement consists of a sequence of icons, which represent objects and operations. Whenever an icon is selected, it will be added to the sequence. Editing functions can be used to modify his query statement. For example, clicking on the object icon will pop up an Object Attributes Set window to let the user reset the attribute values. Spatial functions and parameters can also be modified this way. The user can also remove some conditions or add new conditions into the statement.

- *Command buttons.* The panel containing command buttons is in the top right part of the VSQL window. Commands included ending the editing of the visual query statement, compiling or canceling the query statement, etc.

The visual query statement will be translated to a SSQL statement that will be sent to the query server to be processed further.

### 30.4.3. SSQL, analysis, and execution

The analysis and execution of SSQL statements will be carried out on the query server, while using the catalog functionality of the catalog server in the analyzing and assembling process. First, the query statement will be analyzed against the standard SSQL syntax and grammar. Then, several steps will be carried out serially to analyze and execute the query command. Figure 30.6 gives a block diagram about the SSQL process in the query server. These steps include query routing, SSQL decomposition, sub-query translation, execution, and query results integration, which we will discuss in the following separate sections.

**30.4.3.1. Query routing.** The main task of query routing is to select relevant geospatial data sources from the available ones for answering the query request. Thanks to the functionality of the catalog server, this task will be done through the catalog mapping links that have been built in the catalog association process. The result of the query routing process is a query catalog tree comprising instance nodes created from the client catalogs, system catalogs, and resource catalogs maintained in the catalog server database.

As mentioned in the previous section, every feature in the query statement has a corresponding node in the application-related client catalog tree. The first step is to build an instance catalog tree with these nodes so that the created instance catalog tree is a minimum sub-tree of the client catalog tree, with all the feature nodes referred to in the SSQL statement as the only leafs of this tree. We call this tree the *Client Catalog Instance Query Tree*, or CCIQ tree.

The next step is to extend CCIQ tree with system catalogs and resource catalogs, until the final query catalog tree is formed. The first extension is made with the system catalog tree. This is done through extending the leaf node of the CCIQ tree, that is, adding nodes of the system catalog tree as children. The selection of the catalog nodes is according to the mapping links that map the client catalogs to the system catalogs. Similarly, the second extension adds the resource catalog nodes into the tree with the system-resource catalog links. The extension algorithm can be described as:

```
a) For each leaf node <node-I> of CCIQ tree,
   i) find client-system mapping link <map-link> that
      starts from <node-I>;
   ii) find the end node of mapping link <node-j>;
   iii)  add <node-J> into CCIQ Tree as child of <node-I>.
b) While there is any leaf node <node-k> in the above tree
   that has no mapping link starting from it,
   i) in system catalog tree, select all the child nodes of
      <node-k>;
   ii) add these nodes to CCIQ tree as the children of
      <node-k>.
c) For each leaf node <node-I> of the new tree (which will
   have system catalog nodes as its leaf nodes),
   i) find system-resource mapping link <map-link> that
      starts from <node-I>;
   ii) find the end node of mapping link <node-j>;
   iii)  add whole sub-tree with root node <node-J> in
      resource tree into the new tree as one sub-tree of
      node <node-I> in the new tree.
d) Return the finally formed tree, which is the query
   catalog.
```

The first extension includes steps (a) and (b), which actually translates the client catalog schema of the SSQL statement to the system catalog schema. Step (c) is the second extension, which translates the system schema further to various resource schemas of remote spatial data stores. With these three steps, the final query catalog tree is formed.

**30.4.3.2. SSQL decomposition and parallel query planning.** An SSQL statement needs to be decomposed to several sub-SSQL statements sending to correspondent remote geospatial data store wrappers to query for spatial feature data. A sub-query statement concerns only spatial features in one remote resource so that it can be executed in the spatial repository management system of that resource.

The first step is to apply the query conditions of the SSQL statement against the query catalog tree, so that conditions on the user-domain features are translated to conditions on the resource-domain features. The conditions-applying process begins at the root node of the query catalog tree, then is propagated deeply, from parent nodes to child nodes, and ends on the leaf nodes. In the meanwhile, at the propagated child node, parent attribute data types are replaced with child node data types and parent attribute data values are converted into the child node attribute values. Therefore, the user domain feature attributes are first translated to system-domain feature attributes, and then to the respective various resource-domain feature attributes, as are feature names and feature attributes names.

The next step is to generate sub-query statements for all of the related remote geospatial data stores. For each remote data store, there is a sub-tree consisting of catalog nodes only for that resource. This sub-tree is in fact a window specifying the contents of the correspondent remote data source that are to be queried in the SSQL query command. One sub-query statement is derived from each of these sub-trees.

Finally, the sub-query statements are to be scheduled into an order in which the sub-query is to be executed. Some sub-query statements can be simultaneously executed if they are independent of each other; some sub-query statement must be executed after others because the former sub-query uses the result values returned by the later sub-queries. Sub-query planning groups those parallel sub-query statements that can be executed simultaneously and schedules the sub-query groups that will be dispatched serially.

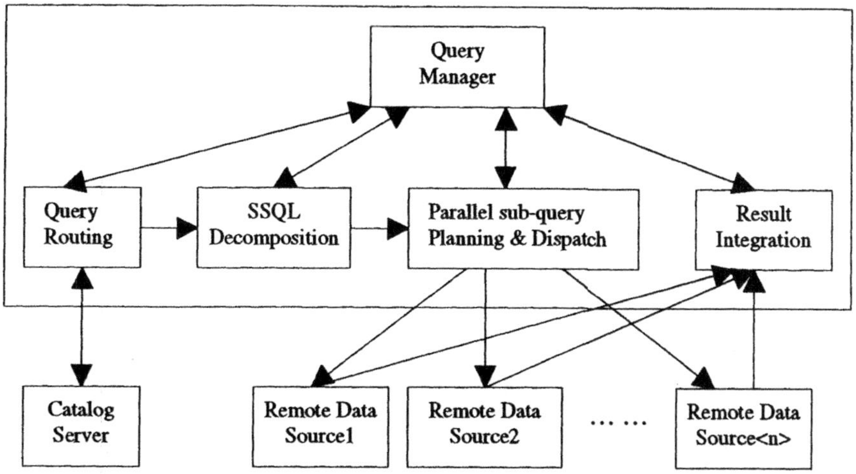


Figure 30.6. Distributed query processing in VGIS query server

**30.4.3.3. Query result integration.** The execution of the sub-queries in various data store wrappers will be data source dependent and of no interest to the readers of this chapter, so they are not included from here. What we discuss here is the packaging of the returned result data of sub-queries, and the integration of the result data according to the user-domain standards and conventions.

First of all, the result data must be converted to the system domain and packaged in a temporary workspace. The conversion is also according to the query catalog tree. From the leaf nodes of the query catalog tree, the features with their attribute values in a child node are converted into the features and their attribute values of the parent node. The conversion continues until all the system catalog nodes have their system-domain feature physical and attribute data successfully calculated.

Then global SSQL query conditions will be imposed on the result features in the system-domain. Spatial relationship verification and algebra functions will be called to check the spatial relationships of spatial features which usually are across the remote data store boundaries and cannot be done on any single remote data source. Unsatisfied features are discarded from the workspace.

Spatial process functions must be called with system-domain spatial feature data. That is because most of the spatial calculation algorithms and spatial semantic concepts used in the implementation of these spatial functions in the query server are all system-domain schema-based, and thus do not function correctly and cannot return correct results for feature data in other domains. For example, positional relationship functions such as *North-of*, *Northwest-of*, etc, are closely related to the coordinate system of the physical data. In one coordinate system, a point has a larger *y* coordinate value than any point south of it; while in another coordinate system, this relation is reversed, i.e., a point has a smaller *y* coordinate value than any point south of it. Therefore for the same two spatial features, if we call North-of to compute the positional relationship in two different coordinate systems, the answer may be different. That is why we insist that the execution of the spatial relationship verification and spatial algebra functions should be in the system internal domain schemas and semantics.

Finally, system-domain feature data are converted to user-domain feature data. This is also done on the query catalog tree. The result data are packed into spatial objects that will be sent back to the user application program to be drawn on the map window, and one query process is finished.

## 30.5. Conclusions

Our objective in VGIS is not only to make interoperable the geospatial features and spatial processes of the heterogeneous geospatial data stores on the Internet environment, but also to satisfy the different requirements of various applications. With the unique characteristics of the VGIS catalog server, the objective of VGIS is well achieved. This is done through the three-level catalog structure of our catalog server, where catalogs are organized into three kinds of catalog trees: client catalog tree, system catalog tree, and resource catalog tree. The run-time query catalog tree derived from SSQL statements facilitates the schema translation and data conversion for a specific query language. So it is possible to build application programs in the user domain while using VGIS server services to access heterogeneous geospatial data stores. VGIS indeed provides a transparent view of geospatial features over the Internet environment.

Our future work will be on the following aspects:

- *Extending VGIS-supported geospatial data stores.* More existing GIS sources need to be added to VGIS. It means that more and more repository wrappers with drivers to access existing GIS data stores and geoprocessing services will be built.

- *Integration with other information systems.* The boundaries between GIS data and other information are becoming vaguer. Integrating geographic data with other information, such as multimedia information, needs more studies of system design and implementation;

- *Database updating and consistency.* We have not discussed remote database updating and consistency maintenance in VGIS, which needs more effort in the next stage of development.

## References

Bloomer J 1991 *Power Programming with RPC.* O'Relly & Associates, Sebastopol, Calif.

Buehler K, McKee L (eds) 1996 *The OpenGIS Guide, Introduction to Interoperable Geoprocessing.* Open GIS Consortium, Wayland, Mass.

LAS Inc (1996) *Open Geographic Datastore Interface (OGDI) Version 1.0, Programmer's Reference.* Defense Research Establishment, Valcartier Scientific Authority, Canada

Liu L, Pu C 1997 An adaptive object-oriented approach to integration and access of heterogeneous information sources. *Distributed and Parallel Databases* 5: 167–205

Muth P 1997 Application specific transaction management in multidatabase systems. *Distributed and Parallel Databases* 5: 357–403

OGIS Project Technical Committee (1997) *The OpenGIS Specification Models*. Open GIS Consortium, Wayland, Mass.

Shklar L, Behrens C, Au E 1997 *New Approaches to Cataloguing, Querying and Browsing Geospatial Metadata. http://usdac2.rutgers.edu/new/new.html*

# 31 INTRODUCTION: EDUCATION, INSTITUTIONS, AND MARKETS

Robin G. Fegeas

In this last section we leave the lower-level technical aspects of interoperability, for the most part, and begin to address organization, education, and infrastructure issues and implications. As we have seen from earlier chapters, the technical challenges facing developers of interoperable GIS components are great. But just as formidable are the challenges to our institutions. Interoperability means changing paradigms not just for the technologist, but for the bureaucrat, the teacher, and the businessperson as well.

There are but five chapters in this last section. All in some manner discuss building infrastructures, including the interplay between the institutional and technological aspects of interoperability. Two chapters deal primarily with building organizational infrastructures to share geographic data. Two others describe a geoservices marketplace infrastructure based on the Internet. The fifth chapter envisions an interoperable educational infrastructure modeled on GIS interoperability itself.

John Evans, in *Organizational and technological interoperability for geographic information infrastructures*, looks at how to build organizational and technological infrastructures for sharing geographic information, based on three case studies and his ongoing prototyping efforts. His focus is on groups that had different goals and tasks, but whose concern for a jointly owned natural resource led them to work together, to coordinate their information activities and to build, with varying degree of success, some permanent underlying mechanism—an infrastructure—to share information. He examines both technological and organizational design choices and contexts; their mutual interaction; and their joint relationship to planning and policy. The case studies and his own experience building an interactive online orthophoto server modeled on the principles of the U.S. National Spatial Data Infrastructure, have important implications for infrastructure-building efforts at several scales.

Volker Gaede, *in Spatial Internet marketplaces from a database perspective*, and Oliver Günther and Rodolf Müller, in *From GISystems to GIServices: spatial computing on the Internet marketplace*, describe Internet marketplaces as a possible model for providing geographic data and services to a broad public. GIS services are seen as especially commercially promising. The vision of interoperating, interchangable GIS modular components (as pursued by the Open GIS Consortium) could be realized in such a marketplace. But as Gaede, Gunther, and Muller outline, much work is needed to build the necessary infrastructure. For example, by contrasting with what is now possible with federated databases, Gaede itemizes numerous differences, increased capabilities, and new concepts.

Ian Heywood, Karen Kemp, and Derek Reeve, in *Interoperable education for interoperable GIS*, outline a vision of geographic information education in 2010. This introduces their exploration of the impact of interoperable GISs on GIS education. They conclude by suggesting that interoperability itself provides a model for the new collaborative methods of teaching that will be necessary. *Institutional* interoperability is envisioned among academics and industry, collaborating to build a GIScience Education Resource Network. The path toward the vision is admittedly fraught with obstacles, but the need for change is clear.

Xavier Lopez, in *Interoperability through organization: digital libraries for distributed geospatial information*, further examines the *organizational dimension* of interoperability. The chapter focuses on the requirements of the geographic information science community for "network-centered organizational frameworks" to facilitate interoperability. The themes of interacting technology and institutions and the need for organizational interdependence and integration introduced in John Evans' chapter is continued. An in-depth discussion of digital libraries provides a good model for the development of interoperable organizational infrastructures.

In summary, the chapters in this final section make an important contribution to our understanding of the institutional as well as technological infrastructure changes that will be necessary in the brave new world of interoperability. In each chapter, we see shifts—new ways of doing business, both literally and figuratively. The new infrastructures, based on collaboration, synergy, egalitarianism, openness, and the ability to adapt (among other qualities), promise untold societal rewards, if only we can change.

# 32 ORGANIZATIONAL AND TECHNOLOGICAL INTEROPERABILITY FOR GEOGRAPHIC INFORMATION INFRASTRUCTURES

John D. Evans

## 32.1. Introduction

What will it take to share geographic information between real-world organizations? Despite progress in networking technology, database design, standards, and organizational wisdom over the last three decades, it's still rare for planners or public managers to share information across organizational boundaries. It's rare even when (as in environmental management for instance) key decision variables are linked by physical pathways (waterways, landforms, habitat) that cross jurisdictions, hierarchies, and other territorial lines. It's rare even for geographic information, despite its cost and its potential for widespread re-use, and has remained rare even as the Internet has come of age in recent years, and as public agencies have been called to increase their efficiency and public accountability. The problem seems to be part technical, part organizational, and often peculiar to the nature of geographic information: complex in structure and interpretation, rich in meaningful inter-relationships, and difficult to understand or use without special-purpose tools.

Part of the answer may be found in geographic information infrastructures—permanent, multi-purpose mechanisms built to help people make use of each other's geographic information. Learning to design such infrastructures, and to deploy and maintain them appropriately within a rapidly evolving technological and organizational context, seems key to effective inter-agency collaboration and sharing of geographic information. This chapter, then, looks at how to build organizational and technological infrastructures for sharing geographic information, based on recent empirical observations and ongoing prototyping efforts (Evans 1997).

The research described here followed a hybrid approach, drawing on social and behavioral perspectives as well as on GIS technology, standards, and networking—as I have summarized in Section 32.2 below. Section 32.3's case studies of inter-organizational information infrastructures highlight essential change processes suggested by recent US experience. Section 32.4 examines the technological *style* needed for such infrastructures, exemplified by a networked service for digital orthophotos built onto the National Spatial Data Infrastructure. Section 32.5 sketches broader implications.

## 32.2. Background and perspectives

*Sharing* can mean a great many things to people, from a simple moral principle (Fulghum 1987) to quite abstract matters of semantics and perception (Frank 1992). Here I have roughly followed Carter's (1992) definition, "the use of multipurpose information by members of a partnership." My focus is on groups that have different goals and tasks, but whose concern for a jointly owned natural resource leads them to work together, to coordinate their information activities, and to build some permanent underlying mechanism—an infrastructure—to share information, rather than rely on *ad hoc*, single-use methods. The next few paragraphs provide a foundation for discussing the technical design of such infrastructures and their deployment and growth within organizations.

### *32.2.1. Technologies for geographic information sharing*

Over the past three decades, a diverse array of research in distributed data networks, standards, and multi-database theory has focused on information sharing.

First, as the Internet has come of age, so have tools for locating, retrieving, filtering, and using networked information, from the file transfer protocol (ftp) to Wide Area Information Servers (Kahle 1991), Gopher (Schwartz et al. 1992), and especially the World Wide Web (Berners-Lee et al. 1992). Frank (1994) sketches the role of data catalogs and navigational tools in several possible scenarios of a future spatial data infrastructure. The Internet and especially the World Wide Web have provided unprecedented levels of connectivity: much of that connectivity is still fairly simple at present (electronic mail and static Web pages), but information standards and distributed data designs promise a rich set of useful interactions among different information systems.

Information standards—shared vocabularies to facilitate communication—have been shifting in emphasis from comprehensive data format and quality specifications (U.S. National Mapping Standards, Spatial Data Transfer Standard) to interface standards that define the interactions with information without changing the information itself. Minimal standards of this second sort are useful in linking autonomous organizations that have different yet established formats, procedures, or quality requirements. Noteworthy efforts include the US Federal Content Standard for Digital Geospatial Metadata (FGDC 1994), the Common Object Resource Broker Architecture (CORBA; Siegel 1996), and the Open Geodata Interoperability Specification (OpenGIS; Buehler and McKee 1996).

Multi-database research pursues these software interactions in further depth, seeking to reconcile different data structures and models (Batini et al. 1986; Litwin et al. 1990; Peckham and Maryanski 1988), and to interpret data semantics (definitions and relationships; Sheth and Larson 1990; Siegel and Madnick 1991).

Geographic information, with its multidimensional and topological nature, presents unique interoperability issues, both theoretical (Morehouse 1990; Nyerges 1991) and practical (Baker and Broadhead 1992). Yet these multi-database concepts and methods are clearly useful for understanding and comparing geographic information sharing systems (Mackay and Robinson 1992; Nyerges 1989).

### 32.2.2. Organizational aspects of geographic information sharing

To many practitioners, the greatest obstacles to information sharing seem to be behavioral, rather than technical (Croswell 1989); topics such as collaboration, consensus building, and coordination provide useful insights for geographic information sharing. Much of this literature follows a *variance* model, in which levels of certain inputs (factors) are seen to lead to performance outcomes by some functional relationship. This literature, reviewed by Grandori and Soda (1995), Pinto et al. (1993), and others, presents a complex set of factors for successful collaboration. For instance, Alexander (1995) finds collaboration to be facilitated by a clear interdependence between organizations, in the form of pooled resources, sequential tasks, or resource transfers—all geared towards improved efficiency and survival in a complex organizational ecology. In the area of geographic information, important factors include appropriate pricing (Rhind 1992) or, conversely, open access (Epstein 1995); negotiation (Obermeyer 1995); a common, super-ordinate goal (Pinto and Onsrud 1995), a killer application (Brodie 1993), data ownership (Carter 1992), and technical expertise (August 1991; Craig 1995).

The variance approach contrasts with a *process* model, in which individual decisions over time explain how the observed outcomes actually come about. Van de Ven and Walker (1984) and others emphasize the power of this model in explaining complex inter-organizational patterns. For instance, Boland and Tenkasi (1995) link shared information to consensus-building through the use of boundary objects (maps, structured narratives, and the like; Harvey, this volume) to help structure one's world-view and relate it to that of another. Granovetter (1972) uses social network theory (strong vs. weak social relationships) to link individual micro-decisions to overall macro-outcomes.

The two kinds of research are not necessarily incompatible; either perspective on organizational research can be harnessed usefully. Nonetheless, I found the process model more appropriate here, given the complexity of the information-sharing question and of the multi-organizational setting. Some attention to process was also helpful as I sought to understand organizational and technical aspects in concert, rather than in isolation from each other.

### 32.2.3. Interdependence of organizational and technological factors

Sharing geographic information is often characterized as either a technological problem or an organizational one. However, in an unsettled, rapidly changing technological and organizational context, it rarely fits cleanly into one or the other category (Evans and Ferreira 1995), but is closely tied instead to the relationship between technological and organizational structures. Markus and Robey (1988) emphasize this two-way relationship, favoring it over *technological determinism* (in which technologies are presumed to have known, inexorable effects on organizations) and *social strategic choice* (in which technologies are seen as inexorably shaped by organizational intentions and actions). Barley (1986) invokes structuration theory (Giddens 1984) to trace the ongoing, recursive influences, triggered by new technologies, between an organization's rules and resources and the behavior of its members. DeSanctis and Poole (1994) emphasize the intertwined nature of technological and behavioral patterns, and Orlikowski (1992) proposes a useful view of technology as a malleable structural property of organizations. In such a model, organizational intentions alone cannot give rise to a given technology, nor can a technology have a fully predictable effect on organizations. Rather, in every phase of a technology's existence—conception, design, deployment, use, evaluation, and modification—the human actors involved mediate both causal effects in essentially unpredictable ways.

Accordingly, this research examines both technological and organizational design choices and contexts; their mutual interaction; and their joint relationship to planning and policy. A focus on interactions implies a dynamic view of information sharing infrastructures, not as a set of fixed, interlocking components but as a chosen direction, or even a style of evolution through an uncertain future..

## 32.3. Lessons from three inter-organizational information infrastructures

The background and perspectives just outlined were helpful in interpreting the experience of several groups of organizations in the US that have undertaken infrastructures for sharing environmental, geographic, and other information. I compared the design and history of these infrastructures, seeking to understand their patterns of growth and change over time, and to document the tangible impacts of inter-agency information sharing. The study's nature, and the complexity of its context, favored a case-study methodology (Yin 1984). I chose three cases in which loosely coupled partnerships among autonomous organizations were motivated by the joint stewardship of a valued natural resource. These cases were the Great Lakes Information Network (GLIN); the Gulf of Maine Environmental Data and Information Management System (EDIMS); and the Pacific Northwest StreamNet and its precursors. For each case, I conducted on-site interviews, reviewed paper and electronic documents, conversed with my study subjects by telephone and electronic mail, and submitted draft summaries to them for review. Qualitative analysis drew on grounded theory (Glaser and Strauss 1967) for inductive hypothesis generation. The next four paragraphs sketch the findings from each case, then draw some overall findings across the cases.

### *32.3.1. The Great Lakes Information Network (http://www.great-lakes.net)*

The Great Lakes Commission, an inter-state compact formed in 1955, began experimenting with the Internet in 1993 to enhance communication among the many groups concerned with the water and air quality and the regional economy of the Great Lakes. The resulting Great Lakes Information Network (Ratza 1996) links the Great Lakes states (Minnesota, Wisconsin, Illinois, Indiana, Michigan, Ohio, Pennsylvania, and New York), the Canadian Province of Ontario, and several Federal agencies and other public and private groups in both the US and Canada.

For its first two years, GLIN saw rapid, mostly unplanned growth in size and usage, thanks to the evangelistic efforts of its founder and director, the tireless support of its technical architect, and the nature of the emerging World Wide Web. It encountered several challenges related to its growth. The first of these was the difficulty of moving from a small, mostly centralized proof of concept to a truly distributed data resource, owned and run by its participating agencies. "The little pilot has become a messy beast!" said GLIN's technical director a year into the project. By that time, he was struggling to support some 400 accounts on a central server (where only 30 had been anticipated), and dozens of other relevant sites were going online all around the region, with no indexing or cataloging scheme in place.

A second challenge for GLIN was its interaction with other information-sharing initiatives in the region. Indeed, several other information-sharing infrastructures were being built for the Great Lakes region at the same time. GLIN's relationship to them was not always clear, which led to some degree of duplication and competition instead of cooperation and coordination. In all, however, GLIN has made impressive strides in equipping public and non-profit organizations throughout the Great Lakes to use the Internet meaningfully in their work.

### *32.3.2. The Gulf of Maine Environmental Data and Information Management System (http://www.gulfofmaine.org)*

In early 1990, three New England states (Massachusetts, New Hampshire, and Maine) and two Canadian Maritime provinces (New Brunswick and Nova Scotia) formed the Gulf of Maine Council on the Marine Environment to facilitate joint use and protection of their shared coastal and marine habitats.

In its initial Action Plan, the Council called for "a common regional protocol allowing for the transfer and periodic updating of data and information" among its participants. The means chosen for this was an Environmental Data and Information Management System, or EDIMS (Brown et al. 1994). Built in the pre-Web years, EDIMS began as an ftp/telnet-accessible directory of the region's available data on coastal and marine resources. This metadata directory, already advanced for its time, was intended as a first step towards a region-wide, distributed set of data resources. However, after an enthusiastic start under a dynamic team leader, funding for EDIMS ran out in late 1993 and the project went dormant for a couple of years—just as the World Wide Web was becoming a household word. When its funding resumed in 1995, EDIMS struggled to reinvent itself within a vastly different networked world, and to redefine its role among a divergent set of organizations and needs. For instance, the Gulf of Maine Council now had many alternatives for electronic communications, and it was now more interested in its online presence than in the technical or scientific issues addressed by the EDIMS databases.

Unlucky timing was only one part of the EDIMS story; another was the limited resources given to building an information infrastructure. With annual budgets on the order of US$50,000 (about one-tenth those of GLIN and StreamNet), dubious support and direction from the Gulf of Maine Council, limited information-systems expertise, and little influence on its partner agencies, the EDIMS project had difficult odds. A third piece of the story was the lack of a clear shared goal among participants. Indeed, few in the region saw themselves as Gulf of Maine citizens, and the region's most obvious trans-boundary resource, the dwindling offshore fishery, was a federal issue, outside the purview of the Gulf of Maine Council. So overall, the EDIMS project was an interesting, early vision of regionally distributed information, but unexpected events and organizational weaknesses prevented it from achieving its goals.

### *32.3.3. The Pacific Northwest StreamNet and its precursors (http://www.streamnet.org)*

Beginning in 1984, the states and Indian tribes of the Pacific Northwest region of the United States (Montana, Idaho, Washington, and Oregon) worked with the US Bonneville Power Administration to build two region-wide rivers information systems to support the management of riparian fisheries and other natural resources.

The first of these, the Northwest Environmental Database (NED), was a geographic database on fisheries, wildlife, and hydroelectric facilities throughout the four-state region. It was built from a detailed Rivers Study in 1985–7, using the US Environmental Protection Agency's River Reach Files to codify the hydrologic network. Its comprehensive, regionally consistent information was key to a landmark policy decision in 1988, which designated 44,000 miles (71,000 km) of the river network as Protected Areas, off-limits to federal hydroelectric development.

NED's regional consistency and maintenance declined after about 1990, but it spurred a second region-wide rivers database, the Columbia River Coordinated Information System (CIS). This database tracked anadromous (i.e., ocean-migrating) species such as salmon and steelhead within the Columbia River Basin, using the same Reach File identifiers. Both systems were built before the Internet was widely available, and they shared digital data by traditional non-networked methods, via coordinators in each state using the common stream identifiers.

Both systems encountered challenges in maintaining region-wide standards and joint usage among agencies with a wide spectrum of technological maturity. In particular, as participating state agencies developed their own GIS capabilities in the early 1990s, they tended to abandon the regionally consistent Reach Files, at a 1:250,000 scale, in favor of more detailed local identifiers at 1:100,000 or larger scales. Anticipating this shift in 1985, NED's designers had planned to update the Reach Files to larger scales within about three years; but usable 1:100,000-scale products didn't become available until nearly ten years later.

Another challenge was maintaining a productive rivalry among regional database and standards efforts. NED and CIS each had proponents and detractors, and some duplication of effort and staffing remained unresolved for several years. In addition, the US Forest Service was heading up regional inter-agency responses to the spotted-owl controversy and related environmental concerns.

In 1996, NED and CIS were merged into an Internet-based repository and data-interchange system known as StreamNet (BPA 1996), with state-of-the-art data querying and interactive map-building functions, and a steering committee (almost a separate shadow organization) that maintains shared ownership of technical and business aspects of the infrastructure. By some accounts at least, StreamNet became regarded as the region's principal source of regional rivers and fisheries data; whether and how this will grow into a fully distributed data resource remains to be seen.

### *32.3.4. Synthesis of the three cases*

From an organizational viewpoint, the three cases illustrate well the importance of a clearly articulated common goal in building and maintaining infrastructures for sharing information. The Great Lakes case succeeded thanks in part to several widely held regional environmental concerns; the Pacific Northwest banded together to harmonize fisheries and hydroelectric development; and in the Gulf of Maine case, the lack of a clear shared goal kept commitment levels low.

Second, all three infrastructures had to define themselves in relation to other, related information activities. Although all three infrastructures were initially conceived as the single or primary gateway to data for their respective regions, none held that simple luxury for long. Furthermore, all three cases found it easy to build a scaffolding, a temporary technical and institutional structure to get started; but moving to a distributed architecture, or to dynamic data rather than fixed brochures, or maintaining an infrastructure over time, were much more difficult. In no case did a comprehensive blueprint exist for assembling the mix of networked technologies, data structures, institutional relationships, and resources needed for an effective infrastructure. Improvised learning and growth were at least as important as any *a priori* set of factors.

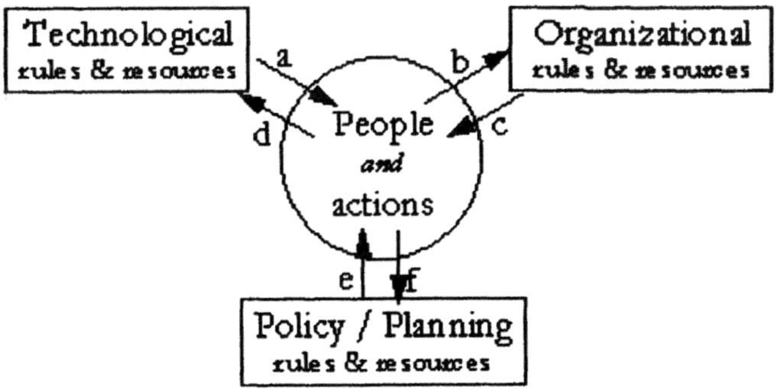


Figure 32.1. Mutual influence of technology, organization, and policy/planning on people and actions.

A structuration model (Figure 32.1) offers one way to interpret this complex set of influences, and to reconcile individual choices and events with broader trends over time. Such a model highlights the mutual influences between people's actions and the structures (that is, the rules and resources) within which they operate. In these cases, the structures of concern are (1) technologies (e.g., data standards and software tools), (2) organizations (e.g., teams and committees, expertise, budgets), and (3) policies and plans (congressional decrees, jurisdictions, laws, etc.). In this model, each of these structures exerts some influence (arrows a, c, e) on the individuals involved, who may respond by trying to alter the various structures that govern their actions (arrows b, d, and f). For instance, in the Pacific Northwest case, a simple data standard led many to collaborate on a shared regional database (arrow a); their actions helped to bring about the Protected Areas consensus (arrow f). In the Great Lakes, one person's initiative and energy did much to build a norm of shared information among the region's various groups (arrow b). In the Gulf of Maine, regional environmental stewardship was a fairly new idea, so collaborative work was difficult (arrow c); and so on.

These reciprocal influence patterns provide a persuasive model of growth and change, and suggest a number of levers for perturbing existing behavior and guiding it towards a particular target—though not with fully certain outcomes, as all of these influences are mediated by human actions. Thus, constructs as complex as information-sharing infrastructures need to be designed and built as living, growing organisms rather than fixed artifacts. This requirement may be especially important when dealing with geographic information systems, for which the technical tools are still evolving rapidly, and which tend to encourage relationships that cut across traditional hierarchies and boundaries (as implied by Chorley 1988 and Goodchild 1992). For organizations, this loosely planned, nimble approach to building and deploying real-world information infrastructures implies a shift from autonomy to interdependence, with policies and plans defined not by authority within a hierarchy, but by persuasion within an organizational ecology (akin to Moore's 1996 *business ecosystem*). This loosely coupled interdependence is an organizational counterpart to the interoperability concept that has become popular in information systems.

Interdependent organizations with loose boundaries and relationships will tend to favor and thrive on certain styles of technology: for instance, modular, layered components, rather than predefined *stovepipe* suites of hardware and software, allow for more improvised learning and change. Such organizations will also tend to blur the traditionally separate technologies of internal data management and external data dissemination. The following section clarifies what such a technology style might look like, using a specific example, a prototype spatial data service on the National Spatial Data Infrastructure.

## 32.4. Lessons from an interactive online orthophoto data service

The organizational case study just described provides insights into the state of the art of information sharing infrastructures as they emerged among government agencies in North America in the mid- to late 1990s. To further clarify these findings, and to ensure their continued relevance in a rapidly changing technological context, a second phase of this study took a different approach, and built a prototype online service for digital orthophotos. (Orthophotos are aerial photographs that have been aligned digitally with a standard geographic coordinate system. In the US and elsewhere, many public agencies have built digital orthophoto series, but have found them difficult to distribute to users due to their large size and the difficulty of integrating them with other maps and GIS datasets.) The orthophoto service was developed as part of the US National Spatial Data Infrastructure (NSDI; National Research Council 1994; Tosta and Domaratz 1997), with support from the Federal Geographic Data Committee (FGDC) and the Massachusetts Executive Office of Environmental Affairs.

The prototype (online at *http://ortho.mit.edu*) provides orthophoto data and standardized metadata through intuitive Web interfaces. It has also proven to be a useful testbed for learning how to give a broad class of users networked access to the orthophotos and metadata; how to distribute the orthophotos efficiently over a slow network; and how to link the online orthophoto service to client-side GIS software.

### *32.4.1. Design goals*

The orthophoto service followed several design goals, within an overall rubric of making digital orthophotos more accessible to a wide audience. The first of these, as mentioned above, was to build onto the National Spatial Data Infrastructure (NSDI): this allowed the project to fit into a broader community and related activities, and to suggest likely futures for the NSDI's Clearinghouse and Framework initiatives. Second, an important part of making the orthophotos more accessible was to provide multi-resolution views. The assumption was that few users wanted entire images (over 60 MB apiece)—that most wanted either low-resolution overviews, or high-resolution snippets of small areas. Third, given that many potential users had thin, dial-up access to the Web, it was important to use network bandwidth sparingly. A fourth design goal was to piece together simple, easily available software components, to facilitate experimenting with the service and adapting it to other data series or operating environments. A final goal was to ease the task of integrating online image data with local maps and common desktop GIS software, so as to begin bridging the gap between local and remote data.

### *32.4.2. Development process*

To accomplish these goals, a process of learning-by-doing was important: that is, a development path that provided early and frequent products, from which more complex systems could be built. The first outcome was a set of metadata files describing the images, which complied with the US federal metadata content standard (FGDC 1994) and were searchable via the Internet's Z39.50 protocol (Kahle 1991). Next was a simple, preliminary user interface to pre-tiled image excerpts, produced by cutting each large orthophoto into 16 equal-size pieces; then cutting each of these pieces in the same way, into snippets of about 80kB each (after compression); and resampling the larger images down to the same size.

This scheme provided an early online presence for the orthophoto service, but its limitations were clear. On the client side, the rigid tiling kept users from precisely specifying their desired viewport; on the server, each orthophoto required an impractical number of files. Extracting sub-image snippets in real-time, using the Web server's Common Gateway Interface (CGI) to run scripts on the server, promised to address both of these limitations. After some experimentation, a modified version of the *pnmcut* freeware utility from the *pbmplus* suite (© Jef

Poskanzer; available at *http://www.acme.com*) proved adequate to the task. This led to a new data-serving architecture based on extracting and compressing image snippets in real time, rather than choosing among static image files.

This new scheme (Figure 32.2) used Unix pipes to control streaming input and output in real time. First, the *pnmcut* utility extracts a user-specified portion of a master image at one of three resolutions. It then sends the sub-image through one of three image compressors/formatters (*cjpeg*, *ppmtogif*, or *pnmtotiff*, all part of the *pbmplus* suite). If requested, a GeoTIFF header is put on the image; and the result is sent back to the client, with a Web protocol (*http*) header.

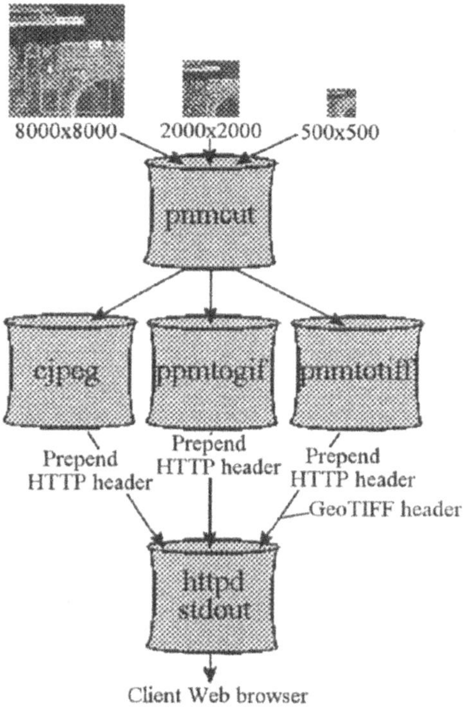


Figure 32.2. Dynamic image extraction and compression.

With this architecture in place, a final step was to build similar real-time services to create customized geographic header files for each snippet, so that users could easily incorporate a downloaded image into a few common desktop GIS software packages.

The resulting service provides an intuitive interface to over 7 GB of digital orthophotos in several popular graphic formats, with encapsulated georeferencing information. Furthermore, it does use the network sparingly (most image views are under 80 kB in size), yet its resizable viewport serves high-bandwidth (or very patient) users as well, with a throughput of about 90kB/s through the real-time data-processing pipes. The server-side scripts are easily modified to suit new orthophoto series, and the maintenance overhead on the server is minimal (two extra files per orthophoto, occupying about 7% additional disk space). Finally, thanks to its standard metadata content and Z39.50 query interface, people across the Internet can discover this service, peruse metadata entries to determine its fitness for their use, and use hypertext links in the metadata to get to the images themselves.

### *32.4.3. Interpretation*

By several measures, the orthophoto service has attracted a large audience. According to the Web server's access log, it has continued to see several thousand hits per day since it was launched in the fall of 1996. The logs also illustrate the variety of sites accessing the server, including educational institutions, city, state, and federal government agencies, and firms involved in construction, engineering, logging, real estate, mapping, and photogrammetry. Anecdotal evidence sheds light on the service's many uses. Boston city government staff, for instance, or architects and engineers working on Boston's Central Artery, found that the orthophotos provided quick, detailed, easily understood views of urban patterns, useful for communicating with neighborhood associations and for correcting and updating street maps.

Although the Massachusetts GIS data center sells bundles of these images on CD-ROMs at low cost, using the orthophoto service's custom-prepared data snippets is often easier than making a telephone call to order the full images. Granted, this shift towards data interdependence wasn't immediate or universal: at least one visitor to the site systematically downloaded a large portion of the image data for future internal use. But as the orthophoto service has proven reliable and remained available, other users have opted to retrieve imagery data only as needed. Furthermore, beyond public data access, a service like this one also lessens the data packaging needed to support internal purposes.

Finally, although most users get to the imagery data through a browser using one particular Web form, requests to the orthophoto server may also be embedded in HTML pages, word-processing documents, or desktop mapping systems. On the server side, data requests could be sent to other servers on the network, in order to tap into adjacent or overlapping data libraries. Thus, this orthophoto service is built to fit into a tapestry of independent, heterogeneous software components that use a few key pieces of vocabulary to interact. This technological interoperability favors, and thrives on, the loosely-coupled, interdependent organizational structures described earlier, and allows for learning as we go into a complex, uncertain future.

## 32.5. Conclusions

For real-world organizations to share geographic information, building the right kinds of infrastructures will be crucial. A few recent real-world examples of such infrastructures highlight the intertwined and unpredictably changing nature of their technological, organizational, and policy environments. These examples also suggest that building effective inter-organizational mechanisms to share geographic and other information, and harnessing them for joint work, may require larger-than-expected organizational shifts, towards loosely-coupled interdependence.

Interoperable organizations of this sort are likely to function in concert with modular technology components, of which the orthophoto service is a seminal example. This service, which prepares custom orthophoto snippets to suit a diverse audience, is tuned to current bandwidths, smoothly integrated with client-side software tools, easily managed and easily replicable. This service has proven popular at least partly because it is designed not merely as a standalone product, but as a component within a larger "tapestry" of interacting software systems, and thus functions well within interdependent, interoperable styles of work.

Together, these findings have important implications for infrastructure-building efforts at several scales. For instance, they suggest that standards efforts may be best spent defining how different software clients, servers, and peer services will interact, and what part of the communication stack to agree upon, rather than what data formats everyone should use. Furthermore, growing and adapting these specifications over time is crucial to stay abreast of a changing environment and user base. These concerns are of particular interest for the future of the National Spatial Data Infrastructure (which will be built from the diverse data services of individual regions and participants); and for its regional and smaller counterparts. Likely impacts go well beyond technology, possibly leading to ephemeral, task-oriented partnerships between large numbers of organizations, and to broader participation and more effective consensus in policy decisions based on the best available information.

## Acknowledgment

This material is based upon work supported by the National Science Foundation, under Grant No. SBR-9507271, and by the 1995 Competitive Cooperative Agreements Program of the Federal Geographic Data Committee. Any opinions, findings, conclusions, or recommendations expressed in this material are my own and don't necessarily reflect the views of the National Science Foundation or of the Federal Geographic Data Committee.

## References

Alexander E R 1995 *How Organizations Act Together: Interorganizational Coordination in Theory and Practice*. Gordon and Breach

August P V 1991 Use vs abuse of GIS data. *Journal of the Urban and Regional Information Systems Association* 3(2): 99–101

Baker T, Broadhead J 1992 Integrated access to land-related data from heterogeneous sources. In Salling M J (ed) *Proceedings of the 1992 Annual Conference of the Urban and Regional Information Systems Association (URISA)*. URISA, Washington DC

Barley S R. 1986 Technology as an occasion for structuring: evidence from observations of CT scanners and the social order of radiology departments. *Administrative Science Quarterly* 31:78–108

Batini C, Lenzerini M, Navathe S 1986 A comparative analysis of methods for database schema integration. *ACM Computing Surveys* 18(4): 323–364

Berners-Lee T J, Cailliau R., Groff J-F 1992 The world-wide web. *Computer Networks and ISDN Systems* 25(4–5): 454
Boland R J, Tenkasi R V 1995 Perspective making and perspective taking in communities of knowing. *Organization Science* 6(4): 350-372
Bonneville Power Administration (BPA) 1996 *StreamNet: Report on the status of salmon and steelhead in the Columbia River basin, 1995.* Bonneville Power Administration, Portland, Ore.
Brodie M L 1993 The promise of distributed computing and the challenges of legacy information systems. In Hsiao D K, Neuhold E J, Sacks-Davis R (eds) *Interoperable Database Systems (DS-5).* North-Holland, Elsevier: 1–31
Brown W S, Garrison K M, MacBeth M 1994 *Prototype Environmental Data and Information Management Systems (EDIMS). http://gulfofmaine.unh.edu/edims/edims_description.html*
Buehler K, McKee L (eds) 1996 *The OpenGIS Guide: introduction to interoperable geoprocessing.* Part I of the Open Geodata Interoperability Specification (OGIS). Open GIS Consortium, Wayland, Mass. *http://www.opengis.org/public/guide.html*
Carter J R 1992 Perspectives on sharing data in geographic information systems. *Photogrammetric Engineering and Remote Sensing* 58(11): 1557–1560
Chorley R 1988 Some reflections on the handling of geographic information. *International Journal of Geographic Information Systems* 2(1): 3–9
Craig, W J 1995 Why we can't share data: institutional inertia. In Onsrud H J, Rushton G (eds) *Sharing Geographic Information.* Rutgers University, New Brunswick, NJ
Croswell P L 1989 Facing reality in GIS implementation: lessons learned and obstacles to be overcome. In Salling M J, Gayk W F (eds) *Proceedings of the 1989 Annual Conference of the Urban and Regional Information Systems Association (URISA).* URISA, Washington DC: 15–35
DeSanctis G, Poole M S 1994 Capturing the complexity in advanced technology use: adaptive structuration theory. *Organization Science* 5(2)
Epstein E F 1995 Control of public information. In Onsrud H J, Rushton G (eds) *Sharing Geographic Information.* Rutgers University, New Brunswick, NJ
Evans J D, Ferreira J 1995 Sharing spatial information in an imperfect world: interactions between technical and organizational issues. In Onsrud H J, Rushton G (eds) *Sharing Geographic Information.* Rutgers University, New Brunswick, NJ
Evans J D 1997 *Infrastructures for Sharing Geographic Information among Environmental Agencies.* PdD Thesis, Massachusetts Institute of Technology. *http://mit.edu/jdevans/thesis*
Federal Geographic Data Committee (FGDC) 1994 *Content Standards for Digital Geospatial Metadata.* Federal Geographic Data Committee, Washington DC
Frank A U 1992 Acquiring a digital basemap: a theoretical investigation into a form of sharing data. *Journal of the Urban and Regional Information Systems Association* 4(1): 10–23
Frank S 1994 The National Spatial Data Infrastructure: designing navigational strategies, *Journal of the Urban and Regional Information Systems Association* 6(1): 37–55
Fulghum R 1987 *All I Really Need to Know I Learned in Kindergarten.* Villard Books, New York: 6
Giddens A 1984 *The Constitution of Society.* University of California, Berkeley
Goodchild M F 1992 Geographic information science. *International Journal of Geographic Information Systems* 6(1): 31-45
Glaser B G, Strauss A L 1967 *The Discovery of Grounded Theory.* Aldine Publishing Co, Chicago
Grandori A, Soda G 1995 Inter-firm networks: antecedents, mechanisms, and forms. *Organization Studies* 16(2): 183–214
Granovetter M S 1972 The strength of weak ties. *American Journal of Sociology* 78(6): 1360–1380

Kahle B 1991 An information system for corporate users: Wide Area Information Servers. *Online Magazine* 15(5): 56–62

Litwin W, Mark L, Roussopoulos N 1990 Interoperability of multiple autonomous databases. *ACM Computing Surveys* 22(3): 267–293

Mackay D S, Robinson V B 1992 *Towards a Heterogeneous Information Systems Approach to Geographic Data Interchange*. Institute for Land Information Management Discussion Paper 92/1. University of Toronto, Mississauga, Ontario

Markus M L, Robey D 1988 Information technology and organizational change: causal structure in theory and research. *Management Science* 34(5): 583–598

Moore J F 1996 *The Death of Competition: Leadership and Strategy in the Age of Business Ecosystems*. HarperCollins, New York

Morehouse S 1990 The role of semantics in geographic data modeling. In *Proceedings of the 4th International Symposium on Spatial Data Handling, Zurich, Switzerland*: 689–698

National Research Council 1994 *Promoting the National Spatial Data Infrastructure through Partnerships*. National Academy Press, Washington DC

Nyerges T L 1989 Schema integration analysis for the development of GIS databases. *International Journal of Geographical Information Systems* 3(2): 153–183

Nyerges T L 1991 Geographic information abstractions: conceptual clarity for geographic modeling. *Environment and Planning A* 23: 1483–1499

Obermeyer N J 1995 Reducing inter-organizational conflict to facilitate sharing geographic information. In Onsrud H J, Rushton G (eds) *Sharing Geographic Information*. Rutgers University Press, New Brunswick, NJ

Orlikowski W J 1992 The duality of technology: rethinking the concept of technology in the context of organizations. *Organization Science* 3(3): 398–427

Peckham J, Maryanski F 1988 Semantic data models. *ACM Computing Surveys* 20(3): 153–189

Pinto J K, Onsrud H J 1995 Facilitators of organizational information sharing: a research framework. In Onsrud H J, Rushton G (eds) *Sharing Geographic Information*. Rutgers University Press, New Brunswick, NJ

Pinto M B, Pinto J K, Prescott J E 1993 Antecedents and consequences of project team cross-functional cooperation. *Management Science* 39(10): 1281–1297

Ratza C A 1996 The Great Lakes Information Network: the region's Internet information service. *Toxicology and Industrial Health* 12(3/4): 557–561

Rhind D W 1992 Data access, charging, and copyright and their implications for geographical information systems. *International Journal of Geographical Information Systems* 6(1): 13–30

Schwartz M F, Emtage A, Kahle B, Neuman B C 1992 A comparison of Internet resource discovery approaches. *Computing Systems: the Journal of the USENIX Association* 5(4): 461–493

Sheth A P, Larson J A 1990 Federated database systems for managing distributed, heterogeneous, and autonomous databases. *ACM Computing Surveys* 22(3): 183–236

Siegel J 1996 *CORBA Fundamentals and Programming*. Wiley, New York

Siegel M D, Madnick S E 1991 Context interchange: sharing the meaning of data. *ACM SIGMOD Record* 20(4): 77–78

Tosta N, Domaratz M 1997 The US National Spatial Data Infrastructure. In Craglia M, Couclelis H (eds) *Geographic Information Research: Bridging the Atlantic*. Taylor & Francis, London

Van de Ven A H, Walker, G 1984 The dynamics of interorganizational coordination. *Administrative Science Quarterly* 29(4): 598–621

Yin R K 1984 *Case Study Research: Design and Methods*. Sage Publications, Beverly Hills, CA

# 33 SPATIAL INTERNET MARKETPLACES FROM A DATABASE PERSPECTIVE

Volker Gaede

## 33.1. Introduction

Internet marketplaces have recently been proposed as a new and interesting model to make data and computational services available to a broad public (Abel 1997; Abel et al. 1997; Günther et al. 1997). Unlike electronic commerce (on-line shopping) where services are only requested over the Internet, Internet Marketplace services are both requested and delivered through the Internet. The key idea of Internet Marketplaces is that providers make a range of services available on the Internet and customers rent them whenever necessary. Such a market model offers a range of advantages over conventional marketplaces and is attractive for both customers and providers (Gaede 1998).

The idea of Internet Marketplaces has been adapted to a range of different application areas including decision support systems (Bhargava et al. 1996;l Jeusfeld and Bui 1997), mathematical software (Günther et al. 1996), and spatial information systems (Abel 1997). Among the different proposals for marketplaces, *spatial Internet Marketplaces* seem to have a somewhat distinct role for the following reasons:

- They address a wide range of different communities of interest (e.g., Earth observation, global positioning, environmental management).
- Geographic space acts as a common underlying domain.
- There is a pressing need for data and computational services.

Advanced spatial applications require access to up-to-date data, specialized hardware and tailor-made software modules. Furthermore, answering complex spatial requests often requires the combination of various disparate and possibly distributed services. For these reasons, spatial Internet Marketplaces may have a much greater commercial potential than other marketplaces.

Before a seamless combination of services is possible, there are many interoperability issues to be resolved. In the area of spatial information systems the recent Open Geodata Interoperability Specification (OGC 1997), proposed by the Open GIS Consortium, specifies a first framework on which to build a spatial Internet Marketplace. Recently, a concrete implementation of a first spatial marketplace design has been proposed in the context of the SMART project currently conducted at CSIRO Mathematical and Information Sciences (Abel et al. 1997). The proposed SMART marketplace consists essentially of two different kinds of services: data and computational services. Services participating in the spatial marketplace have to register with a service, called *registry*. Registered services have to provide a pre-defined interface. Besides the registry service, the SMART marketplace envisages the existence of services for execution and planning. In comparison with other Internet marketplaces, both services are unique to SMART. Execution services accept as input execution plans and safeguard their execution. Planning services, on the other hand, are intended to facilitate service access and to help customers in solving complex problems requiring the combination of several, possibly distributed services. Except for the SMART model, none of the above Internet marketplace proposals caters for automatic planning. Customers in these marketplaces are expected to have full comprehension of the provided services and how to combine them. Obviously, this hampers the usability and therefore a swift dissemination of the services offered.

In marketplaces that support automatic planning, customers ideally state only what they want and the planning service determines an execution plan specifying how this can be computed using existing services. In other words, the planner accepts a declarative specification of customers' requests and determines automatically from which site to acquire data and where to perform certain computations. Such a planning procedure involves identification of suitable services, generating the corresponding requests for each of them, and devising ways to combine the computed results. At first glance, the outlined marketplace planning problem appears to be very similar to execution planning in federated databases, but a more detailed comparison reveals that they are quite different. Roughly speaking, the latter can be considered as a special case of the former. Further results concerning the planning problem can be found in a companion paper (Gaede 1998) where we identify three orthogonal dimensions whose combination yields different marketplace designs. The concrete parameter combination determines whether the marketplace is likely to offer support for planning.

The discussion of the planning problem shows that Internet marketplaces require new languages for specifying requests. While existing database query languages provide a suitable starting point for the design of an Internet marketplace request specification language, none of them provides the functionality required. In the second part of the paper, we argue some of the extensions necessary for Internet marketplaces. Again, further details together with a preliminary definition of a marketplace request specification language can be found in a forthcoming technical report.

Even though most of the results presented in this paper have been derived in the context of the SMART project, we believe that they apply to any Internet marketplace offering similar concepts. In the remainder of the text, we shall therefore not restrict our discussion to spatial aspects but, unless noted otherwise, speak of Internet marketplaces in general. Since databases are obviously closely related to the issues addressed, we use them as a reference point in our analysis. The intentions of this approach are manifold. First, it helps us in getting a better understanding of the differences between the two paradigms. Second, it clearly shows the limits of federated database systems and to what extent their concepts can be carried over to Internet marketplaces.

The chapter is organized as follows. Section 2 briefly reviews related work. In Section 3 we give a definition of planning and discuss how the unique requirements of Internet marketplaces affect the planning problem. This section presents a comparison of execution planning in federated database systems and Internet marketplaces. In Section 4 we argue why current query languages for databases or semi-structured data need to be extended for Internet marketplaces. Section 5 concludes the chapter with a summary and open research issues.

## 33.2. Related work

The integration of autonomous, heterogeneous data resources and computational services is probably one of the major challenges in the future. As for the integration of data resources there exists a large body of literature (for an overview, see for example, Hull 1997). To deal with existing heterogeneity, federated database systems typically adopt an architecture consisting of the following two major components:

- An integrated federated schema which stores the schema information as made available by the component database systems. This information is represented in a data model which is semantically rich enough to represent all concepts of the participating databases. It is also used to formulate global requests.
- Wrappers (translators) are responsible for transforming queries as issued by the federated database system into corresponding requests against the underlying system(s) and translating (and possibly assembling) the result returned.

Figure 33.1 shows a setting consisting of a federated schema and $i$ component systems with their corresponding wrappers.

The central assumption in federated databases is that the export schemata of the participating database systems have been integrated up-front into the federated schema and existing conflicts (e.g., structural, semantical) have been resolved. Much work has been done concerning automatic integration (Buneman et al. 1992a,b; Kim et al. 1995), but typically human intervention is still required. Consequently, this approach does not scale well with the expected number of providers in an Internet marketplace. Developing ontologies (Gruber 1993) capturing the services' semantics might mitigate or even resolve these problems for certain restricted application areas.

As for the integration of computational resources, the area of distributed computing has developed a wide range of different models. Proposals such as the Object Management Architecture (OMA) by the Object Management Group (OMG) (Orfali et al. 1996; Vinoski 1997) or DCE (Bloomer 1995) address this problem at least partially, by providing a framework for application interoperability and resource discovery. The OMA assumes that participating services advertise their interfaces. The OMA also caters for domain independent interfaces (object services) which can be used for example to search for services by their names or properties. The communication between the different services is handled by the Common Object Request Broker. In the field of database systems this work has been complemented by initiatives to facilitate the data exchange among databases. For example, the Object Database Management Group (ODMG) proposed an Object Model with its corresponding Object Definition Language and query language together with mappings to object-oriented languages (Cattell 1994). The goal of the ODMG was, however, not so much interoperability but portability of applications among different object-oriented database systems.

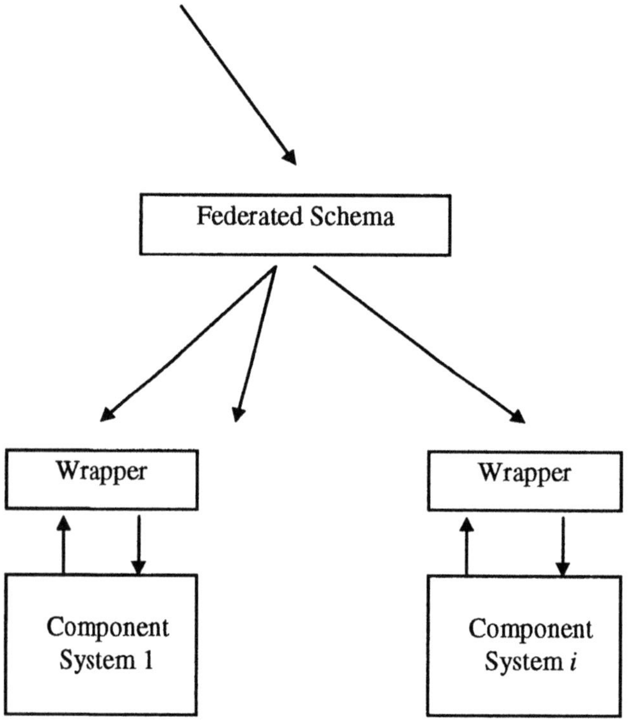


Figure 33.1. Simplified federated database setting consisting of a federated schema and *i* component systems with their corresponding wrappers.

## 33.3. Execution planning

Before we discuss differences of execution planning in Internet marketplaces and federated database systems, let us first clarify the meaning of execution planning. A first definition of execution planning is that a designated component, called *planner*, accepts as an input a declarative request. The planner generates by means of pre-stored or dynamically acquired information automatically an execution plan, i.e., a program which materializes the given request upon execution. The information necessary to generate such an execution plan is often called meta information. This information is of crucial importance for deciding questions such as where to execute certain operations, which algorithm to use for materializing a given request, and so on. Depending on the decisions taken the quality of a plan might differ considerably in terms of execution time, cost, or other criteria. With this abstract definition in mind, let us now look at execution planning in federated database systems.

### *33.3.1. Federated database systems*

In federated database systems, execution planning differs depending on the underlying architectural model. Basically, we can distinguish those models which assume a global integrated schema (federated schema) such as the five-level schema architecture (Sheth and Larson 1990) and those requiring a database-by-database integration. Examples of the latter are the import-export schema architecture (Heimbigner and McLeod 1985) or the federated database architecture (Litwin et al. 1990).

In the presence of a globally integrated schema, users express queries using the global information. The federated planner uses this information to map and possibly decompose the global request GR into a set of local requests $LR_i$, each of which is tailor-made for a particular service. These local requests $LR_i$ are forwarded to the wrappers of the involved services. Typically, the wrapper performs an additional transformation step to map $LR_i$ onto the underlying system. Queries as expressed by the user do not contain information specific to the local services, and from a user's viewpoint the federated system is similar to a non-distributed database. Therefore, it is possible to use conventional database query languages such as SQL for expressing queries in this environment.

In the absence of a global schema such an approach is no longer feasible and the query language has to offer provisions to access the local schema information. Examples of such query languages are MSQL (Litwin et al. 1990) and SchemaSQL (Lakshmanan et al. 1996). Even though queries expressed in one of these languages are more specific to the component database systems, the federated planner still has to perform some decomposition and mapping work. Common to both models, however, is the use of a declarative query language for expressing requests.

A typical database request contains the data sets involved in the query, attributes of interest, and relations between attributes such as equality, greater-than or substring. In addition to these basic operators, there might also be support for the incorporation of more complex, possibly user-defined operators. In general, we note that the set of operators is fairly restricted and often not extensible. To some extent there is also support for transforming attributes of one domain into another one, but such transformations have to be specified explicitly by the user and are not performed automatically by the planner. At query submission time the planner expects that queries are syntactically and semantically correct. This requires among other things that all operations are domain-compatible with respect to the attributes to which they are applied.

### 33.3.2. Comparison: Internet marketplaces and federated databases

Even though planning in Internet marketplaces is expected to be similar in various aspects to planning in federated database systems, there are a number of crucial differences to note. As with federated databases, the complexity of the planning problem and the detailed procedure are determined by the chosen architecture. In the upcoming discussion we assume no particular architectural model, but try to identify the key differences with respect to different requirements imposed by the Internet.

Potential customers in an Internet marketplace are expected to have no or only a limited knowledge of the services available and how to access them. Instead of referring to particular services, Internet marketplace customers simply describe what kind of services they are interested in and a planning service should identify suitable services for answering the given request. For example, a customer should be able to express such requests as: "Intersect the Australian crop map from 1996 with the Australian soil map." Assuming the planner accepts the request specification of the above request as an input, the planning sequence has to include a resource discovery step which aims at identifying services useful for answering the request. As part of this step, the planning system also has to perform some classification of the identified services. Subsequently, it is possible to perform a selection and integration of the equivalence classes. We note that the integration step might necessitate an additional resource discovery step in order to enable the integration (e.g., identify an additional transformation service).

The sheer size, variety, and dynamics of an Internet marketplace make it impossible to hold cost information about the offered services in a global schema. Additionally, services might have special offers for answering certain requests. Hence, for planning it is necessary to contact selected services to acquire cost and quality estimates. This, in turn, requires that the planning service generate an appropriate request prior to contacting individual services. As a result, planning is no longer restricted to a single site as in federated databases but involves multiple sites. Obviously, this makes planning a very costly undertaking.

In summary, we expect that Internet marketplace planning consists of the following steps:

1. *Resource discovery and classification step.* The planning service accepts a customer's request and by means of the information stored in the registry performs (if possible) a first selection of qualifying services, that is, identifies classes of equivalent services. If the registry's information is not adequate or insufficient, the planning service contacts other registries or services directly.
2. *Estimation and integration step.* In the next step the planner generates requests for each of these equivalence classes and contacts the various services within each class to acquire further information (e.g., cost estimates, precision). In case two service classes are incompatible, the planning service tries to identify a transformation service to make them compatible, i.e.,

conducts another resource discovery step. Clearly, the cost induced by such a transformation service has to be added to the overall cost.

3. *Selection step.* On the basis of the returned information, the planning service ranks eligible services and generates a set of alternative plans. The result of this step depends on the chosen marketplace architecture (Gaede 1998).

In a spatial Internet marketplace we expect a large number of providers to be competing in the same market segment by offering equivalent or closely related services. Besides this inherent replication of services, providers might decide for reasons of availability or global presence to offer their services at multiple distributed sites. This adds further to service replication. As a result, the planning service has to select services among a possibly wide array of equivalent services. Compared to federated databases this kind of replication is not restricted to data but also carries over to computational services. The variety of equivalent services is likely to be significantly larger than in federated databases. While such a setting is desirable from a user's point of view, it complicates planning considerably. Without additional information about services it is hard (if not impossible) for the planner to discern them. Even though execution cost can be used as a first criterion to distinguish among equivalent services, it is hardly sufficient. For example, users might prefer reliable but more expensive services over cheaper services. Therefore, planning in an Internet marketplace must incorporate not only a single cost measure but also possibly a variety of different criteria (e.g., minimal usage of resources, up-to-date information, numerical robustness). This in turn requires that the planning service can access information concerning the quality of services. There is no equivalent concept in federated database systems.

In comparison to federated database systems, Internet marketplaces are expected to be much more dynamic. As new services enter the marketplace other services close down, change their interfaces, or are simply not available due to some problem. A planning service in such an ever-changing environment has to keep track of these changes and adapt gracefully to them. Such an adaptation is hampered by the fact that services in an Internet marketplace are even more autonomous than component databases in a federated database system. Related to the above problem is the innate dynamics of the Internet. For example, down-loading a data set from a German site to Australia in the early afternoon might take only a couple of minutes, while it takes hours in the late evening. Data throughput is of minor importance in federated database systems, since their current use is mostly restricted to local or metropolitan area networks in contrast to networks such as the Internet. Because of the constantly and rapidly changing environment, the planning service also has to cater for dynamic service substitution. Services available at planning time might be down or overloaded at execution time, requiring the use of alternative services. Since planning is considered to be a very expensive task and the likelihood of such problems is comparatively high, we require planning services to incorporate contingency information into the plan.

In database systems, planning is performed at two levels: the logical and the physical level (Graefe 1993). The logical level deals with a restricted set of conceptual operators such as join and projection; it represents the mathematical underpinning of database systems. By manipulating these logical operators, the planner (optimizer) aims at finding a first execution sequence based on these high level concepts. In a next step, each of these logical operators is associated with a physical operator, that is, an algorithm implementing the logical operator. Unless it is possible to develop for Internet marketplaces similar mathematical foundations as for current (relational) databases, there is no such distinction.

In general, federated database systems only support a restricted set of functions, which is typically not extensible. In Internet marketplaces we expect the number of supported functions to be much larger. Furthermore, the marketplace has to provide the capability for providers to extend the set of already existing operator and function definitions by new ones. As for planning, this means that Internet marketplace planners are expected to understand and use the newly introduced concepts. Consequently, planners have to be inherently extensible.

Table 33.1 summarizes the above discussion.

| **Feature** | **FDBMS** | **IM** |
|---|---|---|
| Support for imprecision | no | yes |
| Service replication | low | high |
| # of participating systems | low | high |
| Dynamics | low | high |
| Schema integration | up front | up front/ dynamically |
| Service substitution | no | yes |
| Cost measure | one | multiple |
| Scope | LAN/MAN | WAN |
| Resource discovery | no | yes |
| Extensibility | limited | unlimited |
| Logical/physical Algebra | yes | no |

Table 33.1. Comparison between federated database systems (FDBMS) and Internet Marketplaces (IM) with respect to planning.

The above discussion clearly shows that there are numerous differences between execution planning in Internet marketplaces and federated database system. In fact, the latter turns out to be a special case of the former. Unless we impose severe restrictions on the marketplace most of the assumptions made in database systems do not hold for Internet marketplaces. Having such restrictions is neither in the spirit of the Internet nor in the interest of the marketplace. Experience in other areas, however, shows that support for practically useful planning requires the existence of certain standards. In conclusion, if we do not want to give up on the idea of having some support for planning, the major challenge is to find a good balance between the interests of the service providers and their customers and the necessities of an Internet marketplace which enables planning. In other words, we have to identify a marketplace architecture that offers a good compromise between the interests of the marketplace participants without sacrificing the idea of planning. Gaede (1998) identifies such an architecture.

## 33.4. Request specification language

In the preceding section, we noted that request specification languages for Internet marketplaces need to support new concepts. Simply adopting an existing query language is anything but satisfactory, since none of the existing languages caters for all aspects desirable in an Internet marketplace. As the requirement analysis in Section 4.1 shows, not even query languages for semi-structured data (Abiteboul et al . 1997; Fernandez et al. 1997) cater for all desirable features.

### *33.4.1. Requirement analysis*

Though declarative languages such as SQL have been criticized in the past, there is little doubt that these languages facilitate considerably the interaction with often highly complex systems by abstracting from their underlying technical details. Common to declarative database languages is that they are based on simple concepts and are typically easy to comprehend. With respect to Internet marketplaces we conclude that the availability of a declarative request specification language is a stepping stone for the simple access of services.

In Internet marketplaces we expect the participation of various different services. For example, object-oriented systems are expected to run side by side with hierarchical database systems. As a result, the query language upon which Internet marketplaces are built should not be specific to a particular data or computational model; it should be model independent. Similarly, it should be possible to translate the request specification language easily into the respective language of the underlying system. Therefore, the request specification language should exclude neat but complicated concepts. Precluding complex concepts simplifies the development of interfaces to legacy systems and thus reduces the likelihood of errors.

Adopting a query language such as SQL3 is hardly sufficient for Internet marketplaces for the following reasons. First, the language is geared towards relational databases even though it caters for some object-oriented concepts. Second, SQL in its current form is not extensible and prohibits the integration of new functions and operators. Third, there is an emphasis on data services and a lack of support for other types of services such as computational services. Fourth, though SQL3 is similar in spirit to SQL as a simple, declarative query language, it can often be surprisingly intricate to express simple requests.

Novice users of the Internet marketplace will have only insufficient knowledge of the services available on the marketplace and how to access them. Therefore, an Internet marketplace request specification language has to cater for some imprecision in formulating requests. For example, if users do not have exact knowledge of the service name and the offered functions, the request language should allow them to express the request anyway. It is then the task of the planner to identify qualifying services in a resource discovery step. Furthermore, the request specification language should not force customers to formulate their requests in a service-specific way, but it should be flexible enough to be submitted to a range of services.

A request specification language should allow customers to refer to generic services as well as specific services in a single request. For example, a customer may want to draw her data from a specific service, while not knowing where to perform a subsequent computation. Instead of being forced to identify the computational service, it should be possible to refer to a generic service and leave it up to the planner to choose one.

Another issue related to the design of a request specification language is that of expressiveness. Clearly, such a language should be as expressible as possible, but on the other hand is should guarantee the efficient evaluation of requests. In other words, a request specification language should offer a good balance between expressiveness and data complexity. As experience from deductive database systems shows, often the integration of apparently simple concepts can result in a query language which becomes hard to evaluate in the worst case.

Given that Internet marketplaces are likely to evolve over the years, the query language should facilitate the integration of new concepts. As new technologies, services, and functions become available, it should be possible to integrate and use them seamlessly in the request specification language. This implies that the request specification language should not distinguish between different services, that is, it should not be unique to a service type.

In contrast to database query languages, a request specification language must also allow customers to express preferences for certain kinds of services (e.g., cheap and reliable services). This is particularly important with respect to the identification of appropriate services and service substitution. For example, if the preferred service is not available at execution time, it must be possible to name alternative services in the request. Furthermore, the language should also allow customers to exclude certain services from the list of qualifying services. Similarly, up to a certain level it should be possible to include execution information into the request. Having the possibility to specify the order of service invocation or where the result of a computation should be shipped facilitates planning considerably. Clearly, the above concepts contradict to some extent the declarativity of the language and are clearly intended to be used by the advanced user or the planner.

## 33.5. Conclusions

Internet marketplaces are a new and interesting research direction with huge economical potential. Advantages of such marketplaces include that they can be used as an efficient vehicle for the distribution of information technology whose usage was so far restricted to certain niches; they can be used for building up complex applications drawing from computational and data services residing on spatially distributed sites; and they allow customers to access cutting-edge technology readily. The potential of this approach has been recognized in different application areas, which is reflected by the increasing number of marketplace proposals. Among these proposals, spatial Internet marketplaces have a somewhat distinguished role in that they incorporate a wide range of different communities of interest and geographical space acts as a common underlying domain.

For spatial Internet marketplaces it has been suggested that there is need for planning support, that is, the automatic generation of execution plans given declarative requests. Execution plans for spatial Internet marketplaces can be considered as complex applications integrating services from remote sites to solve a given task.

In the first part of the chapter we discussed the issue of planning in an Internet marketplace and how it differs to planning in federated database systems. From our discussion it becomes clear that the federated database paradigm needs to be revised to support execution planning in Internet marketplaces. For example, the expected number of services makes it impossible to assume that the schema information of the services is integrated and maintained by a single administrator. Furthermore, the federated database model does not allow for service substitution which is considered as being essential in a dynamic environment such as the Internet. To address the specific needs of Internet marketplaces, it is necessary to develop new architectural models that carefully balance the various aspects important for planning. For example, the quality of the information made available by the services is vital for planning. Therefore, it is desirable to extract as much information as possible from the providers. On the other side, registering services should be very simple and ideally require no human intervention while yielding high quality results. Unless it is possible to perform automatic schema integration yielding high quality results, the challenge lies in developing tools permitting a simple registration. For restricted application domains such as spatial Internet marketplaces it seems to be possible to devise such tools. Furthermore, the use of domain knowledge for the automatic integration of registries might prove useful. To the best of our knowledge, no research has been conducted in this direction. Other interesting aspects related to planning include the questions of what kind of information should be stored in the registry and how to maintain it. Issues such as efficient access and domain specific classification are of vital importance for planning. Even though this paper represents a first contribution to a better understanding of planning in Internet marketplaces there are still numerous issues to be addressed (Gaede 1998).

As our analysis of request specification languages for Internet marketplaces shows, existing database query languages have to be modified in order to cater for the specific needs of such marketplaces. Query languages for semi-structured data offer some of the desired functionality (e.g., imprecise queries) but also need to be extended in various ways (e.g., support for preferences). In this chapter, we identified a range of desirable language features. The request specification language currently under development at CMIS, Canberra aims at capturing most of them.

## Acknowledgment

I am very grateful to my colleagues at CSIRO Mathematical and Information Sciences, Canberra, Australia, in particular, David J. Abel, who brought the area of spatial Internet marketplaces to my attention. Financial support from CSIRO, Australia and Humboldt University, Berlin, Germany under grant SFB 373 is also acknowledged.

## References

Abel D 1997 Spatial Internet marketplaces: a grand challenge? *Proc. 5th Int. Symposium on Spatial Databases (SSD'97)*. Springer-Verlag

Abel D J, Gaede V, Taylor K L, Zhou X 1997 *SMART: Towards Spatial Internet Marketplaces*. Technical Report ISS-40. Humboldt-Universität zu Berlin, Germany. *http://www.wiwi.hu-berlin.de/~gaede/smart.ps.gz*

Abiteboul S, Quass D, McHugh J, Widom J, Wiener J L 1997 The LOREL query language for semi-structured data. *Int. Journal on Digital Libraries* 1(1): 68–80

Bhargava H L, Krishnan R, Müller R 1996 *Electronic Commerce in Decision Technologies: A Business Cycle Analysis*. Technical Report. Humboldt-Universität zu Berlin, Germany

Bloomer J 1995 Distributed computing and the OSF/DCE. *Dr. Dobbs Journal* 2

Buneman P, Davidson S, Kosky A 1992a Theoretical aspects of schema merging. *Proc. Extending Database Technology (EDBT'92)*: 152–167

Buneman P, Davidson S, Kosky A, VanInwegen M 1992b A basis for interactive schema merging. *Proc. Hawaii International Conference on System Sciences*

Cattell R G G (ed) 1994 *The Object Database Standard: ODMG-93*. Morgan Kaufman

Fernandez M F, Popa L, Suciu D 1997 A structured-based approach to querying semi-structured data. *Int. Workshop on Database Programming Languages*

Gaede V 1998 *Execution Planning in Internet Marketplaces*. Technical Report. Humboldt-Universität zu Berlin, Germany

Graefe G 1993 Query evaluation techniques for large databases. *ACM Computing Surveys* 25 (2): 73–170

Gruber T R 1993 Toward principles for the design of ontologies used for knowledge sharing. *Workshop on Formal Ontology, Padua*

Günther O, Müller R, Schmidt P, Bhargava H, Krishnan R 1996 *MMM: A WWW-Based Method Management System for Using Software Modules Remotely*. Technical Report 32-1996. Humboldt-Universität zu Berlin.

Heimbigner D, McLeod D 1985 A federated architecture for information management. *ACM Trans. on Office Information Systems* 3(3): 253–278

Hull R 1997 Managing semantic heterogeneity in databases: A theoretical perspective. *Proc. 16th ACM Symposium on Principles of Database Systems (PODS)*: 51–61

Jeusfeld M A, Bui T X 1997 Decision support components on the Internet. *Int. Journal Decision Support Systems*

Kim W, Choi I, Gala S, Scheevel M M 1995 On resolving schematic heterogeneity in multidatabase systems. In Kim W (ed) *Modern Database Systems*. Addison Wesley: 521–550

Lakshmanan L V S, Sadri F, Subramanian I N 1996 SchemaSQL: A language for interoperability in relational multidatabase systems. *Proc. 22nd Int. Conference on Very Large Data Bases (VLDB)*: 239–250

Litwin W, Mark L, Roussopoulos N 1990 Interoperability of multiple autonomous databases. *ACM Computing Surveys* 22(2): 267–293

OGC 1997 *The OpenGIS Guide*. Technical Report 96-001. Open GIS Consortium. *http://www.ogis.org/guide*

Orfali R, Harkey D, Edwards J 1996 *The Essential Distributed Objects Survival Guide*. John Wiley and Sons, New York

Sheth A P, Larson J A 1990 Federated database systems for managing distributed heterogeneous, and autonomous databases. *ACM Computing Surveys* 22(3): 182–236

Vinoski S 1997 CORBA: integrating diverse applications within distributed heterogeneous environments. *IEEE Communications Magazine* 14(2)

# 34 FROM GISYSTEMS TO GISERVICES: SPATIAL COMPUTING ON THE INTERNET MARKETPLACE

Oliver Günther and Rudolf Müller

## 34.1. Introduction

*Question*: Why do people buy a GIS? *Answer*: Because their neighbor has one. Richard Newell of Smallworld Systems told this joke during his keynote speech at the 1997 *Symposium on Spatial Databases*—and he did not only refer to Smallworld customers. The truth behind his joke is that GIS are often greatly underutilized. Many customers use only a small fraction of the functionalities offered by their GIS. Some of them are aware of that: they simply do not care about the remaining features. Others are not: they may thus miss functionalities that are actually there and use complicated ways to reimplement them with the features they know. Yet other users may not use their GIS at all: they bought it because they thought it may help them with their problems but then found out that it does not. Some customers may not even have bothered to look: they bought the GIS andleft it in the package.

To be fair, this can be said not only about GIS but also about many other types of software. Microsoft, for example, estimates that 90% of Excel's functionalities are used by only 10% of its users worldwide. A large part of requests for new functionalities received by Microsoft each day can be answered simply by telling the customer that the requested functionality already exists. What makes the situation somewhat different for GIS, however, is the relatively high price of a GIS license. GIS come in relatively large packages: a single license often costs $3,000 US or more, it requires powerful hardware to run on, and it takes considerable training on the customer's part to use the software in a productive way. For most commercial GIS, potential customers face an all-or-nothing choice. Either they invest a relatively large amount to get the license, the required hardware, and some training—or they do not, in which case they get nothing.

We claim that many potential GIS users would become faithful GIS customers if they could do so at a smaller entry cost. Of course, the lower ticket price would not buy them the whole license indefinitely. But rather than putting a time limit on the license, as is typically done, vendors could limit the functionalities they offer. This could mean in particular that the vendor does not sell a classical system license but selected *services* that perform GIS-typical tasks at the customer's request. Services would typically be located at a site run by the vendor or a third party. Customers would make their data available to the service and retrieve the results from the remote site after the computation has been performed. No special hardware or training would be required on the user's part. Payment schemes would follow this service-oriented approach: rather than buying a whole system, users only pay for the particular usage of the vendor's software.

A direct consequence of such a shift from geographic information *systems* to geographic information *services* would be the rise of an electronic market (Kalakota and Whinston 1996) for spatial data and services. Anybody with Internet access could act as both provider and consumer of related goods.

In Section 2 we discuss electronic markets for information goods in more detail. We evaluate the technology that is currently available to support Internet markets, including user interfaces, communication protocols, and server implementations. Section 3 presents MMM, our own middleware approach to support electronic markets for information goods. Section 4 discusses related work, and Section 5 presents conclusions and open problems.

## 34.2. Electronic markets for information goods

### *34.2.1. Basic concepts*

In an *electronic market for information goods*, providers and consumers trade data and computational services by means of an information network, in particular the Internet (Figure 34.1). Providers advertise their offerings, make available requested data, and perform requested computations. They may also bundle some of their services with those of other providers. Consumers search and select services for a given task, apply them to some data set, and pay providers. Consumers may also integrate services as remote function calls into their local computations and combine services from different vendors. In a GIS context, typical services include data retrieval and conversion, map production, map overlay, spatial access methods, statistical analyses (e.g. multivariate data analysis), and visualization tasks.

Our vision is an Internet-based market for information goods, where anybody with Internet access can participate as consumer or provider. It should be as easy to *register* a service as it is to write a Web page today, and it should be as easy to *use* a service as it is to read a Web page today. With regard to GIS, this vision would most likely translate into an electronic market for GI services, where the number of GI service providers would be considerably greater than the number of GIS vendors today. Moreover, the number of consumers would be much greater than the number of GIS users today. Although the average revenue *per consumer* would be lower than the average revenue per GIS user today, *overall* revenue could well increase considerably. We also expect customer satisfaction to improve because users pay only for those functionalities they really take advantage of.

Which technologies are available today to implement such electronic marketplaces for GI services? Table 34.1 rates technological alternatives in terms of ease of use for consumers and providers; it shows that the situation for content providers is still rather unsatisfactory. For simple documents, providers can either offer them in their native format, which requires additional effort on the consumer's side, or they have to convert their holdings into HTML. The Common Gateway Interface (CGI)—a *de facto* standard to connect dynamic information sources—requires even more advanced skills on the provider's part, especially when used for services that are more complex than simple database access. Non-technical content providers will rarely be willing to go through the required learning effort. A similar argument applies to Java-based solutions: many potential service providers, especially in the research community, would not be willing to spend much time on converting their software to Java. Even the effort required to provide an existing service with an applet-based user interface is often prohibitively high.

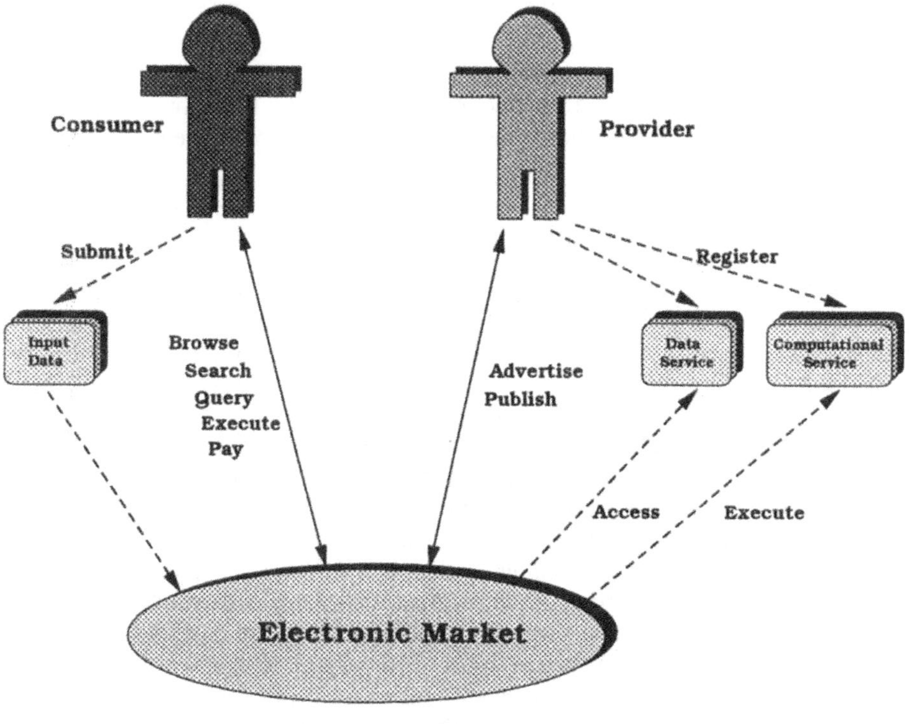


Figure 34.1. Interactions in an electronic market

The evolution of GI service markets thus requires more than application programming with HTML, CGI, and Java. We need appropriate conceptual foundations for the required technologies and their economics, and we need the transfer of these concepts into organized *electronic marketplaces*. A marketplace should free providers from implementing a fully integrated service, thus significantly reducing their market entry cost. Such marketplaces were first proposed for decision support applications (Bhargava et al. 1995a, 1997b).

The Internet technologies underlying such a marketplace can be grouped into four categories: *user interfaces*, *communication*, *connectivity*, and *market infrastructure*.

Any combination of choices from each category defines a middleware (Bernstein 1996) for an information market. For example, if one takes a Web browser as user interface, HTTP as communication protocol, Web sites for connectivity, and Web search engines as market infrastructure, then one obtains the existing Web.

| Format/technology | Consumer | Provider |
|---|---|---|
| HTML | ++ | 0 |
| Native document formats (other than HTML) | 0 | ++ |
| HTML forms + CGI + database access | + | - |
| HTML forms + CGI + general services | 0 | - |
| Java applets (as user interfaces) + general services | 0 | - |
| Java applets (to implement services) | + | -- |

Table 34.1. Consuming and providing Web content today (++ = very easy, -- = very difficult)

The remainder of this section briefly reviews technologies in each category, concentrating on their roles in an electronic marketplace. Bhargava and Krishnan (1998) give more implementational details in the application context of operations research and decision support.

### *34.2.2. User interfaces*

User interfaces for Internet services include simple Web browsers, programs embedded in a Web browser, or stand-alone programs that embed communication protocols for remote services. Among the most important technologies for browser-embedded GUIs are markup document languages (HTML, HTML forms, XML), JavaScript, plug-ins, Active-X components, and Java applets.

These technologies differ in terms of complexity for both consumer and provider. Writing applets with the Java Application Windowing Toolkit, for example, is far more complex than writing HTML forms. Of course, the higher complexity also buys more functionality. Most significant is that all solutions but plain HTML allow client-side computing, that is, the downloading of a software module to a client with subsequent execution. One may use this option, for example, to validate user input before it is sent to the remote service, or one may download a whole service and execute it locally. Other differences exist with respect to the ways downloaded programs can access client resources (such as files), and with regard to communication protocols. Plug-ins, Active-X, and Java applets can use any combination of protocols discussed in the next section, while the simpler approaches are more limited.

Note that user interfaces to GI services would not be restricted to browser-embedded solutions. One could also envision new lightweight GI systems that embed transparent access to an Internet-based market of spatial data and computational services.

### *34.2.3. Communication*

We distinguish four categories of communication protocols: stream socket protocols (e.g., libraries that implement TCP/IP-based sockets between client and server), Internet application protocols (HTTP, FTP, SMTP, or proprietary telnet solutions), remote function calls (RPC, DCE, etc.), and distributed object environments (RMI, CORBA, COM/OLE, or proprietary solutions).

These categories are not mutually exclusive. In fact, the latter three categories are all based on socket protocols. From the programmer's perspective, however, they provide quite different functionalities. The first two are particularly appropriate if byte strings (e.g., images or formatted ASCII documents) are exchanged between client and server. The latter two are more appropriate if communication relies on the exchange of instances of data types.

In the last category, RMI by JavaSoft only works between Java clients and servers. It may use stream sockets *or* HTTP for communication, which allows it to work through firewalls. Microsoft's COM/OLE is supposed to work across different programming languages (but only on Windows platforms). The Common Object Request Broker Architecture (CORBA) provides a language-independent communication layer between distributed objects across operating systems and programming languages (currently supporting C, C++, Java, Smalltalk, and ADA).

GI services may use any of these communication protocols in various combinations, depending on the particular application requirements. Remote function calls or method invocation mechanisms may be less suitable for performance-sensitive applications since data have to be passed through several protocol layers. Data confidentiality and integrity could likely be implemented by secure sockets under any of these protocols. An authentication service, on the other hand, may require higher-level communication protocols, like a CORBA object service.

### *34.2.4. Connectivity*

The term *connectivity* refers to the different ways to wrap existing software in order to make it accessible via one of the communication protocols listed above. Choices include:

- CGI scripts, which process a request (typically encoded as an HTML form) by invoking a program, generating an HTML form to present the result, and forwarding it to the requesting client;
- server-side scripts, which are embedded in pre-compiled HTML documents and interpreted at the server before the document is forwarded to the client;
- callable libraries, which are invoked via an RPC or DCE daemon;
- object wrappers, which give the software system an object-oriented interface.

An essential design decision concerns the question whether the response of the system should depend on previous requests, that is, whether communication should be *stateful*. The first three solutions do not maintain state *per se*, although one may mimic state on top of their stateless protocols. In CGI, for example, this could be done by depositing *cookies* at the client's site. A cookie is a string that contains some state information for the server. The server may later retrieve it and use it to restore the state reached at the end of a previous session.

Another important characteristic of an information market is topology. The classical Web solution is to contact one service at a time and get the result back to the browser. In a market, however, it may be desirable to advise services to get their input from other services. This requires wrappers for services that can communicate with each other.

Finally, wrappers should be adaptive with respect to the service they encapsulate. Simple wrapping techniques, such as CGI scripts, may have to be changed with every service modification (e.g., in order to change the HTML form used to present the result). More advanced techniques, such as server-side scripts, can adapt automatically.

### 34.2.5. Market infrastructure

The term *market infrastructure* refers to services that are shared between providers and consumers. They can be divided into four categories: repositories, serving as yellow pages for market content; data transformation services; support for service registration; and execution planning.

The need for *repositories* is motivated by the complexity of describing a service. Indexing techniques as used by Web search engines may fail, since they are based on full-text analysis. Relationships between different services, such as *belong to the same domain* or *solve an equivalent problem*, need to be made explicit. The repository may also serve for certification purposes by implementing ratings, annotations, and reviews of services. Repositories may simply be the result of an integration of services. As an example, consider the TSIMMIS architecture (Garcia-Molina et al. 1995; Papakonstantinou et al. 1995) of wrappers and mediators (Figure 34.2). The mediator between the application and the wrapper is able to decompose a query and spawn subqueries to the wrappers. The wrapper then translates the incoming queries into requests to the encapsulated legacy system. While there is actually considerably more functionality provided in TSIMMIS, this simple description already shows a simple market infrastructure where the mediator realizes a repository implicitly.

*Data transformation* services are required as long as there is no single data interchange format in the chosen application domain. In this case, data transformation at market level lowers the entry barriers for providers because they can register existing services in their original form, without adapting them to other data formats. If a market has already converged towards a common language to define formats (e.g., XML) or interfaces (e.g., CORBA's Interface Definition Language), data transformation may still be necessary at the semantic level. Consider, for example, two tables of geographic distances that are both represented in XML but whose semantics are defined differently. Interoperation between these two tables requires some mapping between them.

Support for service registration can come in different flavors. One way consists of wrapper frameworks that interoperate with the repository as well as with data transformation services. The latter aspect is particularly important since it enables interoperation with registered services, thus creating *positive externalities* (i.e., the benefit of participation increases with the number of participating consumers and providers). Another kind of support consists of component frameworks (such as CORBA's vertical common facilities or SunSoft's Java beans) that may serve as starting points for the implementation of a service. Interoperation with Web browsers could then be based on systems like CorbaWeb (Merle et al. 1996), Web* (Almási and Jagannathan 1995), and W3Objects (Ingham et al. 1995). A third kind of support applies to the positioning of a service in a given classification scheme. Finally, support could come in the form .of *execution agents* to be created from service descriptions (Bhargava et al. 1997a).

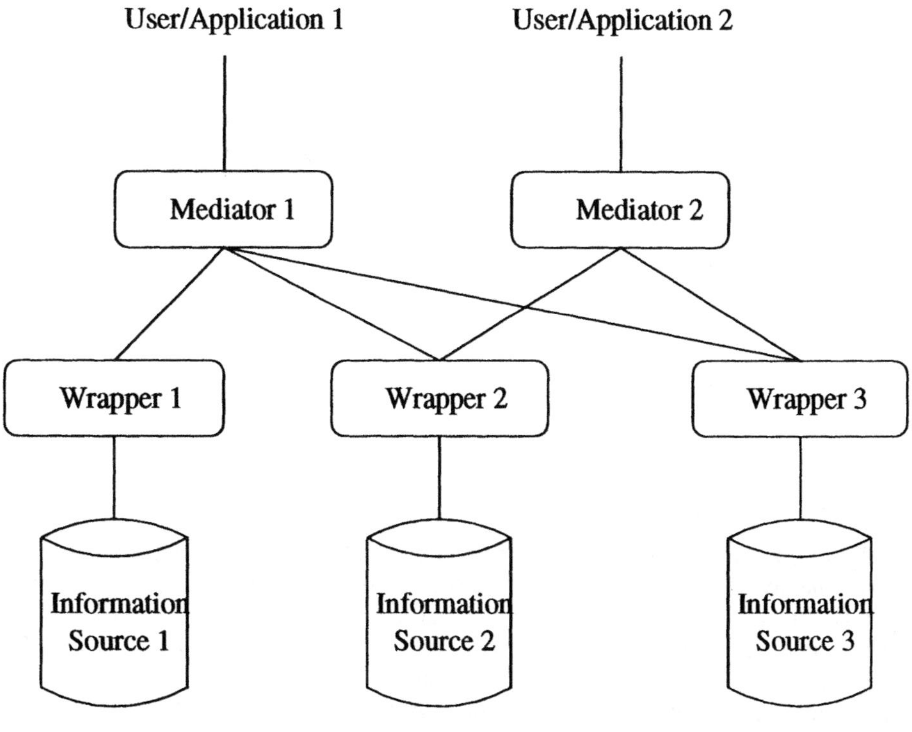


Figure 34.2. Mediators and wrappers in the TSIMMIS architecture

Execution planning in GIS marketplaces has been proposed by Abel et al. (1997). They define a plan as an acyclic digraph. Its source nodes are data services or data provided by the user, interior nodes represent computational services, and target nodes are results from services. Arcs relate to data flow between data sources, computational services, and data targets. Execution planning may take place manually by means of scripting languages (Becker 1996) or graph editors, or it could be done automatically. Automatic execution planning would generate a plan from a declarative request (Gaede 1998). A typical example of plan generation comes up in the context of interactive map making (e.g., on a WWW map server). Depending on the user's specifications, the system has to determine appropriate data sources, retrieve the required data sets, possibly perform some translation between formats, align the data sets, and perform a map overlay. In addition, one may have to perform some complex aggregation on the associated non-spatial data.

### 34.2.6. Summary

Figure 34.3 summarizes the different solutions for user interfaces, communication, and connectivity. Increasing distance from origin represents higher complexity and functionality. A region spawned by intervals on the three axes describes a middleware for which a market infrastructure could provide value-added services.

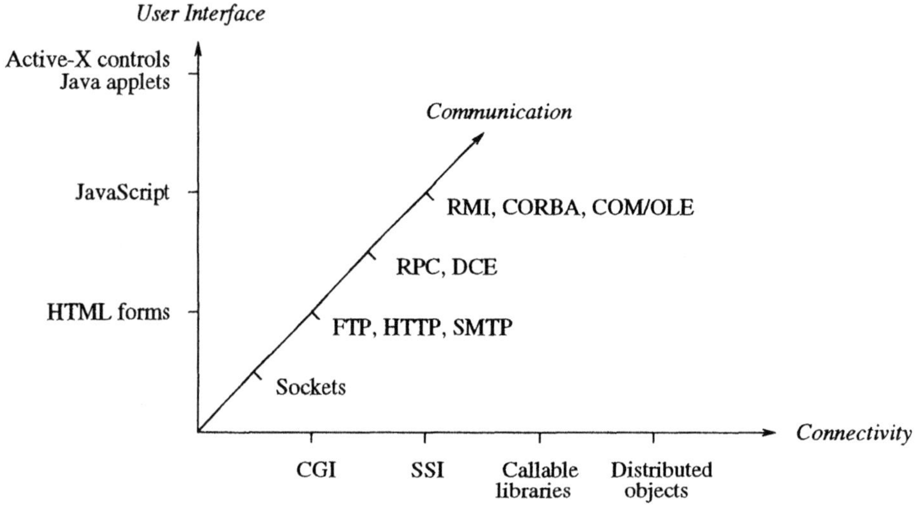


Figure 34.3. Technologies for electronic marketplaces

While one could argue for maximum functionality in each dimension, it is not clear whether this would stimulate market participation because it may have a negative impact on interoperability with other technologies used in the market. Take, for example, an infrastructure that heavily relies on CORBA. The fact that CORBA environments implement binary data interchange formats excludes the corresponding data from being processed with other tools. A better approach might be to agree on the exchange of documents written in a markup language defined in XML. Such documents can be processed by many tools and are also readable by humans. NAOMI, the North American Openmath Initiative (*http://naomi.math.ca/*) and a follow-up to the European Openmath project (*http://www.openmath.org/*), is currently defining such languages for mathematical data. As part of these projects, *the Java OpenMath Library* (*http://pdg.cecm.sfu.ca/openmath/lib/*) implements Java classes that are instantiated by XML documents. Applications using these classes may communicate directly with each other via RMI or CORBA, while communication with other applications could be based on the exchange of XML documents via TCP/IP sockets.

Overall it can be said that most of the technology to create GI services is available. Other than pure implementation issues, however, we identify five major problem areas that need to be addressed before our vision of an electronic market for GI services could become reality:

- user interfaces that are simple enough even for the occasional user;
- formats for metadata to describe data and computational services, and to match them with consumers' problems;
- data transport mechanisms that are reliable, efficient, and secure;
- data interchange formats that interoperate with customers' desktop data representation;
- business models that address related legal, marketing, and pricing issues (Bhargava et al. 1997b, c).

## 34.3. MMM: A method management system

We recently presented our MMM (Method ManageMent) system, a distributed computing infrastructure that supports a business model and electronic marketplace as described above (Günther et al. 1997). MMM is a collection of middleware services that facilitate Web-based access to software modules. In particular, MMM supports providers and consumers in:

- browsing, searching, and querying for relevant services;
- executing services in (possibly proprietary) execution environments;
- transferring and converting input and output data;
- state maintenance to support exploratory data analysis;
- registering new services (publication, check-in, interface definition).

Early prototypes of MMM were based on traditional CGI technology. Consumers entered their requests by means of HTML forms, which were then translated into CGI and forwarded to MMM. In more recent versions, we have replaced the CGI component by a Java applet. A prototype is available at *http://mmm.wiwi.hu-berlin.de*. A CORBA-based reimplementation is currently under consideration (Perrochon and Müller 1998).

As illustrated in Figure 34.4, MMM maintains a registry of services in the form of data and method repositories. They contain two kinds of meta objects (Gayer 1998). *Data service objects* (DSOs) serve to wrap and describe *data* services. *Method session objects* (MSOs) wrap and describe *computational* services. Data and method repositories are organized along subject-specific taxonomies. In MMM's initial application domains—statistical computing and operations research—these are based on the *Guide to Available Mathematical Software* (GAMS) taxonomy (Boisvert 1989).

Meta objects may be stored in different formats, such as XML documents or relational database tables. Figure 34.6 gives an XML example of an MSO for a

Matlab routine, which resides on a Princeton WWW server. The metadata includes all the information that is necessary for the MMM infrastructure to place the method in the given taxonomy, to query a consumer for appropriate input data, to retrieve the routine via the Internet, and to execute it on the consumer's data by a Matlab execution service. The provider of such a routine only has to create an MSO in order to register it as a service in the MMM marketplace. Registration can either be done by editing an XML document and submitting it to MMM, or by going through a registration session with a Java applet.

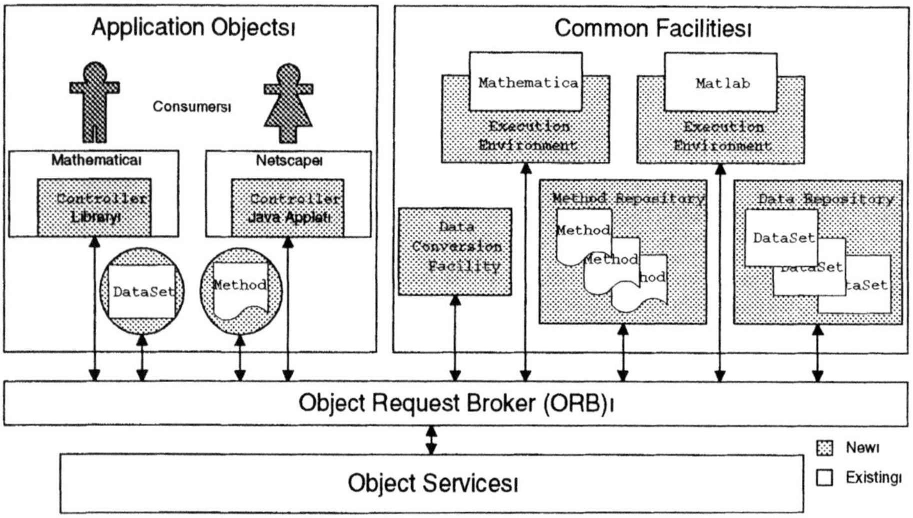


Figure 34.4. MMM architecture in OMG terminology

When a consumer decides to run the Matlab script, the MMM infrastructure uses distributed object technology to create a temporary session object, which resides remotely. In CORBA terms, one creates an *application object* that uses middleware services and common facilities to support its operations (e.g., billing and authorization facilities). In order to generate a user interface to the session object, the Java applet creates a dynamic stub. On method invocation, the session object uses in turn stubs of MMM common facilities to negotiate the execution (Figure 34.5). A session object for the Matlab routine described in Figure 34.6 would first invoke an operation on an object that implements a Web client to download the implementation from Princeton. Then it would use a stub of the execution engine to install the implementation temporarily and transmit the user's input data. Finally, it would trigger execution and return the result to the applet.

In order to locate relevant services, one can browse or search the MMM registry. This includes simple queries in the style of a WWW search engine (e.g., `time series AND estimation`), as well as more complex SQL queries (e.g., `SELECT plan FROM MSOclass m WHERE category = "estimation"`).

With regard to the taxonomy presented in Section 34.2, MMM combines different choices for its user interface, communication, and connectivity. MMM's *user interface* is a Java applet that implements generic layout functions for a set of data types (basic types like string or integer, and abstract data types like records or tables). The applet uses type information and dynamic stubs to create input masks on the fly for the operations offered. For *communication*, MMM gives objects the choice between stream sockets and method invocation via stubs. For example, when an MSO transmits the script for execution, it can either pass it as parameter value in a method call, or it can create a stream socket from which the execution service reads the script via plain TCP/IP. *Connectivity* is realized by a framework of object-oriented wrappers, which may in turn encapsulate other protocols to contact services on the Internet. An example is provided by the Web Connector object in Figure 34.5 that encapsulates the HTTP protocol. Other wrappers encapsulate connectivity to engines that can interpret routines in specific languages, like the Matlab routine in the example.

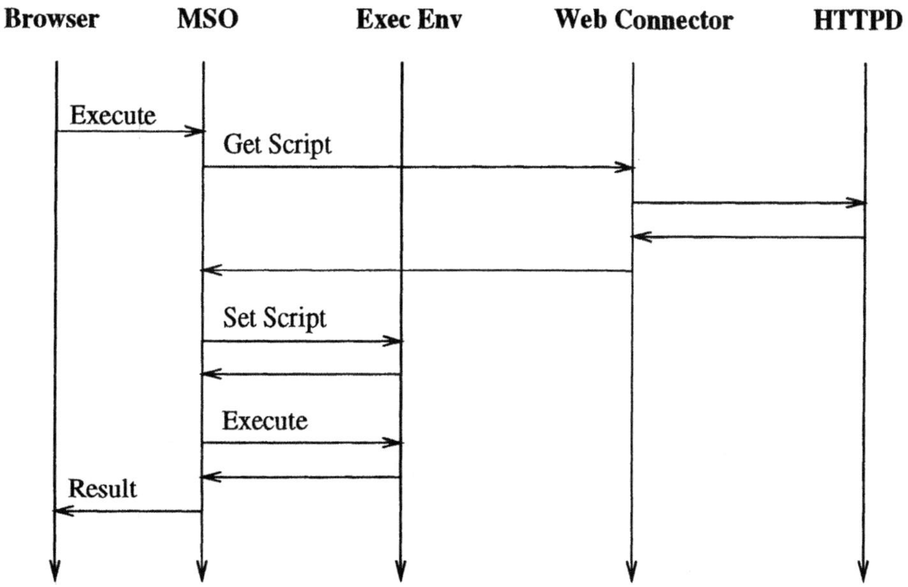


Figure 34.5. A vertical time sequence diagram illustrating method invocation in MMM

## 34.4. Related work

MMM is cooperating closely with two other projects that have related objectives: DecisionNet and SMART. *DecisionNet* (Bhargava et al. 1995a, b) is an organized electronic market for decision support technologies. The market infrastructure consists of agents that support consumers and providers in transactions. The decision technologies themselves reside on provider machines distributed across the Internet. DecisionNet is accessible at *http://dnet.sm.nps.navy.mil/*.

*SMART* (Abel 1997; Abel et al. 1997) is another Internet marketplace model with an emphasis on spatial data and related algorithms. Like MMM and DecisionNet,

SMART is based on the asynchronous communication between service providers and consumers. It offers *query services* to obtain data from a provider, *function services* to model computational tasks (such as conversion between representations), *planning services* to combine and coordinate different tasks, and *execution services* to execute a plan on behalf of a customer.

In 1994, an international group of GIS users and vendors founded the Open GIS Consortium (OGC), which has quickly become a powerful interest group to promote open systems approaches to geoprocessing (Kottman, this volume; *http://www.opengis.org*). The OGC defines itself as a "membership organization dedicated to open systems approaches to geoprocessing." It promotes an *Open Geodata Interoperability Specification* (OGIS), which is a computing framework and software specification to support interoperability in the distributed management of geographic data. OGC seeks to make geographic data and geoprocessing an integral part of enterprise information systems.

An interesting application of some of the technologies described in this paper has recently been presented by Bhargava and Tettelbach (1998). They implemented a waste disposal planner for the Monterey area by combining geographic data about waste disposal points with data about which types of waste are taken at what price. Providers of waste disposal services can register their location and prices online. Users may employ the service for simple look-up or for the computation of an optimal disposal strategy. The system combines a form-based user interface that operates via CGI with a relational database.

Olsen and Associates, a Swiss company founded in 1985, operates a commercial decision support service on the World Wide Web (*http://www.olsen.ch*). The site features a complete set of models, library resources, and customer support for financial market traders. The underlying decision technology consists of mathematical models to compute directional forecasts and timing indicators and to provide trading support for applications in investment and risk management.

The non-commercial Network-Enabled Optimization System (NEOS; Czyzyk et al. 1997) of Argonne National Laboratory can also be regarded as an electronic marketplace for computational services. NEOS allows users to solve complex optimization problems via a Web interface (*http://www-c.mcs.anl.gov/home/otc/Server/neos.html*) and a proprietary client that users may install locally. The CGI-based system gives access to a large number of related services provided by research institutions in the US and Canada.

## 34.5. Conclusions and open problems

Many of the functions performed by GIS seem to be amenable to a business model that is fundamentally different from the one we see today. At present, GIS users typically own the hardware and software they use. They pay license and maintenance fees to various vendors and they have to train their staff in using the system. The alternative would be a service-oriented approach where users make their input data available to some GI service center that performs the necessary computations remotely and sends the results back to the user. Customers pay only for that particular usage of the GIS technology—without having to own a GIS. The system is open for anybody to participate as consumer or provider of services. We believe that with such an approach the number of consumers will be much greater than the number of GIS users today. We also expect customer satisfaction to increase.

Our MMM system is one example of a middleware to implement such an electronic marketplace and associated business model. MMM's architecture is based on object request broker technology. Its support for consumers includes features for browsing, searching, and querying. Support for providers concentrates on the registration of new methods and on the related management of metadata.

Before electronic marketplaces for spatial services will become commonplace, however, several critical issues will have to be resolved.First, the development of appropriate licensing and (micro-)payment schemes is still work in progress. It is crucial for the success of Internet marketplaces that service providers can be certain to collect fees from all customers that use their services (directly or indirectly). Systems like MMM are independent of any particular payment scheme. They can in principle be combined with any of the major systems. The main problem is economical, not technical: what prices are consumers willing to pay, such that it is still attractive for potential providers to enter the market?

Second, one needs ways to guarantee confidentiality for the input and output data transmitted. If the service in question requires the input data in its original form, only organizational measures are possible to identify trustworthy service providers that maintain confidentiality. In many cases, however, a service can be performed on some transformed version of the data (see Figure 34.7). The traveling salesman problem is a typical example of this case: in order to solve a given problem $P$, one can first apply a series $T$ of geometric transformations (such as rotations or scalings) to $P$, then ask the service provider to solve the resulting problem $T(P)$. Once the service provider has found a solution $S$ of $T(P)$, one obtains the solution of the original problem $P$ by applying $T^{-1}$, the inverse of $T$, to $S$. This way the service provider never has access to the input and output data in its original form. It is an open research problem to define problem classes for which this kind of transformation is possible, that is, where $S(P) = T^{-1}(S(T(P)))$ or, more generally, for which there is *some* transformation (not necessarily $T^{-1}$) to generate $S(P)$ from $S(T(P))$.

Third, there is the related problem of *certification* of services. How can we make sure that a service does what it promises to do, in the desired quality? Some providers may already have acquired a reputation in their respective field. Others, however, may have to submit proofs of their competence and trustworthiness before customers decide to use their services.

Fourth, the usage and configuration of services should not be too complicated. Many GIS vendors pride themselves on the *turnkey* nature of their systems: setup efforts are minimal, and one can start using the system shortly after purchase. This may not always be the case in a digital marketplace type of situation, where users have to select and combine the services they need before they can use them.

```
<OBJECT><MSO><ID>ADJPH</ID>
<GENERAL>
<AUTHOR><PERSON> ... </PERSON></AUTHOR>
<PUBLISHER><PERSON> ... </PERSON></PUBLISHER>
<CREATION> ... </CREATION>
</GENERAL>
<ACCESS location="extern" decompression="none" tarfile="">
http://www.aos.princeton.edu/WWWPUBLIC/tapio/arfit/adjph.m</ACCESS>
<SIGNATURE>
<CALL>adjph(x)</CALL>
<INPUT> <PARAMETER> <ID>x</ID> ...
<DESCRIPTION>complex matrix</DESCRIPTION>
</PARAMETER> </INPUT>
<OUTPUT> <PARAMETER> <ID>ox</ID>...
<DESCRIPTION>complex matrix</DESCRIPTION>
</PARAMETER> </OUTPUT>
</SIGNATURE>
<ENVIRONMENT> <ENGINE><MATLAB></MATLAB></ENGINE>
</ENVIRONMENT>
<CONTENT>
<TYPE><MSO_TYPE> <SCRIPT format="ascii" language="matlab"></SCRIPT>
</TYPE></MSO_TYPE>
<DESCRIPTION>
<URL>http://www.aos.princeton.edu/WWWPUBLIC/tapio/arfit/</URL>
%     ox=ADJPH(x) produces a complex matrix ox
%     obtained by multiplying column vectors in x
%     with phase factors exp(i*phi) such that the
%     real and imaginary parts of column vectors in ox
%     are orthogonal and the norm of the real part is
%     at least the norm of the imaginary part.
</DESCRIPTION>
<KEYWORDS>
<KEYWORD>estimation</KEYWORD>
<KEYWORD>decomposition</KEYWORD>
<KEYWORD>autoregressive model</KEYWORD>
</KEYWORDS>
<ABSTRACT>
<URL>http://www.aos.princeton.edu/WWWPUBLIC/tapio/papers/arfit.html</URL>
<URL>http://www.aos.princeton.edu/WWWPUBLIC/tapio/papers/arfit_alg.html</URL
>
</ABSTRACT>
</CONTENT>
</MSO></OBJECT>
```

Figure 34.6. A method session object (MSO) for a Matlab routine

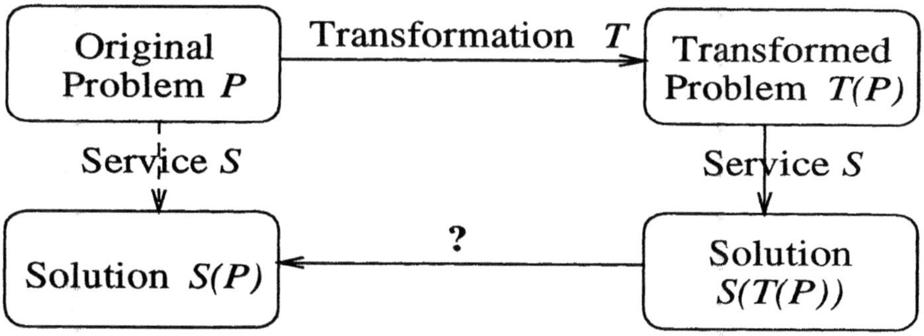


Figure 34.7. Possible transformation of a problem *P* to maintain data confidentiality

## Acknowledgments

Support from the German Research Society (DFG SFB 373/A3 and GRK 316) is gratefully acknowledged. We would also like to thank the members of the MMM team for their contributions to this paper: Matthias Asche, Brian Fox, Kathrin Gayer, Louis Perrochon, and Gerrit Riessen.

## References

Abel D J 1997 Spatial Internet marketplaces: A grand challenge. In Scholl M, Voisard A (eds) *Advances in Spatial Databases*. Lecture Notes in Computer Science 1262. Springer-Verlag, Berlin/Heidelberg/New York

Abel D J, Gaede V, Taylor K L, Zhou X 1997 SMART: Towards spatial Internet marketplaces. Technical Report SIS-1997-01. CSIRO, Canberra

Almási G, Jagannathan V 1995 Integrating the WWW and CORBA-based environments. *Proceedings of the Fourth World Wide Web Conference. http://webstar.cerc.wvu.edu/www4/www4.html*

Becker P 1996 An embeddable and extendable language for large-scale programming on the Internet. *Proceedings of the 16th International Conference on Distributed Computing Systems (ICDCS'96)*

Bernstein P 1996 Middleware: A model for distributed system services. *Communications of the ACM* 39(2): 86–98

Bhargava H K, Krishnan R 1998 The World Wide Web: Opportunities for OR/MS. *INFORMS Journal on Computing*

Bhargava H K, Tettelbach C G 1998 A Web-based decision support system for waste disposal and recycling. *Computers, Environment and Urban Systems*

Bhargava H K, King A S, McQuay D S 1995a DecisionNet: An architecture for modeling and decision support over the World Wide Web. In Bui T X (ed) *Proceedings of the Third International Society for Decision Support Systems Conference.* International Society for DSS, Hong Kong II: 541–550

Bhargava H K, Krishnan R, Kaplan D 1995b On customized access to a WWW-based distributed network of decision support services. In Bui T X (ed) *Proceedings of the Third International Society for Decision Support Systems Conference*. International Society for DSS, Hong Kong II: 551–560

Bhargava H K, Krishnan R, Casey M, Kaplan D, Roehrig S, Müller R 1997a Model management in electronic markets for decision technologies: A software agent approach. In Sprague R (ed) *Proceedings of the 30th Hawaii International Conference on System Sciences*

Bhargava H K, Krishnan R, Müller R 1997b Decision support on demand: Emerging electronic markets for decision technologies. *Decision Support Systems* 19(3): 193–214

Bhargava H K, Krishnan R, Müller R 1997c Electronic markets for decision technologies: A business cycle analysis. *Journal on Electronic Commerce* 1(4): 109–127

Boisvert R F 1989 The guide to available mathematical software advisory system. *Math. and Comp. in Simul* 31: 453–464

Czyzyk J, Mesnier M P, More J J 1997 The networked-enabled optimization system (NEOS) server. Preprint MCS-P615-1096. Mathematics and Computer Science Division, Argonne National Laboratory

Gaede V 1998 Execution planning in Internet marketplaces. Discussion paper, Sonderforschungsbereich 373. Humboldt-Universität zu Berlin

Garcia-Molina H, Hammer J, Ireland K, Papakonstantinou Y, Ullman J, Widom J 1995 Integrating and accessing heterogeneous information sources in TSIMMIS. *Proceedings of the AAAI Symposium on Information Gathering*: 61–64

Gayer K 1998 Metaobject repositories for managing services in electronic marketplaces. Discussion paper, Sonderforschungsbereich 373. Humboldt-Universität zu Berlin

Günther O, Müller R, Schmidt P, Bhargava H, Krishnan R 1997 MMM: A WWW-based approach for sharing statistical software modules. *IEEE Internet Computing* 1(3)

Ingham D, Little M, Caughey S, Shrivastava S 1995 W3Objects: Bringing object-oriented technology to the Web. *World Wide Web Journal. http://www.w3.org/pub/WWW/Journal/1/ingham.141/paper/141.html*

Kalakota R, Whinston A B 1996 *Frontiers of Electronic Commerce*. Addison-Wesley, Reading, Mass.

Merle P, Gransart C, Geib J-M 1996 CorbaWeb: A generic object navigator. *Proceedings of the Fifth International World Wide Web Conference, Paris*. INRIA. *http://www5conf.inria.fr/fichhtml/papers/P33/Overview.html*

Papakonstantinou Y, Garcia-Molina H, Widom J 1995 Object exchange across heterogenous information sources. *Proceedings of the 11th International Conference on Data Engineering*: 251–260

Perrochon L, Müller R 1998 Integrated scientific computing in global networks. In El-Rewini H (ed) *Proc. 31st Hawaii Int. Conf. on System Sciences* VII: 628–636

# 35 INTEROPERABLE EDUCATION FOR INTEROPERABLE GIS

D. Ian Heywood, Karen K. Kemp, and Derek E. Reeve

## 35.1. The vision

It is 2010 and geographic information (GI) is everywhere. Geographic information systems (GIS) interoperability has revolutionized the way vendors, data providers, and application developers deliver GI solutions. GI is embedded in a very wide range of information systems and decision support products from real estate sales through military applications to personal navigation. These all rely on a common global spatial data infrastructure that provides scale-independent data constantly updated. GI is available whenever and wherever it is needed. GIS have become so thoroughly interoperable that they have all but disappeared as distinct products.

GI education has also radically changed. The one-size-fits-all GIS courses of the late 20$^{th}$ century have gone. In their place has emerged a diversity of courses, each targeted upon the needs of differing groups of students. There remain, of course, some true GIS modules designed for the system developers, analysts, programmers and researchers whose careers will be devoted to developing future GIS technologies. For most students, including most Geography students, the focus of their learning is upon the use of digital geographic information, not upon the technologies which deliver this information. Most students now use geographic information, they don't study GIS. The ubiquity of digital geographic information and the extreme ease of use of geo-processing tools have allowed geographical analysis to spread way beyond its traditional confines within the spatial sciences. Consider the following scenario...

*Monday, 11 October 2010 Department of Archaeology, University Somewhere. It is two hours before Module 4 (Spatial Analysis) of ARCH 333 (Field Work in Archaeology) starts and though it is the 21$^{st}$ century, last minute preparation is still pretty common in academia. Some 100+ students will attend the class this week. Thirty of these will be in the local classroom on campus; the remainder will tune in through their satellite-linked home or office conferencing systems. Another 2000 will view reruns of the lecture over the communications network in the coming months. Of course, it won't really be a lecture,* *more of a multimedia experience, but the term seems to have stuck since the bad old days of talk and chalk.*

*Last minute preparation is possible because most of the material for the module already exists and can be found in GERN (the GIScience Education Resource Network). This resource base is a globally-distributed collection of GI educational materials created, maintained, and distributed by educators, software developers, application specialists, commercial publishers, public organizations, and others. It contains educational materials of many different types, including data sets, multimedia case studies, hands-on exercises, self-study problems, and lesson plans for different groups of learners. All the contents display a quality rating from an independent organization and are indexed by theme, education level, learning style, and language. All that the educator needs to work out is what the students should learn from any particular learning experience—the difficult part of education is still curriculum development!*

*Our professor begins by reviewing her own course syllabus and comparing this with those for similar courses being offered by other education providers around the world. After making decisions about what the students need to learn, the professor selects a learning framework appropriate for her students' level and the selected learning objectives. She then begins to fill in the elements, choosing:*

- *a basic content outline prepared by an international group of experts,*
- *a related case study prepared by the local land management agency,*
- *a spatial analysis applet from a major software vendor*
- *a background article describing the analytical procedures contained within the applet,*
- *a cheat sheet outlining the essential elements of the procedures,*
- *a sound bite from an archaeologist currently using these tools in a dig in Mexico who has some concerns about their use, and*
- *a set of paper-and-pencil problems to reinforce learning.*

*All of these elements are easily located in the resource network even though they reside in locations as diverse as Beijing, Pretoria, and Toronto. They all snap together easily, interoperably, into the learning framework. One hour later, the professor reviews her selections, slightly modifies some of the elements within a couple of the components, and, finally, clicks the submit button. She is returned a notice that an academic charge of $14.73 has been calculated for the use of the proprietary materials she selected and this amount is deducted from her course materials account. An additional charge of $2.95 will be deducted from each student's individual course account when he or she accesses the materials. A reasonable price, muses the professor, compared to the cost of the textbooks she used to buy in the late 20$^{th}$ century. Her course is now ready for presentation.*

This vision may or may not come true. It is clear, however, that a paradigm shift is occurring in GIS technology. The monolithic GIS packages of today are giving way to new forms of geo-processing based upon interoperable principles. In a few years time, GIS simply will not be like it has been for the last twenty years. This presents an enormous challenge for GIS educators. There is a need to reconsider what is taught, how it is taught, and who should be taught.

This chapter explores the implications for GIS education of the emerging new breeds of interoperable GIS. It starts by reflecting upon the unfortunate effects that the non-interoperability of present GIS products has had on traditional GIS teaching. Interoperability will ultimately remove many of these frustrations, but the pace at which the new technologies are being introduced will mean that the transition period for GIS teachers will be difficult. This problem is explored by reviewing four current approaches to GIS education. The chapter concludes by suggesting that interoperability itself provides an apt metaphor for the new collaborative methods of teaching that will be necessary to cope with the demands of introducing interoperable GIS to future students.

## 35.2. GIS education in the non-interoperable present

The non-interoperability of GIS products during the last two decades has imposed a number of unfortunate characteristics upon GIS education:

**Conceptual diversity**: During the last twenty years the relational and object models have provided common conceptual bases upon which students of conventional database technologies could ground their understanding. *Tuple*, *relation*, and *method* each have a universal meaning. SQL has provided a universal language across RDBMS products. Contrast this with the position of GIS students. The wide diversity of GIS products means there has been relatively little commonality to which GIS students could cling. GIS products have differed in their models of space, their terminology, their storage structures and their application languages. For GIS students and their teachers, this has meant that GIS education has been built upon an insecure, shifting framework of ideas and terminology.

**Product specificity**: As a consequence of this conceptual diversity, many teachers have retreated into a product-specific view of GIS. Adopting a particular vendor's software, terminology, and models greatly simplifies the teaching process. Assuming the chosen vendor has a reasonable market presence, this approach also offers the pragmatic advantage of providing a reasonable guarantee of subsequent employment to students. The academic dangers of this approach are evident. Limitations in the functionality of the chosen system constrain what is taught. The students invest their energies in learning what is special about system X's view of geography and its computational idiosyncrasies, with the result that students often have a better grasp of specific software than of the underlying geographical principles.

**Limited transferability of knowledge**: The product specificity of GIS has meant that knowledge gained on one system has not always been readily transferable to another. A database lecturer could, hand on heart, tell a class that SQL knowledge gained on one RDBMS would largely be applicable to another. A GIS lecturer could not make the same claim. Transferring, say, ESRI's AML (Arc Macro Language) knowledge to a Smallworld environment would not be straightforward. From the viewpoint of society, the amount of un-learning and re-learning which GIS incompatibilities have imposed is grossly inefficient.

**The software focus**: Because GIS products have been idiosyncratic in their models, terminology, and interfaces, many GIS have had steep learning curves associated with them. Add to this the other aspects of non-interoperability in GIS—problems with data integration and data sharing—and introductory GIS courses have a tendency to degenerate into beat-the-software exercises: "We'll take ten weeks to show you how to use this software and how to get data into it and at the end you might be able to do something useful with it". This focus on mastering the system has left little time for understanding the wider business and research contexts within which GISs are used.

**GIS is seen as difficult**: Because of the characteristics outlined above, GIS has been seen as an advanced topic most often taught as a penultimate- or final-year option at degree level and as a specialist course at postgraduate level. Rather than being presented as a tool that all students interested in spatial analysis should be able to use, it has been the preserve of specialist students. Professors now expect that students coming into higher education will have mastered word processors and spreadsheet programs. Yet the notion that GIS should be so easy to use that it should be introduced at school level is only now beginning to gain currency.

## 35.3. Getting from here to there: GIS education in an interoperable era

It seems very likely that the new age of interoperable geo-processing will ameliorate many of the unfortunate characteristics of current GIS education. GIS now seems sufficiently mature that a common model of geo-processing, most probably based on the work of the Open GIS Consortium (OGC), is feasible and that it will be supported by the most significant vendors (Buehler and McKee 1996). Based upon this common model, terminology will become standardized. Interoperability at the data level will lessen the attention that needs to be paid to this element of GIS work in courses. Standard GIS languages, such as that embedded in SQL3/MM, will increase the transferability of GIS knowledge from system to system. In short, there will be less need to concentrate upon the *mechanics* of GIS which will free more time for lecturers to focus upon the *purposes* and *contexts* of digital geographical processing. The Promised Land, it would seem, is at hand. The transitional phase, however, during which GIS education will need to adjust to the new demands which interoperable GIS will impose, will create a period of discomfort for teachers and their institutions. Achieving the Promised Land is often more troublesome than the prophets imply.

In addition, pressures for change in GIS educational programs will arise not only from changes *within* GIS but also from external influences. If the trend watchers (Denning 1996; Dertouzos 1997; Tapscott 1996; Twigg and Oblinger 1996) are to be believed, the new generation of GIS courses will emerge in an educational environment that features:

- the convergence of work and learning as society moves towards an information economy;

- the encouragement of lifelong learning and the impact which lifelong learning will have on the demographic composition, and expectations, of the student body;
- the increasing demand for accountability and efficiency in higher education;
- the adoption of new teaching methodologies; and
- the emergence of for-profit educational institutions and vendor-sourced educational products.

Within the next few years, therefore, not only will the GIS curriculum change but so too will the methods used to teach it, the institutional framework within which it is taught and, indeed, the characteristics of the students who want to learn.

### *35.3.1. Increase in technological expertise of true GIS students*

A recent discussion between representatives from the GIS vendor community suggests that current MSc GIS programs are failing to produce graduates with sufficient technical knowledge to allow them to be appointed as software engineers, developers, and researchers within their companies. The GIS companies prefer to recruit mainstream computer science graduates for such positions, appointing GIS graduates to sales and support roles. "'I'm afraid it's actually easier at the moment to take a computer scientist and teach them the geographical area than the other way round" (Coote, quoted by Heywood 1997:108). If an aim of GIS education has been to supply GIS vendors with technologically competent recruits, the evidence is that so far there has been only limited success. The drive towards interoperability probably will raise further the level of technological expertise required by students who aspire really to understand what goes on behind the glass in GIS. Therefore, unless new educational arrangements are made, it seems likely that the gulf between what the GIS software companies require for their technical positions and what GIS educational programs can supply will become even wider.

If GI *systems* science is defined as the body of knowledge that is necessary to master the technology of GIS, a critical question for the next few years therefore will be: what are the most efficient arrangements for teaching this body of knowledge? The trend within the technology itself perhaps hints at the answer. The technologies upon which interoperability is based—object orientation, distributed computer platforms, three-tier architectures, middleware, etc.,—are mainstream technologies which are being driven forward by major IT companies such as Microsoft and Oracle. The GIS vendors, via organizations like OGC, are importing and adapting these technologies for geo-processing. RDBMS vendors such as Oracle, Sybase, Informix, and IBM now provide spatial data extenders that can be plugged into their mainstream DBMS products. It seems clear that GIS, as a technology, is being incorporated into the mainstream IT industry. "You could go so far as to say that the GIS industry as it has been historically defined is going to be gobbled up by the mainstream IT infrastructure… I get the sense at the moment that Microsoft have bigger fish to fry than spatial data but it is only a matter of time." (Stafford, quoted by Anonymous 1997: 15).

If GIS, as a technology, is moving into the mainstream of IT, perhaps teaching GI *Systems* might also be efficiently relocated into Information Systems degree programs. Locating GIS as a sub-discipline of Information Systems would attract a different student intake than present GIS programs and would provide a computer science, rather than a geographical, milieu. Perhaps, in future, Information Science professors might teach the technology of GIS, with a few guest lectures from Geography faculty, rather than the other way round as at present? Already some GIS programs are emerging which are based in Computer Science departments, for example, Texas A&M University in the United States, Keele University in the UK and Curtin University of Technology in Australia.

Clearly an issue for the next few years will be to determine the institutional arrangements that will most efficiently produce the GIS programmers and developers who vendors will need to help them develop interoperable GIS products. It will be interesting to watch the turf wars that will emerge between the disciplines that currently claim GIS as their own.

### *35.3.2. Develop GI analysis courses.*

Kuiper (1992) observes that by the late 1980s a divergence had emerged within the mainstream database industry between information specialists, that is, people who could analyze the information needs of their organizations and use databases to provide information products, and the computer scientists whose role it is to work at the systems and code level. He estimated that the ratio is probably about 20:1 in favor of the information specialists and argued that separate educational programs needed to be developed for the two streams of students.

Clearly the GIS courses discussed above will provide the hard-core, technically competent GI computer scientists required by the GIS developers, but Kuiper is probably correct when he suggests that this will be a minority specialism. Most students who choose to study GIS as a specialist subject will probably follow a business analyst approach. The curricula for such GI analysis modules will emphasize the human aspects of GIS: understanding organizations, understanding project analysis, and project planning skills, as well as providing a basic grounding in the technologies and their applications. The emphasis in many GIS courses in future will be upon implementing GI strategies for organizations and specific application areas, not on the core technologies.

### *35.3.3. Embedded geo-processing for the majority*

As GIS technology becomes ever more integrated into desktop computing, for most students GIS as an optional module title will disappear. They simply will not need to do GIS. A wide diversity of courses which have geo-processing embedded within them will emerge. Rather than spending time studying the mechanics of using GIS packages, students will focus upon the value of geographical analysis in their own domains. In such courses, there may be a need to devote a few lectures to outlining the basics of geo-processing as an enabling technology but the focus will be upon the *usefulness* of the technology and the dos and don'ts of working with geographical data, not the technology itself. Nobody suggests that all students need to understand the algorithms and data structures that are built into spreadsheets before they can use them. It is expected, however, that the student understand the nature of the application for which the spreadsheet is being used and the underlying mathematical and statistical principles that an application might employ. If the hope is that interoperable GIS products will eventually become as ubiquitous and easy to use as spreadsheets, soon there will be little justification for dragging many students through the mysteries of geo-processing. Most students in the future will use GIS products, they won't study them. Most students do not need to know what goes on behind the glass. If GIS interoperability needs to be taught, other than to specialist students, then GIS interoperability will have failed.

The market for GI-based courses will extend far beyond the traditional spatial sciences. If accountants, lawyers, managers, and advertising executives are going to have interoperable GIS functions on their desktops, someone will need to teach them why they should want to press their geographical analysis buttons, and what the results mean from having done so. Having a spreadsheet on your computer does not make you an accountant. Equally, having map-based functions will not automatically make accountants and managers into geographers. Interoperability might well mean that the demand for geography professors to act as conventional GI systems trainers declines within the next few years and the market for professors to act as geography trainers increases. These professors, however, may not be plying their trade within geography departments.

There will clearly be a segmentation and specialization within GIS education during the coming years. For the minority of GIS students there will need to be an increase in the level of technical expertise provided compared to present GIS programs. For the majority, the emphasis upon the technology of GIS will decrease, allowing more time to be devoted to the use of the technology, the application of GI for broad-based problem solving and the societal and organizational changes this may bring about.

### *35.3.4. Resourcing the change*

A very serious problem for institutions and individual professors is the rapidity with which new forms of geo-processing technology are being introduced. The GIS industry is one of an increasing number of high technology industries in which the conventional flow of information from academia to the commercial mainstream is being reversed. There is a real danger that the technology being proposed by OGC and being implemented now by vendors will outpace the ability of educational programs to change. While the mainstream industry forges ahead with new interoperable software, educational programs, unless capable of financing rapid change, will continue to introduce students to the concepts embedded in the last-but-one generation of GIS software. A significant percentage of GI knowledge, particularly as it relates to technology, becomes outdated within six months. How can institutions afford to re-equip their GIS laboratories with new products at the same rate as the vendors produce them? In this respect, it should also be noted that the specification for a lab to demonstrate interoperability may exceed that of most present-day GIS labs. Most GIS practical classes are currently based upon relatively cheap, desktop packages using small, static, pre-packaged data sets. To introduce students to the full vision of interoperability, as it is currently conceived, educators will need to have access to large-scale, heterogeneous geodata resources providing live data and downloadable program components. How many educational institutions will independently be able to fund such infrastructures?

For individual professors, the rate of change is impossibly fast. It is impossible for an individual teacher to keep pace with all the changes. The technologies change faster than lecture notes. On the other hand, it might be argued that university teachers in GIS should not aspire always to use the most recent software or chase the most recent trends. It is often asserted that there is a distinction to be made between training, in which the vendors introduce operatives to their latest products, and education, wherein university professors focus upon the eternal truths of spatial analysis. In practice, however, many students have a keen eye upon the currency of what they are taught and, frankly, expect to be taught on the systems they that hope to use in employment.

## 35.4. Interoperable GIS education

The expectation is that during the next decade, GIS educators are going to be faced with an enormous task. The very subject matter that is taught, GIS, is being radically transformed. GIS products are emerging at ever-increasing rates. A wider range of students than previously will require GI knowledge. Some students will need to master aspects of advanced computer science; some, perhaps from the traditionally non-spatial disciplines, might need to be introduced to the basic concepts of mapping. The university environments within which professors teach will become more challenging.

The conclusion of many current GIS educators, and indeed of many other educationalists in other fields, is that no individual or institution will in future be able to cope independently. It is argued, therefore, that the ingrained model of competition between universities—for funds, for students, for kudos—will need to be replaced by one of collaboration. A need is foreseen to develop *interoperable* educational programs, in which educational *objects* authored by staff from many institutions, and not necessarily exclusively the traditional educational institutions, can be snapped together to meet the needs of any particular group of students. Within the GIS educational community, this need for wide-ranging collaboration to meet the challenges of the next few years has been emphasized in the priorities of the University Consortium for Geographic Information Science (see *http://www.ucgis.org*) and the conference summaries from the Second and Third International Symposia of GIS in Higher Education (*http://www.ncgia.org/gishe*)

The vision at the start of this chapter contains one model by which interoperable GIS education might be facilitated. It is envisioned that by 2010 the GERN (GIScience Educational Resource Network) will act as an online, international repository of GIS educational objects. These objects will consist of a wide range of materials—data sets, multimedia case studies, hands-on exercises, self-study problems, applets providing geo-processing algorithms, exam questions, lesson plans—all indexed by theme, education level, learning style, language, etc. Just as Guenther (this volume) and Gaede (this volume) anticipate mechanisms by which conventional interoperable GIS applets will soon be rented via the Web by GIS analysts, the GERN would allow GIS teachers to rent educational objects. The role of individual institutions will be to assure the quality of the educational objects they adopt and to validate the qualifications their students obtain (Heywood et al. 1998). Objects will be deposited into the GERN by a wide range of authors, including conventional higher educational instructors, software and data vendors, mapping agencies, local and private sector GIS project developers, independent researchers, publishers, etc.

Developments like the GERN might presently seem Utopian and indeed there remain many obstacles to overcome and issues to address. Some of these are listed in Tables 35.1 and 35.2. On the other hand some of the building blocks are already falling into place, including models for collaboration between educational institutions and between individual educators; industry provision of educational materials; and the development of generic on-line flexible learning environments. These are discussed in greater detail in the following sections.

| |
|---|
| What kinds of resources are needed in a GI/S education resource base? |
| What are the required components of GI/S educational resource objects? |
| How can resources be categorized according to different GI/S learning requirements (i.e., system-specific, application-centered)? |
| How to handle obsolescence of GI/S knowledge? Is a system of archiving and versioning needed? |
| How to develop strategies for meta GI/S education infrastructures? |
| How to formalize GI/S knowledge for educational purposes? |
| How to establish common terminology for GI/S? |
| How can quality be evaluated for GI/S education products and what mechanisms can assist individual educators to assess it? |

Table 35.1: GIS-specific issues to be addressed in the development of a GERN

| |
|---|
| How to build and promote distributed learning resources? |
| How to develop future-proof learning resources and systems? |
| How to handle the not-invented-here syndrome? |
| How to handle disciplinary chauvinism? |
| How to assess and control the quality of distributed resources? |
| How to carry out localization of global materials? How to globalize locally-developed materials? |
| How to develop value-added products from global resources? |
| How to collaborate with publishers and other commercial enterprises including data providers and vendors? |
| How to manage academic credit and financial return for contributions to an education resource base? |
| How to handle multiple languages? |

Table 35.2: Generic educational obstacles to be addressed in the development of a GERN

### *35.4.1. Institutional collaboration*

Within the GIS community there are already several well-established models for sharing educational materials.

**The NCGIA Core Curriculum** (*http://www.ncgia.org*) released originally in 1990 consists of over 1000 pages of lecture notes and related materials contributed by academics around the world. Over 1400 copies were distributed and the materials have been translated into several languages. Although now out-of-print and out-of-date, the continued demand for these materials encouraged NCGIA to begin work on a new Core Curriculum in 1996. Intriguingly, this new Curriculum might be regarded as a GERN in embryo. It is being developed with contributions from authors around the world who prepare the on-line materials and act as editors and peer reviewers. With strict format and editorial control at the NCGIA, the materials are both accredited and consistent in quality (Kemp 1997). The result will be a dynamic resource base that can be used, in this case free of charge, to help determine course content and to locate educational materials for GIS courses at different levels and with considerable content variation.

**The UNIGIS network of universities** (*http://www.unigis.org*), originally launched by three UK institutions in 1992, has now grown into a network of some 14 institutions located in 10 countries across the globe which collaborate to deliver postgraduate GIS education by distance-learning methods. The initial impetus behind the UNIGIS network was a recognition that individually none of the participating institutions had the resources to produce sufficient high-quality learning materials to be able to mount a successful program. By pooling materials and sharing development costs it has been possible to achieve this goal. Although initially the UNIGIS materials were delivered primarily in paper form, increasingly the emphasis is being put on developing a digital UNIGIS resource base similar to that imagined in the GERN. The UNIGIS partnership has shown that it is possible for otherwise competitive institutions to find regulatory and financial mechanisms that allow co-operation.

**The Geographer's Craft project** (*http://www.utexas.edu/depts/grg/gcraft/contents.html*) also provides an example of the willingness of professors and students to adopt materials that are not locally authored. Traditionally, there may have been a reluctance to use materials authored at another institution, the not-invented-here syndrome, but this reluctance now seems to be receding. The Geographer's Craft materials began as an experimental introductory course in geographical techniques at the University of Texas Geography Department. The project has since developed into a comprehensive set of on-line course materials, including an electronic textbook and laboratory manual, which are linked to by faculty around the world, at no cost, for use by their own students. In autumn 1996, logs showed that over 100,000 files were being accessed each week from the Web site, only 3,000-5,000 of these by students from the University of Texas (Foote 1997). Foote suggests that the very high level of interest in the materials and the links to the site from the homepages of other educators clearly indicate the potential for collaboration in the development of Web resources and distance education.

### *35.4.2. Industry participation*

The traditional distinction between vendor training and university education has always been, in practice, questionable in GIS. Within the universities, many GIS modules have in effect been training courses based on a particular vendor's software. Conversely, the vendors have long been aware of the need to promote general GIS education as a means of growing the market for their products. In future, this blurring between the two camps will need to be carried further. A GERN would depend as much on vendors depositing their applets, data sets, and training materials as much as on contributions from traditional academic authors.

In this respect, ESRI's development of a Virtual Campus provides an indication of the sort of contributions vendors might make in future to a GERN. The ESRI Virtual Campus (*http://www.esri.com*) has been developed to help overcome the problem that not enough people understand the concepts behind GIS, a situation which ESRI feels limits widespread use of the technology. By providing a virtual campus, ESRI hopes to overcome some of the barriers to attending traditional training courses that are imposed by cost and location. The campus is intended to provide a learning resource center for existing and potential customers as well as other interested individuals who wish to explore and develop skills using ESRI's software products. In addition to courses, this campus has *cafes*, faculty who hold office hours, and certificates of satisfactory completion based on student assessment. Since these courses are either free or low cost, this high-quality hands-on GIS training program is being used to supplement or even replace some elements of the laboratory program in some traditional GI courses. For example, the UNIGIS program requires students to complete the ESRI-accredited ARCVIEW training modules provided through the campus. Only when UNIGIS students have received their ESRI certificate will they be awarded UNIGIS credits for the ARCVIEW part of their program. In addition to building a virtual campus, ESRI is also in the process of developing its own GI knowledge base which like the GERN would be populated with lecture notes, exercises, application case studies, and reading materials drawn from the ESRI user community (Miller, personal communication). By 2010, it is likely that a GERN could contain links to similar knowledge bases maintained by all the major vendors, so that course builders could download materials and applets from whichever vendor seemed most appropriate for a particular need.

### *35.4.3. Online learning environments*

The ESRI virtual campus also exemplifies the emergence of the on-line educational environments that might be necessary for GERN-like developments. Of course, ESRI is not alone by any means in pursuing the concept of a virtual university. Several state-funded virtual universities are now being developed in the US and elsewhere (see for example the Western Governors' University at *http//www.westgov.org/ smart/vu/vu.html/*).

To support such virtual universities and other distance-learning environments, major digital infrastructures are needed. For example, in November 1994, Educom, a non-profit consortium of higher education institutions in the US, launched the National Learning Infrastructure Initiative (NLII). The NLII is in response to a common need among educational institutions for a non-proprietary, Internet-based instructional management system (IMS) to provide the means to customize and manage the instructional process while at the same time integrating content from multiple publishers in distributed—or virtual—learning environments. The IMS project was formed as a catalyst for the development of a substantial body of instructional software, the creation of an on-line infrastructure for managing access to learning materials and environments, the facilitation of collaborative and authentic learning activities, and the certification of acquired skills and knowledge (see *http://www.imsproject.org*). Other online learning shells such as WebCT from the University of British Columbia (*http://homebrew1.cs.ubc.ca/webct/*) provide similar, though local, facilities.

Clearly by 2010, indeed possibly already, the software tools will be available for GIS authors to design and implement efficiently on-line educational objects ranging in scope from short case-study descriptions to examination questions to full course modules. The role of GERN will be to advertize the availability of these objects to the community of GIS teachers.

## 35.5. Towards the promised land

To make all this happen there are a number of issues which will need to be pursued. There is much work to be done at a technical level. Following the OpenGIS model, we will need to focus upon developing metadata standards which will allow the authors of educational GI objects to define with some precision the objectives, contents, levels, and intended outcomes of their objects, so that potential users can evaluate their usefulness for their courses. There may well need to be a testing procedure to ensure that educational objects comply to an agreed standard to enforce interoperability. The granularity at which GI educational objects are best defined—exercise, concept, topic, course, etc.—will need to be established. Perhaps more than in many other disciplines, there will need to be mechanisms to allow individual objects to customize themselves to localized requirements—the educators' version of polymorphic objects! Imagine a general purpose educational object which explains coordinate systems but which, depending upon context, presents itself in English using Ordnance Survey data, or in Spanish using Spanish examples. A tracking mechanism that automatically monitors the sequences in which educational objects are commonly assembled could generate valuable de facto GI curricula. Methods for handling assessments and student queries will need to be addressed.

It is important to emphasize, however, that GIS educators will not be alone in developing the technical basis for educational objects. On the contrary, as the example of IMS above suggests, there are a number of super-disciplinary projects which are already exploring the technicalities of interoperable education. Just as the OpenGIS consortium is presently ensuring that *geography* is properly treated within computer industry standards such as DCOM and CORBA, the immediate agenda for GIS educators should be to liaise with the educational interoperability projects to ensure that GI requirements are properly built into their specifications.

Even when the technical challenges involved with delivering educational interoperability are resolved, this will not in itself be sufficient to guarantee that GI educational interoperability will succeed. Technical standards and facilities can provide only mechanisms by which educational objects might be delivered. Without academic *content*, such delivery mechanisms will be redundant. It is necessary, therefore, to consider the incentives which might motivate faculty to pour their hard-earned knowledge into GI educational objects. Why should academics share their knowledge in a collaborative, interoperable manner? Why should educational institutions, whose stock-in-trade might be considered to be the knowledge of their staff, allow these staff to make their teaching materials available to other institutions? Here we enter the fascinating, and still fuzzy, area of *institutional* interoperability.

The phrase *Educational Object Economy* (EOE) is being used to describe the bases upon which individuals and organisations might trade in interoperable education objects (Apple 1998). Possibly, a conventional market economy in educational objects might develop. As envisaged in the scenario at the beginning of this chapter, educators might simply pay a rental fee each time they borrow an object from the GERN system. Where the authors of educational objects are conventionally employed university faculty staff, the proportions of these fee payments which are retained by authors and by their institutions would doubtless be the subject of local negotiations. It is important to remember, however, that in an EOE, objects could be contributed by non-academic authors. GI practitioners could well encapsulate their expertise within educational objects and release them into the EOE. There should be no presumption that in the future EOE the generation of content will remain the exclusive preserve of conventional academics. The EOE will be an open market in which the success of an object will depend upon the willingness of others to buy it. Alternatively, a barter EOE may evolve, with authors, or institutions, gaining the right to borrow objects by depositing their objects within GERN-like clearinghouse systems.

The EOE model assumes that educational interoperability will be based upon a trading relationship of some kind. There are, however, already examples of educational materials being shared on an altruistic basis. Within GIS, the NCGIA's core curriculum and the University of Texas's Virtual Geography Department provide excellent examples of the willingness of academics to contribute voluntarily to projects in which they believe. As educational interoperability develops, it may be that the incentives and kudos associated with authoring an educational object increase. Being the author of one of the top ten most borrowed objects in the GERN system may well become one of the most significant achievements available to GI faculty and be recognized as such by university promotion boards.

In all probability, a mixed EOE will emerge. Within the software industry, freeware, shareware, and fully commercial products happily co-exist. Similarly, educational objects might be provided with varying degrees of commercialism. The central point is that universities and individuals will need to explore models which provide sufficient incentives to them to ensure that GERN-like clearinghouses are populated with sufficient high-quality educational objects to make them worthwhile.

## 35.6. Conclusion

Clearly nobody really knows what GIS education will look like by 2010. The only certainty is that it will not be at all like it is at present.

Using our admittedly cloudy crystal ball it is anticipated that present trends towards interoperability will continue to the extent that GIS as a separate technology will disappear from superficial view. Geo-processing will be embedded within information systems almost universally. Mirroring this trend, formal teaching of GIS, that is, the technology of interoperable geo-processing, will retrench into specialist courses, probably taught as part of Information Systems programs. GIS as a specialism will disappear from the curricula of many degree programs. Conversely, courses which assume the availability of digital geographic information and which use geo-processing objects to explore application areas will blossom not only in the traditional spatial sciences but across many disciplines. The rapidity of change, and growing pressures within the educational system, suggest that the only way that educators will be able to keep pace will be by collaboration. Present embryonic experiments in on-line collaborative GIS teaching and learning will continue to expand and develop, and may possibly formalise into something similar to the vision of the GIScience Educational Resource Network (GERN). It seems to us that interoperable GIS provide an apt metaphor for the future of GIS education. Only by creating interoperational educational objects will each educator be able to keep pace with the emerging technology of GIS.

To make all this happen, there is an urgent need for a mechanism by which academics and industry partners can obtain career advancement or other rewards for contributions to projects like the GERN. By 2010, possibly the highest accolade a GIS professor will be able to claim will be that he or she authored one of the ten most frequently rented GIS objects from the GERN network. Equally from an industry perspective, the developers of case studies or software applets that are accredited to GERN might expect to be rewarded for promoting the activities of the company on a global scale.

*But then, again, we could be entirely wrong!*

## References

Anonymous 1997 Who's who in GIS Number 4: Smallworld Systems. *Mapping Awareness* 11(10): 15

Apple 1998 *The Educational Object Economy. http://trp.research.apple.com*

Buehler K, McKee L 1996 *The OpenGIS Guide Introduction to Interoperable Geoprocessing: Part I of the Open Geodata Interoperability Specification).* OGIS TC Document 96-001

Denning P 1996 Business designs for the new university. *Educom Review* 31(6): 20–30

Dertouzos M 1997 *What Will Be: How The New World of Information Will Change Our Lives.* Piatkus, London

Foote K 1997 The Geographer's Craft: teaching GIS on the Web. *Transactions in GIS* 2(2): 137–150

Heywood D I 1997 *Beyond Chorley: Current Geographic Information Issues.* Association for Geographic Information, London

Heywood D I, Cornelius S C, Cremers P H 1998 Developing a virtual campus for UNIGIS: an international distance learning programme for geographical information professionals. In Anonymous (eds) *Bringing Information Technology into Education.* Dortrecht, Kluwer

Kemp K K 1997 The NCGIA Core Curricula in GIS and Remote Sensing. *Transactions in GIS* 2(2): 181–190

Kuiper K 1992 *Those Idiots in the Computer Room: Computer Myth Conceptions Spelled out in Plain English.* MacAdam House Publishing, Portland

Tapscott D 1996 *The Digital Economy: Promise and Peril in the Age of Networked Intelligence* McGraw-Hill, New York

Twigg C A, Oblinger D G 1996 *The Virtual University: A report from a Joint Educom/IBM Roundtable.* EDUCOM, Washington DC, viewed on the WWW on February 26, 1998 at *http://www.educom.edu/nlii/VU.html*

# 36 INTEROPERABILITY THROUGH ORGANIZATION: DIGITAL LIBRARIES FOR DISTRIBUTED GEOSPATIAL INFORMATION

Xavier R. Lopez

## 36.1. Introduction

The rapid development of communication and computing technology is changing the way scientific information is created, disseminated, managed, and used. Distributed computing, in particular, is increasingly being used as the foundation for scientific information management, collaboration, and publishing. This new foundation comes at a time when scientific budgets can rarely afford the creation of the large, vertically integrated scientific data centers which would otherwise be necessary to manage the explosion of information being produced today. Distributed computing is, therefore, being used to create a new scientific information infrastructure, and digital libraries are beginning to play a central role in emerging information systems.

The creation of digital libraries heralds a fundamentally new organizational structure. Building upon a distributed computing model, digital libraries will support the networked access, dissemination, and combining of scientific information. Because geographic information scientists routinely collaborate on a global scale, they are likely to be at the forefront of developments in globally integrated digital libraries. Thus, they will be among the first both to reap the benefits of this new organizational structure, and to encounter the problems associated with implementing it. Geoscientists will first have to solve the problems of interoperability between data sets, platforms, and organizations to experience the boost in scientific innovation, productivity, and return on investment that this new structure promises.

Organizations have evolved their own systems, legacy databases, and applications to serve internal needs. This has resulted in data models and applications uniquely tailored to meet specific internal requirements. However, as organizations recognize the importance of sharing information across organizational and computing environments, the issue of data interoperability becomes paramount. The activities of the Open GIS Consortium, the Federal Geographic Data Committee (FGDC), and CENDI, a U.S. interagency working group addressing scientific and technical information, are national efforts to promote technical interoperability across government and industry sectors. These facilitating efforts recognize the growing importance of interorganizational interaction and consensus to redefine technical requirements.

Traditionally, government data suppliers have operated under centralized and hierarchical structures. Both government and scientific data suppliers have also traditionally served bounded communities of users with unique semantic and conceptual requirements (e.g., military, federal agencies, transportation agencies). This hierarchical framework has profoundly influenced system architecture, resulting in closed, proprietary, and centralized geoprocessing. These developments have also hindered data interoperability between platforms and organizations.

The interoperability of geographic information across systems and platforms is also an *organizational* issue. Increasingly there is an urgent need to access distributed information from many organizations to address problems or provide services. Disaster relief, environmental monitoring, interagency coordination, joint force deployment, and the provision of integrated geospatial mapping services over the Internet all require interorganizational access of data. This need to access information resources across bureaucratic and hierarchical boundaries calls for new organizational processes that permit open network exchanges.

This chapter examines the *organizational dimension* of interoperability for geographic information. The chapter focuses on the requirements of the geographic information science community to examine possible network-centered organizational frameworks to facilitate interoperability. It addresses each of the following themes in sequential order:

- Institutional and technical forces of change
- Interoperability challenges in geoprocessing
- Interoperability through organizational integration
- Digital libraries as interoperable organization systems

## 36.2. Institutional and technical forces of change

The convergence of computing and communications is powerful, enabling, if not driving, the emergence of digital libraries. Figure 36.1 illustrates recent institutional and technical forces of change leading to the formation of network organizations, such as digital libraries. While the financial justifications for implementing digital libraries are generally recognized (i.e., reduced storage, dissemination, infrastructure, and transaction costs), another important feature is that such highly-structured organizations enable the global exchange of digital resources and expertise.

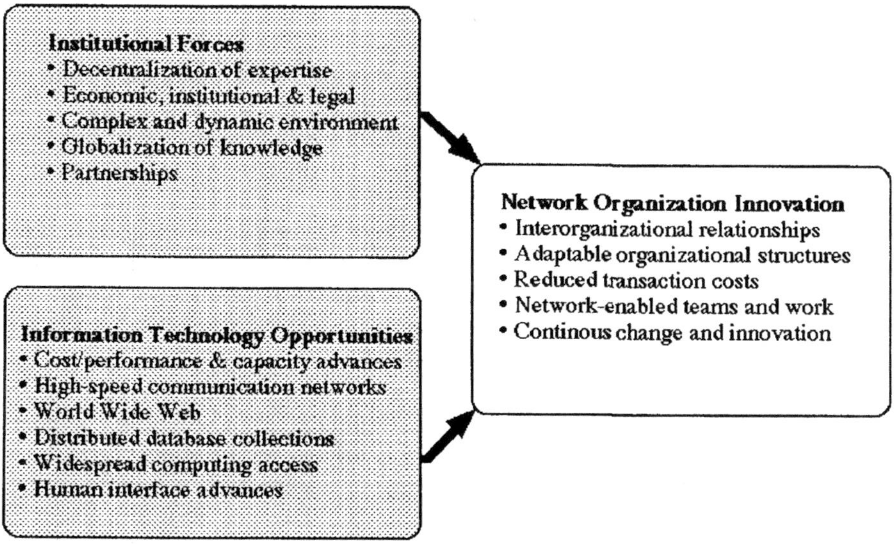


Figure 36.1. Institutional and technical forces of change

Drawing from the experiences in industrial systems, it is likely that changes in networking technologies may ultimately re-shape our concept of libraries, data centers, and similar information-based services. Electronic bookstores, such as *Amazon.com*, already use web-based supply chains to coordinate on-line demand for books. Moreover, software publishers and digital content providers increasingly use the Internet and Web to sell and give away products. Digital libraries are poised to extend this publishing and information exchange model to handle databases, imagery, and peer-reviewed digital content to suit for-profit and not-for-profit objectives.

Information exchange is not limited to numerical datasets. Increasingly, tacit knowledge and expertise held at the individual unit level must be amplified to the organizational level through explicit knowledge creating processes (Nonaka and Takeuchi 1995). Knowledge generation is enabled by structuring group processes and technologies (collaborative systems, video-conferencing) to transform tacit knowledge into explicit information. In doing so, a new organizational design is created. By linking competencies or pooling resources along an inter-organizational network of suppliers, partners, and intermediaries, a digital library can create alliances that pool knowledge resources and increase strategic flexibility. We are creating a new world, one that demands new metaphors for data management and dissemination systems. Previous metaphors of organization, such as the machine, are ill-suited for today's complex, boundary-spanning, networked environment.

## 36.3. Interoperability challenges in geoprocessing

The interoperability issues of the geographic information community are both *technical* and *organizational* in nature. As defined by Litwin et al. (1990) and paraphrased by the UCGIS: "interoperability generally refers to a bottom-up integration of pre-existing systems and applications that were not intended to be integrated but are systematically combined to address problems that require multiple DBMS and application programs" (UCGIS 1996: 1). Without consistent and open standards, the benefits of ongoing investments in spatial data may not be realized. Standards or interoperability capabilities are necessary for data models, content data, feature delineation, data collection, geo-referencing, indirect positioning, data quality, metadata, and data transfer (Guptill 1994). Information technology can only enhance organizational performance if it matches emerging business process needs.

In response, organizations are quickly moving away from proprietary standards and protocols, towards more open system platforms and interoperable processes. To reap the full benefits of communication and information technologies, networks, middleware, applications, and data need to be interoperable. Efficiency and effectiveness will likely increase where organizations embrace open technology infrastructures and agree to standard information processing conventions and business principles. The adoption of interoperable data models and modular software components, along with open network standards and protocols, permit concurrent and autonomous development of applications (Buehler and McKee 1996). Process and component design can be enhanced by *embedding coordination* between distributed component activities and other activities required to fully develop a new product, eliminating the need for much, if not all, of the overt managerial coordination of development activities.

However, geoprocessing must be thought of as more than simply the disparate uses, exchanges, and transformations of spatial data. To serve both the mass consumer and specialized user needs, geoprocessing would benefit by incorporating the metaphor of a virtual supply chain—pooled geospatial data components that can be assembled virtually to generate timely solutions. Effective communication and transfer of geographic information requires that organizations resolve interoperability of data models and components across organizational boundaries and applications. Over time, interorganizational alliances and partnerships, in combination with market forces, will generally establish *de jure* or *de facto* standards. However, distributed computing standards and protocols alone cannot support the complex supply chain of interactions that will increasingly be needed to support web-enabled geoprocessing.

A new business model is needed to keep an organization's business system synchronized with its information requirements. Organizations today must deal with increasing amounts of rapidly changing geospatial and attribute data. To deal with the increasing volume of data and the decreasing time frame for responding to requests, users want their information systems to become an integral part of business processes, rather than simply reflecting a static point in time. This increasingly integral use of information (and the systems that manage it) means that information systems must evolve at the same pace as the organization. The need to access information resources across bureaucratic and hierarchical boundaries calls for new interorganizational structures and strategies that permit the two-way exchange of information.

## 36.4. Interoperability through organizational integration

In today's networked economy, a single organization cannot meet the needs of a whole industry. No one data supplier or intermediary can do it alone. The whole community must be viewed as a *market* whereby the competitors and partners provide substitutable products and services that are differentiated by quality, price, and features. In many cases network-enabled strategic alliances provide a competitive advantage. This places a premium on *interoperable business processes and services* that integrate data, add value, and exchange geoprocessing services along an electronic supply chain. Major developments in computer-assisted design (CAD), computer-integrated manufacturing (CIM), electronic data interchange (EDI), and electronic commerce (EC) have demonstrated the extent to which product development, integration, and dissemination now use object components and virtual supply chains. It is urgent that managers of geospatial digital libraries consider these virtual work processes.

### *36.4.1. Systemic interdependencies*

Network organizations are more than networked information systems. A fundamental criterion for understanding these emerging forms of organization is recognition of their network characteristics. Figure 36.2 illustrates three structural approaches for addressing the systemic interdependencies in network organizations, focusing on the so-called Web, Star, and Chain network configurations (Kumar and Dissel 1996).

The first structure for addressing systemic interdependencies is the web. Readers will notice that the web structure represents the relations generated by today's Internet and World Wide Web. The web structure provides two-way communication, and enables intensive use of technologies including text, images, video, sound, telephony, and transactions. However, the structurability of the Web is low, resulting in high interoperability conflicts.

The *star* configuration attempts to address the lack of structure in the conventional web system. Through the use of standards, rules, and membership, customized technologies can service the needs of a particular network or user community. Such a community might be a global network of observatories and measurement stations cooperating to form a digital library to support global change research. The star network explicitly structures technology and content to support a community with well-defined information requirements. In doing so, interoperability conflicts are resolved through consensus-based standards.

The third configuration is commonly known as a *chain*. Chain structures are routinely used to enable long, linked supplier processes, as is increasingly common in electronic data interchange (EDI) and electronic commerce (EC) applications. Because typical supply chain interactions are subject to periodic change, these structures require more flexibility than a typical star structure. They also require a higher level of standards and protocol development than is common in a web structure.

| **Type of Interdependence** | **"Web"** Reciprocal Interdependence | **"Star"** Pooled Interdependence | **"Chain"** Sequential Interdependence |
|---|---|---|---|
| **Configuration** | 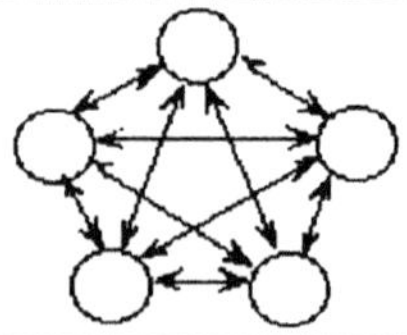 | 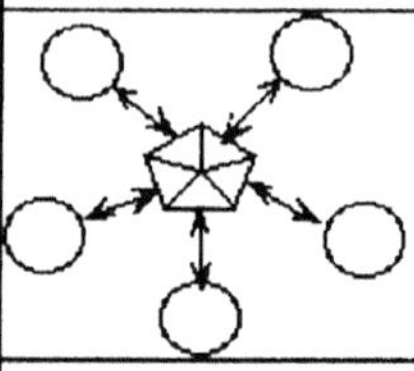 |  |
| **Coordinating Mechanisms** | Standards, Rules Schedules, Plans & Mutual Adjustment | Standards & Rules | Standards, Rules Schedules, & Plans |
| **Technologies** | Intensive | Mediating | Long-linked |
| **Structurability** | Low | High | Medium |
| **Interoperability Conflicts** | High | Low | Medium |
| **Examples of Application and Technologies** | Intranets, Extranets Video Conferencing | Shared Databases Digital Libraries, Virtual Organizations | Electronic Commerce EDI Applications Voice Mail, Fax |

(Adapted from Kumar & Dissel 1996)

Figure 36.2. Independence, structure, and interoperability conflicts

It may now become obvious to the reader that the application of these structures is not mutually exclusive. Network organizations such as digital libraries will employ multiple structures to support diverse functions. For example, electronic supply-chains can be embedded within a star structure to support activities that require sequential interdependence (e.g. electronic publishing and transactions). Moreover, the star pattern is likely to be embedded in the web structure of the Internet and WWW. The result is a complex organizational structure with multiple dimensions, each dimension well suited for carrying out specific functions.

The challenge of exploiting new technological opportunities, therefore, lies in adopting meaningful organizational structures to guide implementation. While advances in networking, computing, and geoprocessing are precipitating fundamental changes in organizational strategy and design, the resulting structures need to be flexible enough to accommodate change. Digital libraries are emerging as an organizational form that is responsive to managing the rapid growth of scientific information in a fast-changing environment.

## 36.5. The digital library as an interoperable organization system

Digital libraries provide a meaningful framework for integrating information resources and competencies from multiple organizations to deliver a synergistic service that is greater than its parts (Lopez 1997). While they may retain many of the important characteristics of a traditional research library (patron support, content management, archiving, inter-system transfer and sharing, specialized services), they will also exhibit important differences from traditional libraries and data centers. First, a digital library will focus not on physical collections but on digital content—databases, imagery, component objects, and video. To be of value as a supplier of government data, they must increasingly add some form of value (indexing, searching, enhancing) to government information resources (databases, imagery), which are increasingly available through the Web.

### *36.5.1. Definition*

There has been much discussion of the role that digital libraries may play in overcoming interoperability constraints and harmonizing the transfer of open geospatial data. However, the concept of a digital library suffers from semantic ambiguity, multiple interpretations, and imprecise definitions, and so must be clearly defined before further discussion can be productive. A digital library is defined as a coordinated set of heterogeneous actors or organizations which interact along an electronic communication network to develop, add value to, disseminate, and archive electronic information and related specialist services. It is characterized by its flexibility, which balances centralization with decentralization. The chief structural characteristic of a digital library is a high degree of data integration across formal boundaries. This definition highlights the importance of organizational relations in a digital library.

Because organizational relations are the key to digital library success, relationships between library actors must be defined and reinforced through interorganizational norms, consensus, and trust. Contractual arrangements and rigid, hierarchical rules alone will not prove to be adaptable and innovative enough to respond to changing technologies and client demands. A digital library must be effective in coordinating resources across horizontal boundaries and geographic space. Finally, many useful and difficult-to-imitate resources may lie outside a digital library's immediate sphere, but still be available to it if integrated along an electronic network. The focus of this network alliance is on the common pooling of competencies (data, services) to generate a synergy that is greater than its parts.

### *36.5.2. Affordances of digital library design*

As a purposive organization, the digital library directs concerted action toward a common goal. Interoperation of information and communications along electronic networks opens up new opportunities for innovation in digital library processes and strategy. A digital library integrates resources and competencies needed to deliver services into a coherent operational unit. While a digital library will retain many of the important service functions of a library (patron support, content management, archiving, inter-system transfer and sharing, specialized services), it will also exhibit important differences from traditional libraries and data centers (Table 36.1).

First, a digital library will shift from focusing on physical collections to digital content—databases, imagery, component objects, and video. A digital library will also decouple storage, organization, and access from a centralized site. The organizational structure of a conventional library is vertical and deterministic, while the structure of a digital library is flat and adaptive—responding to the requirements and resources of its community. It increasingly adds value (indexing, searching, enhancing) to electronic information resources in ways that bring synergy to the user. Access to singular databases alone is not enough. Increasingly, enhanced authoring will be supported, whereby origination, annotation, and collaboration take place on-line. In keeping with the characteristics of their networked environment, digital libraries must be capable of responding to the challenges and opportunities generated by the rapidly changing and uncertain environment in which they operate.

| | University Library | Digital Library |
|---|---|---|
| Information Object | book/journal article | databases and imagery |
| Hierarchy | vertical/closed | flat/open |
| Boundary | deterministic/closed | organic/adaptive |
| Strategy | product-centered | process-centered |
| Value-added | sum of parts | synergistic |
| Change processes | gradual/predictable | adaptive/uncertain |

Table 36.1. Comparative structural and process assumptions

Other notable basic characteristics of a digital library include:

- *Distributed Resources, Central Coordination*: Digital libraries provide fundamental *intermediary* processes between clients and suppliers. Success depends on the organization's capacity to coordinate data supply chains, intermediation and disintermediation, and other value-adding processes. Successful positioning along a network of dependencies is determined by the ability to enroll the support of others who recognize the value of the library's contribution. While decentralization permits flexibility, it also imposes coordination costs that must be taken into account.

- *Inter-organizational Interdependence*: The digital library is an organizational structure embedded in a network of relationships. The activities and purposes of digital libraries are interdependent with those of other suppliers, partners, and intermediaries. Its success lies in its ability to have other organizations conform to common data interoperability standards, while contributing to new knowledge about markets, information resources, and supply chain processes. While interorganizational interdependence imposes heavy coordination costs, it also generates tremendous potential for new product development, market creation, and organizational learning.

- *Leveraging Complementary Competencies*: A digital library increases flexibility and generates value by integrating resources and activities along a global supply network. Complementarity is a measure of the processes of a unit within a network. It is concerned with the ease with which combinations with a specific resource may, in general, be carried out. Complementarity within a network facilitates synergy.

- *Network Externalities*: Digital libraries are enabled by advances such as the Internet and World Wide Web. The skyrocketing growth of the Internet demonstrates how the adoption and acceptance of common standards and protocols can generate positive network externalities. Each additional unit added to the network increases the value of the whole. Likewise, by adopting common geoprocessing data models, network standards, and protocols, digital libraries can minimize costs of singular investments to users while maximizing the unit value of supplier and intermediary investments.

- *Dynamic Network Structure*: A digital library's organizational relations and structure result from dynamic interactions with other organizations and their environment. Strategic managerial action is necessary to guide an organizational mission that is in tune with community demands, changes in technology, and resource opportunities. This calls for a flexible and evolving network structure that is reinforced with a stable core of partner/client relationships.

- *Long-term Strategy*: Investments are needed to secure and sustain a start-up period. Developing the necessary competencies to sustain a digital library over the long haul will take time. One viable area of focus will be to support emerging requirements of the geoscience community. Additional concerns include: assessment of user needs, data development and integration, strategic alliance formation, and data licensing. Concerted efforts to establish and maintain long-term partnerships with key government, commercial, and academic suppliers, and with other organizations sharing common objectives are key.

The organizational forms of network environments are emerging from the opportunistic alliance of suppliers and partners that provide *best of the breed* products and services. The ability to generate *virtual* value-added networks requires strategic integration of suppliers, partners, and intermediaries into one coherent and tightly coupled whole (Figure 36.3). *Virtual* vertical integration is therefore a direct outcome of the creation of digital libraries. This super-organizational structure balances flexibility with stability and centralization with decentralization to provide low-cost, high-quality service.

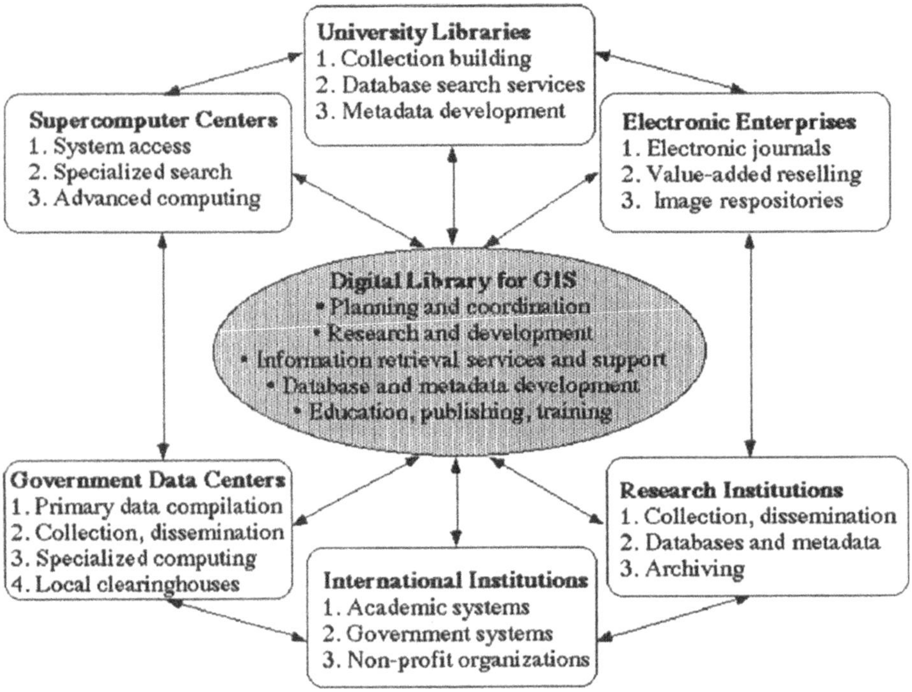


Figure 36.3. Digital library for geographic information science

Organizational relations are crucial to a digital library. Digital libraries can play an instrumental role in overcoming current impediments to interoperability, by working at the organizational level to harmonize the adoption of open geospatial data. Such a role reduces transaction costs, stimulates component generation, and provides a standard platform for new components and applications. On one level, digital libraries provide a technical infrastructure to meaningfully translate geographic information across system platforms with minimal loss. On another level, they can serve the long-term interoperability requirements of unique user communities, by actively engaging in the evolution of technical standards.

In a sense, the organizational challenge of maintaining peer to peer communication and relationships is analogous to the technical challenges of transferring geographic information across computing platforms. Open data models can reduce transaction costs, stimulate component generation, and provide a standard platform for new components and applications. Contractual arrangements and hierarchical rules also facilitate data interchange between the digital library and suppliers and the digital library and clients. However, static agreements and protocols, alone, may not provide the needed flexibility to respond to changing internal and external requirements and opportunities. Increasingly, human relations are necessary to (re)define common objectives and establish consistent work processes. Such relations are reinforced through interorganizational norms, consensus, and trust, thereby coordinating resource integration across institutional boundaries and geographic space. In doing so, the network self-organizes, responding to its community demands.

### *36.5.4. Towards a modular organizational form*

While this chapter has focused exclusively on network relations for digital libraries, further work that closely examines the components of a digital library is needed. Daft and Lewin (1993) have suggested the evolution of a new form of *modular organization* that uses "interconnected coordinated self-organizing processes" to support continuous learning and flexible responses to a changing environment. Modularity of function is a central hallmark of vertically integrated industries and bureaucracies (Weber 1947). Modularity is also used in object-oriented software engineering to manage complexity and develop portable, efficient, and extensible applications (Szyperski 1998). An object orientation model may provide a useful metaphor for describing digital library creation and management, applying concepts such as components, behaviors, frameworks, and functions to the network organization domain.

In a digital library environment, data object repositories and their associated organizations can be considered to be components. Organizations have associated *behaviors*. Likewise, the interorganizational logic and processes for building a digital library are analogous to the frameworks through which the library performs its *functions*. These frameworks are reusable in that they can be used to produce customized digital libraries. Such modular organizational structures are compelling and may help to better structure distributed knowledge and expertise.

## 36.6. Summary

In today's fast-changing, networked society and economy, closed and rigidly hierarchical organizational structure may not be optimally suitable for diverse geospatial information services. Flexible and open organizational processes and structures that permit the smooth flow of raw and enhanced information along computing networks are needed. Digital libraries provide an efficient and effective means to overcome current technical and organizational impediments to interoperable geoprocessing. These network systems attempt to resolve interoperability problems not simply through technical means, but by harmonizing information handling practices at the organizational level. This is a promising development which can lead to decreased user cost, increased unit returns on organizational investments, and an accelerated pace of innovation and development in the geographic information community.

## References

Buehler K, McKee L (eds) 1996 *The Open GIS Guide.* Open GIS Consortium, Wayland, Mass

Daft R L, Lewin A Y 1993 Where are the theories of the 'new' organization forms? An editorial essay. *Organization Science* 4(4): i–iv

Guptill S C 1994 Spatial data standards. *Government Information Quarterly* 11(4): 387–401

Kumar K, van Dissel H G 1996 Sustainable collaboration: managing conflict and cooperation in interorganizational systems. *MIS Quarterly* 20(3): 279–299

Litwin W, Mark L, Roussopoulos N 1990 Interoperability of multiple autonomous databases. *ACM Computing Surveys* 22(3): 267–293

Lopez X R 1997 The network as organization: digital libraries for spatial information. *Proceedings of the First Assembly and Retreat of the University Consortium for Geographic Information Science(UCGIS), Bar Harbor, ME June 15–20*

Nonaka I, Takeuchi H 1995 *The Knowledge-Creating Company: How Japanese Companies Create the Dynamics of Innovation.* Oxford University Press, New York

Szyperski C 1998 *Component Software: Beyond Object-Oriented Programming.* Addison-Wesley, Reading, Mass

University Consortium for Geographic Information Science (UCGIS) 1996 *Interoperability of Geographic Information. http://www.ucigs.org/*

Weber M 1947 *The Theory of Social and Economic Organization.* Oxford University Press, New York

# EPILOG

One of our objectives in developing this book was to bring practice and theory together, and to build a more productive partnership between them. We argued at the outset that practice in GIS had generally evolved in the absence of theory. Much theoretical progress has been made in the past few years, reflected in the proceedings of the COSIT (International Conference on Spatial Information Theory), SSD (International Symposium on Large Spatial Databases), SDH (International Symposium on Spatial Data Handling), and other series of conferences, the journals (*International Journal of Geographical Information Science*, *Geoinformatica*, etc.), and books. But GIS is a very practical subject, and a wide gap still exists between the large and growing population of practitioners, and the relatively small community of theoreticians. Nowhere is this more apparent than in the first efforts at developing interoperation between GISs. Yet the time seems ripe, with the appearance of the Open GIS Consortium and other groups, for a stronger effort to bridge the gap between theory and practice. We hope this book will be viewed as part of that effort.

We are impressed at the power and ingenuity of the many prototypes and systems described in this book. The field is very fast-moving, and it is hard enough to keep up with developments reported in the literature. It is even harder to keep up with the development of prototypes, many of which never appear in the literature, so we hope the collection presented in this book will be of value both through the descriptions of their designs, and also through the lessons that have been learned. At the same time they illustrate the complexity of the problem of GIS interoperation, and the great range and diversity of issues that fall under that general rubric.

One issue that is almost conspicuously absent from this book is uncertainty. It is impossible to determine location on the Earth's surface with infinite accuracy, and other forms of uncertainty derive from the inherent vagueness of definition of many geographical terms and variables, and from the failure of a producer of information to describe the process of production fully to users. Perhaps uncertainty can be treated like any other attribute of data, as a metadata field; but recent research suggests that the story is far more complex, and that consideration of uncertainty often forces a reassessment of the entire process of data modeling, and thus impacts interoperation (see, for example, Guptill and Morrison 1995; Heuvelink 1998).

After assembling the book, one is tempted to ask what the world of GIS might look like in five or ten years, as the efforts described here come to fruition and influence future efforts. What progress is likely to be made in interoperating GISs, and will the hopes expressed here evolve into standard practice? We offer the following predictions:

- Hardware speed will continue to improve, as will network connectivity;
- There will be greater emphasis on geospatial information, as improved GIS and other geographic information technologies facilitate greater awareness of our global and local surroundings;
- There will be greater competition in the GIS industry, as the market for software products matures, and vendors will see interoperation as a significant marketing edge;
- GISs will be easier to use in future, as vendors adopt the user interface concepts that have been proven in other applications, and users come to expect them in all computing environments. It is, however, questionable whether the typical point-and-click interactions with the most common software components are the appropriate types of interactions with GIS (Egenhofer and Kuhn 1998);
- More aspects of geographic information processing will be hidden from the user, including issues of geodetic datum and map projection, and certain details of discretization;
- We will not make much progress towards a universal language of geographic information, but theory will advance, driven in part by more general advances in computer, cognitive, and information science, and there will be a degree of convergence within the GIS industry on common languages and specifications;
- Interoperating GIS will enable a transition from a one-size-fits-all GIS to information technologies that are specific to domains and tasks. We see already such diversification with specialized software systems that are inherently geospatial, but not necessarily considered GISs. Examples are car navigation systems or GPS-based personal tracking devices. The missing link that interoperation should provide in the future is the reuse of software components designed for such specialized software systems, and the ability to connect and integrate them with other information services.

Progress in GIS interoperation depends in no small part on the cohesiveness of the field, and the degree to which geographic information science can be recognized as a specialty in its own right. We are encouraged by the signs of an emerging discipline, including the growing list of books, journals, and conferences, new university programs, and the success in the U.S. of the University Consortium for Geographic Information Science, and hope that similar efforts to UCGIS will be successful elsewhere.

## References

Egenhofer M J, Kuhn W 1998 Interacting with GIS. In Longley P A, Goodchild M F, Maguire D J, Rhind D W (eds) *Geographical Information Systems: Principles, Techniques, Management, and Applications*. Wiley, New York

Guptill S C, Morrison J L (eds) 1995 *Elements of Spatial Data Quality*. Elsevier, New York

Heuvelink G B M 1998 *Error Propagation in Environmental Modelling*. Taylor and Francis, London

# CONSOLIDATED REFERENCES


Abadi M, Cardelli L 1996 *A Theory of Objects*. Springer-Verlag, New York

Abel D J 1997 Spatial Internet marketplaces: A grand challenge. In Scholl M, Voisard A (eds) *Advances in Spatial Databases*. Lecture Notes in Computer Science 1262. Springer-Verlag, Berlin/Heidelberg/New York

Abel D J, Gaede V, Taylor K L, Zhou X 1997 *SMART: Towards Spatial Internet Marketplaces*. Technical Report ISS-40. Humboldt-Universität zu Berlin, Germany. *http://www.wiwi.hu-berlin.de/~gaede/smart.ps.gz*

Abel D J, Gaede V, Taylor K L, Zhou X 1997 SMART: Towards spatial Internet marketplaces. Technical Report SIS-1997-01. CSIRO, Canberra

Abel D, Yap S, Ackland R, Cameron M, Smith D, Walker G 1992 Environmental decision support system project: an exploration of alternative architectures for geographical information systems. *International Journal of Geographical Information Systems* 6(3): 193–204

Abiteboul S, Quass D, McHugh J, Widom J, Wiener J L 1997 The LOREL query language for semi-structured data. *Int. Journal on Digital Libraries* 1(1): 68–80

Abler R 1987 The National Science Foundation National Center for Geographic Information and Analysis. *International Journal of Geographical Information Systems* 1: 303–326

Absil van de Kieft I, Kok B 1997 The development of a geo metadata service for the Netherlands. *Third Joint European Conference & Exhibition on Geographical Information (JEC-GI'97)*: 1165–1176

Adali S, Subrahmanian V S 1994 Amalgamating knowledge bases, II: Distributed mediators. *International Journal of Intelligent and Cooperative Information Systems* 3(4): 349–383

Adobe 1990 *Postscript Language Reference Manual*. Second Edn. Addison-Wesley, Reading

Akrich M, Latour B 1992 A summary of convenient vocabulary for the semiotics of human and nonhuman assemblies. In Bijker, W, Law J (eds) *Shaping Technology/Building Society*. MIT Press, Cambridge, Mass.: 259–264

Albrecht J 1994 Universal elementary GIS tasks—beyond low-level commands. In Waugh T C and Healey R G (eds) *Sixth International Symposium on Spatial Data Handling*: 209–22

Alexander E R 1995 *How Organizations Act Together: Interorganizational Coordination in Theory and Practice*. Gordon and Breach

Alexander-Tomlinson 1997 *Queensland Land Information Strategy (QLIS) Benefit Study. Volume II: Situation Report (Assessment of Government and Industry Spatial Information Requirements)*

Almási G, Jagannathan V 1995 Integrating the WWW and CORBA-based environments. *Proceedings of the Fourth World Wide Web Conference. http://webstar.cerc.wvu.edu/ www4/www4.html*

Alms R, Balovnev O, Breunig M, Cremers A B, Jentzsch T, Siehl A 1997 Space–time modeling of the Lower Rhine Basin supported by an object-oriented database. *Physics and Chemistry of the Earth.*

Ambite J L, Knoblock C A 1995 Reconciling distributed information sources. *Working Notes of the AAAI Spring Symposium on Information Gathering in Distributed Heterogeneous Environment, Palo Alto*

AMIRA P431 1995 *A Data Dictionary of Geological Objects.* Australian Mineral Industries Research Association LTD, Melbourne

Anonymous 1997 Who's who in GIS Number 4: Smallworld Systems. *Mapping Awareness* 11(10): 15

ANSI/NISO 1995 *Information Retrieval: Application Service Definitions and Protocol specification: Z39.50.* American National Standards Institute, New York

Apple 1998 *The Educational Object Economy. http://trp.research.apple.com*

Arctur D, Hair D, Timson G, Martin E P, Fegeas R 1998 Issues and prospects for the next generation of the spatial data transfer standard (SDTS). *International Journal of Geographical Information Science* 12(4): 403–426

Arens Y, Knoblock C A, Shen W 1996 Query reformulation for dynamic information integration. In Wiederhold G (ed) *Intelligent Integration of Information.* Kluwer Academic Publishers: 11–42

Armstrong M P, Marciano R J 1996 Local interpolation using a distributed parallel supercomputer. *International Journal of Geographical Information Systems* 10(6): 713–730

Arya M, Cody W, Faloutsos C, Richardson J, Toga A 1994 QBISM: extending a DBMS to support 3D medical images. *Proceedings of the 10th International IEEE Conference on Data Engineering, Houston, Texas*: 314–325

ASTM (American Society for Testing and Materials) 1985 *Annual Book of ASTM Standards.* (Section 4: *Construction,* Vol. 04.08, *Soil and Rock; building stones*). ASTM, Philadelphia

Aufaure-Portier M A, Trépied C 1996 A survey of query languages for geographic information systems. *Proc. 3rd Int. Workshop on Interfaces to Database Systems, Edinburgh*

August P V 1991 Use vs abuse of GIS data. *Journal of the Urban and Regional Information Systems Association* 3(2): 99–101

Baker T, Broadhead J 1992 Integrated access to land-related data from heterogeneous sources. In Salling M J (ed) *Proceedings of the 1992 Annual Conference of the Urban and Regional Information Systems Association (URISA).* URISA, Washington DC

Balovnev O, Breunig M, Cremers A B 1997 From GeoStore to GeoToolKit: The second step. *Proceedings of the 5th International Symposium on Spatial Databases.* Lecture Notes in Computer Science 1262. Berlin, Springer: 223–37

Barley S R. 1986 Technology as an occasion for structuring: evidence from observations of CT scanners and the social order of radiology departments. *Administrative Science Quarterly* 31:78–108

Batini C, Lenzerini M, Navathe S 1986 A comparative analysis of methodologies for database schema integration. *ACM Computing Surveys* 18(4): 323–64

Bayardo R, Bohrer W, Brice R, Cichocki A, Fowler G, Helal A, Kashyap V, Ksiezyk T, Martin G, Nodine M, Rashid M, Rusinkiewicz M, Shea R, Unnikrishnan C, Unruh A, Woelk D 1997 Semantic integration of information in open and dynamic environments. *Proceedings of the 1997 ACM International Conference on the Management of Data (SIGMOD), Tucson*: 195–206

Beard K, Smith T 1998 A framework for meta-information in digital libraries. In Sheth A, Klas W (eds) *Multimedia Data Management: Using Metadata to Integrate and Apply Digital Media*. McGraw Hill: 341–365

Becker P 1996 An embeddable and extendable language for large-scale programming on the Internet. *Proceedings of the 16th International Conference on Distributed Computing Systems (ICDCS'96)*

Beckmann N, Kriegel H-P, Schneider R, Seeger B 1990 The R*-tree: an efficient and robust access method for points and rectangles. *Proceedings of the 1990 ACM SIGMOD International Conference on Management of Data, Atlantic City*: 322–31

Bédard Y, Proulx M J, Létourneau F, Moulin B, Maamar Z 1997 Géorépertoires sur Internet: Outils puissants de démocratisation des données à référence spatiale. *Proceedings of Géomatique VI: Un monde accessible*. Canadian Institute of Geomatics, Montreal: 23–37

Beech D 1997 Data semantics on the information superhighway. In Meersman R, Mark L (eds) *Database Application Semantics*. Chapman and Hall

Bennett D A 1997 Managing geographical models as repositories of scientific knowledge. *Geographical and Environmental Modelling* 1(2): 115–133

Bennett D A 1997 A framework for the integration of geographic information systems and modelbase management. *International Journal of Geographical Information Systems* 11(4): 337–357

Berners-Lee T J, Cailliau R., Groff J-F 1992 The world-wide web. *Computer Networks and ISDN Systems* 25(4–5): 454

Bernstein P 1996 Middleware: A model for distributed system services. *Communications of the ACM* 39(2): 86–98

Bernstein P, Hadzilacos V, Goodman N 1987 *Concurrency Control and Recovery in Database Systems*. Addison-Wesley

Bertin J 1983 *Semiology of Graphics. Diagrams, Networks, Maps*. The University of Wisconsin Press, Madison

Bhargava H K, King A S, McQuay D S 1995a DecisionNet: An architecture for modeling and decision support over the World Wide Web. In Bui T X (ed) *Proceedings of the Third International Society for Decision Support Systems Conference*. International Society for DSS, Hong Kong II: 541–550

Bhargava H K, Krishnan R 1998 The World Wide Web: Opportunities for OR/MS. *INFORMS Journal on Computing*

Bhargava H K, Krishnan R, Casey M, Kaplan D, Roehrig S, Müller R 1997a Model management in electronic markets for decision technologies: A software agent approach. In Sprague R (ed) *Proceedings of the 30th Hawaii International Conference on System Sciences*

Bhargava H K, Krishnan R, Kaplan D 1995b On customized access to a WWW-based distributed network of decision support services. In Bui T X (ed) *Proceedings of the Third International Society for Decision Support Systems Conference*. International Society for DSS, Hong Kong II: 551–560

Bhargava H K, Krishnan R, Müller R 1997b Decision support on demand: Emerging electronic markets for decision technologies. *Decision Support Systems* 19(3): 193–214

Bhargava H K, Krishnan R, Müller R 1997c Electronic markets for decision technologies: A business cycle analysis. *Journal on Electronic Commerce* 1(4): 109–127

Bhargava H K, Tettelbach C G 1998 A Web-based decision support system for waste disposal and recycling. *Computers, Environment and Urban Systems*

Bhargava H L, Krishnan R, Müller R 1996 *Electronic Commerce in Decision Technologies: A Business Cycle Analysis*. Technical Report. Humboldt-Universität zu Berlin, Germany

Birkhoff G 1945 Universal algebra. *Proceedings of First Canadian Mathematical Congress, Toronto University*: 310–326

Birkhoff G, Lipson J D 1970 Heterogeneous algebras. *Journal of Combinatorial Theory* 8: 115–133

Bishr Y 1997 *Semantic Aspects of Interoperable GIS*. PhD Thesis, Publication Series Number 56. ITC, Enschede, The Netherlands

Bishr Y 1998 Overcoming the semantic and other barriers to GIS interoperability. *International Journal of Geographic Information Science* 12(4): 299–314

Bishr Y, Molenaar M, Radwan M 1996 Spatial heterogeneity of federated GIS in a client/server architecture. In *Proceedings, GIS'96, Vancouver, Canada*

Black A, Hutchinson N, Jul E, Levy H, Carter L 1987 Distribution and abstract types in Emerald. *IEEE Transactions of Software Engineering* SE-13(1)

Bloomer J 1991 *Power Programming with RPC*. O'Relly & Associates, Sebastopol, Calif.

Bloomer J 1995 Distributed computing and the OSF/DCE. *Dr. Dobbs Journal* 2

Bode T, Breunig M, Cremers A B 1994 First experiences with GeoStore, an information system for geologically defined geometries. *Proceedings IGIS'94*. Lecture Notes in Computer Science 884. Springer, Berlin: 35–44

Boisvert R F 1989 The guide to available mathematical software advisory system. *Math. and Comp. in Simul* 31: 453–464

Boland R J, Tenkasi R V 1995 Perspective making and perspective taking in communities of knowing. *Organization Science* 6(4): 350-372

Boll S, Klas W, Sheth A 1998 Overview on using metadata to manage multimedia data. In Sheth A, Klas W (eds) *Multimedia Data Management: Using Metadata to Integrate and Apply Digital Media*. McGraw-Hill Publishers: 1–23

Bonneville Power Administration (BPA) 1996 *StreamNet: Report on the status of salmon and steelhead in the Columbia River basin, 1995*. Bonneville Power Administration, Portland, Ore.

Booch G 1996 *Object Solutions: Managing the Object-Oriented Project*. Addison-Wesley, Menlo Park, California

Bordie M L 1992 The promise of distributed computing and the challenge of legacy information systems. *Proceedings of the IFIP WG2.6 Database Semantics Conference on Interoperable Database Systems, Lorne, Australia, 16-20 November*: 1–31

Boursier P, Dbouk M, Kvedarauskas D 1996 Querying and visualizing geographical data sets with dynamic maps. *Proc. 2nd ACM Workshop on New Paradigms for Information Visualization and Manipulation, Rockville*

Boursier P, Kvedarauskas D, Iris S, Guilloteau A 1995 MAGIC TOUR: the integration of multimedia and geographic information technologies in an authoring system. *Proc. 1st Int. Conference on Spatial Multimedia and Virtual Reality (SMVR '95), Lisbon*

Boursier P, Mainguenaud M 1992 Spatial query languages: extended SQL vs. visual languages vs. hypermaps. *Proc. 5th Int. Symposium on Spatial Data Handling (SDH '92), Charleston*

Brassel K, Bucher F, Stephan E M, Vckovski A 1995 Completeness. In Guptill S C, Morrison J L (eds) *Elements of Spatial Data Quality*. Elsevier Science, Amsterdam: 81–108

Breitbart Y, Garcia-Molina H, Silverschatz A 1992 Transaction management issues in a failure-prone multidatabase system environment. *VLDB Journal* 1(1): 1–39

Breunig M, Bode T, Cremers A B 1994 Implementation of elementary geometric database operations for a 3D GIS. *Proceedings of the 6th International Symposium on Spatial Data Handling, Edinburgh*: 604–17

Breunig M, Perkhoff A 1992 Data and system integration for geoscientific data. *Proceedings of the Fifth International Symposium on Spatial Data Handling, Charleston, South Carolina*: 272–81

Brodie M L 1993 The promise of distributed computing and the challenges of legacy information systems. In Hsiao D K, Neuhold E J, Sacks-Davis R (eds) *Interoperable Database Systems (DS-5)*. North-Holland, Elsevier: 1–31

Brown W S, Garrison K M, MacBeth M 1994 *Prototype Environmental Data and Information Management Systems (EDIMS). http://gulfofmaine.unh.edu/edims/edims_description.html*

Buehler K 1994 *OGIS Geodata Model Overview*. OGIS project documentation 94-019. Open GIS Consortium

Buehler K 1996 *The Open Geodata Interoperability Specification Part II: Abstract Specification*. Draft version. Open GIS Consortium, Wayland, Mass, USA

Buehler K, McKee L (eds) 1996 *The OpenGIS Guide, Introduction to Interoperable Geoprocessing*. Open GIS Consortium, Wayland, Mass. *http://www.opengis.org/public/guide.html*

Buneman P, Davidson S, Kosky A 1992a Theoretical aspects of schema merging. *Proc. Extending Database Technology (EDBT'92)*: 152–167

Buneman P, Davidson S, Kosky A, VanInwegen M 1992b A basis for interactive schema merging. *Proc. Hawaii International Conference on System Sciences*

Burrough P A 1986 *Principles of Geographical Information Systems for Land Resources Assessment*. Clarendon Press, Oxford

Burrough P A 1997 Environmental modeling with geographical information systems. In Kemp Z (ed.) *Innovations in GIS 4*. Taylor and Francis, London: 143–53

Bushmann F, Meunier R, Rohnert H, Sommerlad P, Stal M 1996 *A System of Patterns*. Wiley, England

Buttenfield B P, Ganter J H 1990 Visualization and GIS: What should we see? What might we miss? *Proceedings of the 4th International Symposium on Spatial Data Handling, Zurich*: 307–316

Câmara G 1995 *Models, Languages and Architectures for Geographic Databases*. PhD Thesis, National Institute of Spatial Research (INPE), Brazil (in Portuguese)

Câmara G, Casanova M A, Hemerly A S, Magalhães G C, Medeiros C M B 1996 *Anatomy of Geographic Information Systems*. Computing Institute, University of Campinas, Brazil (in Portuguese)

Camara G, Freitas U M, Souza R, Casanova M, Hemerly A, Medeiros C 1994 A model to cultivate objects and manipulate fields. *Proceedings 2nd ACM Workshop on Advances in GIS*: 20–28

Campbell J B 1983 *Mapping the Land: Aerial Imagery for Land Use Information*. Association of American Geographers, Washington DC

Cardelli L, Wegner P 1985 On understanding types, data abstraction, and polymorphism. *ACM Computing Surveys* 17(4): 471–522

Carter J R 1992 Perspectives on sharing data in geographic information systems. *Photogrammetric Engineering and Remote Sensing* 58(11): 1557–1560

Casner S M 1991 A task-analytic approach to the automated design of graphic presentations. *ACM Transactions on Graphics* 10(2): 111–151

Cattell R G G (ed) 1994 *The Object Database Standard: ODMG-93*. Morgan Kaufman

Cattell R G G, Barry D K (eds) 1997 *The Object Database Standard: ODMG-2.0*. Morgan Kaufmann: 251–5

CEN-META 1996 CEN/TC 287. Geographic information—data description—metadata. Technical Report draft prEN 12657. CEN, Brussels

CEN-QUERY 1997 CEN/TC 287. Geographic information—processing—query and update. Technical Report draft prEN 12660. CEN, Brussels

CEN-SCHEMA 1996 CEN/TC 287. Geographic information—data description—spatial schema. Technical Report final draft prENV 12160. CEN, Brussels

Central Imagery Office 1995 *Standards Profile for Imagery Access, Version 2*. CIO-2020. Central Imagery Office, Vienna, VA

CEN-TRANSF 1997 CEN/TC 287. Geographic information—data description—transfer. Technical Report draft prEN 12658. CEN, Brussels

CEOS 1996 *Catalogue Interoperability Protocol (CIP) Specification Release B.* CEOS/WGISS/PPT/CIP-B, European Space Agency/ESRIN, Frascati, Italy

Chappell D 1996 *Understanding ActiveX and OLE.* Microsoft Press, Redmond

Chen F, Hearst M, Kimber D, Kupiec J, Pedersen J, Wilcox L 1998 Metadata for mixed-media access. In Sheth A, Klas W (eds) *Multimedia Data Management: Using Metadata to Integrate and Apply Digital Media.* McGraw Hill: 319–340

Cheung W, Hsu C 1996 The Model-Assisted Global Query System for multiple databases in distributed enterprises. *ACM Transactions on Information Systems* 14(4): 421–470

Chorley R 1988 Some reflections on the handling of geographic information. *International Journal of Geographic Information Systems* 2(1): 3–9

Clark University 1998 *IDRISI Software Homepage. http://www.idrisi.clarku.edu/*

Clarke K C 1990 *Analytical and Computer Cartography.* Prentice Hall, Englewood Cliffs, NJ

Clément G, Larouche C, Gouin D, Morin P, Kucera H 1997 OGDI: Toward interoperability among geospatial databases. *SIGMOD Record* 26(3)

Clementini E, Felice P D, Van Oosterom P 1993 A small set of formal topological relationships suitable for end user interaction. In Abel D, Ooi B C (eds) *Advances in Spatial Databases, Proceedings of the Third International Symposium on Spatial Data Handling.* Lecture Notes in Computer Science No. 692. Springer Verlag, New York: 277–295

Cleveland W S 1993 *Visualizing Data.* Hobart Press, Summit NJ

Clinton W 1994 Coordinating geographic data acquisition and access: the National Spatial Data Infrastructure, Executive Order 12906. *Federal Register* 59(71): 17671–4. *http://www.fgdc.gov/Communications/GenInfo/execord.html*

Cook S, Daniels J 1994 *Designing Object Systems: Object-Oriented Modelling with Syntropy.* Prentice Hall

CORBA 1995 *The Common Object Request Broker: Architecture and Specification.* Version 2.0. Object Management Group, Framingham

Couclelis H 1992 People manipulate objects (but cultivate fields): beyond the raster-vector debate in GIS. In Frank A U, Campari I, Formentini U (eds) *Theories and Methods of Spatio-Temporal Reasoning in Geographic Space.* Lecture Notes in Computer Science No. 639. Springer Verlag, Berlin: 65–77

Craig, W J 1995 Why we can't share data: institutional inertia. In Onsrud H J, Rushton G (eds) *Sharing Geographic Information.* Rutgers University, New Brunswick, NJ

Crawley S, Davis S, Indulska J, McBride S, Raymond K 1997 Meta information management. *Formal Methods for Open Object-based Distributed Systems (FMOODS) Conference, Canterbury, UK, July 1997*

Croswell P L 1989 Facing reality in GIS implementation: lessons learned and obstacles to be overcome. In Salling M J, Gayk W F (eds) *Proceedings of the 1989 Annual Conference of the Urban and Regional Information Systems Association (URISA).* URISA, Washington DC: 15–35

Czyzyk J, Mesnier M P, More J J 1997 The networked-enabled optimization system (NEOS) server. Preprint MCS-P615-1096. Mathematics and Computer Science Division, Argonne National Laboratory

Daft R L, Lewin A Y 1993 Where are the theories of the 'new' organization forms? An editorial essay. *Organization Science* 4(4): i–iv

Daniel R, Lagoze C, Payette S 1998 A metadata architecture for digital libraries. *Proceedings of the Forum on Research and Technology Advances in Digital Libraries (ADL'98), Santa Barbara*: 276–288

Dao S, Perry B 1996 Information mediation in cyberspace: scalable methods for declarative information networks. In Wiederhold G (ed) *Intelligent Integration of Information.* Kluwer Academic Publishers: 43–62

Data Engineering Group 1998 *DIAL Release Version 1.0.* Raytheon STX Corporation, Greenbelt, Maryland. *http://hops.stx.com/dialhome.html*

De Oliveira J L, Pires F, Medeiros C B 1997 An environment for modeling and design of geographic applications. *GeoInformatica* 1: 29–58

Defense Mapping Agency 1983 *Geodesy for the Layman (DMA TR 80-003).* Defense Mapping Agency, Washington, DC

Defense Mapping Agency 1996 *Global Geospatial Information & Services (GGI&S) and Data Architecture and Gateway Services (DAGS).* Solicitation released on June 27, 1996

Denning P 1996 Business designs for the new university. *Educom Review* 31(6): 20–30

Department of the Interior 1987 *Digital Line Graphs from 1:100,000-Scale Maps—Data Users Guide 2.* U.S. Geological Survey, Reston, VA

Department of the Interior 1990 *Standards for Digital Line Graphs, Part 3: Attribute Codes.* U.S. Geological Survey, Reston, VA

Department of the Interior 1997 *DLG-3 SDTS Transfer Description, Draft.* *ftp://sdts.er.usgs.gov/pub/sdts/datasets/tvp/dlg3/*

Dertouzos M 1997 *What Will Be: How The New World of Information Will Change Our Lives.* Piatkus, London

DeSanctis G, Poole M S 1994 Capturing the complexity in advanced technology use: adaptive structuration theory. *Organization Science* 5(2)

DeWitt D J, Kabra N, Luo J, Patel J M, Yu J-B 1994 Client–server Paradise. *Proceedings of the 20th VLDB Conference, Santiago, Chile*

DiBiase D 1996 Rethinking laboratory education for an introductory course on geographic information. *Cartographica* 33(4): 61–71

DND-Canada and Geomatics Canada 1995 Canadian Geomatics CD-ROM. *GEOMATICS 1995 Conference, Ottawa, Canada, June 11–15*

Dozier J 1992 How Sequoia 2000 addresses issues in data information systems for global change. Technical Report 94-14. Sequioa 2000, University of California, Berkeley

Drew P, King R, McLeod D, Rusinkiewicz M, Silberschatz A 1993 Report of the workshop on semantic heterogeneity and interoperation in multidatabase systems. *SIGMOD Record* 22(3): 47–56

Drew P, Ying J 1998 Metadata management for geographic information discovery and exchange. In Sheth A, Klas W (eds) 1998 *Multimedia Data Management: Using Metadata to Integrate and Apply Digital Media.* McGraw Hill: 89–121

Duda K A 1998 *Database Query and Map Generation via the WWW : Online Resources and Examples. http://www.ssc.msu.edu/~geo/wwwgis.html*

Dykes J A 1997 Exploring spatial data representation with dynamic graphics. *Computers & Geosciences* 23(4): 345–70

Egenhofer M J 1991 Reasoning about binary topological relations. *Proceedings of the Second International Symposium on Spatial Databases*: 143–60

Egenhofer M J 1997 Spatial relations: models and interface. *Tutorial in 5th International Symposium on Spatial Databases*

Egenhofer M J, Frank A, Jackson J 1991 A topological data model for spatial satabases. *Proceedings SSD'89.* Lecture Notes in Computer Science 409. Springer, Berlin: 271–85

Egenhofer M J, Franzosa R D 1991 Point set topological spatial relations. *International Journal of Geographic Information Systems* 5(2): 161–174

Egenhofer M J, Kuhn W 1998 Interacting with GIS. In Longley P A, Goodchild M F, Maguire D J, Rhind D W (eds) *Geographical Information Systems: Principles, Techniques, Management, and Applications.* Wiley, New York

Egenhofer M J, Richards J 1995 Visual map algebra: a direct-manipulation user interface for GIS. In Spaccapietra S, Jain R (eds) *Visual Database Systems 3: Visual Information Management, Proceedings of Third IFIP 2.6 Working Conference on Visual Database Systems.* Chapman & Hall, London: 235–53

Egenhofer M, Mark D 1995 Naive geography. In Frank A, Kuhn W (eds) *Spatial Information Theory—A Theoretical Basis for GIS, International Conference COSIT '95, Semmering, Austria*. Lecture Notes in Computer Science 988. Springer-Verlag, Berlin: 1–15

Eliassen F, Nicol J R 1996 Supporting interoperation of continuous media objects. *Theory and Practice of Object Systems* Special Issue on Distributed Object Management Systems 2(2)

Elmagarmid A (ed) 1992 *Database Transaction Models for Advanced Applications*. Morgan-Kaufmann

Elmagarmid A, Pu C (eds) 1990 Heterogeneous databases. Special Issue. *ACM Computing Surveys* 22 (3)

Elmagarmid A, Sheth A, Rusinkiewicz M (eds) 1998 *Heterogeneous Distributed Databases*. Morgan Kaufmann Publishers

Elmasri R, Navathe S B 1994 *Fundamentals of Database Systems*. Benjamin-Cummings

Epstein E F 1995 Control of public information. In Onsrud H J, Rushton G (eds) *Sharing Geographic Information*. Rutgers University, New Brunswick, NJ

Eriksson H-P, Penker M 1998 *UML Toolkit*. John Wiley, New York

Evans J D 1997 *Infrastructures for Sharing Geographic Information among Environmental Agencies*. PdD Thesis, Massachusetts Institute of Technology. *http://mit.edu/jdevans/thesis*

Evans J D, Ferreira J 1995 Sharing spatial information in an imperfect world: interactions between technical and organizational issues. In Onsrud H J, Rushton G (eds) *Sharing Geographic Information*. Rutgers University, New Brunswick, NJ

Ewusi-Mensah K 1997 Critical issues in abandoned information systems development projects. *Communications of the ACM* 40(9): 74–80

Farley J 1998 *Java Distributed Computing*. O'Reilly, Sebastopol

Federal Geographic Data Committee (FGDC) 1994 *Content Standards for Digital Geospatial Metadata*. Federal Geographic Data Committee, Washington DC

Federal Geographic Data Committee 1997 *Content Standards for Digital Geospatial Metadata, Version 2.0)*. Federal Geographic Data Committee, Washington, DC

Fegeas R G, Casio J L, Lazar R A 1992 An overview of FIPS 173, the spatial data transfer standard. *Cartography and Geographic Information Systems* 19(5): 278–293

Fernandez M F, Popa L, Suciu D 1997 A structured-based approach to querying semi-structured data. *Int. Workshop on Database Programming Languages*

Ferreira J 1992 User interfaces for geographic information systems. In Mark D M, Frank A U (eds) *User Interfaces for Geographic Information Systems: Report of the Initiative 13 Specialist Meeting*. Technical Report 92-3. National Center for Geographic Information and Analysis, Santa Barbara: 63–66

Fingerman P W 1997 *HDF-EOS 2.00 Version Description Document (VDD) for the ECS Project, Version 1.00* (814-RD-009-001). Hughes Information Technology Systems, Upper Marlboro, MD

Finin T, Weber J, Wiederhold G, Genesereth M, Fritzson R, McKay D, McGuire J, Pelavin R, Shapiro S, Beck C 1993 *Specification of the KQML Agent-Communication Language*. Working paper, ARPA Knowledge Sharing Initiative

Flanagan D 1996 *Java in a Nutshell*. O'Reilly & Associates, Sebastopol

Floyd C 1985 On the relevance of formal methods to software development. In Ehrig C F H, Nivat M, Thatcher J (eds) *Formal Methods and Software Development*. Springer-Verlag 2: 1–11

Foote K 1997 The Geographer's Craft: teaching GIS on the Web. *Transactions in GIS* 2(2): 137–150

Frank A 1987 Towards a spatial theory. Proceedings of International Geographic Information Systems (IGIS) Symposium: The Research Agenda: 215–227

Frank A 1992 Qualitative spatial reasoning about distances and directions in geographic space. *Journal of Visual Languages and Computing*: 343–71

Frank A U 1992 Acquiring a digital base map: a theoretical investigation into a form of sharing data. *Journal of the Urban and Regional Information Systems Association* 4(1): 10–23

Frank A U 1996 The prevalence of objects with sharp boundaries in GIS. In Burrough P A, Frank A U (eds) *Geographic Objects with Indeterminate Boundaries*. Taylor & Francis, London: 29–40

Frank A U, Campari I (eds) 1993 *Spatial Information Theory, A Theoretical Basis for GIS*. Lecture Notes in Computer Science No. 716. Springer Verlag, Berlin

Frank A U, Kuhn W 1995 Specifying open GIS with functional languages. In Egenhofer M, Herring J (eds) *Advances in Spatial Databases: 4th International Symposium on Large Spatial Databases, SSD'95, Portland*. Springer-Verlag, New York: 184–195

Frank A, Kuhn W 1995 *Spatial Information Theory—A Theoretical Basis for GIS, International Conference COSIT '95, Semmering, Austria*. Springer-Verlag, Berlin

Frank A U, Mark D M 1991 Languages issues for GIS. In Maguire D J, Goodchild M F, Rhind D W (eds) *Geographic Information Systems: Principles and Applications*. Longman, London: 147–63

Frank S 1994 The National Spatial Data Infrastructure: designing navigational strategies, *Journal of the Urban and Regional Information Systems Association* 6(1): 37–55

Fujimura J H 1992 Crafting science: standardized packages, boundary objects, and 'translation'. In Pickering A (eds) *Science as Practice and Culture*. University of Chicago Press, Chicago: 168–211

Fulghum R 1987 *All I Really Need to Know I Learned in Kindergarten*. Villard Books, New York: 6

Gaede V 1998 *Execution Planning in Internet Marketplaces*. Technical Report. Humboldt-Universität zu Berlin, Germany

Gahegan M N 1996 Specifying the transformations within and between geographic data models. *Transactions in GIS* 1(2): 137–52

Gahegan M N, Ehlers M 1997 A framework for the modeling of uncertainty in an integrated geographic information system. To appear in *ISPRS International Workshop on Dynamic and Multi-Dimensional GIS*, Hong Kong

Gallagher J, Milkowski G 1995 Data transport within the distributed oceanographic data system. *4th International World Wide Web Conference Proceedings*: 691–700.

Garcia-Molina H, Hammer J, Ireland K, Papakonstantinou Y, Ullman J, Widom J 1995 Integrating and accessing heterogeneous information sources in TSIMMIS. *Proceedings of the AAAI Symposium on Information Gathering, Stanford, California*: 61–64

Gayer K 1998 Metaobject repositories for managing services in electronic marketplaces. Discussion paper, Sonderforschungsbereich 373. Humboldt-Universität zu Berlin

Genesereth M R, Ketchpel A P 1994 Software agents. *Communication of the ACM* 37(7): 48–53

Genesereth M, King R (eds) 1995 *Reference Architecture, Intelligent Integration of Information*. Stanford University and University of Colorado. *http://logic.stanford.edu /achitecture/reference.html*

Georgakopoulos D, Rusinkiewicz M 1994 Using tickets to enforce the serializability of multidatabase transactions. *IEEE Transactions On Knowledge and Data Engineering* 6(1): 166–180

Gersmehl P J 1985 The data, the reader, and the innocent bystander—a parable for map users. *Professional Geographer* 37(3): 329–34

Giddens A 1984 *The Constitution of Society*. University of California, Berkeley

Glaser B G, Strauss A L 1967 *The Discovery of Grounded Theory*. Aldine Publishing Co, Chicago

Goh H, Madnick S, Siegel M 1994 Context interchange, overcoming the challenges of large-scale interoperable database systems. In *Third International Conference on Information and Knowledge Management*: 337–46

Goldstein J, Roth S F 1994 Using aggregation and dynamic queries for exploring large data sets. *Proceedings of CHI'94, Boston*. ACM Press: 23–29

Gonçalves P, Diogo P 1994 Geographic information systems and cellular automata: A new approach to forest fire simulation. *Proceedings of the European Conference on Geographical Information Systems (EGIS 94), Paris*. EGIS Foundation: 702–712

Goodchild M F 1990 Keynote address: spatial information science. *Proceedings of the 4th International Symposium on Spatial Data Handling, Zurich*: 3–12

Goodchild M F 1992 Geographic information science. *International Journal of Geographic Information Systems* 6(1): 31-45

Goodchild M F 1992 Geographical data modeling. *Computers and Geosciences* 4(18): 401–408

Goodchild M F 1995 Sharing imperfect data. In Onrud H J, Rushton G (eds) *Sharing Geographic Information*. Center for Urban Policy Research, New Brunswick, NJ: 413–425

Goodchild M F, Egenhofer M J, Fegeas R 1997 *Interoperating GISs: Report of the Specialist Meeting*. National Center for Geographic Information and Analysis, Santa Barbara. *www.ncgia.org*

Goodchild M F, Gopal S (eds) 1989 *Accuracy of Spatial Databases*. Taylor & Francis, London

Goodchild M F, Proctor J 1997 Scale in a digital geographic world. *Geographical and Environmental Modelling* 1(1): 5–23

Goodchild M, Egenhofer M, Fegeas R 1997 *Interoperating GISs: Report of a Specialist Meeting Held under the Auspices of the Varenius Project, Panel on Computational Implementations of Geographic Concepts*. National Center for Geographic Information and Analysis, Santa Barbara

Gouin D, Morin P, Clément G, Larouche C 1997 Solving the geospatial data barrier. *Geomatica* 51(3): 278–287

Graefe G 1993 Query evaluation techniques for large databases. *ACM Computing Surveys* 25 (2): 73–170

Grandori A, Soda G 1995 Inter-firm networks: antecedents, mechanisms, and forms. *Organization Studies* 16(2): 183–214

Granovetter M S 1972 The strength of weak ties. *American Journal of Sociology* 78(6): 1360–1380

Gray H, Chaudhuri S, Bosworth A, Layman A, Reichart D, Venkatrao M 1997 Data cube: A relational aggregation operator generalizing group-by, cross-tab, and sub-totals. *Data Mining and Knowledge Discovery* 1(1): 29–53

Gray P M D, Preece A, Fiddian N J, Gray W A, Bench-Capon T J M, Shave M J R, Azarmi N, Wiegand M, Ashwell M, Beer M, Cui Z, Diaz B, Embury S M, Hui K, Jones A C, Jones D M, Kemp G J L, Lawson E W, Lunn K, Marti P, Shao J, Visser P R S 1997 KRAFT: knowledge fusion from distributed databases and knowledge bases. *Database and Expert System Applications (DEXA '97), Toulouse, France*

Grosky W, Fotouhi F, Jiang Z 1998 Using metadata for the intelligent browsing of structured media objects. In Sheth A, Klas W (eds) *Multimedia Data Management: Using Metadata to Integrate and Apply Digital Media*. McGraw Hill: 123–148

Grossley D, Boston T 1996 A generic map interface to query geographical information using the World Wide Web. *World Wide Web Journal* 1: 723–35

Gruber T 1991 The role of a common ontology in achieving sharable, reusable knowledge bases. *Proceedings of the 2nd International Conference on Principles of Knowledge Representation and Reasoning, Cambridge*: 601–602

Gruber T R 1993 Toward principles for the design of ontologies used for knowledge sharing. *Workshop on Formal Ontology, Padua*

Gruber T R 1995 Toward principles for the design of ontologies used for knowledge sharing. *International Journal of Human and Computer Studies* 43(5/6): 907–928

Guarino N 1995 Formal ontology, conceptual analysis and knowledge representation. *International Journal of Human and Computer Studies* 43(5/6): 625–640

Guarino N 1998 Formal ontology and information systems. *Proceedings of the 1st International Conference on Formal Ontology in Information Systems [FOIS'98], Torino*: 3–15

Günther O, Müller R, Schmidt P, Bhargava H, Krishnan R 1996 *MMM: A WWW-Based Method Management System for Using Software Modules Remotely*. Technical Report 32-1996. Humboldt-Universität zu Berlin.

Günther O, Müller R, Schmidt P, Bhargava H, Krishnan R 1997 MMM: A WWW-based approach for sharing statistical software modules. *IEEE Internet Computing* 1(3)

Günther O, Riekert W–F 1993 The design of GODOT: an object-oriented geographic information system. *Bulletin of the Technical Committee on Data Engineering* 16(3): 4–9

Günther O, Voisard A 1998 Metadata in geographic and environmental data management. In Sheth A, Klas W (eds) *Multimedia Data Management: Using Metadata to Integrate and Apply Digital Media*. McGraw Hill: 57–87

Gupta A, Jain R 1997 Visual information retrieval. *Communications of the ACM* 40(5): 70–79

Guptill S C 1994 Spatial data standards. *Government Information Quarterly* 11(4): 387–401

Guptill S C, Morrison J L (eds) 1995 *Elements of Spatial Data Quality*. Elsevier, New York

Güting R H 1988 Geo-relational algebra: A model and query language for geometric database systems. In Schmidt J W, Ceri S, Missiko M (eds) *Advances in Database Technology. Proceedings of the International Conference on Extending Database Technology (EDBT' 88)*. Lecture Notes in Computer Science No. 951. Springer-Verlag, Berlin: 506–527

Guttag J V, Horning J J, Wing J M 1985 *Larch in Five Easy Pieces*. Digital Equipment Corporation, Systems Research Center

Guttman A 1984 R-trees: a dynamic index structure for spatial searching. *Proceedings of ACM SIGMOD Conference, Boston*: 47–57

Habermann T, Miller T 1994 *The FREEFORM Tutorial*. NOAA National Geophysical Data Center, Boulder, Colorado. *ftp://ftp.ngdc.noaa.gov/Access_Tools/FREEFORM*

Hadzilacos T, Tryfona N 1996 Logical data modeling for geographical applications. *International Journal of Geographical Information Systems* 10(2): 179–203

Hallgren T, Carlsson M 1995 Programming with Fudgets. In Jeuring J, Meijer E (eds) *Advanced Functional Programming*. Lecture Notes in Computer Science 925. Springer-Verlag, Berlin: 137–182

Hamilton G 1997 *JavaBeans*. Sun Microsystems, Inc., Mountain View, California. *http://splash.javasoft.com/beans/docs/beans.101.pdf*

Hamilton G, Cattell R, Fisher M 1997 *JDBC Database Access with Java: A Tutorial with Annotated Reference*. Addison-Wesley

Han J 1997 Spatial data mining and spatial data warehousing. *Tutorial Notes of the International Symposium on Spatial Databases (SSD'97), Berlin*

Harvey F 1997 Improving multi-purpose GIS design: participative design. In Hirtle S, Frank A U (eds) *Spatial Information Theory: A Theoretical Basis for GIS*. Springer Verlag, Berlin: 313–328

Harvey F, Chrisman N R 1998 Boundary objects and the social construction of GIS technology. *Environment and Planning A*

Hearnshaw H M, Unwin D J 1994 *Visualization in Geographical Information Systems*. John Wiley and Sons, New York

Heimbigner D, McLeod D 1985 A federated architecture for information management. *ACM Transactions on Office Information Systems* 3(3): 253–278

Herring J, Egenhofer M J, Frank A U 1990 Using category theory to model GIS applications. *Proceedings of the 4th International Symposium on Spatial Data Handling, Zurich*: 820–829

Heuvelink G B M 1998 *Error Propagation in Environmental Modelling*. Taylor and Francis, London

Heywood D I 1997 *Beyond Chorley: Current Geographic Information Issues*. Association for Geographic Information, London

Heywood D I, Cornelius S C, Cremers P H 1998 Developing a virtual campus for UNIGIS: an international distance learning programme for geographical information professionals. In Anonymous (eds) *Bringing Information Technology into Education*. Dortrecht, Kluwer

Heywood D I, Oliver J, Tomlinson S 1995 Building an exploratory multicriteria modeling environment for spatial decision support. In Fisher P (ed.) *Innovations in GIS 2*. Taylor & Francis, London: 127–36

Hirano S 1996 What's HORB?. *http://ring.etl.go.jp/openlab/horb/doc/what.htm*

Hirtle S, Frank A 1997 *Spatial Information Theory—A Theoretical Basis for GIS, International Conference COSIT '97, Laurel Highlands, Pennsylvania*. Springer-Verlag, Berlin

Hoare C A R 1969 An axiomatic basis for computer programming. *Communications of the ACM* 12(10): 576–580

Hobbes T 1651 *Leviathan, or, the Matter, Form, and Power of a Common-Wealth Ecclesiatical and Civil*. Andrew Crooke, London

Houstis C, Nikolaou C, Marazakis M, Patrikalakis N, Sairamesh J, Thomasic A 1997 THETIS: Design of a data repositories collection and data visualization system for coastal zone management of the Mediterranean Sea. *D-Lib Magazine, The Magazine for Digital Library Research,* November. *http://www.dlib.org/dlib/november97/thetis/11thetis.html*

Hsiao D, Neuhold E, Sacks-Davis R (eds) 1993 *Interoperable Database Systems (DS-5)*. North-Holland

Hsu C 1996 *Enterprise Integration and Modeling: the Metadatabase Approach*. Kluwer Academic Publishers, Boston

Hudak P, Peyton Jones S L, Wadler P L and others 1992 Report on the functional programming language Haskell, Version 1.2. *SIGPLAN Notices 27*

Hughes Information Technology Corp 1996 *Release-B Science Data Processing Segment (SDPS) Database Design and Database Schema Specifications for the ECS Project*. Document No. 311-CD-008-001

Huhns M, Singh M 1997 Ontologies for agents. *Internet Computing* 1(6): 81–83

Hull R 1997 Managing semantic heterogeneity in databases: A theoretical perspective. *Proc. 16th ACM Symposium on Principles of Database Systems (PODS)*: 51–61

Hurson A R, Lee B, Shirazi R, Wang M 1990 A program allocation scheme for data flow computers. *Proceedings of the 1990 International Conference on Parallel Processing*: I415–I423

Ingham D, Little M, Caughey S, Shrivastava S 1995 W3Objects: Bringing object-oriented technology to the Web. *World Wide Web Journal. http://www.w3.org/pub/WWW/Journal/1/ingham.141/paper/141.html*

INGRES 1995 *CA-OpenIngres. Object Management Extension User's Guide, release 1.1*

Inmon W H 1996 *Building the Data Warehouse*. Wiley Computer Publishing, New York

INTERLIS 1990 *INTERLIS: A data exchange mechanism for land information systems*.Version 1.0e

ISO 1992 *The EXPRESS Language Reference Manual*. ISO TC 184. International Standards Organization

International Standards Organization 1997 *Geographic Information—Metadata, Version 2.0*. ISO/TC211 Working Group 3, Working Document No. 1997-01-20

Java 1996 *JavaBeans—Component APIs for Java. http://www.javasoft.com*

JavaServer Product Group 1997 *The Java Servlet API*. Sun Microsystems Inc., Mountain View, California. *http://jserv.javasoft.com/products/java-server/documentation/webserver 1.0.2 servlets/api.html*

Jeusfeld M A, Bui T X 1997 Decision support components on the Internet. *Int. Journal Decision Support Systems*

Jones M P 1991 *An Introduction to Gofer*. Yale University

Jones M P 1994 *The Implementation of the Gofer Functional Programming System*. Report No. YALEU/DCS/RR-1030. Yale University

Jones M P 1995 Functional programming with overloading and higher-order polymorphism. In Jeuring J, Mejer E (eds) *Advanced Functional Programming*. Lercture Notes in Computer Science 925. Springer-Verlag, Berlin: 97–136

Jung V 1995 Knowledge-based visualization design for geographic information systems. *Proceedings of the 3rd ACM International Workshop on Advances in Geographic Information Systems, Baltimore*. ACM Press: 101–108

Kahle B 1991 An information system for corporate users: Wide Area Information Servers. *Online Magazine* 15(5): 56–62

Kalakota R, Whinston A B 1996 *Frontiers of Electronic Commerce*. Addison-Wesley, Reading, Mass.

Kambayashi Y, Rusinkiewicz M, Sheth A (eds) 1991 *Proceedings of the RIDE-IMS'91: International Workshop on Interoperability in Multidatabase Systems*. IEEE Computer Society

Kashyap V 1997 *Information Brokering over Heterogeneous Digital Data: A Metadata Based Approach*. PhD Thesis, Rutgers University

Kashyap V, Shah K, Sheth A 1995 Metadata for building the MultiMedia Patch Quilt. In Jajodia S, Subrahmanian V S (eds) *Multimedia Database Systems: Issues and Research Directions*. Springer-Verlag: 297–319

Kashyap V, Sheth A 1994 Semantics based information brokering. *Proceedings of the 3rd International Conference on Information and Knowledge Systems*: 363–370

Kashyap V, Sheth A 1995 *Semantic and Schematic Similarities Between Database Objects: A Context-based Approach*. Technical Report TR-CS-95-001. Department of Computer Science, University of Georgia, Athens

Kashyap V, Sheth A 1996 Schematic and semantic similarities between database objects: a context-based approach. *The Very Large Databases Journal* 5(4): 276–304

Kashyap V, Sheth A 1998 Semantic heterogeneity in global information systems: the role of metadata, context and ontologies. In Papazoglou M, Schlageter G (eds) *Cooperative Information Systems: Current Trends and Directions*. Academic Press: 139–178

Kemp K K 1997 The NCGIA Core Curricula in GIS and Remote Sensing. *Transactions in GIS* 2(2): 181–190

Kent W 1991 The breakdown of the information model in multi-database systems. *SIGMOD Record* 20(4): 10–15

Kernighan B W, Ritchie D M 1988 *The C Programming Language*. Prentice Hall, Englewood Cliffs, NJ

Kerschberg L 1997 Knowledge rovers: cooperative intelligent agent support for enterprise information architectures. *Proceedings of International Workshop on Cooperative Information Agents '97 (CIA'97)*. Lecture Notes in Artificial Intelligence. Springer Verlag: 77–100

Khang J, McLeod D 1998 Dynamic classificational ontologies: mediation of information sharing in cooperative federated database systems. In Papazoglou M, Schlageter G (eds) *Cooperative Information Systems: Current Trends and Directions*. Academic Press: 179–203

Kim W (ed) 1995 *Modern Database Systems: The Object Model, Interoperability and Beyond*. Addison Wesley

Kim W, Choi I, Gala S, Scheevel M 1993 On resolving schematic heterogeneity in multidatabase systems. *Distributed Parallel Databases International Journal* 1: 251–279

Kim W, Choi I, Gala S, Scheevel M M 1995 On resolving schematic heterogeneity in multidatabase systems. In Kim W (ed) *Modern Database Systems*. Addison Wesley: 521–550

Kim W, Lochovsky F H 1989 *Object-Oriented Concepts, Databases, and Applications*. Addison-Wesley

King D J, Wadler P 1993 Combining Monads. In Ayr S, Launchbury J, Sansom P M (eds) *Proceedings of the Functional Programming Workshop Glasgow 1992*. Springer-Verlag, Berlin

King J L, Kraemer K L 1993 Models, facts, and the policy process: The political ecology of estimated truth. In Goodchild M, Parks B O, Stayaert L T (eds) *Environmental Modeling with GIS*. Oxford University Press, New York: 353–360

Kiyoki Y, Kitagawa T, Hayama T 1998 A metadatabase system for semantic image search by a mathematical model of meaning. In Sheth A, Klas W (eds) *Multimedia Data Management: Using Metadata to Integrate and Apply Digital Media*. McGraw Hill: 191–222

Koch T, Ardo A, Brummer A, Lundberg S 1996 *The Building and Maintenance of Robot-Based Internet Search Services*. Technical Report, Lund University

Kosters G, Pagel B, Six H 1996 The GeoOOA tool and its interface to open software developments for GIS. *Proceedings of the 4th ACM Workshop on GIS, Rockville*. ACM Press: 165–173

Kosters G, Pagel B, Six H 1997 GIS-application development with GeoOOA. *International Journal of Geographical Information Science* 11(4): 307–335

Kraak M J 1996 Integrating multimedia in geographical information systems. *IEEE Multimedia* 3(2): 59–65

Kuhn W 1994 Defining semantics for spatial data transfers. In Waugh T C, Healey R G (eds) *Sixth International Symposium on Spatial Data Handling*: 973–87

Kuiper K 1992 *Those Idiots in the Computer Room: Computer Myth Conceptions Spelled out in Plain English*. MacAdam House Publishing, Portland

Kumar K, van Dissel H G 1996 Sustainable collaboration: managing conflict and cooperation in interorganizational systems. *MIS Quarterly* 20(3): 279–299

Labrou Y, Finin T 1997 *A Proposal for a New KQML Specification*. Technical Report TR CS-97-03, University of Maryland

Lagoze C, Lynch C, Daniel R 1996 The Warwick Framework: A container architecture for aggregating sets of metadata. Technical Report TR96-1593. Cornell University, Department of Computer Science. *http://cs-tr.cs.cornell.edu:80/Dienst/UI/2.0/Describe/ncstrl.cornell/TR96-1593*

Lakshmanan L V S, Sadri F, Subramanian I N 1996 SchemaSQL: A language for interoperability in relational multidatabase systems. *Proc. 22nd Int. Conference on Very Large Data Bases (VLDB)*: 239–250

Langran G 1992 *Time in Geographic Information Systems*. Taylor & Francis, London

Larouche C, Clément G, Moore T 1996 Adapting GIS technology for decision-makers. *Proceedings of the 1996 Canadian Conference on GIS, Ottawa, Canada, June 4-10*

LAS 1996 *Open Geographic Datastore Interface (OGDI), Version 1.0: Programmer's Reference*. Logiciels et Applications Scientifiques Inc

LAS 1996 *GRASSLAND User's Guide for Windows 95 and Windows NT, Version 1.1*. Logiciels et Applications Scientifiques Inc

Lassila O, Swick R R (eds) 1997 *Resource Description Framework (RDF) Model and Syntax*. WWWC Draft Specification (WD-rdf-syntax-971103). *http://www.w3.org/Metadata/RDF/Group/WD-rdf-syntax/*

Latour B 1988 *The Pasteurization of France*. Harvard University Press, Cambridge, Mass.

Laurini R, Thompson D 1992 *Fundamentals of Spatial Information Systems*. Academic Press, London

Lee J, Madnick S, Siegel M 1996 Conceptualizing semantic interoperability: a perspective from the knowledge level. *International Journal of Cooperative Information Systems* 5(4): 367–393

Lenz H J, Shoshani A 1997 Summarizability in OLAP and statistical data bases. *Proceedings of 9th International Conference on Statistical and Scientific Database Management*. IEEE Computer Society Press

Létourneau F 1998 Analyse du potentiel de l'approche entrepôt de données pour l'intégration des métadonnées d'un ensemble de géorépertoires disponibles sur internet. M.Sc. thesis, Dept. of Geomatics Sciences, Laval University, Québec, Canada

Létourneau F, Bédard Y, Proulx M J 1997 Data warehouse-based system for the optimized selection of spatial data (SOS-DS: Système optimal de sélection des données spatiales couplé à un data warehouse). *D-Lib Magazine: The magazine of Digital Library Research. http://www.dlib.org/dlib/march97/laval/03letourneau.html*

Levy A Y, Srivastava D, Kirk T 1995 Data model and query evaluation in global information systems. *Intelligent Information Systems* 5(2): 121–143

Litwin W, Boudenant J, Esculier C, Ferrier A, Glorieux A, La Chimia, J, Kabbaj K, Moulinoux C, Rolin P, Stangret C 1982 SIRIUS: systems for distributed data management. In Schneider H-J (ed) *Distributed Data Bases*. North-Holland, Netherlands: 311–66

Litwin W, Mark L, Roussopoulos N 1990 Interoperability of multiple autonomous databases. *ACM Computing Surveys* 22(2): 267–293

Liu L, Pu C 1997 An adaptive object-oriented approach to integration and access of heterogeneous information sources. *Distributed and Parallel Databases* 5: 167–205

Locke J 1731 *Essay Concerning Human Understanding*. 10th Edition. Edward Symon, London

Lopez X R 1997 The network as organization: digital libraries for spatial information. *Proceedings of the First Assembly and Retreat of the University Consortium for Geographic Information Science(UCGIS), Bar Harbor, ME June 15–20*

Maamar Z 1997 Contribution à la résolution des problèmes d'interopérabilité des systèmes, une méthode de conception par frameworks orientés-agents logiciels. PhD thesis, Computer Science Department, Laval University, Québec, Canada

Maamar Z, Moulin B 1997 An agent-based approach for intelligent and cooperative systems. *Proceedings of IEEE Knowledge and Data Engineering Exchange Workshop '97 (KDEX'97), Newport Beach*: 19–26

Maamar Z, Moulin B 1997 C-foal: une méthode de conception par frameworks orientés-agents logiciels de systèmes multiagents interopérables. *Proceedings of Third International Symposium on Programming and Systems '97 (ISPS'97, Algiers*: 374–396

Maamar Z, Moulin B, Bédard Y, Babin G 1997 Software agent-oriented frameworks meet georeferenced digital library interoperability. *D-Lib Magazine: The magazine of Digital Library Research. http://www.dlib.org/dlib/september97/laval/09laval.html*

MacEachren A M 1994 *Some Truth with Maps: A Primer on Symbolization and Design*. Association of American Geographers, Washington DC

MacEachren A M 1995 *How Maps Work: Representation, Visualization, and Design*. Guilford Press, New York

Mackay D S, Robinson V B 1992 *Towards a Heterogeneous Information Systems Approach to Geographic Data Interchange*. Institute for Land Information Management Discussion Paper 92/1. University of Toronto, Mississauga, Ontario

Mackinlay J 1986 Automating the design of graphical presentation of relational information. *ACM Transactions on Graphics* 5(2): 110–41

MacLane S, Birkhoff G 1967 *Algebra*. Macmillan, New York

Manola F 1998 *Towards a Web Object Model*. Object Services and Consulting, Inc. *http://www.objs.com/OSA/wom.htm*

Mark D M 1993 Toward a theoretical framework for geographic entity types. In Frank A U, Campari I (eds) *Spatial Information Theory, A Theoretical Basis for GIS*. Lecture Notes in Computer Science No. 716. Springer Verlag, Berlin: 270–283

Mark D M, Frank A U 1992 *User Interfaces for Geographic Information Systems: Report on the Initiative 13 Specialist Meeting*. Technical Report 92-3. National Center for Geographic Information and Analysis, Santa Barbara

Markus M L, Robey D 1988 Information technology and organizational change: causal structure in theory and research. *Management Science* 34(5): 583–598

McCauley J D, Navulur K C S, Engel B A, Srinivasan R 1996 Serving GIS data through the World Wide Web. Proceedings 3rd International Conference/Workshop on Integrating GIS and Environmental Modeling, Santa Fe, New Mexico. National Center for Geographic Information and Analysis, Santa Barbara

McKee L, Buehler K (eds) 1996 *The Open GIS Guide*. Open GIS Consortium, Inc, Wayland, Massachusetts. *http:www.opengis.org/guide*

McKellar D 1995 The Global Geospatial Information and Services Initiative. *GEOMATICS 1995 Conference, Ottawa, Canada, June 11-15*

Medeiros C B, Pires F 1994 Databases for GIS. *SIGMOD Record* 23(1): 107–15

Medyckyj-Scott D, Hearnshaw H M (eds) 1993 *Human Factors in Geographical Information Systems*. Belhaven Press, London

Meersman R 1997 An essay on the role and evolution of data(base) semantics. In Meersman R, Mark L (eds) *Database Application Semantics*. Chapman and Hall

Meersman R, Mark L (eds) 1997 *Database Application Semantics*. Chapman and Hall

Mena E, Kashyap V, Illarramendi A, Sheth A 1998 Domain specific ontologies for semantic information brokering on the global information infrastructure. *Proceedings, International Conference on Formal Ontology in Information Systems (FOIS'98), Torino*: 269–283

Mendelzon A, Mihaila G, Milo T 1997 Querying the World Wide Web. *Journal of Digital Libraries* 1(1): 68–88

Mendelzon A, Mihaila G, Raschid L, Tomasic A 1997 Locating and accessing heterogeneous data sources. *Proceedings of CASCON'97*

Meredith P 1995 Distributed GIS: If its time is now, why is it resisted? In Onrud H J, Rushton G (eds) *Sharing Geographic Information*. Center for Urban Policy Research, New Brunswick, NJ: 7–21

Merle P, Gransart C, Geib J-M 1996 CorbaWeb: A generic object navigator. *Proceedings of the Fifth International World Wide Web Conference, Paris*. INRIA. *http://www5conf.inria.fr/fichhtml/papers/P33/Overview.html*

Mesrobian E, Muntz R, Shek E, Nittel S, LaRouche M, Kriguer M 1996 OASIS: An open architecture scientific information system. *Workshop on Research Issues in Data Engineering (RIDE 96), New Orleans*. IEEE Computer Society

Meyer B 1988 *Object Oriented Software Construction*. Prentice-Hall, New York

Microsoft Corporation 1994 *Microsoft ODBC 2.0 Programmer's Reference and SDK Guide: for Microsoft Windows and Windows NT*. Microsoft Press, Redmond, Washington.

Milner R 1978 A theory of type polymorphism in programming. *Journal of Computer and System Sciences* 17: 348–375

Mimick I S, Quass D, Mumick B S 1997 Maintenance of data cubes and summary tables in a warehouse. *SIGMOD Record* (ACM Special Interest Group on Management of Data), 26(2): 100–110

Monmonier M 1996 *How to Lie with Maps*. The University of Chicago Press, Chicago

Montgomery G E, Schuch H C 1993 *GIS Data Conversion Handbook*., GISWorld, Ft. Collins

Moore J F 1996 *The Death of Competition: Leadership and Strategy in the Age of Business Ecosystems*. HarperCollins, New York

Morehouse S 1990 The role of semantics in geographic data modeling. In Brassel K, Kishimoto H (eds) *Fourth International Symposium on Spatial Data Handling*. Department of Geography, University of Zurich, 2: 689–98

Moreno J 1994 Fire in the Mediterranean region: implications of land-use changes in a global change context. In Viegas D (ed) *Proceedings of 2nd International Conference on Forest Fire* Research, Coimbra: 61–2

Morgenstern M 1996 Integrating diverse information resources: capturing meaning in a sea of data. In Terashima N, Altman E (eds) *Advanced IT Tools*. Chapman & Hall, London: 471–478

Morris B 1988 CARTO-NET: graphic retrieval and management in an automated map library. *Special Libraries Association Geography and Map Division Bulletin* 152: 19–35

Moulin B, Chaib-draa B 1996 An overview of distributed artificial intelligence. In O'Hare G M P, Jennings N R (eds) *Foundations of Distributed Artificial Intelligence*. Wiley, New York: 3–55

Mudumbai S 1997 *ZEBRA: Customizable, Extensible Metadata-based Access to Federated Image Repositories*. M.S. Thesis, Department of Computer Science, University of Georgia

Murray B S 1994 Intelligent information presentation systems. *The Knowledge Engineering Review* 9(3): 269–286

Muth P 1997 Application specific transaction management in multidatabase systems. *Distributed and Parallel Databases* 5: 357–403

National Aeronautics and Space Administration 1997 *Directory Interchange Format (DIF) Writer's Guide, Version 5.0a*. NASA Goddard Space Flight Center, Global Change Data Center, Code 902

National Biological Service 1997 *National Biological Information Infrastructure Metadata Standard*. National Biological Service, Washington, DC

National Imagery and Mapping Agency 1996 *Geospatial Metadata: DoD Geospatial Data Standardization Project Report, Vol. 3*. National Imagery and Mapping Agency

National Imagery and Mapping Agency 1997 Geospatial Information Integrated Product Team (GI IPT) Geospatial Information Infrastructure GII 97 Requirement—SOL BAA. *Commerce Business Daily* (January 17–23): A6–A8

National Research Council 1994 *Promoting the National Spatial Data Infrastructure through Partnerships*. National Academy Press, Washington DC

Nebert D D 1997 *Z39.50 Application Profile for Geospatial Metadata or GEO*. Federal Geographic Data Committee, Reston. *http://www.fgdc.gov/Clearinghouse/Tutorials/GEO21a.html*

NEN1878 1993 Exchange format for data about spatial objects that are related to the earth surface (in Dutch). Nederlands Normalisatie-instituut

NEN3610 1995 Terreinmodel vastgoed. Terms, definitions and general rules for the classification and coding of spatial objects related to the earth surface (in Dutch). Nederlands Normalisatie-instituut

Neumann L J, Star S L 1996 Making infrastructure: the dream of common language. *PDC'96, Participatory Design Conference*. Cambridge, Mass. 1: 231–240

Neves N, Silva J P, Gonçalves P, Muchaxo J, Silva J, Câmara A 1997 Cognitive spaces and metaphors: A solution for interacting with spatial data. *Computers & Geosciences* 23(4): 483–8

NGDC 1997 *GeoVu Software Package*. NOAA National Geophysical Data Center, Boulder, Colorado. *http://www.ngdc.noaa.gov/seg/geovu/*

NIST 1992 *Spatial Data Transfer Standard (SDTS)*. Federal Information Processing Standards Publication No.173. Computer Systems Laboratory, National Institute of Standards and Technology, Gaithersburg, MD

Nittel S, Muntz R R, Mesrobian E 1997 GeoPOM: a heterogeneous geoscientific persistent object system. *Proceeding of 9th Int'l Conference on Statistical and Scientific Database Management, Washington*

Nittel S, Yang J, Muntz R R 1996 Mapping a common geoscientific object model to heterogeneous spatial data repositories. *Proceedings of the Fourth ACM Workshop on Advances in GIS*: 51–8

Nonaka I, Takeuchi H 1995 *The Knowledge-Creating Company: How Japanese Companies Create the Dynamics of Innovation*. Oxford University Press, New York

NPR3611 1997 Practical guidance for NEN 1878 and NEN 3610 (in Dutch). Bijlage B - Ketting-topologie. Nederlands Normalisatie-instituut

Nyerges T L 1989 Schema integration analysis for the development of GIS databases. *International Journal of Geographical Information Systems* 3(2): 153–183

Nyerges T L 1991 Geographic information abstractions: conceptual clarity for geographic modeling. *Environment and Planning A* 23: 1483–1499

O2 Technology Inc 1995 *Technical Overview of the O2 System*. Technology Technical Report No. 9. O2 Technology Inc, Palo Alto, CA

Obermeyer N J 1995 Reducing inter-organizational conflict to facilitate sharing geographic information. In Onsrud H J, Rushton G (eds) *Sharing Geographic Information*. Rutgers University Press, New Brunswick, NJ

Object Design 1995 *C++ Programming with ObjectStore v4.0*. Object Design, Inc., Burlington

Object Management Group (OMG) 1997 *http://www.omg.org*

Object Management Group 1997 *CORBA 2.0/IIOP Specification*. OMG technical document formal/97-02-25. http://www.omg.org/corba/corbiiop.htm

Objectivity 1996 *Using Objectivity/C++, Version 4*. Objectivity Inc

ODBC 1997 *Microsoft ODBC 3.0: Software Development Kit and Programmer's Reference*. Second Edition. Microsoft Press

OGC 1997 *The OpenGIS Guide*. Technical Report 96-001. Open GIS Consortium. *http://www.ogis.org/guide*

OGIS Project Technical Committee (1997) *The OpenGIS Specification Models*. Open GIS Consortium, Wayland, Mass.

Oliveira J L, Cereja N, Medeiros C M B 1997 Intermediate interface model for geographic information systems. *Proceedings of GIS Brasil 97* (in Portuguese)

Oliveira J L, Medeiros C M B 1995 *A Direct Manipulation User Interface for Querying Geographic Databases*. Technical Report DCC-95-08. Computer Science Department, University of Campinas, Brazil

Oliveira J L, Medeiros C M B 1996 *User Interface Issues in Geographic Information Systems*. Technical Report IC-96-06. Institute of Computing, University of Campinas, Brazil

OMG ORBTF 1992 *Common Object Request Broker Architecture*. Object Management Group, Framingham

Online Computer Library Center, Inc 1997 *Dublin Core Metadata Element Set: Reference Description*. Office of Research and Special Projects, Dublin, Ohio. *http://www.oclc.org:5046/research/dublin_core/*

Open GIS Consortium 1996 *Request for Information: OGIS Catalog Service Interfaces*. OGC Request 3, Open GIS Services Working Group (August 30, 1996). Open GIS Consortium, Wayland, MA

Open GIS Consortium 1996 *The OpenGIS Abstract Specification: An Object Model for Interoperable Geoprocessing, Revision 1*. Project Document No. 96-015R1. Open GIS Consortium

Open GIS Consortium 1997 *OpenGIS Simple Features Specification for CORBA*. Revision 0. Open GIS Consortium, Inc

Open GIS Consortium 1998 *The OpenGIS Abstract Specification Model*. TC Document 98-101. Open GIS Consortium Inc.

Orbix 1997 *Orbix Programming Guide*. IONA Technologies Ltd

Orfali R, Harkey D, Edwards J 1996 *The Essential Distributed Objects Survival Guide*. John Wiley and Sons, New York

Orlikowski W J 1992 The duality of technology: rethinking the concept of technology in the context of organizations. *Organization Science* 3(3): 398–427

Ouksel A, Naiman C 1993 Coordinating context building in heterogeneous information systems. *Journal of Intelligent Information Systems* 3: 151–183

Ousterhout J K 1994 Tcl and the Tk Toolkit. Addison-Wesley

Paepcke A, Chang C, Garcia-Molina H, Winograd T 1998 Interoperability for digital libraries worldwide. *Communications of the ACM* 41(4): 33–43

Papakonstantinou Y, Garcia-Molina H, Widom J 1995 Object exchange across heterogenous information sources. *Proceedings of the 11th International Conference on Data Engineering*: 251–260

Papazoglou M, Schlageter G (eds) 1998 *Cooperative Information Systems: Current Trends and Directions*. Academic Press

Park J 1997 Geographic information systems and problem solving environment. *Crossroads* 4(1): 3–8

Pascoe R T, Penny J P 1995 Constructing interfaces between (and within) geographical information systems. *International Journal of Geographical Information Systems* 9(3): 275–91

Peckham J, Maryanski F 1988 Semantic data models. *ACM Computing Surveys* 20(3): 153–189

Peloux J, Rigaux P 1995 A loosely coupled interface to an object-oriented geographic database. In Hartmanis J, Goos G, van Leeuwen J (eds) *Spatial Information Theory (International Conferences COSIT, Semmering, Austria)*: 123–37

Perrochon L, Müller R 1998 Integrated scientific computing in global networks. In El-Rewini H (ed) *Proc. 31st Hawaii Int. Conf. on System Sciences* VII: 628–636

Peterson J, Hammond K, Augustsson L, and others 1996 *Report on the Functional Programming Language Haskell, Version 1.3*. Yale University Research Report YALEU / DCS/ RR-1106. *http://haskell.cs.yale.edu/haskell-report/haskell-report.html*

Peterson J, Hammond K, Augustsson L, and others 1997 *The Haskell 1.4 Report*. *http://haskell.org/report/index.html*

Peuquet D J 1988 Representations of geographic space: towards a conceptual synthesis. *Annals of the Association of American Geographers* 78: 375–94

Peyton Jones S L, Wadler P 1993 Imperative functional programming. *Proceedings of the ACM Symposium on Principles of Programming Languages (POPL), Charleston*. ACM: 71–84

Pinto J K, Onsrud H J 1995 Facilitators of organizational information sharing: a research framework. In Onsrud H J, Rushton G (eds) *Sharing Geographic Information*. Rutgers University Press, New Brunswick, NJ

Pinto M B, Pinto J K, Prescott J E 1993 Antecedents and consequences of project team cross-functional cooperation. *Management Science* 39(10): 1281–1297

Pissinou N, Makki K, Park E 1993 Towards the design and development of a new architecture for geographic information systems. *Proceedings of the Second International Conference on Information and Knowledge Management*: 565–73

Proulx M J, Bédard Y, Létourneau F, Martel C 1997 Catalogage des données spatiales sur le world wide web: Concepts, analyse des sites et présentation d'un géorépertoire personnalisable georep. *Revue Internationale de Géomatique* 7(1): 7–32

QLIS 1996 *QLIS Project Report: Technology Architecture Research & Development Project.* Queensland Land Information Council, State of Queensland Department of Natural Resources, Australia

Ram S (ed) 1991 Heterogeneous distributed database systems. Special Issue. *IEEE Computer* 24(12)

Raper J, Livingstone D 1995 Development of a geomorphological spatial model using object-oriented design. *International Journal of Geographical Information Systems* 9(4): 359–84

Rational 1998 *UML Notation Guide.* Rational Software Corp

Ratza C A 1996 The Great Lakes Information Network: the region's Internet information service. *Toxicology and Industrial Health* 12(3/4): 557–561

Reese G 1997 *Database Programming with JDBC and Java.* O'Reilly, Sebastopol

Reiter R (in preparation) *Knowledge in Action: Logical Foundations for Describing and Implementing Dynamical Systems*

Rew R 1997 Georeferencing with Java: An example of executable metadata. *http://www.unidata.ucar.edu/cgi-bin/mfs/01/staff/russ/MapGeoGrid/ams98.html*

Rhind D W 1992 Data access, charging, and copyright and their implications for geographical information systems. *International Journal of Geographical Information Systems* 6(1): 13–30

Rigaux P 1995 *Interfaces Graphiques pour Bases de Données Spatiales: Application à la Représentation Multiple.* PhD thesis, Conservatoire National des Arts et Metiers, France

Roth S M, Mattis J 1990 Data characterization for intelligent graphics presentation. *Proceedings of SIGCHI'90: Human Factors in Computing Systems Conference, Seattle.* ACM Press: 193–200

Rothermel R 1972 *A Mathematical Model for Predicting Fire Spread in Wildland Fuels.* Res. Pap. INT-115, U.S. Department of Agriculture, Forest Service, Intermountain Forest and Range Experiment Station

Rothermel R 1983 *How to Predict the Spread and Intensity of Forest Fire and Range Fires.* Gen. Tech. Rep. INT-143, U.S. Department of Agriculture, Forest Service, Intermountain Forest and Range Experiment Station

Rubin J 1996 *Handbook of Usability Testing: How to Plan, Design and Conduct Effective Tests.* John Wiley & Sons, New York

Rumbaugh G, Blaha J, Premelani W, Eddy F, Lorensen W 1991 *Object-Oriented Modeling and Design.* Prentice Hall

Russell S J, Norvig P 1995 *Artificial Intelligence: A Modern Approach.* Prentice Hall, Engleweed Cliffs, NJ

Samet H 1990 *Design and Analysis of Spatial Data Structures.* Addison-Wesley, Reading

Schek H–J, Sheth A, Czjedo B (eds) 1993 *Proceedings of the RIDE-IMS'93: International Workshop on Interoperability in Multidatabase Systems.* IEEE Computer Society

Scholl M, Rigaux P 1995 Multi-scale partitions: applications to spatial and statistical databases. In Egenhofer M J, Herring J (eds) *Proceedings of the 4th International Symposium on Spatial Databases (SSD'95).* Lecture Notes in Computer Science No. 951. Springer-Verlag, Berlin: 170–183

Scholl M, Voisard A 1989 Thematic map modeling. In Buchman A, Günther O, Smith T R, Yang Y F (eds) *Design and Implementation of Large Spatial Databases, First Symposium SSD'89.* Lecture Notes in Computer Science No. 409. Springer-Verlag, Berlin: 167–192

Scholl M, Voisard A 1992 *Geographic Applications: An Experience with $O_2$.* In Bancilhon F, Delobel C, Kanellakis P (eds) 1992 *Building an Object-Oriented Database System: The Story of $O_2$.* Morgan Kaufmann: 585–618

Schwartz M F, Emtage A, Kahle B, Neuman B C 1992 A comparison of Internet resource discovery approaches. *Computing Systems: the Journal of the USENIX Association* 5(4): 461–493

Schweitzer P 1996 A document type declaration for formal metadata. In *Tools for Creation of Formal Metadata*. U.S. Geological Survey, Reston. *http://geochange.er.usgs.gov/pub/tools /metadata/tools/doc/dtd.html*

Schweitzer P 1997 A compiler for formal metadata. In *Tools for Creation of Formal Metadata*. U.S. Geological Survey, Reston. *http://geochange.er.usgs.gov/pub/tools /metadata/tools/doc/mp.html*

Scott D S 1997 Logic and programming languages. *Communications of the ACM* 20(9)

Sellis T K, Roussopoulos N, Faloutsos C 1987 The R+-tree: a dynamic index for multi-dimensional objects. *Proceedings of the 13th International Conference on Very Large Data Bases, Brighton*: 507–18

Senay H, Ignatius E 1994 A knowledge-based system for visualization design. *IEEE Computer Graphics and Applications* 14(6): 36–47

Sengupta R, Bennett D A, Wade G A 1996 Agent mediated links between GIS and spatial modeling software using a model definition language. *Proceedings of GIS/LIS '96*. American Congress on Surveying and Mapping, Bethesda, MD: 295–309

Sgouros T 1996 DODS User Guide Version 1.0. *http://dods.gso.uri.edu/DODS/user/guide-html/guide.html*

Shah K, Sheth A 1998 Logical information modeling of Web-accessible heterogeneous digital assets. *Proceedings of the Forum on Research and Technology Advances in Digital Libraries (ADL'98), Santa Barbara*: 266–275

Shah K, Sheth A, Mudumbai S 1997 Black box approach to image feature manipulation used by visual information retrieval engines. *Second IEEE Metadata Conference*

Shekhar S, Coyle M, Goyal B, Liu D, Sarkar S 1997 Data models in geographic information systems. *Communications of the ACM* 40(4): 103–111

Sheth A (ed) 1991 Semantic issues in multidatabase systems. Special Issue. *SIGMOD Record*

Sheth A 1987 *Heterogeneous Distributed Databases: Issues in Integration*. Tutorial Notes, 3rd International Conference on Data Engineering

Sheth A 1995 *Multidatabase Interoperation: Perspective of Researchers and Practitioners*. Tutorial Notes, 11th International Conference on Data Engineering, Taiwan

Sheth A 1996 Bellcore's ADAPT/XHarness system for managing information on Internet and intranets. *Proceedings of the 22nd International Conference on Very Large Data Bases, Bombay, India*: 585

Sheth A 1997 Data semantics: What, where, and how? In Meersman R, Mark L (eds) *Database Application Semantics*. Chapman and Hall: 601–610

Sheth A P, Larson J A 1990 Federated database systems for managing distributed heterogeneous, and autonomous databases. *ACM Computing Surveys* 22(3): 182–236

Sheth A, Kashyap V 1993 So far (schematically) yet so near (semantically). In Hsiao D, Neuhold E, Sacks-Davis R (eds) 1993 *Interoperable Database Systems (DS-5)*. North-Holland: 283–312

Sheth A, Kashyap V 1996 Media-independent correlation of information: what? how? Proceedings of First IEEE Metadata Conference

Sheth A, Klas W (eds) 1998 *Multimedia Data Management: Using Metadata to Integrate and Apply Digital Media*. McGraw Hill

Shklar L, Behrens C, Au E 1997 *New Approaches to Cataloguing, Querying and Browsing Geospatial Metadata*. *http://usdac2.rutgers.edu/new/new.html*

Shklar L, Behrens C, Basu C, Yeager N, Au E 1997 *New Approaches to Cataloguing, Querying and Browsing Geospatial Metadata*. Paper presented at the 2nd IEEE Metadata Conference. NOAA, Silver Springs, Maryland, Sept. 16–17, 1997

Shklar L, Sheth A, Kashyap V, Shah K 1995 InfoHarness: Use of automatically generated metadata for search and retrieval of heterogeneous information. *Proceedings of CAiSE-95*: 217–330

Shneiderman B 1992 *Designing the User Interface: Strategies for Effective Human–Computer Interaction*. Addison-Wesley, New York

Shoham Y 1991 Varieties of context. In Lifschitz V (ed) *Artificial Intelligence and Mathematical Theory of Computation: Papers in Honor of John McCarthy*. Academic Press, Boston: 393–408

Shoham Y 1993 Agent-oriented programming. *Artificial Intelligence* 60: 51–104.

Siegel J 1996 *CORBA Fundamentals and Programming*. Wiley, New York

Siegel M D, Madnick S E 1991 Context interchange: sharing the meaning of data. *ACM SIGMOD Record* 20(4): 77–78

Sims R A, Corns I G W, Klinka K (eds) 1996 *Global to Local: Ecological Land Classification*. Kluwer, Dordrecht, The Netherlands

Smith T R, Park K K 1992 Algebraic approach to spatial reasoning. *International Journal of Geographical Information Systems* 6(3): 177–192

Smith T R, Ramakrishnan R, Voisard A 1992 Object-based data model and deductive language for spatio-temporal database applications. In Gambosi G, Scholl M, Six H-W (eds) *Geographic Database Management Systems*. Springer Verlag: 79–102

Snyder P 1987 *Map Projections, A Working Manual*. Geological Survey Professional Paper 1395. U.S. Government Printing Office, Washington, DC

St. Laurent L, Kucera H, Keighan E 1997 Geospatial data warehousing: an architecture. *Proceedings of the International Symposium: Geomatics in the Era of Radarsat (GER'97), May 25-30, Ottawa, Canada*

Stafford P, Powell J 1995 COM: A model problem solver. *Microsoft Developer Network News* 4(2): 1–12

Star S L (ed) 1995 *The Cultures of Computing*. Blackwell Publishers/The Sociological Review, Oxford

Star S L, Griesemer J R 1989 Institutional ecology, 'translations', and boundary objects: amateurs and professionals in Berkeley's Museum of Vertebrate Zoology. *Social Studies of Science* 19: 387–420

Stevens W T 1990 *UNIX Network Programming*. Prentice Hall, Englewood Cliffs, New Jersey

Stoy J 1977 *Denotational Semantics*. MIT, Cambridge

Sun Microsystems Inc. 1997 *Java Remote Method Invocation (RMI)*. Sun Microsystems Inc., Mountain View, California. *http://java.sun.com/products/jdk/rmi/index.html*

Sycara K, Decker K, Pannu A, Williamson M, Zeng D 1996 Distributed intelligent agents. *IEEE Expert-Intelligent Systems and Their Applications* 11(6): 36–45 *http://www.cs.cmu.edu/*

Szyperski C 1998 *Component Software: Beyond Object-Oriented Programming*. Addison-Wesley, Reading, Mass

Tapscott D 1996 *The Digital Economy: Promise and Peril in the Age of Networked Intelligence* McGraw-Hill, New York

Tomlin C D 1990 *Geographic Information Systems and Cartographic Modeling*. Prentice Hall, Englewood Cliffs, NJ

Tosta N, Domaratz M 1997 The US National Spatial Data Infrastructure. In Craglia M, Couclelis H (eds) *Geographic Information Research: Bridging the Atlantic*. Taylor & Francis, London

Tsichritzis D C, Klug A (eds) 1978 *The ANSI/SPARC DBMS framework—Report of the study group on data base management systems*. Volume 3. ANSI, USA

Tufte E R 1983 *The Visual Display of Quantitative Information*. Graphics Press, Cheshire

Tukey J W 1977 *Exploratory Data Analysis*. Addison-Wesley, Reading Mass.

Tweedie L 1997 Characterizing interactive externalizations. *Proceedings of CHI'97, Atlanta*: 375–82

Twigg C A, Oblinger D G 1996 *The Virtual University: A report from a Joint Educom/IBM Roundtable*. EDUCOM, Washington DC, viewed on the WWW on February 26, 1998 at *http://www.educom.edu/nlii/VU.html*

UML 1977 *Universal Modeling Language (UML) Document Set*. Rational Software Corporation, Santa Clara. *www.rational.com/uml*

University Consortium for Geographic Information Science (UCGIS) 1996 *Interoperability of Geographic Information*. *http://www.ucigs.org/*

Van de Ven A H, Walker, G 1984 The dynamics of interorganizational coordination. *Administrative Science Quarterly* 29(4): 598–621

van Engelen V W P, Ting-Tian W (eds) 1995 *Global and National Soils and Terrain Digital Databases - SOTER. Procedures Manual*. Revised edition. International Soil Reference and Information Centre, Wageningen, The Netherlands

Vckovski A 1998 *Interoperable and Distributed GIS*. Taylor & Francis, London (in press)

Vckovski A 1998 Guest editorial: Special Issue, Interoperability in GIS. *International Journal of Geographical Information Science* 12(4): 297–298

Ventura J 1992 *Propagação de Fogos em áreas Florestais*. Instituto Superior Técnico, Lisboa

Versant 1993. *Versant Reference Manual*. Versant Object Technology, Inc., Menlo Park, California.

Vijlbrief T, van Oosterom P 1992 The GEO++ system: An extensible GIS. *Proceedings of the 5th International Symposium on Spatial Data Handling, Charleston, South Carolina*: 40–50

Vinoski S 1997 CORBA: integrating diverse applications within distributed heterogeneous environments. *IEEE Communications Magazine* 14(2)

Voisard A 1991 Towards a toolbox for geographic user interfaces. *Proceedings of the Second Symposium on Spatial Database Systems*. Springer Verlag Lecture Notes in Computer Science 525: 75–97

Voisard A 1994 Designing and integrating user interfaces of geographic database applications. *Proceedings of the 1994 ACM Workshop on Advanced Visual Interfaces, Bari, Italy*: 133–42

Voisard A 1995 Mapgets: a tool for visualizing and querying geographic information. *Journal of Visual Languages and Computing*

Wade G A, Bennett D A 1997 GNS: A distributed hierarchical topical nomenclature for geographic data. *Proceedings of GIS/LIS '97*. American Congress on Surveying and Mapping Bethesda, MD

Wade G A, Bennett D A, Sengupta R 1997 An interactive distributed architecture for geographical modeling. *Proceedings of Auto-Carto 13*. American Congress on Surveying and Mapping, Bethesda, MD: 307–316

Walker D R F, Newman I A, Medyckyj-Scott D J, Ruggles C L N 1992 A system for identifying datasets for GIS users. *International Journal of Geographical Information Systems* 6(6): 511–27

Weber M 1947 *The Theory of Social and Economic Organization*. Oxford University Press, New York

Wegener M 1996 Spatial models and GIS. Presentation for the ESF-GISDATA and NSF-NCGIA 1996 Summer Institute in Geographic Information held at Villa Borsig, Berlin, 24 July – 1 August 1996

Weisbuch G 1991 *Complex Systems Dynamics*. Lecture Notes Volume in the Santa Fe Institute Series in the Science of Complexity. Addison-Wesley

Wesseling C G, Van Deursen W P A, Burrough P A 1996 A spatial modeling language that unifies dynamic environmental models and GIS. *Proceedings, Third International Conference/Workshop on Integrating GIS and Environmental Modeling, Santa Fe, NM, January 21-26, 1996*. National Center for Geographic Information and Analysis, Santa Barbara, CA. *http://www.ncgia.ucsb.edu/conf/SANTA_FE_CD-ROM/main.html*

Widom J 1995 Research problems in data warehousing. *Proceedings of the 4th International Conference on Information and Knowledge Management (CIKM'95)*

Wiederhold G 1989 *The Architecture of Future Information Systems*. Computer Science Department, Stanford University

Wiederhold G 1992 Mediators in the architecture of future information systems. *IEEE Computer* 25(3): 38–49

Wiederhold G (ed) 1996 *Intelligent Integration of Information*. Kluwer Academic Publishers

Wiederhold G 1997 Value-added mediation in large-scale information systems. In Meersman R, Mark L (eds) *Database Application Semantics*. Chapman and Hall

Wiener J, Gupta H, Labio W, Zhuge Y, Garcia-Molina H, Widom J 1996 System prototype for warehouse view maintenance. *Proceedings of the ACM Workshop on Materialized Views: Techniques and Applications*: 26–33

Wirth N 1976 *Algorithms + Data Structures = Programs*. Prentice Hall, Englewood Cliffs, NJ

Wirth N 1996 Recollections about the development of Pascal. In Bergin T J, Gibson R G (eds) *History of Programming Languages II*. ACM Press, New York

Wolfram S 1986 Cellular automaton fluids 1: Basic theory. *Journal of Statistical Physics* 45: 471–526

Worboys M F 1992 A generic model for planar geographic objects. *International Journal of Geographical Iinformation Systems* 6(5): 353–372

Worboys M F 1994 *GIS: A Computing Perspective*. Taylor & Francis, London

Worboys M F, Hearnshaw H M, Maguire D J 1990 Object-oriented data and query modelling for geographical information systems. In *Proceedings of the 4th International Symposium on Spatial Data Handling, Zurich*: 679–88

Yin R K 1984 *Case Study Research: Design and Methods*. Sage Publications, Beverly Hills, CA

Yuan M 1996 Temporal GIS and spatio-temporal modeling. *Proceedings of Third International Conference/Workshop on Integrating GIS and Environmental Modeling, Santa Fe*. National Center for Geographic Information and Analysis, Santa Barbara (CD-ROM)

Zhan F B, Buttenfield B P 1995 Object-oriented knowledge-based symbol selection for visualizing statistical information. *International Journal of Geographical Information Systems* 9(3): 293–315

Zhang A, Elmagarmid A K 1993 A theory of global concurrency control in multidatabase systems. *VLDB Journal* 2(3): 331–60

# INDEX

GPSR Compliance

*The European Union's (EU) General Product Safety Regulation (GPSR) is a set of rules that requires consumer products to be safe and our obligations to ensure this.*

*If you have any concerns about our products, you can contact us on ProductSafety@springernature.com*

In case Publisher is established outside the EU, the EU authorized representative is:

Springer Nature Customer Service Center GmbH
Europaplatz 3
69115 Heidelberg, Germany

**Batch number: 10755248**

Printed by Printforce, the Netherlands